Transportation Engineering

An Introduction

SECOND EDITION

C. Jotin Khisty
Illinois Institute of Technology, Chicago

B. Kent Lall
Portland State University, Portland

Prentice-Hall International, Inc.

Acquisitions editor: William Stenquist
Production editor: Rose Kernan
Editor-in-chief: Marcia Horton
Cover designer: Bruce Kenselaar
Director of production and manufacturing: David W. Riccardi
Managing editor: Bayani Mendoza De Leon
Manufacturing buyer: Julia Meehan
Interior designer/Compositor: Preparé / Emilcomp srl

© 1998, 1990 by Prentice-Hall, Inc.
Simon & Schuster / A Viacom Company
Upper Saddle River, New Jersey 07458

The author and publisher of this book have used their best efforts in preparing this book. These efforts include the development, research, and testing of the theories and programs to determine their effectiveness. The author and publisher make no warranty of any kind, expressed or implied, with regard to these programs or the documentation contained in this book. The author and publisher shall not be liable in any event for incidental or consequential damages in connection with, or arising out of, the furnishing, performance, or use of these programs.

Printed in the United States of America

10 9 8 7 6 5 4 3 2 1

ISBN 0-13-861527-6

Prentice-Hall International (UK) Limited, London
Prentice-Hall of Australia Pty. Limited, Sydney
Prentice-Hall Canada Inc., Toronto
Prentice-Hall Hispanoamericana, S.A., Mexico
Prentice-Hall of India Private Limited, New Delhi
Prentice-Hall of Japan, Inc., Tokyo
Simon & Schuster Asia Pte. Ltd., Singapore
Editora Prentice-Hall do Brasil, Ltda., Rio de Janeiro
Prentice-Hall, Inc., Upper Saddle River, New Jersey

I dedicate this book to the memory
of my father, Dr. B. R. Khisty,
physician and surgeon, philosopher and teacher,
who introduced me to the "music of the spheres."

—C. Jotin Khisty

A dedication from both of us
To our students, past and present, with whom we continue to learn.

—C. Jotin Khisty
B. Kent Lall

Contents

Preface to the Second Edition

The challenges in transportation engineering education remain just as valid as at the time of the first edition. A trend toward offering two courses in undergraduate civil engineering programs is continuing. There is talk in some circles of proposing a bachelor's degree in transportation engineering as the diverse background required by the professionals expands with additional applications of computer science, electronics, and computer engineering in Intelligent Transportation Systems. The second edition should serve as a good model and a platform on which to build further knowledge.

It is heartening to see that the pace of research has picked up with increased support from the National Cooperative Highway and Transit Research Programs (NCHRP and NCTRP). A better understanding of traffic operations and highway capacity issues led to major changes in the Highway Capacity Manual during 1994. More changes are in the works and would be reflected in future publications. The text includes all changes introduced in the *Highway Capacity Manual* during 1994 and some that are proposed for 1997. Similarly, the text accommodates the recent *Policy on Geometric Design of Highways and Streets* by the American Association of State Highway and Transportation Officials (AASHTO). All chapters have undergone some rewriting and a few are completely revised. Additional exercises are included at the end of the chapters along with some reorganization of the existing ones to better serve the reader. Increasing use of software to applications in transportation engineering is recognized and appropriately emphasized.

The authors are grateful to many colleagues and friends who have suggested improvements. The revision, though enjoyable, has been painstaking and a labor of love. The authors deeply acknowledge the support and understanding shown by their families. Several of our students have directly assisted in the revisions. Notable contributions were made by Titus Reynolds, Philip Taylor, and P. S. Sriraj. Admirable support was provided by Bill Stenquist, Joe Scordato, Rose Kernan and Meg Weist of Prentice Hall.

C. Jotin Khisty
B. Kent Lall

Preface to the First Edition

The main purpose of this book is to cover the major areas of transportation engineering, planning, and management at an introductory level. The contents of this book are intended for use primarily at the junior or senior undergraduate level in the civil engineering curriculum, and at the graduate level in the disciplines of urban geography, economics, public administration, and city and regional planning. Professionals working directly or indirectly in the field of transportation would also find this book useful. For them, the book is intended to give sufficient background and sources to references for further elucidation, should this be desired. Informed laymen and elected officials wishing to gain a quick understanding of the technical implications of a particular transportation-related problem, method, or procedure would hopefully find the text helpful also.

This textbook would also prove useful for self-study, for both the beginning student as well as for those mature students inclined to review and integrate information. Numerous worked examples are provided in every chapter to reinforce the contents, and exercises of varying complexity are to be found at the end of each chapter. An instructor's manual is available.

Transportation engineering and planning have been developed to a large extent by the joint efforts of engineers, planners, economists, geographers, mathematicians, physical scientists, and social scientists. Transportation is a multidisciplinary area of study, which has created several problems in teaching a required course (of courses), particularly in the undergraduate civil engineering (CE) program. Some of these problems are lack of suitable, moderately priced, relatively self-contained, introductory textbooks; general deficiency among students in areas such as microeconomics and statistics, which are needed to comprehend transportation problems; lack of understanding of the systems approach necessary to address socioeconomic issues connected with transportation; lack of appreciation of the multivariable, open-ended, conflict-ridden, value-laden nature of real-world problems; and presentation of the principles of transportation from a modally oriented point of view (Khisty, 1986, 1987).

The questions stemming from these problems are: What constitutes transportation engineering education for an undergraduate CE curriculum? What do employers expect from a CE undergraduate? How should the course be developed so that it addresses the needs of a relatively large number of CE students who in all probability do not foresee the possibility of pursuing further studies in transportation, and at the same time stimulates a relatively small number of students who may develop an active interest in transportation?

Although the master's degree is considered by most educators and practitioners as the degree of specialization in transportation, only a small percentage of undergraduates elect to pursue the master of science in civil engineering (MSCE) with a major in transportation. This is not surprising. When industry pays an individual with a BSCE a respectably good starting salary, it deprives the young engineer of any significant motivation to acquire an advanced degree. This posture is changing.

Proper grounding in the principles of transportation is essential because the entry-level BSCE in federal, state, and local government, as well as in construction, design, and consulting firms, may have had only one required course in transportation engineering. Over the years, there has been a running debate about what to include in this required course (or courses) because no two teachers seem to have identical views as to what transportation engineering topics should be taught to aspiring civil engineers.

Not too long ago, the author of a transportation textbook conducted a survey of professors teaching transportation to determine the content of a transportation course(s) that should be included as a requirement in a CE curriculum (Wright, 1983). To follow up on the results of this survey, the writer conducted another survey to identify the views of transportation practitioners working for departments of transportation, counties, cities, and private firms. The practitioners were asked to evaluate the importance of 30 topics that could possibly be included in a required course in transportation for CE students. Table P-1 shows 20 topics by rank on a 5-point scale. It also includes, as a comparison, the 10 topics that received the highest scores awarded by transportation educators. There is little doubt that there is a high congruence in the expectation of educators and practitioners in prioritizing the topics (Khisty, 1986).

Conversations with practitioners interviewed in my survey resulted in the following general observations, views, and suggestions with respect to enriching a required course in transportation:

- Students should be given the opportunity to tackle open-ended problems, defending their solutions or conclusions with short narratives.
- Students should be given every opportunity to tackle real-life problems. This could be in the form of one or more projects done individually or in a group. The group project idea should be encouraged because it provides students with a realistic experience in team dynamics.
- The ability to solve problems with incomplete or redundant data should be impressed on students through appropriate examples and class assignments.
- The fundamental principles underlying transportation should be emphasized.

TABLE P-1   TRANSPORTATION TOPICS

Topics	Practitioners (N = 50)		Educators (N = 51)	
	Score	Rank	Score	Rank
Geometric Design of Highways	4.80	1	4.62	2
Vehicle Operating Characteristics	4.72	2	4.34	5
Highway Capacity Studies	4.69	3	4.28	6
Intersection Design	4.58	4	4.00	8
Transportation Planning	4.44	5	3.96	9
Traffic Control Devices	4.32	6	4.38	4
Economics of Transportation	4.20	7	—	—
Land-Use/Transportation Interaction	4.18	8	—	—
Evaluation Techniques	4.13	9	3.90	10
Transportation Systems Management	4.06	10	—	—
Description of Transport System	4.04	11	4.72	1
Traffic Flow Characteristics	4.04	12	4.54	3
Traffic Safety	4.00	13	4.22	7
Contracting Procedures	3.92	14	2.30	—
Specifications	3.80	15	—	—
Operational Characteristics of Modes	3.80	16	—	—
Mass Transit	3.79	17	—	—
Airport Planning	3.63	18	—	—
Human Powered Transport	3.50	19	—	—
History of Transportation	3.41	20	—	—

Source: Khisty, 1986.

- To do justice to such topics as pavement design, construction methods, maintenance of facilities, and so forth, it would be best to address these topics in courses other than the required course.

On the basis of these results, I have framed the contents of this textbook to focus on clarity of exposition, topical coverage, technical content, and pedagogical elements. The 16 chapters in this text correspond closely to the ones indicated in the table. Some of the text's special characteristics are as follows:

- The material is built on ideas, concepts, and observations that students are likely to be most familiar with, e.g., roads, streets, highways, buses, bicyclists, pedestrians, and so on.

- The organization of the book and individual chapters has been carefully planned for easy transition from one area to another.

- While numerical problem solving has been emphasized where appropriate, the need to substantiate these numerical results, buttressed by proper explanations and discussions, has been duly illustrated. Several exercises at the end of chapters are the open-ended type questions requiring creativity and critical thinking.

- The latest manuals, codes, reports, and practices have been incorporated, e.g., *Highway Capacity Manual*, 1985; and *A Policy on Geometric Design of Highways and Streets*, 1984.

This text is a partially multimodal work in that it deals primarily with highways and the people who use them—motorized, nonmotorized, private, and public. A separate chapter on public transport deals to a limited extent with the rail mode. No attempt is made to describe transportation engineering as it relates to air transport, water transport, or pipelines. The results of the survey described earlier amply justifies the choice of topics.

The first three introductory chapters set the stage for the rest of the book; they are crucial and fundamental. Chapters 4 through 9 are traffic engineering–related, Chapter 10 deals with public transport, and Chapter 11 through 14 are planning-related. The last two are on evaluation and safety. A brief description of each chapter follows:

Chapter 1, "Transportation As a System," introduces the student to the field of transportation engineering, planning, and management. It provides an overview of transportation systems characteristics, hierarchies, and classifications.

Chapter 2, "Transportation Economics," covers the most elementary ideas in economics useful to the transportation engineer. Most of these principles are applied to problems taken up in later chapters.

Chapter 3, "The Land-Use/Transportation System," illustrates the basic interdependence between land use and transportation. It is a critical chapter for students to comprehend and one that introduces a myriad of basic concepts underlying this relationship.

Chapter 4, "Vehicle and Human Characteristics," describes how human beings, as vehicle operators, passengers, and pedestrians, interact with vehicles and the transportation facilities they use. This chapter synthesizes several topics connected with the human element, the vehicle, and the enviroment.

Chapter 5, "Traffic Flow Characteristics," examines the uninterrupted flow of vehicles moving individually or in groups on roadways or tracks, subject to constraints imposed by human behavior or vehicle dynamics. The fundamental equations of vehicular flow are derived by taking into consideration safety, speed requirements, and capacity.

Chapter 6, "Geometric Design of Highways," deals with proportioning of the physical elements of highways, such as vertical and horizontal curves, lane widths, and cross sections. The 1984 edition of *A Policy on Geometric Design of Highways and Streets*, published by the American Association of State Highway and Transportation Officials (AASHTO), is the principal source of reference.

Chapter 7, "Highway Capacity," involves the quantitative evaluation of a highway section, such as a freeway, multilane, or two-lane to carry traffic. The procedures and methodologies contained in the 1985 *Highway Capacity Manual* published by the Transportation Research Board (TRB) are used in this chapter.

Chapter 8, "Intersection Control and Design," deals with at-grade intersections and the traffic signals, signs, and markings needed to regulate, guide, warn, and channel traffic. The design of traffic signals is an important part of this chapter.

Chapter 9, "At-Grade Intersection Capacity and Level-of-Service," covers the analysis of intersections based on the procedures spelled out by the 1985 *Highway Capacity Manual*. Analysis is done at two levels: the operational and planning levels. Unsignalized intersections are also considered.

Chapter 10, "Public Passenger Transportation," describes those modes of passenger transportation open for public use, such as bus, light rail, and rail-rapid transit. Beginning with the historical development of urban transportation, the chapter includes a classification of mass transport systems and their capabilities to carry passengers. The operational designs of a simple rail and a bus system are also explained.

Chapter 11, "Urban Transportation Planning," presents the traditional four-step sequential process of travel forecasting. The chapter initially explains the general organization and philosophy of long- and short-range (TSM) planning, currently followed in the developed and developing world.

Chapter 12, "Local Area Traffic Management," deals with problems and solutions related to existing neighborhoods and their possible expansions and renovations. The planning and design of pedestrian and bicycle facilities and parking and terminal facilities are considered in detail.

Chapter 13, "Energy Issues Connected with Transportation," provides an introduction to techniques for energy planning and energy conservation.

Chapter 14, "TSM Planning," introduces the reader to the short-range component of transportation systems. Transportation Systems Management (TSM) covers a broad range of potential improvement strategies focussing on nonfacility and low-capital-cost operations.

Chapter 15, "Evaluation of Transportation Improvements," covers the basic techniques of benefit-cost analysis of alternative proposals, including cost-effective and multicriteria evaluation.

Chapter 16, "Transportation Safety," describes the Highway Safety Improvement Program (HSIP). It begins by examining the nature and characteristics of accidents by type, severity, contributing circumstances, and environmental conditions. Methods of identifying hazardous locations are also discussed.

This textbook is designed for use in engineering, city planning, and management courses. Although the emphasis in these courses may differ to some extent, a combination of chapters can be chosen for each course to suit specific objectives. Table P-2 may be used as a guide for structuring a course outline. Two courses are indicated for engineering students. The first is assumed to be a mandatory course, and the second may be an elective. Planning and management courses can cover the first three introductory chapters followed by the planning-related chapters, 10 through 16. The traffic engineering chapters may be omitted or briefly scanned.

TABLE P-2

Topics	1	2	3	4
1. Transportation As a System	x		x	x
2. Transportation Economics	x		x	x
3. The Land-Use/Transportation System	p	x	x	x
4. Vehicle and Human Characteristics	x			p
5. Traffic Flow Characteristics	x			p
6. Geometric Design of Highways	x		p	p
7. Highway Capacity	x		p	p
8. Intersection Control and Design	p	x		
9. At-Grade Intersection Capacity and Level-of-Service	p	x		
10. Public Passenger Transportation	p	x	x	x
11. Urban Transportation Planning	p	x	x	x
12. Local Area Traffic Management	p	x	x	x
13. Energy Issues Connected with Transportation		x	x	x
14. TSM Planning-Framework	p	x	x	x
15. Evaluation of Transportation Improvement	x		x	x
16. Transportation Safety	p	x		x

1 = Engineering 1 course (mandatory)
2 = Engineering 2 course (mandatory/elective)
3 = Planning (graduate)
4 = Management (graduate)
x = entire chapter: p = partial chapter

A companion textbook, *Laboratory and Field Manual for Transportation Engineering* (Prentice Hall, 1991) supplements this textbook for those students taking a lab course. My colleague Dr. Michael Kyte and I are the authors.

Although the initial chapters of this book were written and rewritten in Pullman, Washington, serious attempts to put several of the crucial chapters together were done at the University of Washington, Seattle, where I spent the 1984–1985 academic year on a sabbatical. I appreciate the interaction with colleagues at the University of Washington—Jerry Schneider, Nancy Nihan, Scott Rutherford, Stephen Ritchie, Joe Mahoney, Jimmy Hinze, and Sandor Veress—that proved most beneficial. Although Bob Davis of Prentice Hall at Seattle was instrumental in encouraging me to submit parts of the manuscript for possible consideration to Prentice Hall, it was Doug Humphrey, Senior Engineering Editor, who steered me through the prepublication process. It has been a delight to work with him. Also, Ms. Marianne Peters, the production editor, deserves a special word of thanks for her patience and guidance.

In writing this book I have been constantly reminded of the debt I owe to instructors, colleagues, and students, in India, Germany, and the United States. They have influenced my own views on several aspects of transportation engineering and planning. I am indebted to several individuals who reviewed various chapters of the manuscript and offered invaluable suggestions: J. D. Gupta, Michael Kyte, B. Kent Lall, Martin Lipincki, and Thomas Mulinazzi. A special word of thanks is due to Dr. Surinder Bhagat, Chairman of the Civil and Environmental Engineering Department, who provided encouragement and support.

From among my students, I particularly take great pride in mentioning the following who helped me in a number of ways: A. Alzahrani, M. Y. Rahi, Ping Yi, and Morgan Wong. Morgan deserves my special thanks for organizing and proofreading the chapters and spending many hundreds of hours typing the original manuscript. He has proved to be a dedicated and indispensable person during this long period of writing. Lastly, I thank my wife, Lena, for her constant encouragement and support. To each of those who helped in the preparation of this book, I express my deepest gratitude and appreciation. And like the rug weavers of yesteryear who chose their yarns and natural dies from several sources, I alone am responsible for the final design and product.

Finally, I would be especially grateful for suggestions, criticisms, and corrections that might improve this text book. A solutions manual is now available from the publisher.

REFERENCES

KHISTY, C. J. (1986). Undergraduate Transportation Engineering Education, in *Transportation Research Record #1101*, Transportation Research Board, National Research Council, Washington, DC, pp. 1–3.

KHISTY, C. J. (1987). Urban Planning Education for Civil Engineers, *ASCE Journal of Urban Planning and Development*, Nov., pp. 54–60.

WRIGHT, P. H. (1983). *Transportation Engineering Education: An Author's View*, paper presented at the Annual ASCE Conference, Houston, TX.

C. Jotin Khisty
B. Kent Lall

Chapter 1

Transportation As a System

1. INTRODUCTION

The importance of transportation in world development is multidimensional. For example, one of the basic functions of transportation is to link residence with employment and producers of goods with their users. From a wider viewpoint, transportation facilities provide the options for work, shopping, and recreation, and give access to health, education, and other amenities.

The field of transportation can be compared to a mansion with several stories, many chambers, and scores of connections. I would like to take the reader on a short tour of this mansion just to acquaint him or her with some of its characteristics. One of the prerequisites for accompanying me on this trip is to have an open mind. Almost everyone will have had several years of personal experience as a *user* of the transportation system, such as a car driver, a bus passenger, an elevator user, a frequent flyer, or just a sidewalk user. A very small fraction of readers may be involved in providing transportation services, such as a student who partially earns her livelihood by driving the morning express bus for the local transit company. Naturally, almost every person will tend to acquire his or her own personal viewpoint. No two persons can expect to come to the same conclusion about a problem confronting transportation even though they are each known to be highly objective and rational. Try as hard as you can to approach the field of transportation and its myriad problems with an open mind, free of presumptions and prejudice. Like food, shelter, clothing, and security, transportation is an integral part of human culture. Movement in a broad sense offers both inherent joy and pleasure and pain, suffering, and frustration. These factors will assume even greater importance in the years ahead.

1

1.1 National and Individual Involvement

Everybody is involved with transportation in so great a variety of ways that a mere listing of these ways would take us by surprise. Ultimately, all human beings are interacting over distance and time, and this interaction in itself creates involvement. To understand the theory of transportation, one must examine its relationship to various social, economic, and political institutions, and we undertake such an examination in this book (Wolfe, 1963).

The role of transportation in the day-to-day life of Americans can be appreciated just by reading the following (U.S. D.O.T., 1994):

> The transportation system includes about 200 million automobiles, vans, and trucks operating on about 4 million miles of streets and highways; over 100,000 transit vehicles operating on those streets, as well as more than 7,000 miles of subways, street car lines, and commuter railroads; 275,000 airplanes operating in and out of 17,000 airports; 18,000 locomotives and 1.20 million cars operating over 113,000 miles of railroads; 20 million recreational boats, 31,000 barges, and over 8,000 ships, tugs, and other commercial vessels operating on 26,000 miles of waterways; and 1.50 million miles of intercity pipelines.

Travel consumes roughly an hour of an average person's day, and roughly one-sixth of household expenditures. Americans make nearly a thousand trips per year per person, covering a distance of about 15,000 miles annually. It is estimated that households, businesses, and governments spend over $1 trillion to travel 3.8 trillion miles and to ship goods 3.5 trillion ton-miles each year. In summary, transportation accounts for 12% of Gross Domestic Product. The Bureau of Transportation Statistics (BTS) of the U.S. Department of Transportation issues an annual report indicating the state of the transportation system and its consequences, and this information is of vital importance to all students of transportation engineering and planning.

1.2 Progress in Transportation

The principles of transportation engineering have been evolving over many millennia. Human beings are known to have laid out and used convenient routes as early as 30,000 B.C. Although it was traders and migrants who opened up most major routes of communication, the military has generally been responsible for improving the status of early routes built by civilians. The first wheeled military vehicles were developed around 2500 B.C., and since then, significant resources have been devoted by rulers and their builders to constructing and maintaining communication routes in the form of roads (Lay, 1986).

Steady progress has since been maintained in providing the highway and street network (which forms the stationary component of the transportation system), in providing vehicles for moving people and goods over this network (which comprises the dynamic part), and in enhancing the ability of drivers (or controllers) to operate the vehicles. Basically, it is these three major interacting components that are to be studied critically in this book (Lay, 1986).

Before bicycles and motor vehicles came into fashion, vehicle speeds seldom exceeded 10 miles per hour (mph). Naturally, a surface of compacted broken stone made an ideal pavement surface, even for the solid iron wheels then in use. Today, the American highway system consists of about 4 million miles of high-class streets and highways, classified by function, into a series of interconnected networks, which translates to 1 mile of road for every square mile of land. This level of coverage provides access to almost every part of the nation by road. The centerpiece of the highway development program in the United States is the 42,000-mile freeway system, considered to be one of the greatest public works achievement since the dawn of history. In urban areas, the thrust has been in constructing complicated freeway interchanges, pedestrian and bicycle facilities, and high-occupancy vehicle and bus lanes. In recent years, the paucity of funds for new construction has placed the accent on maintenance, rehabilitation of pavements, and pavement management systems (NCPWI, 1986).

Vehicles (and pseudovehicles) have been in use since human beings learned to walk. People who traveled on foot could manage between 10 and 25 miles per day. It is claimed that the Incas were able to transmit messages at the rate of 250 miles per day by using fast runners over short stretches, thus achieving speeds of about 10 mph. Horses, on the other hand, could make almost 40 miles per day.

By the late 1840s, the horse-drawn street car appeared in a number of cities, operating at an average speed of about 4 mph. It was not until the 1880s that electrically propelled transportation was introduced. By the beginning of World War I, the electric street car had already had a major impact on the growth and structure of the city (Gray and Hoel, 1979).

The entire picture for transportation changed in 1885 with Daimler and Benz's introduction of the gasoline-powered internal-combustion engine. Within the last 100 years, the motor vehicle has revolutionized private transportation all over the world. Before the appearance of the motor vehicle, vehicle speeds seldom exceeded 10 mph. The car soon changed the situation, and for purposes of safety and efficiency, traffic signals were introduced at intersections (Lay, 1986).

Some of the most outstanding technological developments in transportation have occurred in the preceding 100 years (Karaska and Gertler, 1978):

- The first pipelines in the United States were introduced in 1825.
- The internal-combustion engine was invented in 1866.
- The first automobile was produced in 1886 (by Daimler and Benz).
- The Wright brothers flew the first heavier-than-air machine in 1903.
- The first diesel electric locomotive was introduced in 1921.
- Lindbergh flew over the Atlantic Ocean to Europe in 1927.
- The first diesel engine buses were used in 1938.
- The first limited-access highway in the United States (the Pennsylvania Turnpike) opened in 1940.
- The Interstate Highway system was initiated in 1950.
- The first commercial jet appeared in 1958.

- Human beings landed on the moon in 1969.
- The use of computers and automation in transportation grew dramatically through the 1960s and 1970s and continues today.
- Microcomputers have revolutionized our capabilities to run programs in the 1980s.

1.3 The Urban System and Transportation

In 1850, there were four cities in the world with more than 1 million people, and in 1950, there were about a hundred cities of this size. But what is most shocking is that by the year 2000, there will be over 1000 cities of this magnitude. Naturally, smaller cities will grow into bigger ones, and these in turn will form megalopolises (Bell and Tyrwhitt, 1972).

Several architects, planners, and engineers have developed matrices and frameworks to represent and understand the urban scene. In the mid-1950s, C. A. Doxiadis, a Greek city planner, gave a new meaning to the science of human settlements and attempted to represent it in the form of a grid. This matrix, called the *ekistic grid*, incorporates a spectrum of the range of human settlements (Figure 1-1). The abscissa of the grid shows population figures ranging from a single human being through an ecumenopolis consisting of approximately 30 billion people. Notice that the units on this horizontal axis generally increase in logarithmic progression by multiples of between 6 and 7, and this progression has been observed by other regional scientists (Bell and Tyrwhitt, 1972).

The five elements shown on the ordinate are nature, man, society, shell, and networks. Nature represents the ecological system within which the city must exist. Man and society are constantly adapting and changing, and in turn molding the city to be a

Community Scale	I	II	III	I	II	III	IV	V	VI	VII	VIII	IX	X	XI	XII
	1	2	3	4	5	6	7	8	9	10	11	12	13	14	15
	Man	Room	Dwelling	Dwelling group	Small neighbourhood	Neighbourhood	Small town	Town	Large city	Metropolis	Conurbation	Megalopolis	Urban region	Urbanized continent	Ecumenopolis
Nature															
Man															
Society															
Shells															
Networks															
Synthesis															
Population t (thousands) m (millions)	1	2	4	40	250	1.5t	7t	50t	300t	2m	14m	100m	700m	5,000m	30,000m

Ekistic logarithmic scale

Figure 1-1 Ekistic Grid (Bell and Tyrwhitt, 1972)

satisfactory environment. The built environment is represented by the shell, which is the traditional domain of the architectural, planning, and engineering professions. Highways, railroads, pipelines, telephones—indeed, the entire gamut of communications—provide the element of networks. To cater to the demands of faster and cheaper communication in the face of the growth of settlements, we are constantly devising means of substituting travel by communication. The sum total of all the elements and their interactions is represented by synthesis. Thus, the dimensions of this grid embrace not only the current situation but also the past and future. The primary advantage in looking at the "forest as well as the trees" is in understanding universal issues as well as local ones. Another, equally important issue is the need to understand the meaning of city structure and the factors determining it (Thomson, 1977). What are the elements that form the basic structure of society? How do these elements relate to one another, interact, and function? What techniques are available to understand and predict what is likely to happen in the future? These are some of the questions usually asked by professionals and citizens. The answers, if they exist at all, are complex and often contradictory.

2. THE FIELD OF TRANSPORTATION ENGINEERING

The desires of people to move and their need for goods create the demand for transportation. People's preferences in terms of time, money, comfort, and convenience prescribe the mode of transportation used, provided of course that such a mode is available to the user.

The Institute of Transportation Engineers (1987) defines transportation engineering as "the application of technological and scientific principles to the planning, functional design, operation, and management of facilities for any mode of transportation in order to provide for the safe, rapid, comfortable, convenient, economical, and environmentally compatible movement of people and goods." Traffic engineering, a branch of transportation engineering, is described as "that phase of transportation engineering which deals with planning, geometric design, and traffic operations of roads, streets, and highways, their networks, terminals, abutting lands, and relationships with other modes of transportation."

3. THE PRACTICE OF TRANSPORTATION ENGINEERING

Transportation engineering involves a diversity of basic activities performed by such specialists as policymakers, managers, planners, engineers, and evaluators. Figure 1-2 illustrates these activities in the context of some of the transportation modes in current use. Several fringe and developing modes in this figure have not been identified. Whereas airways, conveyors, highways, pipelines, railways, and waterways are comparatively commonplace, we need to explain the last three modes listed. When two or more modes are combined to provide utility and service to the public, the combination is known as a multimodal system. Exotic systems are those modes that are not yet

	POLICY MAKING	ADMIN-MANAGEMENT	PLANNING	ANALYSIS, SYNTHESIS AND DESIGN	CONSTRUCTION	OPERATIONS	MAINTENANCE	TESTING AND EVALUATION
AIRWAYS								
CONVEYORS								
HIGHWAYS								
PIPELINES								
RAILWAYS								
WATERWAYS								
MULTI-MODAL								
EXOTIC								
QUASI-TRANSPORT								

Figure 1-2   Transportation As a System
(Khisty, 1983)

being used commercially but that have been tested in a pilot project. Air-cushioned vehicles fall into this category. Transportation substitutes such as the telephone (as used widely in teleconferencing) and facsimile transmission of documents by wire and radio can be considered as quasi-transport (Hay, 1977).

4. THE NATURE OF TRANSPORTATION ENGINEERING

Transportation engineering is a multidisciplinary area of study and a comparatively new profession that has acquired theoretical underpinnings, methodological tools, and a vast area of public and private involvement. The profession carries a distinct societal responsibility. A wide comprehensive training in transportation is therefore the desirable goal of all transportation education (Khisty, 1981; Hoel, 1982).

Because of the multidisciplinary content of transportation engineering, we find that concepts drawn from the fields of economics, geography, operations research, regional planning, sociology, psychology, probability, and statistics, together with the customary analytical tools of engineering, are all used in training transportation engineers and planners.

Figure 1-3 illustrates, in a general way, the interdisciplinary breadth and the depth of involvement of transportation engineering. Most specialization in transportation engineering occurs at the graduate level; undergraduates receive an overall general view of the elements of transportation engineering (Wegman and Beimborn, 1973). Traditionally, the upper left-hand part of this figure represents the "soft" side of transportation engineering, and the lower right-hand side, representing pavement design, bridge engineering, and drainage, may be looked on as the "hard" side of transportation. However, there is no definite demarcation between the two (Khisty, 1985, 1986, 1987).

5. TRANSPORTATION VIEWED AS A SYSTEM

The systems approach represents a broad-based, systematic approach to problems that involve a system. It is a problem-solving philosophy used particularly to solve complex problems (Wortman, 1976).

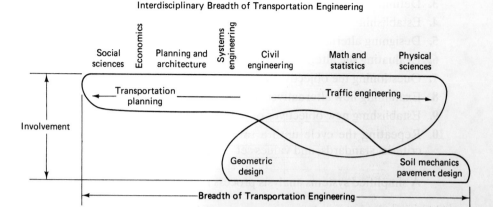

Figure 1-3 Conceptual Outline of Interdisciplinary Training for Engineering Students (Khisty, 1981: Wegman and Beimborn, 1973)

A *system* is a set of interrelated parts, called *components*, that perform a number of functions in order to achieve common goals. *System analysis* is the application of the scientific method to the solution of complex problems. *Goals* are desired end states. Operational statements of goals are called *objectives*; these should be measurable and attainable. Feedback and control are essential to the effective performance of a system. The development of objectives may in itself involve an iterative process. Objectives will generally suggest their own appropriate *measures of effectiveness* (MOEs). An MOE is a measurement of the degree to which each alternative action satisfies the objective. Measures of the benefits forgone or the opportunities lost for each of the alternatives are called *measures of costs* (MOCs). MOCs are the consequences of decisions. A criterion relates the MOE to the MOC by stating a decision rule for selecting among several alternative actions whose costs and effectiveness have been determined. One particular type of criterion, a *standard*, is a fixed objective: the lowest (or highest) level of performance acceptable. In other words, a standard represents a cutoff point beyond which performance is rejected (Cornell, 1980).

With reference to communities, we often find a set of irreducible concepts that form the basic desires and drives that govern our behavior. To these desires, the term *values* is assigned. Values form the basis for human perception and behavior. Because values are shared by groups of people with similar ties, it is possible to speak of *societal* or *cultural values*. Fundamental values of society include the desire to survive, the need to belong, the need for order, and the need for security.

A *policy* is a guiding principle or course of action adopted to forward progress toward an objective. Evaluating the current state of a system and choosing directions for change may be considered as policymaking.

STEPS IN SYSTEM ANALYSIS

1. Recognizing community problems and values
2. Establishing goals

3. Defining objectives
4. Establishing criteria
5. Designing alternative actions to achieve steps 2 and 3
6. Evaluating the alternative actions in terms of effectiveness and costs
7. Questioning the objectives and all assumptions
8. Examining new alternatives or modifications of step 5
9. Establishing new objectives or modifications of step 3
10. Repeating the cycle until a satisfactory solution is reached, in keeping with criteria, standards, and values set

A simplified systems analysis process is shown in Figure 1-4.

6. TRANSPORTATION POLICYMAKING

Transportation planners and engineers recognize the fact that transportation systems constitute a potent force in shaping the course of regional development. Transportation encompasses a broad set of policy variables, and the planning and development of transportation facilities generally raises living standards and enhances the aggregate of community values.

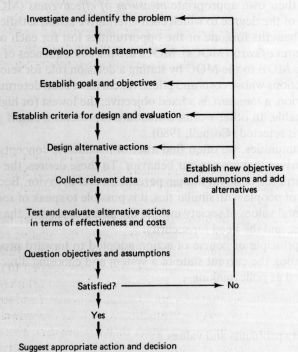

Investigate and identify the problem

Develop problem statement

Establish goals and objectives

Establish criteria for design and evaluation

Design alternative actions

Collect relevant data

Establish new objectives and assumptions and add alternatives

Test and evaluate alternative actions in terms of effectiveness and costs

Question objectives and assumptions

Satisfied? ──────────▶ No

Yes

Suggest appropriate action and decision

Figure 1-4 The System Analysis Process

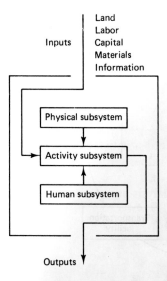

Figure 1-5 Transportation System Model: Transportation Processor (NHI, 1980)

Engineers and planners also recognize that most transportation issues can be characterized by great size, breadth, complexity, diversity, cost, and uncertainty. An example of a transportation system model is shown in Figure 1-5. It consists of inputs, such as land, labor, and capital, fed into three subsystems: the physical subsystem, the activity subsystem, and the human subsystem. The *physical subsystem* consists of vehicles, pavements, tracks, rights-of-way, terminals, and other manufactured or natural objects. The *activity subsystem* includes riding, driving, traffic control, and so on. These activities interface with the *human subsystem*—individuals and groups of people who are involved with the physical and activity subsystems. Outputs from the system include the movement of people and goods and improvement or deterioration of the physical environment.

7. MOVEMENT AND TRANSPORTATION

A city can be considered as a locational arrangement of activities or a land-use pattern. The location of activities affects human beings, and human activities modify locational arrangements. Interaction between activities is manifested by the movement of people, goods, and information.

The reason that people and goods move from one place to another can be explained by the following three conditions: (1) *complementarity*, the relative attractiveness between two or more destinations; (2) the desire to overcome distance, referred to as *transferability*, measured in terms of time and money needed to overcome this distance and the best technology available to achieve this; and (3) intervening opportunities to competition among several locations to satisfy demand and supply. How people and goods move from an origin to a destination is a matter of mode choice (a person might choose to take the bus downtown rather than use her car). This decision is made depending on such attributes as time, speed, efficiency, costs, safety, and

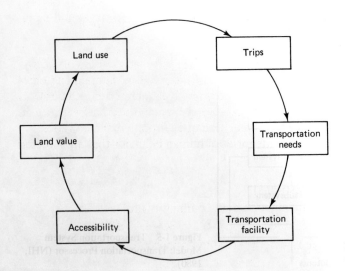

Figure 1-6   Land Use/Transportation Cycle

convenience. Geographers describe a *trip* as an event and *travel* as a process (Abler et al., 1971).

A simple connection between land use and transportation is shown in Figure 1-6. Land use is one of the prime determinants of movement and activity. This activity, known as *trip generation*, will dictate what transportation facilities, such as streets and bus systems, will be needed to move traffic. When such additional facilities have been provided, the system will naturally have increased accessibility.

A change in accessibility will determine the change, if any, in the value of land, and this change may affect the way the land is currently used. If such a change does occur (e.g., a residential neighborhood changes to a commercial area), the trip generation rate (e.g., the number of trips per acre of land) will change and a whole cycle of changes will result. Note that this cyclical process is a highly simplified representation of reality, and also that other land market forces are not shown in the figure. It does, however, illustrate the fundamental interconnection of transportation and land use.

8. OVERVIEW OF TRANSPORTATION SYSTEMS CHARACTERISTICS

The physical plant of most transportation systems consists of four basic elements:

1. *Links:* the roadways or tracks connecting two or more points. Pipes, beltways, sealanes, and airways can also be considered as links.
2. *Vehicles:* the means of moving people and goods from one node to another along a link. Motorcars, buses, ships, airplanes, belts, and cables are examples.
3. *Terminals:* the nodes where travel and shipment begins or ends. Parking garages, off-street parking lots, loading docks, bus stops, airports, and bus terminals are examples.

4. *Management and labor:* the people who construct, operate, manage, and maintain the links, vehicles, and terminals.

These four elements interact with human beings, as users or nonusers of the system, and also with the environment. The behavior of the physical, human, and environmental subsystems is highly complex because it involves interaction of people as drivers, riders, and nonriders, using vehicles of differing character and performance on links with diverse physical characteristics in a myriad of environmental conditions.

Researchers have identified nine categories of human behavior that are affected by transportation:

1. Locomotion (passengers, pedestrians)
2. Activities (e.g., vehicle control, maintenance, community life)
3. Feelings (e.g., comfort, convenience, enjoyment, stress, likes, dislikes)
4. Manipulation (e.g., modal choice, route selection, vehicle purchase)
5. Health and safety (e.g., accidents, disabilities, fatigue)
6. Social interaction (e.g., privacy, territoriality, conflict, imitation)
7. Motivation (positive or aversive consequences, potentiation)
8. Learning (e.g., operator training, driver education, merchandising)
9. Perception (e.g., images, mapping, sensory thresholds)

Similarly, there are at least 11 properties of the physical environment that have a direct impact on human behavior. Details of these environments follow (Khisty, 1983):

1. *Spatial organization.* This dimension often includes the shape, scale, definition, bounding surfaces, internal organization of objects and society, and connections to other spaces and settings. Indeed, this is the dimension that most people are referring to when they talk about the physical environment. The degree of dispersion, concentration, clustering, and proximity of facilities is also included.

2. *Circulation and movement.* This property includes people, goods, and objects used for their movement—cars, trains, highways, and rails—and also the forms of regulating them, such as corridors, portals, turnstiles, and open spaces.

3. *Communication.* Both explicit and implicit signals, signs or symbols communication, required behavior, responses, and meanings are covered by this dimension; in essence, these are the properties of the environment that give users information and ideas.

4. *Ambience.* This dimension usually includes such items as microclimate, light, sound, and odor. Those features of the environment that are critical for maintaining the physiological and psychological functioning of the human organism are included. For instance, a passenger in a bus enjoys the comfort of being protected from the weather but may be subjected adversely by the noise level and vibrations prevalent in the bus.

5. *Visual properties.* The environment as it is perceived by its users is generally implied by this property and includes color, shape, and other visual modalities.

6. *Resources.* The physical components and amenities of a transportation system—paths, terminals, and vehicles—could be included. The measures of these resources could embrace such dimensions as the number of lanes or the square footage of the terminals.

7. *Symbolic properties.* The social values, attitudes, and cultural norms that are represented or expressed by the environment fall into this category.

8. *Architectonic properties.* This refers to the sensory or aesthetic properties of the environment.

9. *Consequation.* This is that characteristic of the environment that strengthens or weakens behavior. Measures of consequation include such items as costs, risks, and congestion.

10. *Protection.* Safety factors in general are implied in this category.

11. *Timing.* All the items mentioned before are scheduled in time and some of them fluctuate with various cyclical rhythms, such as daily, weekly, or hourly timings.

Figure 1-7 illustrates the impact of the environment on aspects of human behavior relevant to transportation. Further, individual differences among the population that use and provide transportation should also be considered. These dimensions include age, ethnicity, income, car ownership, economic status, health, and skills. Also, some of the basic ingredients that are embraced by transportation designs are as follows (Khisty, 1983): safety, security, convenience, continuity, comfort, system coherence, and attractiveness. The variables contained in these lists and figures may appear overwhelming, but they do set forth a systems approach to the interconnection between transportation and human behavior.

Transportation systems can be evaluated in terms of three basic attributes:

1. *Ubiquity:* the amount of accessibility to the system, directness of routing between access points, and the system's flexibility to handle a variety of traffic condi-

Environmental Aspects \ Human Behavior	Activities	Locomotion	Social Interaction	Feelings	Perception	Motivation	Health and Safety	Learning	Manipulation
Spatial Organization		X	X					X	
Circulation and Movements	X	X	X				X	X	
Communication	X	X	X					X	
Ambience	X			X		X	X		X
Visual Properties					X				X
Resources	X			X			X	X	X
Symbolic Properties	X	X			X			X	X
Architectural Properties	X	X	X		X		X	X	X
Consequation		X				X		X	X
Protection							X	X	
Timing		X	X						

Figure 1-7 Relation between Aspects of Transportation and Their Effects on People (Khisty, 1983, p. 43)

tions. Highways are very ubiquitous compared to railroads, the latter having limited ubiquity as a result of their large investments and inflexibility. However, within the highway mode, freeways are far less ubiquitous than local roads and streets.

2. *Mobility:* the quantity of travel that can be handled. The capacity of a system to handle traffic and speed are two variables connected with mobility. Here again, a freeway has high mobility, whereas a local road has low mobility. Water transport may have comparatively low speed, but the capacity per vehicle is high. On the other hand, a rail system could possibly have high speed and high capacity.

3. *Efficiency:* the relationship between the cost of transportation and the productivity of the system. Direct costs of a system are composed of capital and operating costs, and indirect costs comprise adverse impacts and unquantifiable costs, such as safety. Each mode is efficient in some aspects and inefficient in others.

Table 1-1 summarizes the basic characteristics of major transportation modes.

9. TRANSPORTATION SYSTEMS, HIERARCHIES, AND CLASSIFICATION

The classification of transportation modes into different operational systems or functional classes is useful in understanding the complexity of the total transportation system. For example, the emergence of functional classification as the predominant method of grouping highways has helped engineers to communicate with economists, sociologists, planners, and administrators more effectively.

A series of distinct travel movements are recognizable in most trips. On a highway system, for example, these movements are a main movement along a freeway, a transition to an arterial via a freeway off-ramp, then further movement along an arterial where traffic is distributed and later collected via a collector, finally accessing a terminal (a garage or on-street parking lot). Further movement of the passenger may be as a pedestrian on a sidewalk of a local street, and finally to his or her destination.

Note that similar movement is likely to have occurred when the trip started to the point when the passenger got on the freeway facility. Figure 1-8 shows this hierarchy of movement. The inadequacy of parts of the hierarchy to accommodate each trip movement is one of the reasons that systems fail or become obsolete.

An illustration of a functionally classified rural highway network is shown in Figure 1-9, and Figure 1-10 shows a functionally classified suburban street network. Closely connected to the concept of highway classification are the concepts of accessibility and mobility to which reference has been made in this chapter. Figure 1-11 illustrates the connection between access, mobility, and the functional classification of streets and highways. Freeways and arterials have a high level of mobility because they allow high speeds but do not provide sufficient accessibility to adjoining property. Local streets, on the other hand, offer the maximum accessibility, but users find the mobility on such streets rather poor because of the slow speeds. A city or region must therefore try to provide the right proportion of freeways, arterials, collectors, and local

TABLE 1-1 OVERVIEW OF MAJOR TRANSPORTATION SYSTEMS

System	Ubiquity	Mobility	Efficiency	Mode	Passenger service	Freight service
Highways	Very high: land owners have direct access to a road or street. Direct routing limited by terrain and land use.	Speeds are limited by human factors and speed limits. Capacity per vehicle is low, but many vehicles are available.	Not high as regards safety, energy, and some costs.	Truck	Negligible	Intercity, local, farm to processing and market centers. Small shipments; containers
				Bus	Intercity and local	Packages (intercity)
				Automobile	Intercity and local	Personal items only
				Bicycle	Local; recreational	Negligible
Rail transport	Limited by large investment in route structure. Also constrained by terrain.	Speed and capacity can be higher than for highway modes.	Generally high, but labor costs may result in low cost efficiency.	Railroads	Mostly < 300 miles and suburban commuters	Intercity. Mostly bulk and oversized shipments; containers
				Rail transit	Regional, intracity	None
Air transport	Airport costs reduce accessibility. Excellent opportunity for direct routing.	Speeds are highest, but capacity per vehicle is limited.	Fairly low as regards energy and operating costs.	Air carriers	Mostly > 300 miles and across bodies of water	High-value freight (no bulk) on long hauls; containers
				General aviation	Intercity; business, recreation	Minor
Water transport	Direct routing and accessibility limited by availability of navigable waterways and safe ports.	Low speed. Very high capacity per vehicle.	Very high: low cost, low energy use. Safety varies.	Ships	Cruise traffic. Ferry service	Bulk cargos, especially petroleum; containers
				Barges	None	Bulk cargos, especially petroleum, containers
				Hovercraft	Ferry service	Minor
Continuous-flow systems	Limited to few routes and access points.	Low speeds. High capacity.	Generally high: low-cost energy use.	Pipelines	None	Liquids, gases, and slurries on short and long hauls
				Belts	Escalators and belts for short distances	Bulk materials handling, mostly < 10 miles
				Cables	Lifts and tows for short distances in rough terrain	Materials handling in rough terrain

Source: Homburger and Kell, 1988.

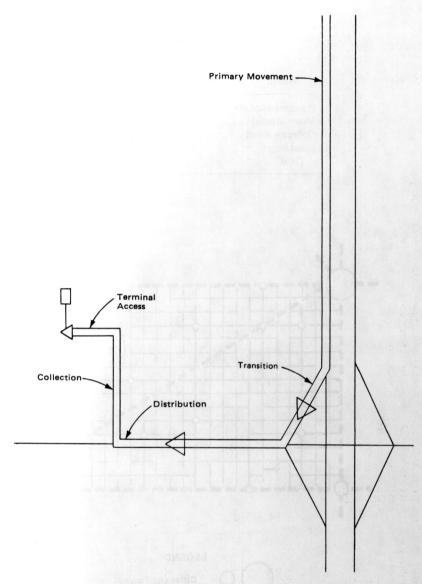

Figure 1-8 Hierarchy of Movement (AASHTO, 1984)

roads to offer a balanced system of accessibility and mobility for the convenience of its inhabitants.

Typical distribution of functional streets and highways. A typical urban area street distribution is shown in Table 1-2. The urban principal arterials serve the major activity centers, such as universities, shopping centers, and stadiums, and also the highest-traffic-volume corridors. Notice that they carry a high proportion of the

TABLE 1-2   TYPICAL DISTRIBUTION OF URBAN
FUNCTIONAL SYSTEMS

	Range	
System	Travel volume (%)	Mileage (%)
Principal arterial	50	5
Minor arterial	25	10
Collector street	5	10
Local street	20	75
Total	100	100

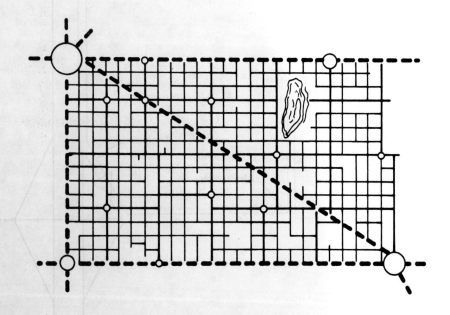

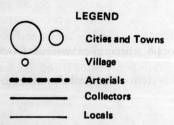

LEGEND

◯ ◯ **Cities and Towns**

○ **Village**

▬ ▬ ▬ ▬ ▬ ● **Arterials**

───── **Collectors**

───── **Locals**

Figure 1-9   Schematic Illustration of a Functionally Classified Rural Highway Network (AASHTO, 1984)

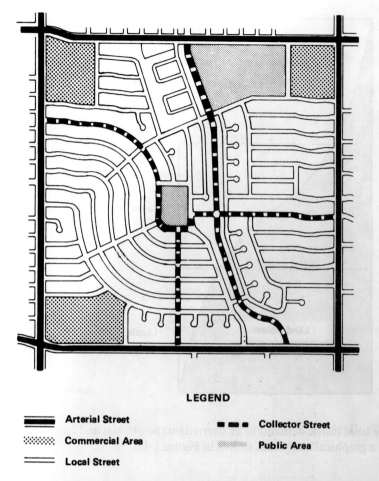

LEGEND

━━━━ **Arterial Street**   ■ ■ ■ **Collector Street**

░░░░ **Commercial Area**   ▒▒▒▒ **Public Area**

═══ **Local Street**

Figure 1-10   Schematic Illustration of a Portion of a Suburban Street Network (AASHTO, 1984)

total urban area travel on a low mileage. Urban minor arterials accommodate trips of moderate length at fairly high speeds and connect the principal arterial system with the collector streets. The collector street system provides both land access service and traffic circulation within residential neighborhoods and commercial and industrial areas. Collector systems interconnect the minor arterial system with the local street system. Direct access to abutting lands is provided through a local street system.

Achievement of a balanced system of streets is an important goal in order to best serve the varying types and number of trips that occur in a metropolitan region. This balance is based on an analysis of total mileage (vehicle-miles of travel, or VMT) occurring in each class. For example, no more than 10% of the VMT occurs on local streets, and collectors take no more than 20% of the VMT. This leaves about 70% of

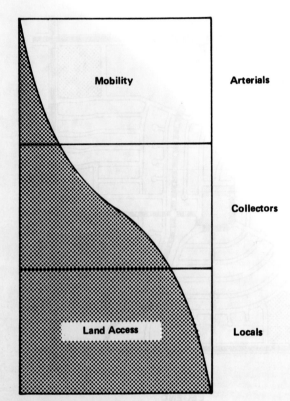

Figure 1-11   Relationship of
Functionally Classified Systems in
Service Traffic Mobility and Land Access
(AASHTO, 1984)

the total traffic demand to be carried on the arterial and freeway systems. An example
of a graphical analysis is shown in Figure 1-12.

10. COMMUNICATIONS, TRANSPORTATION, AND TRANSPORT GAPS

It has been hypothesized that in the United States, three modes of transportation
dominate the overall hierarchy of transportation available to people: walking for short
distances, the car for medium distances, and the airplane for long distances. An overall transportation hierarchy exists that includes the network hierarchies and urban
hierarchies.

Planners are well aware of the "refusal" distance of an average pedestrian who
uses the street system, which is generally about 400 meters (m) or $\frac{1}{4}$ mile. Beyond 400
m, the majority of pedestrians demand some type of mechanical system to transport
them. If, for instance, a pedestrian wants to cover a distance 10 times greater than 400
m (i.e., 4 kilometers (km) or 2.5 miles), he will not ordinarily agree to spend 50 minutes
walking, although he may very well have the time to spare. He will generally seek a
faster means of transport. There is ample evidence to show that the trip maker's choice

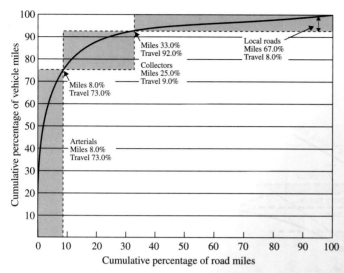

Figure 1-12 labels:

Miles 33.0%
Travel 92.0%

Local roads
Miles 67.0%
Travel 8.0%

Collectors
Miles 25.0%
Travel 9.0%

Miles 8.0%
Travel 73.0%

Arterials
Miles 8.0%
Travel 73.0%

Figure 1-12   Vehicle-Miles of Travel by Street Class

of mode is not based on cost alone, but also on travel time. Subconsciously, distance is connected with time.

It has also been observed that as science and technology have given people the ability to travel at higher and higher speeds, we have readily used these higher speeds, despite their costs, in preference to lower speeds. Figures 1-13 and 1-14 indicate that the demand for speed depends on the distance traveled. Long distances mean greater

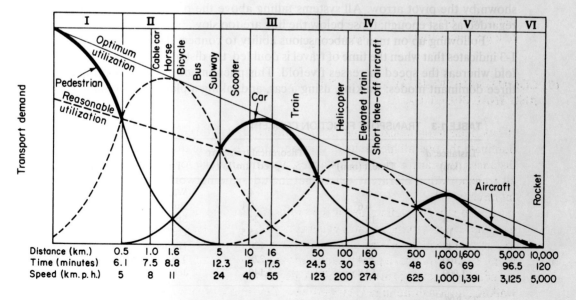

Figure 1-13   Transportation Gaps (Abler et al., 1971)

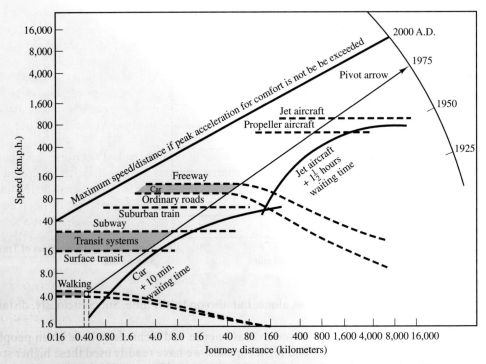

Figure 1-14 Demands for Speed Depend on Distance Traveled (Abler et al., 1971)

speed. Theoretically, with time, all forms of transportation move at higher speeds, as shown by the pivot arrow. All systems falling above the pivot at a point in time are regarded as fast enough; those below the line are too slow.

Following up on man's subconscious ability to connect distance and time, Table 1-3 indicates that when the time of travel is doubled, the distance covered increases ten-fold whereas the speed increases fivefold. This phenomenon generally produces the three dominant modes: walking, using a car, and taking a plane. At the same time, it

TABLE 1-3 TRANSPORT FUNCTION CONCEPT

Distance, d (km)	Time, t (min)[a]	Theoretical transport speed (km/hr)	Transport alternative
0.4	5	4.8	Walking
1	6.6	9.1	Bus (town center)
4	10	24	Streetcar or bicycle
10	13.2	45.5	Car (urban or suburban)
40	20	120	Highway
100	26.4	228	Train or aeroplane
1,000	52.8	1,140	Jet

[a] $t = 6.6d^{0.3}$.
Source: Kolbuszewski, 1979.

produces pronounced "transportation gaps." Of course, these figures, shown in Figure 1-14, are not uniform for people of different levels of economic development. One can easily dismiss the notion of a transportation gap on the grounds that evidently there is no market available for a mode of this description in the transportation hierarchy. On the contrary, it may be advantageous to understand the real needs of the transportation user and the boundary conditions that society and the environment have imposed on different modes in this hierarchy. With the growing shortages and uncertainties of energy supply, the identification and assessment of transportation gaps are indeed crucial.

How can the transportation gaps be filled, particularly the gap between the pedestrian and the car user? Society's spatial organization may have to be altered to fill these gaps if prevailing transportation modes and networks do not provide the answer. There is also the possibility of a modal shift from actual trip making through conventional transport to artificial trips through telecommunication. This relationship is shown in Figure 1-15. Artificial trips will undoubtedly result in cutting down the traffic movement, saving millions of gallons of gasoline and decreasing the risks of accidents (Abler et al., 1971).

The progressive elimination of time from transportation is probably one of the greatest achievements of the later twentieth century, and it is the electronic devices in use that have made this possible. Table 1-4 illustrates speeds of various modes in orders of magnitude to the base 4.5. It must be noted that transmission of thoughts, ideas, and concepts, as well as the general area of information, can be done at the speed of light, but the physical movement of people and goods is still constrained generally to the speed of a jet plane.

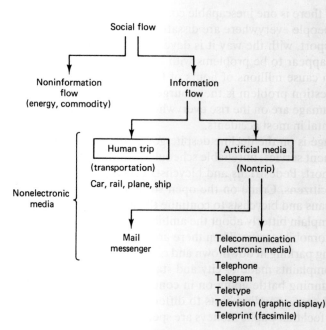

Figure 1-15 Relationship between Human Trip and Artificial Media with Social Flow of Information (Kolbuszewski, 1979)

TABLE 1-4 SPEEDS AND MODES

Mode	Speed (mph)	Factor (approx.)
Army (on foot)	0.5	$4.5^0/2 = 0.5$
Wagon, coach, sailing ship	2.25	$4.5^1/2 = 2.25$
Early steamship	10	$4.5^2/2 = 10.13$
Car, train	45	$4.5^3/2 = 45.56$
Propeller plane	200	$4.5^4/2 = 205.03$
Jet plane	900	$4.5^5/2 = 922.64$
Missile	4,000	$4.5^6/2 = 4,152$
Orbiting satellite	18,000	$4.5^7/2 = 18,683$
Telecommunication	Speed of light	$4.5^{14}/2 = 1.396 \times 10^9$

Source: Wolfe, 1963, p. 120.

11. TRANSPORTATION AND TRANSPORTATION-RELATED PROBLEMS

So far, nothing has been said directly about the transportation problems confronting developed and developing countries. Because urban and rural transportation affects the economic and social efficiencies of regions and cities, and because almost everyone makes use of some form of transport, the operation of transportation systems is a major topic on the agenda of politicians and managers.

Solving transportation problems has become one of the chief tasks confronting city and state governments in this country and abroad. Today's transportation problems have arisen despite enormous annual expenditures on various transportation facilities and systems.

Thomson (1977) says: "If there is one inescapable conclusion from a study of the world's major cities, it is that people everywhere are dissatisfied, often to the point of public protest, with their transport, with the way it is developing and the effects it is having on their cities." There appear to be problems with traffic movement, particularly during peak hours, which cause millions of hours of total delay to the system's users. Coupled with this congestion problem is the scourge of accidents. Fatal accidents, injuries, and property damage are on the rise everywhere, although there seems to be little that is really accidental in most accidents.

Public transportation usage is on the decline despite governmental intervention and high subsidization. Infrequent service, unreliable schedules, and rising fares do little to popularize public transport. Pedestrians and bicyclists constantly complain of being treated as second-class citizens. Crime on the open street does not encourage even health-conscious pedestrians and bicyclists to continue their mode of transportation. Neighborhood groups complain bitterly about the ambient noise and pollution of the atmosphere because of automobile traffic. Then there are the constant complaints regarding curtailment of existing parking in downtown and commercial areas. In recent years, one also hears about complaints made to city and state government regarding equity problems. There is a running battle going on in connection with facility construction and the benefits and corresponding costs to different segments of society. Poor neighborhoods generally feel that public moneys are spent on providing the maximum benefits to the rich at the expense of the poor. Problems in transportation are often solved on a purely economic basis. Energy prices seem to be constantly on the rise and the public appears to blame the government for not intervening in curbing automobile ownership. Citizens in large metropolitan areas claim that city, state, and federal governments are scared to implement road pricing schemes such as the one implemented in Singapore to reduce congestion through area licensing (and increased car occupancy).

One wonders why there appears to be this apathy on the part of government institutions to do something positive. One of the reasons for this state of affairs is of course the "openness" of transportation systems, which makes it difficult, if not impossible, for governments to apply technological alternatives to solve transportation problems. This "openness" is a symptom of the highly intricate connections the transport industry (including the automobile manufacturers) have with everything else. It is this intertwining with the rest of the economy that almost precludes interfering with the entire system (Mitric, 1977).

Be that as it may, the problems one sees with urban transportation emphasize the importance of examining public policy more closely. Owen (1976) suggests that whatever policymakers do in future, they will have to keep the following points at the backs of their minds:

- The tremendous dependence on the automobile and relation of this dependency to urban form and the location of people and their jobs
- The evolution of a public transportation system capable of serving the entire urban area effectively
- The capability of government and its policies to provide a transportation system that is equitable to both car owners and the carless
- The combination of new technologies and effort to design a more satisfying urban environment in the long run
- Complexities of new problems due to the uncertainty of energy supplies
- Solving urban transportation problems through the public and private sectors; and the cost implications of alternative federal policies

12. TRANSPORTATION AND SUSTAINABILITY

Figure 1-16 is a simple model showing basic interactions between three essential components: persons and/or goods that need to be transported; motor vehicles or rail cars or water vessels that are used to move people and/or goods; and the infrastructure, which includes a variety of fixed installations such as roads and streets, railroads, pipelines, canals, airports, and harbors. The *activity system* comprises the movements of persons and goods between two or more points or positions in space relative to the infrastructure. This activity system can be conceived as a market for movement. The size of a market is measured in terms of trade and travel at the macro level, and the size, the type, and the frequency of shipments are important

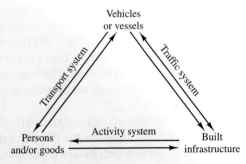

Figure 1-16   A Basic Model Connecting Vehicles/Vessels, Persons/Goods, and the Built Infrastructure

factors at the micro level. The *transport system* consists of persons and goods needing some kind of vehicle or vessel (besides using their own personal power) to move them from one position to another. Each movement is a transport service. The demand (or need) for services is matched by an equivalent supply of services by vehicles and their operators on roads, tracks, and paths. Consequently, vehicles are looked on as means of transport. In the *traffic system*, actual physical movement of transport is realized in space and time, assuming that people and goods move together with the means of transport (i.e., vehicles) along physical networks. Each vehicle (or a set of vehicles physically connected to one another) is seen as a traffic unit, and the resulting flow is normally measured as the number of vehicles per unit of time on a specific link of the infrastructure, and, in effect, utilizing (or occupying) a certain part of the infrastructure for a specific time.

In the last two decades, there has been a coordination of the three subsystems through the communication, information, and transportation revolution, and an attempt is being made to move people and goods by the right mode, in the right quantity, by the best available quality, to the right place at the right time. For example, the interaction between the vehicle and infrastructure subsystems is being investigated for a suitable "control" system. Currently, there are a range of very large research programs going on: PROMETHEUS and DRIVE in Europe, SSVS (Super Smart Vehicle System) in Japan, IVHS (Intelligent Vehicle Highway System) in the United States, all representing attempts to add different kinds of information and control functions to the vehicle-infrastrusture system.

These vast developments in transportation have all come at a price. For example, air pollution, noise pollution, and degradation of the land have brought humankind to realize that there are limits to "progress." Looming threats to the world's climate and the undermining of other global commons are making the transition to sustainability stronger. The 1987 Brundtland Report defines sustainable development as development that satisfies present needs without compromising the capacity of future generations to satisfy their needs (World Commission on Environment and Development, 1987).

The concept of sustainable development is now receiving worldwide attention. It encompasses both a mandate to stop irreversible damage to the ecosystem and recognition that the pursuit of environmental goals must accommodate the human needs of the planet's burgeoning population, both now and in the future. The concept of sustainability ties together economic, environmental, social, and energy goals, all of which affect and are affected by transportation.

13. EMERGING TRANSPORTATION TECHNOLOGIES

During the last decade, a broad spectrum of advanced technology, ranging from in-vehicle components to advanced traffic management systems, has come together and is called Intelligent Transportation Systems (ITS). Although many of these technologies are currently in the experimental or limited real-world application stage, the U.S. D.O.T. intends to deploy ITS technology across the United States in 75 of the largest metropolitan areas within the next 10 years.

The urgency for using ITS technology stems from the fact that in the past 10 years, there has been a 30% increase in traffic, resulting in Americans losing 2 billion hours a year to gridlock (equivalent to losing $20 billion a year). It has also been estimated that businesses lose $40 billion a year to traffic congestion. It is envisaged that ITS technology will reduce the congestion problem considerably.

There are at least nine components of ITS; a short description of each follows.

1. Smart traffic signal control systems sense heavy traffic flows at road intersections and adjust the timing of signals automatically to accommodate the flows.

2. Freeway management systems meter vehicles entering freeways with signals on on-ramps. Although this technology has been in use for a long time, the system is being refined by linking it with other detection systems, for example, accident surveillance systems.

3. Transit management systems help managers to control and monitor the movements of transit vehicles and adjust schedules accordingly.

4. Incident management systems (IMS) detect and manage nonrecurrent traffic congestion caused by random unpredictable incidents, such as traffic accidents, lane blockages, and hazardous material spills resulting in major traffic congestion for considerable time periods. IMS has been in practice for quite some time.

5. Electronic toll collection on bridges and roads has been in operation on over a dozen sites.

6. Electronic fare payment systems have also been in practice for a long time but have to be refined further.

7. Emergency response allows emergency vehicles to control traffic lights at intersections. The emergency vehicle driver can hold the green phase until the intersection is cleared.

8. Travel information systems provide traffic information to users so that they can adjust their travel plans based on what they learn.

9. Route guidance systems are based on Global Positioning System (Satellite) technology and assist motorists with distance and direction information to selected destinations.

14. INTERMODAL SURFACE TRANSPORTATION EFFICIENCY ACT OF 1991 (ISTEA)

On December 18, 1991, President Bush signed the Intermodal Surface Transportation Efficiency Act of 1991, providing authorizations for highways, highway safety, and mass transportation for the next 6 years. Total funding of about $155 billion was available in fiscal years (FY) 1992–1997.

The purpose of the Act is clearly enunciated in its statement of policy: "to develop a National Intermodal Transportation System that is economically efficient, environmentally sound, provides the foundation for the Nation to compete in the global economy and will move people and goods in an energy efficient manner."

The provisions of the Act reflect these important policy goals. Some of the major features include the following:

- A National Highway System (NHS), consisting primarily of existing interstate routes and a portion of the Primary System, is established to focus federal resources on roads that are the most important to interstate travel and national defense, roads that connect with other modes of transportation, and are essential for international commerce.
- State and local governments are given more flexibility in determining transportation solutions, whether transit or highways, and the tools of enhanced planning and management systems to guide them in making the best choices.
- New technologies, such as intelligent vehicle highway systems and prototype magnetic levitation systems, are funded to push the nation forward into thinking of new approaches in providing twenty-first-century transportation.
- The private sector is tapped as a source for funding transportation improvements. Restrictions on the use of federal funds for toll roads have been relaxed and private entities may even own such facilities.
- The Act continues funds for mass transit.
- Highway funds are available for activities that enhance the environment, such as wetland banking, mitigation of damage to wildlife habitat, historic sites, activities that contribute to meeting air quality standards, a wide range of bicycle and pedestrian projects, and highway beautification.
- Highway safety is further enhanced by a new program to encourage the use of safety belts and motorcycle helmets.
- State uniformity in vehicle registration and fuel tax reporting is required. This will ease the recordkeeping and reporting burden on businesses and contribute substantially to increased productivity of the truck and bus industry.

SUMMARY

1. Transportation engineering is a very diverse field. It embraces planning, functional design, operation, and the management of facilities for different modes of transportation.

2. Good transportation provides for the safe, rapid, comfortable, convenient, economical, and environmentally compatible movement of people and goods.

3. Transportation engineering is practiced by policymakers, managers, planners, designers and engineers, operating and maintenance specialists, and evaluators.

4. Transportation engineering is a multidisciplinary field drawing on more established disciplines to provide its basic framework, such as economics, geography, and statistics.

5. The system approach as a problem-solving philosophy has been applied successfully in transportation engineering.

6. The reason why, when, and how people and goods move is a complex issue.

7. Land use is one of the prime determinants of movement and activity. A cyclical process connecting transportation and land-use activities provides fundamental answers to land-use patterns and transportation needs over time.

8. The physical plant of most transportation systems consists of basic elements: links, vehicles, and terminals.

9. There is an interconnection between human behavior and transportation. Several properties of the physical environment, such as spatial organization and physical ambience, have a direct impact on human behavior.

10. At least three basic attributes of transportation systems can be used for purposes of evaluation: ubiquity or accessibility, mobility, and efficiency.

11. Although transportation systems can be classified in different ways, the emergence of functional classification is useful for engineers.

12. There is ample evidence to show that people choose a mode of transportation not purely on the basis of cost, but also on the basis of time. This phenomenon has produced the pronounced transportation gaps we notice in developed countries.

13. The connection between transportation and sustainability is recognized worldwide.

14. Emerging advanced technologies are gaining ground.

REFERENCES

ABLER, R., J. ADAMS, and P. GOULD (1971). *Spatial Organization*. Prentice-Hall, Englewood Cliffs, NJ.

AMERICAN ASSOCIATION OF STATE HIGHWAY AND TRANSPORTATION OFFICIALS (AASHTO) (1984). *A Policy on Geometric Design of Highways and Streets*, AASHTO, Washington, DC.

BELL, G., and J. TYRWHITT (1972). *Human Identity in the Urban Environment*, Penguin Books, Middlesex, England.

CARTER, E. C., and W. S. HOMBURGER (1978). *Introduction to Transportation Engineering*, Reston Publishing, Reston, VA.

CORNELL, A. H. (1980). *The Decision-Maker's Handbook*, Prentice-Hall, Englewood Cliffs, NJ.

GRAY, G. E., and L. A. HOEL (Eds.) (1979). *Public Transportation: Planning, Operations, and Management*, Prentice-Hall, Englewood Cliffs, NJ.

HAY, WILLIAM W. (1977). *An Introduction to Transportation Engineering*, 2nd Ed., John Wiley, New York.

HOEL, L. A. (1982). Transportation Education in the United States, *Transport Reviews*, Vol. 2, No. 3.

HOMBURGER, W.S., and J. H. KELL (1988). *Fundamentals of Traffic Engineering*, 11th Ed., University of California, Berkeley.

INSTITUTE OF TRANSPORTATION ENGINEERS (ITE) (1987). *Membership Directory*, ITE, Washington, DC.

ISTEA (1991). *Intermodal Surface Transportation Efficiency Act of 1991*, U.S. D.O.T., Washington, DC.

KARASKA, G. J., and J. B. GERTLER (Eds.) (1978). *Transportation, Technology, and Society: Future Options,* Clark University Press, Worcester, MA.

KHISTY, C. J. (1981). Challenges in Teaching Design Courses in Transportation Engineering, *Proceedings of the American Society for Engineering Education.*

KHISTY, C. J. (1983). *Assessing the Built Environment for Pedestrians through Behavior Circuits*, Transportation Research Record 904. National Academy of Sciences, Washington, DC.

KHISTY, C. J. (1985). *Is Urban Planning Education Necessary for Civil Engineers?* Transportation Research Record 1045, National Academy of Sciences, Washington, DC.

KHISTY. C. J. (1986). *Undergraduate Transportation Engineering Education*, Transportation Research Record 1101, National Academy of Sciences, Washington, DC.

KHISTY. C. J. (1987). Urban Planning Education for Civil Engineers, *ASCE Journal of Urban Planning and Development,* Nov., pp. 54–60.

KOLBUSZEWSKI. J. (1979). How Decisions in Transportation Safety Will Ultimately Be Taken, in *Transportation Safety*, The Institute for Safety in Transportation Proceedings, San Diego.

LAY, M. G. (1986). *Handbook of Road Technology,* Vol. 1, Gordon and Breach, New York.

MITRIC, S. (1977). *Transportation Engineering Notebook,* Ohio State University, Columbus.

NATIONAL COUNCIL ON PUBLIC WORKS IMPROVEMENT (NCPWI) (1986). *Defining the Issues*, NCPWI, Washington, DC.

NATIONAL HIGHWAY INSTITUTE (NHI) (1980). *Highway Safety and Traffic Study Program*, Washington, DC.

OWEN, W. (1976). *Transportation for Cities*, The Brookings Institution, Washington, DC.

THOMSON, J. M. (1977). *Great Cities and Their Traffic,* Penguin Books, Middlesex, England.

U.S. D.O.T. (1994). *Transportation Statistics, Annual Report, 1994*, U.S. D.O.T., Bureau of Transportation Statistics, Washington, DC.

WEGMAN, F. J., and E. A. BEIMBORN (1973). *Transportation Centers and Other Mechanisms to Encourage Interdisciplinary Research and Training Efforts in Transportation*, Highway Research Record 462, National Academy of Sciences, Washington, DC.

WOLFE R. I. (1963). *Transportation and Politics*, D. Van Nostrand, Princeton, NJ.

WORLD COMMISSION ON ENVIRONMENT AND DEVELOPMENT (1987). Our Common Future, Oxford University Press, Oxford, UK.

WORTMAN. R. H. (1976). Applications of Systems Concepts, in *Transportation and Traffic Engineering Handbook*, ed. John E. Baerwald, Prentice-Hall, Englewood Cliffs, NJ.

EXERCISES

1. What are the basic differences between transportation engineering and traffic engineering?

2. The intersection of two streets is controlled by a four-way stop-sign system (each approach has a stop sign). Over the years, the traffic at this intersection has increased by 50%, result-

ing in inordinate delays to motorists. List all the alternative actions an engineer can take to improve the situation. What would happen if the "do nothing" alternative is adopted?

3. A city adopted the following goals in a public meeting: (1) develop a balanced, integrated, multimodal transportation system that serves the existing and future transportation needs of the city and provides a convenient choice among modes for the trip to work, school, shopping, personal business, and recreational purposes; (2) provide an economical transportation system that will optimize the use of citizens' personal and tax money, reduce fuel consumption, and protect the natural environment. For each of these two goals, provide (a) a set of at least three objectives, matching each goal; (b) a set of at least two policies accompanying each goal; (c) a set of measures of effectiveness that could be used to judge the success of each objective.

4. With the introduction of high-technology into almost every facet of safety, what major changes do you foresee happening in (a) urban form, (b) transportation, and (c) quality of life of the average citizen?

5. Interview 20 students (at random) on your campus and determine the mode of transport they use to commute to school and the number of trips they make every day for different purposes (reasons) per week and their approximate length. Tabulate your results in a meaningful way. Determine the cost of these trips (out-of-pocket expenses). Write a short report about this interview and summarize your results.

6. What are the major characteristics of the automobile that have made it the predominant mode of travel in the United States?

7. Suppose that you are attending a university where you have the choice of walking, bicycling, taking a bus, riding with a friend, or driving. Make a matrix in which your criteria for choice between these modes of travel are listed in the rows and your rough scores of alternatives are listed in columns. Which mode would you choose?

8. A survey of a region in the state of Washington yielded the results shown in the table. Plot these figures with cumulative percentage of miles on the horizontal axis and cumulative percentage of vehicle-miles on the vertical axis. What conclusions do you draw from examining your plot?

	Rural areas		Urban areas	
	Miles %	Vehicle-miles %	Miles %	Vehicle-miles %
Local roads/streets	65	5	75	15
Collectors	20	20	5	5
Minor arterials			15	60
Arterials	15	75		
Principal arterials			5	20

9. Do you think that the transportation gaps can be filled with suitable modes by, say, the year 2020? Support your answer with convincing facts.

10. What would you suggest that federal, state, and local governments ought to do to induce filling up the transportation gap?

11. Refer to Table 1-3.
    (a) Convert the distance column (km) and theoretical transport speed column (km/hr) into miles and miles per hour, respectively.

(b) What are the values of a and b in the expression $t = ad^b$ if time t is measured in minutes and distance d is measured in miles?

12. What lesson have we, as a nation, learned in the last three decades with respect to our transportation system?

13. For each of the transportation problems mentioned in this chapter, suggest at least two ways of mitigating the problem. Are all these "solutions" compatible? Will society accept these solutions? If not, why?

14. Identify problems connected with each of the following transportation systems in a large metropolitan area: (a) street network, (b) transit system, (c) travel information system, and (d) pedestrian facility system.

15. Why has the concept of sustainability gained such prominence in the United States? From your reading of the newspapers and watching the news on TV, do you feel that the federal, state, and local (city) governments are serious about sustainability? Discuss.

16. What are the major advantages and disadvantages of adopting various forms of Intelligent Transportation Systems (ITSs)? How will the rich and poor sectors of society be affected by ITSs?

Chapter 2

Transportation Economics

1. THE SCOPE OF TRANSPORTATION ECONOMICS

The reason that people and goods move from one place to another is explained in Chapter 1. It was also demonstrated that this movement or flow is possible because transportation systems, including their networks (roads, streets, rail lines) can handle this flow. In this chapter, the interaction between transportation demand (e.g., the desire to make trips, with the ability to pay for it) and transportation supply (e.g., the availability of traffic lanes to make the trips) are examined. In short, some elementary ideas of economics as applied to transportation are studied.

"Economics is the study of how people and society end up choosing, with or without the use of money, to employ scarce productive resources that could have alternative uses to produce various commodities and distribute them for consumption, now and in the future, among various persons and groups in society. It analyzes the costs and benefits of improving patterns of resource allocation" (Samuelson, 1976).

Economists conveniently divide the broad area of economics into two main streams. *Microeconomics* concerns itself with the study of economic laws as affecting a firm on a small scale. It deals with the economic behavior of individual units such as consumers, firms, and resource owners. *Macroeconomics* is the study, on the national and international scale, of the wealth of society. It deals with the behavior of economic aggregates such as gross national product and the level of employment (Mitchell, 1980).

Transportation economics is a branch of applied microeconomics. However, it faces a number of special problems and characteristics and for several reasons, the simple laws of market economics cannot be applied to transportation economics. For example, the demand for transportation is derived; it is not a demand for its own sake. Also, each trip is unique in time and space. Technological differences among different modes and economies of scale tend to create problems in dealing with transportation economics. Government intervention in transportation also creates problems in analyzing issues (Stubbs et al., 1980).

Planning, designing, constructing, operating, and maintaining transportation facilities represent annual commitments of hundreds of billions of dollars, yet engineers, planners, and policy analysts who are responsible for transportation work often have little or no formal training or education in economics (Wohl and Hendrickson, 1984).

The area covered by transportation economics is very large, and therefore the reader is urged to refer to standard books on this and applied subjects for an in-depth understanding. The topics covered in this chapter are highly selective and have been included to provide readers with an introduction. You are encouraged to refer to some of the standard texts listed at the end of this chapter, particularly de Neufville and Stafford (1971) and Mansfield (1975), and more recent works by these authors.

Included in this chapter are the basic concepts of demand and supply functions that are fundamental to understanding, designing, and managing transportation systems. Much of the work conducted in the field of transportation is devoted to specifying and estimating performance function (e.g., demand and supply).

2. TRANSPORTATION DEMAND

The demand for goods and services, in general, depends largely on consumers' income and the price of the particular good or service relative to other prices. For example, the demand for travel depends on the income of the traveler. The choice of the travel mode depends on several factors, such as the purpose of the trip, the distance traveled, and the income of the traveler (Stubbs et al., 1980).

A demand function for a particular product represents the willingness of consumers to purchase the product at alternative prices. A demand function shows, for example, a number of passengers willing to use a commuter train at different price levels between a pair of origins and destinations, for a specific trip, during a given period. The term *price* stands for all outlays *perceived* by the traveler for a given trip. For example, the price for the trip could be out-of-pocket costs (the train fare); travel time including access, waiting, and in-vehicle time; comfort; safety; convenience; reliability; and several other tangible and intangible factors. Most of the components of the perceived price for travel can be measured and expressed in monetary units. This synthetic "price" is sometimes called a *generalized price*.

A linear demand function or travel is shown in Figure 2-1 for a given pair of origin and destination points, at a specific time of day and for a particular purpose. Such a demand function is useful for predicting travel over a wide range of conditions. This demand function assumes a particular level and distribution of income, population, and socioeconomic characteristic. Note that it is an aggregate demand curve, representing the volume of trips demanded at different prices by a group of travelers. Functionally,

$$q = \alpha - \beta p \tag{1}$$

where q is the quantity of trips demanded, and α and β are constant demand parameters. The demand function is drawn with a negative slope expressing a familiar situation where a decrease in perceived price usually results in an increase in travel, although this is not always true.

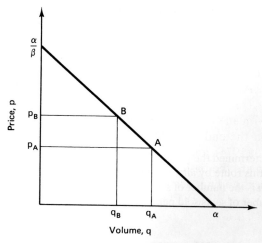

Figure 2-1 Typical Demand Function.

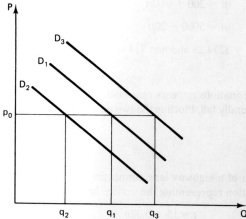

Figure 2-2 Shifted Demand Curves.

Figure 2-2 shows a series of shifted demand curves, representing changes in the quantity of travel due to variables other than the perceived price. Naturally, at a price p_0 one could expect different quantities q_1, q_2, and q_3 as the demand curve changes from D_1 to D_2 and D_3. If the curve shifts upward (from D_1 to D_3), it indicates an increase in trips.

It is important to distinguish short-run changes in the quantity of travel due to price changes represented by movement along a single demand curve, as shown in Figure 2-1 from long-run changes due to activity or behavioral variables, represented by shifts in demand functions (as shown in Figure 2-2).

3. DEMAND, SUPPLY, AND EQUILIBRIUM

We have seen that the demand function is a relation between the quantity demanded of a good and its price. In a similar manner, the supply function (or service function) represents the quantity of goods a producer is willing to offer at a given price, for example, bus seats at a given price, and tons of wheat at a given price. If the demand and supply functions for a transportation facility are known, then it is possible to deal with the concept of equilibrium.

Equilibrium is said to be attained when factors that affect the quantity demanded and those that determine the quantity supplied result in being statically equal (or converging toward equilibrium). A simple example will serve to illustrate equilibrium between demand and supply.

Example 1

An airline company has determined the price of a seat on a particular route to be $p = 200 + 0.02n$. The demand for this route by air has been found to be $n = 5000 - 20p$, where p is the price in dollars, and n is the number of seats sold per day. Determine the equilibrium price charged and the number of seats sold per day.

Solution

$$p = 200 + 0.02n$$

$$n = 5000 - 20p$$

These two equations yield $p = \$214.28$ and $n = 714$ seats.

Discussion

The logic of the two equations appears reasonable. If the price of an airline ticket rises, the demand would naturally fall. Plotting the two equations to scale may help to visualize the equilibrium price.

Example 2

The travel time on a stretch of a highway lane connecting two activity centers has been observed to follow the equation representing the service function.

$$t = 15 + 0.02v$$

where t and v are measured in minutes and vehicles per hour, respectively. The demand function for travel connecting the two activity centers is $v = 4000 - 120t$.

(a) Sketch these two equations and determine the equilibrium time and speed of travel.

(b) If the length of the highway lane is 20 miles, what is the average speed of vehicles traversing this length?

Solution (see Figure 2-E2)

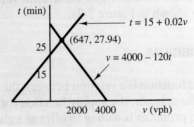

Figure 2-E2. Static Equilibrium of Demand and Supply.

$$t = 15 + 0.02v$$

$$v = 4000 - 120t$$

Therefore,

$$v = 647 \text{ vehicles/hour}$$

$$t = 27.94 \text{ minutes}$$

$$\text{Speed} = (20 \times 60)/27.94 = 42.95 \text{ mph}$$

Discussion

It is customary to plot price, time, or price units on the vertical axis and quantity units on the horizontal axis.

4. SENSITIVITY OF TRAVEL DEMAND

Knowledge of the functional form of travel demand can be used to forecast changes in the volume of travel caused by specified changes in price in the short run. A useful descriptor for explaining the degree of sensitivity to a change in price (or some other factor) is the elasticity of demand (e_p).

If $q = \alpha - \beta p$ (see Eq. 1),

and e_p = percentage change in quantity of trips demanded
that accompanies a 1% change in price   (2)

then $e_p = \dfrac{\delta q/q}{\delta p/p} = \dfrac{\delta q}{\delta p} \times \dfrac{p}{q}$

where δq is the change in the number of trips that accompanies δp, the change in price.

$$\text{Arc price elasticity} = \frac{\delta q}{\delta p} \frac{p}{q} = \frac{Q_1 - Q_0}{P_1 - P_0} \frac{(P_1 + P_0)/2}{(Q_1 + Q_0)/2} \qquad (3)$$

where Q_0 and Q_1 represent the quantity of travel demanded corresponding to prices P_0 and P_1, respectively.

For a linear demand function, we can determine the elasticity with respect to price by taking the derivative

$$e_p = \frac{\delta q}{\delta p} \frac{p}{q} = \frac{-\beta p}{q} \qquad (4)$$

or after substitution for p, using the equation

$$e_p = 1 - \frac{\alpha}{q} \qquad (5)$$

Example 3

An aggregate demand function is represented by the equation

$$q = 200 - 10p$$

where q is the number of trips made, and p is the price per trip. Find the price elasticity of demand when

$$q = 0, \quad q = 50, \quad q = 100, \quad q = 150, \quad q = 200 \text{ trips}$$

corresponding to

$$p = 20, \quad p = 15, \quad p = 10, \quad p = 5, \quad p = 0 \text{ cents}$$

Solution

$$e_p = 1 - \frac{\alpha}{q} \qquad \text{where } \alpha = 200 \tag{5}$$

$$e_0 = 1 - \frac{200}{200} = -0$$

$$e_5 = 1 - \frac{200}{150} = -0.133$$

$$e_{10} = 1 - \frac{200}{100} = -1$$

$$e_{15} = 1 - \frac{200}{50} = -3$$

$$e_{20} = 1 - \frac{200}{0} = -\infty$$

These elasticities are shown in Figure 2-E3(a).

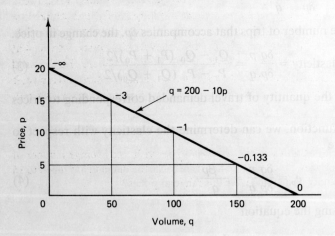

Figure 2-E3. (a) Demand Function Showing Elasticities at Various Volumes.

When the elasticity is less than −1 (i.e., more negative than −1), the demand is described as being elastic, meaning that the resulting percentage change in quantity of trip making will be larger than the percentage change in price. In this case, demand is relatively sensitive to price change. However, when the elasticity is between 0 and −1, the demand is described as being inelastic or relatively insensitive. These ranges are shown in Figure 2-E3(b).

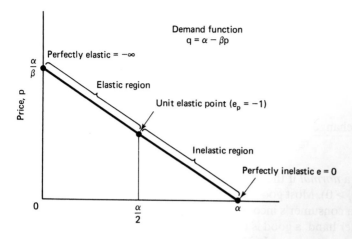

Figure 2-E3. (b) General Case of a Linear Demand Function Showing Elasticities.

Discussion

From Figure 2-E3(a), it is obvious that when the price per trip is 20 cents, no trips are made. Also, when nothing is charged per trip, 200 trips are made. Notice that the price elasticity for this transportation system varies from 0 to ∞, with unit elasticity when $p = 10$.

A linear demand curve has several interesting properties. Notice, as one moves down the demand curve, the price elasticity of demand becomes smaller (i.e., more inelastic). In fact, the elasticity at a given point equals the length of the demand line segment below the point divided by the length of the line segment above it. Another point to note is that the slope of the line is constant, but the *elasticity changes* from ∞ at the top, where the demand line intersects the vertical axis, to zero, where the demand line intersects the horizontal axis. Because elasticity changes along the demand curve, it is essential to specify over what range of prices or quantity the elasticity was measured.

Example 4

When the admission rate to an amusement park was $5 per visit, the average number of visits per person was 20 per year. Since the rate has risen to $6, the demand has fallen to 16 per year. What is the elasticity of demand over this range of prices?

Solution

$$\text{Arc price elasticity, } e_p = \frac{-\Delta Q}{\Delta P} \frac{(P_1 + P_0)/2}{(Q_1 + Q_0)/2} = \frac{4}{1} \frac{(5.5)}{(18)} = -1.22 \quad \text{(elastic)}$$

Discussion

There are problems connected with arc price elasticity because such an elasticity will differ from point elasticity, the difference increasing as ΔP or ΔQ increases.

5. FACTORS AFFECTING ELASTICITIES

5.1 Income Elasticities

Income elasticities have a special significance in transportation engineering and are denoted by

$$e_i = \frac{\% \text{ change in quantity of good demanded}}{\% \text{ change in income}} \tag{6}$$

A good is considered to be a *normal* if the demand for the good goes up when a consumer's income increases ($e_i > 0$). Most goods are normal. A good is a *superior* good if it goes up in demand when a consumer's income increases *and* its share in income also goes up ($e_i > 1$). On the other hand, a good is inferior if the demand for the good goes down when a consumer's income goes up. In North America, an automobile is considered as a superior good, whereas spending money on traveling by mass transit is often considered as an inferior good. Gourmet food is a superior good while cheap beer is an inferior good.

5.2 Price Elasticities

In general, consumers buy more than otherwise of a good when the price goes down and buy less than otherwise when the price goes up. Some factors that affect price elasticity are as follows:

1. If a consumer spends a substantial percentage of income on, say, transportation, the more willing will he or she be to search hard for a substitute if the price of transportation goes up.

2. The narrower the definition of a good, the more substitutes the good is likely to have, and thus the more elastic its demand will be. For example, the demand for Toyotas is more elastic than the demand for automobiles and the demand for automobiles is more elastic than the demand for transportation.

3. If consumers find out that the price and availability of substitutes are easy, the more elastic the demand will be. Advertising plays an important role in making available substitutes to consumers. In the same context, the more time consumers have to find substitutes, the more elastic demand becomes.

4. Those goods that consumers consider to be "necessities" usually have inelastic demands, whereas goods considered by consumers to be "luxuries" usually have elastic demands. For instance, eyeglasses for a consumer are a necessary good, with few substitutes, whereas a vacation trip to Europe is a luxury good with several substitutes.

5.3 Elasticity and Total Revenue

It is possible to tell what the total revenue (price × output) of a firm is likely to be if the price of a unit changes. Here

$$e = \frac{\% \text{ change in quantity (units) demanded}}{\% \text{ change in price}} \tag{7}$$

If $e > 1$, price and total revenue are *negatively* related (or demand is elastic); therefore, an increase in price will reduce total revenue, but a decrease in price will increase total revenue.

If $e < 1$, price and total revenue are *positively* related (or demand is inelastic), in which case, an increase in price will increase total revenue, and a decrease in price will decrease total revenue.

If $e = 1$, total revenue will remain the same whether the price goes up or down.

Example 5

A bus company's linear demand curve is $P = 10 - 0.05Q$, where P is the price of a one-way ticket, and Q the number of tickets sold per hour. Determine the total revenue along the curve.

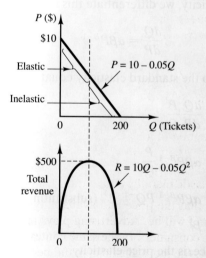

Figure 2-E5.   Total Revenue Curve.

Solution

$$P = 10 - 0.05Q$$

$$R = Q(10 - 0.05Q) \qquad \text{where } R = \text{total revenue}$$

$$R = 10Q - 0.05Q^2$$

$$dR/dQ = 10 - (0.05 \times 2)Q$$

and this is equal to zero when R is maximum.
Therefore, $Q = 100$ when R is 500 (maximum).

Discussion

Starting from a price of $10 at near zero tickets sold and decreasing the price eventually to half ($5), the revenue steadily increases to a maximum of $500/hour (over the elastic portion). After that, the revenue decreases as the price further decreases and finally approaches near zero, when the demand approaches 200 (over the inelastic portion).

6. KRAFT DEMAND MODEL

We occasionally come across a demand function where the elasticity of demand for travel with respect to its price is essentially constant. The demand function for such a situation corresponds to the equation

$$Q = \alpha(p)^\beta \tag{8}$$

where α and β are constant parameters of the demand function. To prove that this function has a constant elasticity, we differentiate this function with respect to price:

$$\frac{dQ}{dP} = \alpha\beta P^{\beta-1}$$

and substitute the result into the standard elasticity equation (Eq. 2).

$$e_p = \frac{dQ}{dP}\frac{P}{Q}$$

$$= \alpha\beta P^{\beta-1}\frac{P}{Q}$$

$$= \alpha\beta P^{\beta-1} PQ^{-1} \quad \text{(substituting } Q \text{ from Eq. 8)}$$

$$= \beta$$

Thus, β, the exponent of price, is the price elasticity.

Example 6

The elasticity of transit demand with respect to price has been found to be equal to -2.75, which means that a 1% increase in transit fare will result in a 2.75 decrease in the number of passengers using the system. A transit line on this system carries 12,500 passengers per day, charging 50 cents per ride. The management wants to raise the fare to 70 cents per ride. What advice would you offer to management?

Solution

$$Q = \alpha P^\beta \tag{8}$$

$$12{,}500 = \alpha(50)^{-2.75}$$

$$\alpha = 12{,}500 \times 50^{2.75}$$

$$= 5.876 \times 10^8$$

$$Q = 5.876 \times 10^8 \times P^{-2.75}$$

An increase in fare from 50 to 70 cents will attract a demand of

$$Q = 5.876 \times 10^8 \times (70)^{-2.75} = 4955$$

Therefore, the increase in fare from 50 to 70 cents (a 40% increase) is likely to reduce the patronage on this line from 12,500 passengers per day to 4955 (a 60.36% decrease). In terms of revenue, the results are as follows:

$$50 \text{ cents/rider} \times 12{,}500 \text{ passengers} = \$6250$$

$$70 \text{ cents/rider} \times 4955 \text{ passengers} = \$3468.50$$

$$\text{Loss in revenue} = \$3406$$

Advice to management would be not to increase the fare.

Discussion

In general, it has been observed that when the price is elastic (e.g., -2.75), raising the unit price will result in total loss, but lowering price will result in total gain. The converse is also true; if the price is inelastic, raising the unit price will result in total gain, whereas lowering the unit price will result in total loss. Students may be helped by graphing these cases to discover why this is so.

Example 7

The demand function for transportation from the suburbs to downtown in a large city is as follows:

$$Q = T^{-0.3} C^{-0.2} A^{0.1} I^{-0.25}$$

where

$Q = $ number of transit trips

$T = $ travel time on transit (hours)

$C = $ fare on transit (dollars)

$A = $ cost of automobile trip (dollars)

$I = $ average income (dollars)

(a) There are currently 10,000 persons per hour riding the transit system, at a flat fare of $1 per ride. What would be the change in ridership with a 90-cent fare? What would the company gain per hour?

(b) By auto, the trip costs $3 (including parking). If the parking charge were raised by 30 cents, how would it affect the transit ridership?

(c) The average income of auto riders is $15,000 per year. What raise in salary will riders require to cover their costs in view of the change in parking charge noted in part (b)?

Solution

(a) This is essentially a Kraft model. The price elasticity of demand for transit trips is

$$\frac{\delta Q/Q}{\delta C/C} = -0.2$$

This means that a 1% reduction in fare would lead to a 0.2% increase in transit patronage. Because the fare reduction is $(100 - 90)/100 = 10\%$, one would expect an increase of 2% in patronage. Patronage would now be $10,000 + (10,000 \times 0.2) = 10,200$.

$$10,000 \text{ passengers at } \$1.00/\text{ride} = 10,000$$

$$10,200 \text{ passengers at } \$0.90/\text{ride} = 9180$$

The company will lose $820 per hour.

(b) The automobile price cross-elasticity of demand is 0.1, or

$$\frac{\delta Q/Q}{\delta A/A} = 0.1$$

This means that a 1% rise in auto costs (including parking) will lead to a 0.1% rise in transit trips. A $0.30 rise is 10% of $3. Therefore, a 10% rise in auto cost would raise the transit patronage by 1%—from 10,000 to 10,100 riders.

(c) The income elasticity should be looked at first:

$$\frac{\delta Q/Q}{\delta I/I} = -0.25$$

which means that an income raise of 1% will result in a 0.25% decrease in transit patronage, or $\delta Q/Q = 1\%$, from part (b). Therefore,

$$\frac{1\%}{\delta I/I} = -0.25$$

and

$$\frac{\delta I}{I} = \frac{1\%}{-0.25} = 0.04 = 4\%$$

So a 4% increase in income would cover a 30-cent increase (or 10% increase) in auto cost. If the average income is $15,000, a $600 raise in salary would change the minds of those auto drivers who were planning to ride the transit system.

7. DIRECT AND CROSS ELASTICITIES

The effect of change in the price of a good on the demand for the *same* good is referred to as *direct* elasticity. However, the measure of responsiveness of the demand for a good to the price of another good is referred to as *cross* elasticity.

When consumers buy more of good A when good B's price goes up, we say that good A is a substitute for good B (and good B is a substitute for good A). For example, when the price of gasoline goes up, travelers tend to use more transit. On the other hand, when consumers buy less of good A when good B's price goes up, we say that good A is a complement to good B. In general, complementary goods are ones that are used together. Thus, for example, when the price of downtown parking goes up, the demand for driving a car downtown goes down (and the demand for an equivalent trip by transit or by taxi to downtown goes up).

Goods are substitutes when their cross elasticities are positive, and goods are complements when their cross elasticities of demand are negative.

Example 8

A 15% increase in gasoline costs has resulted in a 7% increase in bus patronage and a 9% decrease in gasoline consumption in a midsized city. Calculate the implied direct and cross elasticities of demand.

Solution

Let

$$P_0 = \text{price of gas before}$$

$$P_1 = \text{price of gas after}$$

$$Q_0 = \text{quantity of gas consumed before}$$

$$Q_1 = \text{quantity of gas consumed after}$$

Then for *direct elasticity*:

$$Q_0 \text{ (gas)} \times 0.91 = Q_1 \text{ (gas)}$$

$$P_0 \text{ (gas)} \times 1.15 = P_1 \text{ (gas)}$$

$$e = (\Delta Q/Q)/(\Delta P/P) = [-0.09/(1 + 0.91)]/[0.15/(1 + 1.15)] = -0.361$$

And,

$$B_0 = \text{bus patronage before}$$

$$B_1 = \text{bus patronage after}$$

Then for *cross elasticity*:

$$B_0 \text{ (bus)} \times 1.07 = B_1 \text{ (bus)}$$

$$P_0 \text{ (gas)} \times 1.15 = P_1 \text{ (gas)}$$

$$e = (\Delta B/B)/(\Delta P/P) = [0.07/(1 + 1.07)]/[0.15/(1 + 1.15)] = +0.48$$

8. CONSUMER SURPLUS

Consumer surplus is a measure of the monetary value made available to consumers by the existence of a facility. It is defined as the difference between what consumers might be willing to pay for a service and what they actually pay. A patron of a bus service pays a fare of, say, 50 cents per trip but would be willing to pay up to as much as 75 cents per trip, in which case, his surplus is 25 cents.

The demand curve can be considered as an indicator of the utility of the service in terms of price. The consumer surplus concept is shown in Figure 2-3(a). The area ABC represents the total consumer surplus. Maximization of consumer surplus is indeed the maximization of the economic utility of the consumer. In project evaluation, the use of this concept is common, particularly for transit systems.

In general, a transportation improvement can be measured in terms of the change in consumer surplus. Figure 2-3(b) indicates the case of a street having a traffic supply curve S_1, intersecting a demand curve at E_1. An additional lane has been added, shifting the supply curve to S_2 and therefore intersecting the demand curve at E_2. The change in consumer surplus can be quantified as the area of trapezoid $P_1 P_2 E_2 E_1$, or $(P_1 - P_2)(Q_1 + Q_2)/2$. The consumer surplus is normally defined as the difference between the maximum amount that consumers are willing to pay for a specified quan-

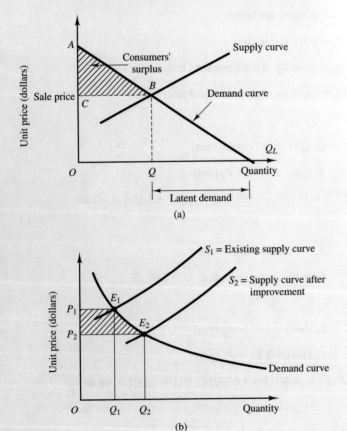

Figure 2-3 (a) Consumer Surplus Concept; (b) Change in Consumer Surplus.

tity of a good rather than going without it. In general, $AOQB$ is equal to the total community benefit, $BCOQ$ is equal to the market value, and ACB is equal to the consumer surplus or net community benefit.

Figure 2-3(a) illustrates an additional concept that is useful to transportation engineers: *latent demand*. Note that travelers between Q and the point of intersection of the demand function and the abscissa do not currently make trips, but would do so if the price per trip were lower than the equilibrium price. The number of such potential travelers is popularly called latent demand. The concept can be used in several ways; for instance, a transit operator hoping to increase transit patronage by introducing a discount rate valid for nonpeak hours may like to investigate latent demand. Indeed, the quantity of trips demanded, if the price of a trip were zero (free transit), would be $Q_L - Q$, as indicated in Figure 2-3(a).

Example 9

A bus company with an existing fleet of one hundred 40-seater buses increases its fleet size by 20% and reduces its fare of $1 to 90 cents per ride. Calculate the change in consumer surplus and the price elasticity of demand. Assume that the existing buses had a load factor of 90% and it is anticipated that the improvement will result in a 95% load factor. Does the company lose money? Assume that all the buses in the fleet are being used during the peak hours. (*Note:* The vehicle load factor is a measure of seat availability, and a load factor of 1.0 means that every seat is occupied.)

Solution

With the existing situation:

$$100 \text{ buses} \times 40 \text{ seats} \times 0.90 \text{ (load factor)} = 3600 \text{ persons/hr}$$

$$\text{Revenue: } 3600 \times 1.00 = \$3600/\text{hr}$$

With the improved situation:

$$120 \text{ buses} \times 40 \text{ seats} \times 0.95 = 4560 \text{ persons/hr}$$

$$\text{Revenue: } 4560 \times 0.90 = \$4104/\text{hr}$$

The company gains $4104 - $3600 = $504/hr.

$$\text{Change in consumer surplus} = \frac{(1.00 - 0.90)(3600 + 4560)}{2}$$

$$= \$408/\text{hr}$$

$$\text{Price elasticity of demand} = \frac{Q_1 - Q_0}{P_1 - P_0} \frac{(P_1 + P_0)/2}{(Q_1 + Q_0)/2}$$

$$= -\frac{960}{0.10}\left(\frac{0.95}{4080}\right) = -2.235$$

Discussion

This is an interesting situation. Even if the number of buses was not increased and the prices were lowered as indicated, resulting in an increase in the load factor, there would be an increase in total revenue by $200 because the price elasticity is elastic (-2.235). Naturally, with more buses being deployed the situation is even better. Consumer surplus is a good way to compare two alternatives.

Example 10

Bob is willing to pay up to $10 to travel by bus once every month to visit his family, $8 to travel twice, and $6 to travel three times for the same purpose. (a) If the price of a bus ticket to visit his family is $7, what is Bob's consumer surplus? (b) If the bus company offers three tickets per month for a flat price of $19, will Bob accept the deal? (c) What is the maximum the bus company should charge for Bob to take the three-ticket offer?

Solution

(a) Bob's individual consumer surplus is $4. [For the first trip, Bob's consumer surplus is $10 − 7 = $3 and for the second trip it is $1, which adds up to $4. He will not go for the third trip.]

(b) Yes, because for three tickets, his consumer surplus is ($10 + 8 + 6 = 24 − 19) = $5, which is better than buying the tickets individually.

(c) At best, Bob will pay $20 for the three-ticket offer, which is the amount that would be the same consumer surplus ($4) as the option of buying the tickets separately ($24 − $20 = $4).

Discussion

Note that the bus company makes more money from Bob with the package deal. Also, note that in this problem we are calculating an individual's consumer surplus, and not an aggregate (or total) consumer surplus, as in example 9.

9. COSTS

It is essential to have a knowledge of costs or the value of a product or service. To establish the true cost of a product, the analyst must determine, for example, where delivery takes place, who pays for transportation, and who pays for insurance and storage.

Before finding average costs, it is convenient to break down costs into fixed costs, variable costs, and total costs. Fixed costs are inescapable costs and do not change with use. If a plant is producing 500 trucks per day and the plant costs $1 million to run whether one truck is produced or a hundred, the fixed cost is $1 million. Naturally, the fixed cost per truck produced will be reduced with increasing production, even though the fixed cost itself remains unchanged. Variable costs, on the other hand, increase with output or production. If, for example, the labor cost for assembling one truck is $1000, it is likely that this labor cost for assembling two trucks is $1900. The total costs of production is the sum of fixed and variable costs and will increase with production. For any particular level of production, the average cost of a single unit (one truck) can be found by dividing the total cost by the number of units corresponding to the total cost.

9.1 Laws Related to Costs

Two concepts related to costs are of importance in transportation. The first, the *law of diminishing returns*, states that although an increase in input of one factor of production may cause an increase in output, eventually a point will be reached beyond which increasing units of input will cause progressively less increase in output. The second, the *law of increasing returns to scale*, states that in practice, the production of units is often likely to increase at a faster rate than the increase of factors of production. This phenomenon may be due to any number of factors, such as technological features or the effects of specialization.

9.2 Average Cost

The mathematical relationship connecting the total cost (C) of a product to the unit cost (c) and magnitude of the output (q) can be written as

$$C = cq = \alpha + \beta(q)$$

where parameter α equals the fixed cost of production, and the function $\beta(q)$ equals the variable cost of production.

The average cost ($\bar{c}$) of each item produced is equal to

$$\bar{c} = \frac{C}{q} = \frac{cq}{q} = \frac{\alpha + \beta(q)}{q} = \frac{\alpha}{q} + \frac{\beta(q)}{q} \tag{9}$$

The relationships of the total and average cost functions are shown in Figure 2-4. Notice that in this particular case, as output q increases, the average cost of production decreases, and then increases at higher levels of production. When the production level reaches q', the average cost is a minimum ($\bar{c}$). *Economy of scale* is defined as a decrease in average cost as output increases. There is an economy of scale for production levels between 0 and q'; beyond q', there is no economy of scale, because the average cost rises. This concept is useful to engineers in deciding whether additional capacity (or production) or growth would result in higher profits.

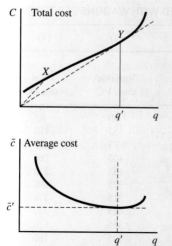

Figure 2-4   Total and Average Cost.

9.3 Marginal Cost

The *marginal cost* of a product is the additional cost associated with the production of an additional unit of output. This is an important concept used in transportation engineering in several ways. An example will clarify this term. The cost of running a train system with variable number of wagons is given in Table 2-1. From this table of costs all other costs can be computed. Column (5) is obtained by dividing the total cost given in column (4) by the number of wagons per train shown in column (1). The marginal cost is calculated by subtraction of adjacent rows of total cost (Morlock, 1978).

The figures in columns (1) to (4) of Table 2-1 are plotted in Figure 2-5(a). Similarly, the figures from columns (5) and (6) are plotted in Figure 2-5(b). Note that the point of minimum cost ($40) occurs at the intersection of the average cost (AC) and marginal cost (MC) curves. Also, the projection of this point to Figure 2-5(a) corresponds to the point where the gradient of the tangent drawn from the origin has the minimum slope.

In general, we can summarize what has been demonstrated in our exercise in the train problem.

$$\text{Total cost} = TC(x) = FC + VC(x) \tag{10}$$

$$\text{Average cost} = AC(x) = \frac{TC(x)}{x} = \frac{FC}{x} + \frac{VC(x)}{x} \tag{11}$$

$$\text{Marginal cost} = MC(x) = TC(x) - TC(x-1) \tag{12}$$

where

$TC = $ total cost

$FC = $ fixed cost

$VC = $ variable cost

TABLE 2-1   COSTS ASSOCIATED WITH WAGONS PER TRAIN

(1)	(2)	(3)	(4)	(5)	(6)
Number of wagons/train	Fixed cost, FC	Variable cost, VC	Total cost, TC	Average cost, AC	Marginal cost/unit, MC
1	55	30	85	85.0	25
2	55	55	110	55.0	20
3	55	75	130	43.3	30
4	55	105	160	40.0	50
5	55	155	210	42.0	70
6	55	225	280	46.7	90
7	55	315	370	52.9	110
8	55	425	480	60.0	130
9	55	555	610	67.8	150
10	55	705	760	76.0	

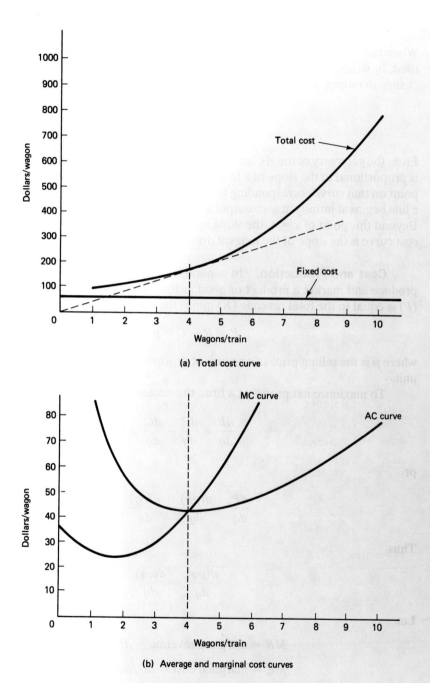

(a) Total cost curve

(b) Average and marginal cost curves

Figure 2-5   Total Cost, Average Cost, and Marginal Cost.

MC = marginal cost

AC = average cost

When the output is a continuous function, the differential form of the marginal cost is used, in which the marginal cost is the rate of change of total cost with respect to a change in output. In this form, the equation is

$$MC(x) = \frac{dTC(x)}{dx} = \frac{dVC(x)}{dx} \tag{13}$$

From the geometry of the AC and MC curves, it may also be noted that the average cost is proportional to the slope of a line connecting the origin of the total cost curve with a point on that curve corresponding to the total output. In our example, the slope of such a line begins at infinity at zero output and then decreases to its lowest point, when $x = 4$. Beyond this point of $x = 4$, the slope increases again. On the other hand, the marginal cost curve is the slope of the tangent drawn at any point on the total cost curve.

Cost and Production.   In general, a private company or firm will continue to produce and market a product or good as long as it is making a profit. The net profit (P) is equal to the total revenue (R) minus the total cost (C).

$$P = R - C = pq - cq$$

where p is the selling price of one unit of product q, and c is the production cost of one unit.

To maximize net profits of a firm, the necessary condition is

$$\frac{dP}{dq} = \frac{dR}{dq} - \frac{dC}{dq} = 0$$

or

$$\frac{dP}{dq} = \frac{d(pq)}{dq} - \frac{d(cq)}{dq} = 0$$

Thus,

$$\frac{d(pq)}{dq} = \frac{d(cq)}{dq}$$

Let

$$MR = \text{marginal revenue} = dR/dq = d(pq)/dq$$

$$MC = \text{marginal cost} = dC/dq = d(cq)/dq$$

Therefore,

$$MR = MC \tag{14}$$

This equation says that to achieve the goal of maximizing profits, the firm should produce where marginal revenue equals marginal cost.

Example 11

A transport company hauling goods by truck has a cost function, $C = 15q^{1.25}$, where C is the total cost of supply q.

(a)   Determine the average cost and the marginal cost.

(b)   Prove that the cost elasticity is 1.25.

(c)   Is there an economy of scale?

Solution

(a) $\bar{c} = \dfrac{C}{q} = \dfrac{15q^{1.25}}{q} = 15q^{0.25}$ which is the average cost

$MC = dC/dq = (15 \times 1.25)\, q^{0.25} = 18.75q^{0.25}$

(b) $e = MC/AC = 18.75q^{0.25}/15q^{0.25} = 1.25$ (see Eq. 15, below)

(c)   Economy of scale does not exist because the average cost increases with increased q.

Cost Elasticity.   The cost elasticity e_c is defined as the ratio of percentage change in cost C to the percentage change in supply q.

$$e_c = \frac{\%\,\Delta \text{ in cost}}{\%\,\Delta \text{ in supply}} = \frac{(\Delta C/C)100}{(\Delta q/q)100} = \frac{q}{c}\frac{\Delta C}{\Delta q}$$

In the limit when $\Delta q = 0$, $e = (q/C)(dC/dq)$. Rearranging terms,

$$e = \frac{dC/dq}{C/q} = \frac{MC}{AC} \tag{15}$$

10. PRICING AND SUBSIDY POLICIES

One of the most difficult problems in the field of urban transportation is the congestion of motor vehicles in the central business districts of our cities. Congestion creates terrible private and social costs, higher operating costs, losses in valuable time, more road accidents, air pollution, discomfort and inconvenience to pedestrians, and noise pollution. Naturally, the choices facing society are to do nothing, to reduce the inconvenience by adding additional lanes, or to restrict road use in the central business district (CBD). It has been observed through long experience that the pursuit of the first two courses is likely to result in further congestion and therefore greater inefficiency.

The third choice offers the possible opportunity to solve the problem through reallocation of resources. Under this choice, we can consider three broad choices. They are, in ascending order of importance: taxes on suburban and dispersed living; subsidies to public transportation; and various methods of increasing the costs of motoring, including road pricing and car parking charges.

One of the best means of restraining congestion is through some method of taxing the motorist. It has long been accepted that introducing an element of marginal cost pricing may be the answer to the problem of congestion.

In most circumstances, it is reasonable to characterize the short-run travel price paid by motorists by an average cost function, as shown in Figure 2-6 and indicated by curve AC. The marginal cost curve MC is also shown. DD is the demand curve. At traffic flows up to OL per hour, the cost per trip is OA, comprising time and operating expenses. Beyond this point flow, the speed of vehicles falls, and therefore the average user cost per trip rises. In the absence of any further interference, the flow will equilibrate at ON vehicles per hour at a cost of OB.

At traffic flows above OL vehicles per hour, each additional vehicle is slowing down the entire flow of traffic and raising the operating costs of all other vehicles in that stream flow. In accord with the pricing principle, the price should equal the marginal cost, to give a flow of OM vehicles. This should be achieved by imposing a tax of GF, thus raising the average cost curve to achieve the optimal traffic flow. The benefit from this action is the reduction in the operating cost of all the remaining vehicles, and the loss is the loss in benefit from the MN trips that cannot be undertaken. This marginal cost pricing policy may result in the most efficient use of the system.

Example 12

The Federal Highway Administration has established the following relationship between travel time on a highway section and the volume [vehicles per hour (veh/hr)] using this highway of length 10 miles:

$$t = 10\left[1 + 0.15\left(\frac{V}{200}\right)^4\right]$$

where t is the travel time for vehicles traveling this section, and V is the volume of traffic on this section (veh/hr). The demand function for this highway is

$$d = 4000 - 100t$$

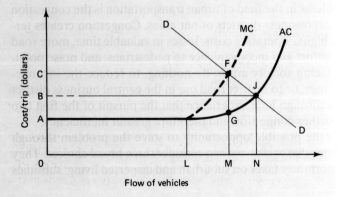

Flow of vehicles

Figure 2-6 Congestion Pricing.

where d is the demand (veh/hr), and t is the time (min). If the highway users value time at \$5 per vehicle per hour, what should be the congestion toll levied on this section of highway?

Solution

The time taken by all vehicles traveling this section is

$$tV = 10\left[1 + 0.15\left(\frac{V}{2000}\right)^4\right]V$$

$$= 10\left[V + \frac{0.15V^5}{(2000)^4}\right]$$

and

$$\text{Marginal time} = \frac{d(tV)}{dV} = 10\left(1 + \frac{0.75V^4}{2000^4}\right)$$

Volume (veh/hr)	Time (min)	Demand $d = 4000 - 100t$	Speed (mph)	Marginal time (min)
1000	10.09	2991	59.46	10.5
1500	10.47	2953	57.31	12.4
2000	11.50	2850	52.17	17.5
2500	13.66	2634	43.92	28.3
3000	27.59	1241	21.75	48.0

$$\text{Toll} = 19.12 - 11.82 \text{ min} = 7.30 \text{ min}$$

$$= 7.30 \times \frac{\$5}{60} = \$0.61$$

$$\text{Length of section} = 10 \text{ miles}$$

$$\text{Toll/mile} = 6.1 \text{ cents}$$

$$\text{Optimum flow} = 2100 \text{ veh/hr}$$

Discussion

Figure 2-E12 shows the personal time (curve a), marginal time (curve b), and the demand curve. Curve a represents the flow on the highway when each tripmaker is only aware of his own personal time. The additional time of adding one extra vehicle to the traffic stream is referred to as the *marginal time*. Note that this additional time is over a distance of 10 miles of highway. The intersection of the marginal time curve *b* with the demand curve at *x* gives the optimal flow (2120 veh/hr). This is the flow on the highway when a trip is made only if the benefit of a trip to the trip maker exceeds the additional time imposed on other trip makers. The ordinate *XZ* indicates the toll that will result in flow conditions represented by curve *b*.

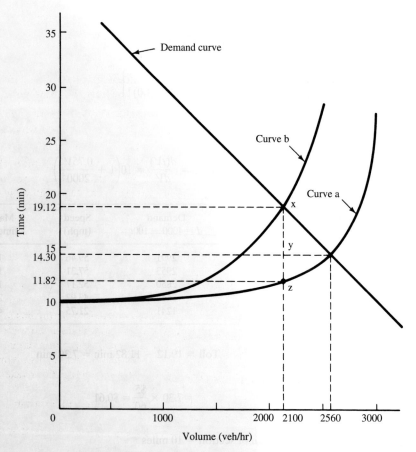

Figure 2-E12.   Personal, Marginal, and Demand Curves.

SUMMARY

Transportation economics is a branch of applied microeconomics. It includes special problems faced by transportation engineers and planners and encountered in the discipline of transportation. Only the basics of this branch of economics are described. Terms such as transportation demand, transportation supply, the sensitivity of travel demand with respect to price and other variables (familiarly referred to as "elasticity"), consumer surplus, total, average, and marginal cost, and pricing and subsidy policies are all briefly described.

The student is urged to read and refer to this chapter from time to time to appreciate the subtleties of the subject. The concepts of price elasticity, marginal cost, demand and supply, and pricing will occur from time to time in other chapters, particularly Chapter 15.

REFERENCES

DE NEUFVILLE, R., and J. H. STAFFORD (1971). *Systems Analysis for Engineers and Managers*, McGraw-Hill, New York.

MANSFIELD, EDWIN (1975). *Microeconomics: Theory and Applications*, 2nd Ed., W. W. Norton, New York.

MITCHELL, ROBERT L. (1980). *Engineering Economics*, John Wiley, New York.

MORLOCK, E. K. (1978). *Introduction to Transportation Engineering and Planning*. McGraw-Hill, New York.

SAMUELSON, PAUL A. (1976). *Economics*, 10th Ed., McGraw-Hill, New York.

STUBBS, P. C., W. J. TYSON, and M. Q. DALVI (1980). *Transportation Economics*, Allen & Unwin, London.

WOHL, MARTIN, and CHRIS HENDRICKSON (1984). *Transportation Investment and Pricing Principles*, John Wiley, New York.

EXERCISES

1. The performance function for a highway connecting a suburb with the business district can be represented by a straight line of the form $t = a + bq$, where t is the travel time in minutes, q is the traffic flow in vehicles per hour, and a and b are constants equal to 10 minutes and 0.01 minute/vehicle-hour, respectively. The demand function, also represented by a straight line, is $q = c + dt$, where c and d are constants equal to 5000 vehicles/hr and -100 vehicles/hour/minute, respectively.

   (a) Find the equilibrium flow (q^*) and the corresponding equilibrium time (t^*) algebraically, and sketch the functions.

   (b) If the length of this highway is 22.5 miles, what is the average speed of vehicles along this highway?

   (c) It is proposed to improve this highway such that constant b is now 0.005. What would be the new values of t^* and q^* and what would be the average speed on this highway?

2. The population growth and added commercial activity have affected the highway link described in Example 2, and improvements planned for this highway are reflected by equations representing the new conditions. The new equations are

$$t_1 = 15 + 0.004v_1$$

$$v_1 = 4333 - 130t_1$$

   (a) Sketch the existing and proposed service and demand functions.

   (b) What are your estimates of the equilibrium time and vehicle flow on the proposed link?

   (c) If the length of this link is 20 miles, what is the average speed over the link?

3. The demand for travel over a stretch of highway is given by the function

$$q = 2000/(t + 1)$$

where q is the travel flow, and t is the travel time (minutes). Plot this demand function and calculate the change in vehicle-hours of travel when travel time increases from 10 to 15 minutes due to road congestion.

4. The demand function for a transit system can be represented by a straight line connecting fare per person and ridership. Observations made on this system resulted in the following: When the fare was $1.50 per ride, the ridership per hour was 2000; when the fare was raised to $2 per ride, the ridership dropped to 1000. What is the equation of the demand function? What would be the patronage if the fare was (a) 50¢ per ride; (b) zero?

5. A bus company is charging a flat rate of 50¢ per ride to any part of the city and has a patronage of 500,000 per day. It has been decided to raise the fare to 60¢ per ride and it is estimated that 470,000 people will ride the buses. Calculate (a) the arc-price elasticity; (b) the possible total gain or loss in total revenue per day.

6. Within certain limits, a bus company has a demand function connecting patronage (Q) and price per ride (P) as follows:

$$Q = 2125 - 1000P$$

where Q is person-trips/day, and P is the price (dollars/ride). The manager has the following options to increase the total revenue: (1) attracting additional riders by rescheduling and rerouting the service and thus changing the demand function to $Q = 2150 - 1000P$ or (2) encouraging more riders onto the system by reducing the fare from $1.30 to $1. What option would you advise the manager to adopt, and give good reasons for doing so.

7. (a) A bus company found that the price elasticity of demand for bus trips during peak hours is -0.40 for small price changes. Management would like to increase the current fare but fears (1) that this action would lead to a reduction in patronage, and (2) that this action would also result in a loss of revenue for the company. Are these fears justified? Discuss your assessment.

(b) If the same situation were to occur in another city where the price elasticity is -1.3, would the fears still be justified?

8. An airline company currently sells a package deal for a $100 return ticket and sells 5000 tickets per week. Because of the high demand, the company raises its fare to $120 per ticket hoping to raise its revenue. If the price elasticity of demand is currently -1.2, what will be the sale of tickets per week and how will this affect total revenue? What conclusions do you draw from this exercise?

9. When the supply of motor scooters falls by 10%, the price of scooters goes up 40%. What is the price elasticity of demand of scooters? What would happen if scooter-sellers raise prices of scooters by 50% because of this reduction of supply?

10. Which of the following pairs of products can be considered as complements and which as substitutes?

       Group A: Car batteries and automobiles
       Group B: Car tires and automobiles
       Group C: Bus travel and airplane travel
       Group D: Hot dogs and hamburgers
       Group E: Horses and carriages
       Group F: Hot dogs and buns

11. Refer to Example 6 in the text. If all the conditions remain the same, but the elasticity is −0.75, what would be the advice you would offer to the management of this transit company?

12. The demand function for automobile travel along a major corridor in a medium-sized city is estimated to be

$$Q = aA^{-2.2}B^{0.13}C^{-0.4}D^{0.75}$$

where

Q = automobile trips per hour (peak-hour)
a = constant
A = travel time by automobile in minutes
B = travel time by bus in minutes
C = average cost by automobile
D = average cost by bus

(a) Justify the signs (+ or −) of the exponents of parameters A, B, C, and D.
(b) Because of congestion likely to occur on this corridor, the travel times of automobiles would increase by 20%, and travel times of buses would increase 10%. At the same time, auto travel costs rise 5% and bus costs decrease 15%. What will be the percentage change in auto traffic?
(c) If the average cost of travel by bus increases by 10%, but the travel time by bus decreases by 10%, what would be the overall percentage increase or decrease in auto travel?

13. A transit company estimates that the cross elasticity of demand between its fast express bus and its ordinary bus is 2. Calculate the effect on the revenue received from the express service if the price of the ordinary bus service is reduced from $75 to $50 while the price of the express bus service remains the same.

14. Latent demand for highway travel is defined as the difference between the maximum number of trips that could be made and the number of trips that are actually made. Given a demand function $q = 1800 - 150t$, what is the time elasticity of demand when $t = 2$ minutes? What is the latent demand at this travel time?

15. A transit authority wants to improve one segment of a light rail system by increasing the peak-hour seating by 20% and also by repricing the fare to achieve full utilization. The existing capacity is 2000 seats/hour at $1/seat, and the price elasticity is −0.75. What additional consumer surplus would be generated by this action? Would the transit authority gain from this action?

16. A bus system consisting of 50 buses (55-seaters) charges $1 per ride. It has been decided to put 10% more buses into service. What should be the new fare per ride in order to achieve full utilization of capacity if the price elasticity of demand for ridership is −0.3? What will be the additional consumer surplus generated by this action?

17. The ferry service from a city to a recreational island is currently served by regular ferries and luxury boat. Five thousand passengers per day use the regular ferries and 7000 use the luxury boat. Travel times (min) and fares ($) are

	TRAVEL TIME (MIN)	FARE ($)
Regular	45	1
Luxury	30	2

The linear arc-time and arc-price elasticities of demand are as follows:

	REGULAR		LUXURY	
	TIME	FARE	TIME	FARE
Regular	−0.03	−0.04	+0.02	+0.05
Luxury	+0.05	+0.02	−0.07	−0.20

(a) If the fare on the luxury boat is raised to $2.50, what will be the effect on ridership?

(b) If the travel time on the luxury boat is reduced to 25 min, what will be the effect on ridership?

(c) If the regular ferries increase their travel time to 50 min, what will be the effect on ridership?

(d) How will the total revenue of the service be affected by parts (a), (b), and (c)?

18. A survey of college students revealed that, in general, they value one train trip a month to a resort at $40, the second the same month at $30, the third at $20, a fourth at $15, and a fifth at $5. The survey found that students would not take more than five trips even if they were free.

(a) If the train tickets cost $25 a trip, how many trips, on average, will a typical student take?

(b) A travel club, charging monthly dues, allows students to travel free. How much, at most, would a student be willing to pay per month as monthly dues?

19. A rapid transit system has estimated the following costs of operation for one of its routes, making a variety of combinations of cars per train:

Number of cars	Fixed cost ($/mi)	Variable cost ($/mi)
1	45	30
2	45	54
3	45	76
4	45	102
5	45	150
6	45	225
7	45	310

Plot the information for each combination, including the total cost, the average cost, and the marginal cost of operating the system. What is the optimum number of cars per train that should be operated?

20. A concrete-mix plant needs to hire a few men at either $5/hour (if semiskilled) or $3.50 per hour (if nonskilled). The following data are available for making a decision:

Number of workers	2	3	4	5	6	7	8
Hours of output (yd^3)	56	120	180	200	210	218	224

Additional fixed costs are $50/hour when semiskilled labor is used and $60/hour when nonskilled labor is used. Determine the number of workers to be hired to minimize production costs. What is the corresponding cost per cubic yard?

21. A section of a highway has the following characteristics:

$$v = 60 - \left(\frac{q}{120}\right)$$

where v is the speed (mph), and q is the stream flow (veh/hr). Further,

$$p = 2.5 \left(3 + \frac{200}{v}\right)$$

where p is the motorist's cost (cents/mile) and

$$D = 3.8 \times \frac{10^4}{p}$$

where D is the demand (vehicles/hour).
(a) Plot the motorist's private and marginal cost curves and the demand curve.
(b) Determine the optimum flow.
(c) Determine the toll that should be charged to limit the flow to the optimal flow.

22. A short section of a freeway has the following characteristics:

$$\text{Average cost (cents/mile/vehicle)} = 8.0 + \left(\frac{350}{28 - 0.008q}\right)$$

where

q = flow rate

t = travel time per mile for a flow rate of q

$$= \frac{1}{28 - 0.008q}$$

If the flow were to increase from 1800 vehicles to 1801 vehicles, calculate the corresponding average and marginal cost per vehicle-mile and the travel time. Discuss your own interpretation of these figures.

23. Ten miles of a congested urban freeway were improved by adding an additional lane, with the hope of relieving congestion and decreasing the average travel time over the 10-mile section. Three years later, after millions of dollars had been expended, and the mayor had just cut the ribbon, motorists found to their surprise that the congestion and travel time over this section was worse than before. Plot two sets of demand and supply functions on a graph, with time and flow on the y and x axes, respectively, to depict this phenomenon. What lessons can be learned from this experience?

24. A small bus company has a cost function of $C = 5 + 7q$, where q is the number of buses, and C is the cost in 1000's.
(a) Determine the average and marginal cost function.
(b) Determine the elasticity function.
(c) Does an economy of scale exist?
(d) Would you recommend providing additional bus capacity based on (c)?

25. Refer to Exercise 24. A similar bus company in a city has a cost function of $C = 7q$. Answer parts (a) through (d) for this city.

Chapter 3

The Land-Use/Transportation System

1. INTRODUCTION

The basic interdependence between land use and transportation was discussed briefly in Chapter 1. It was demonstrated that a piece of land with a particular type of land use produces a certain number of trips. These trips indicate the need for transportation facilities in order to serve the trip-making demand. In turn, the new or improved transportation facilities provide better accessibility. Naturally, the demand to develop this land increases because of its improved accessibility, causing its land value to increase. Eventually, the original land use changes (usually to a higher density), reflecting the state of the land market; and so the cycle continues. Although this is a simplified description of the land-use/transportation cycle, it represents the interactive nature of these two components (Moore, 1975; Deacon et al., 1976; Chapin et al., 1979).

Unfortunately, transportation and land-development decisions have all too often been regarded as distinctly separate issues in analysis, planning, designing, and evaluation (Deacon et al., 1976). In this chapter, we deal with the land-use/transportation interaction by examining the components of the urban system and then scrutinize the criteria for measuring the urban structure. Some selected theories connected with land use are then studied, followed by further discussions on the more complex issues of land use and transportation vis-à-vis population and housing.

This is a critical chapter for students to comprehend, because it covers a large body of background information necessary for understanding the basis of transportation engineering and planning. It is quite possible that students may not appreciate the significance of the content of this chapter on the first reading. Come back and read the chapter after you have advanced through most of this book. It will benefit you immensely.

2. URBAN SYSTEM COMPONENTS

In a democratic society such as the United States, land has historically been used by its owner for whatever purposes the owner saw fit. As society evolved, limitations were placed on such use, particularly if such use was likely to affect neighboring property negatively. These limitations, in effect, maintained other people's rights against detrimental effects. As the need for decent housing, safe streets, and proper sewer and water facilities grew, governmental agencies and private concerns saw the need for planning and regulating the use of land. The origin of the term *land use* comes from agricultural economics. It refers to a parcel of land and the economic use it was then put to—grazing, growing crops, mining, or building.

Land-use planning really can be considered in two contexts. First, it includes all forms of planning. For instance, transportation planning can be considered as a form of land-use planning because it actually consists of planning for that proportion of land used for transportation. Second, land-use planning is a discipline by itself, having its own set of theories and practices (ASCE, 1986). Some of these theories will be examined in this chapter.

Despite the fact that each city and urban area has its own individual characteristics, there is an apparent order or coherence in its overall pattern. Allowing for pockets of chaos here and there, the general arrangement seems to fit together; buildings and land uses appear to be arranged according to some "plan" (Bourne, 1982). One may well ask the question: What is the logic behind this order, despite the chaos? In this and other sections of this chapter, we attempt to provide partial answers to this open-ended question.

There is no unified set of concepts or theories of urban form and spatial structure that has emerged, although several regional scientists and city planners have presented their own theories, hypotheses, and models on this subject. There is, of course, some validity to each approach, but none of them has universal acceptance (Bourne, 1982). The objective here is to identify a consistent body of concepts and ideas to help understand the urban fabric.

It is useful to consider the city (or the urban area) as a system. A number of related elements or components, in combination, make up this city system. Table 3-1 identifies a series of system components and their corresponding elements, within the fabric of the city, defined as a spatial system. The characteristics of each of these components could be described in great detail, but it will suffice here to provide some examples. The nucleus of the city may be considered as the location of the initial settlement, developing in course of time, as the commercial and communications center of the city—usually the central business district (CBD). As the city grows and spreads out, the influence of the CBD may tend to decline as prominent subcenters begin to develop (Bourne, 1982).

Cities have a definable area, with identifiable boundaries, at a certain point in time. Every city exhibits specific types of behavior. That behavior, in the form of growth, change, or decay, is subject to a dominant set of mechanisms that underlie its form and determine the pattern of change that is likely to take place. Cities also have

TABLE 3-1   SYSTEM COMPONENTS IN URBAN SPATIAL STRUCTURE

System components	Corresponding elements in urban spatial structure
1. Nucleus: the point of system origin and the locus of control	1. The initial settlement (e.g., the confluence of two rivers, or a harbor) and the central business district
2. Geometric area and boundaries of the system	2. The geographic extent and limits of the urban area
3. Elements: the parts, units, or bits that form the membership of the system	3. Social groups, land uses, activities, interactions, and institutions
4. Organizational principles: what ties the system together and allocates activities to areas; what "energy" drives the system?	4. The underlying logic or principles of urban structure (e.g., the land market) and the determinants of growth
5. Behavior: how the system acts and changes over time; its routine and nonroutine actions	5. The way the city works; its activity patterns and growth performance
6. Environment: the "external" context that influences that system	6. The source and types of external determinants of urban structure
7. Time path: a trend of evolution and change	7. A development sequence; historical profile of building cycles and transport eras

Source: From *Internal Structure of the City: Readings on Urban Form, Growth and Policy*, 2/e, by Larry S. Bourne. Copyright © 1982 by Oxford University Press, Inc. Reproduced by permission.

an external environment, which may be the immediate territorial hinterland, or regional economy, or the entire array of political, economical, or cultural spheres of which the city is an integral part. In addition, there is an historical sequence of development or a time path connected with a city, such as building cycles or the construction of freeways or rail systems. Such "layers" of development provide what one might term the "personality" of the city (Bourne, 1982).

3. CONCEPTS AND DEFINITIONS (CATANESE AND SNYDER, 1979; BOURNE, 1982; ASCE, 1986; HANSEN, 1986)

A set of simple working definitions and basic concepts is provided for understanding urban form and structure, remembering that in an interdisciplinary topic such as this, everything is related to everything else.

   1. *Urban form.* The spatial pattern or "arrangement" of individual elements—such as buildings, streets, parks, and other land uses (collectively called the *built environment*), as well as the social groups, economic activities, and public institutions, within an urban area, is recognized as the urban form. Figure 3-1(a) represents the broad configuration of land uses.

   2. *Urban interaction.* This is the collective set of interrelationships, linkages, and flows that occurs to integrate and bind the pattern and behavior of individual land uses, groups, and activities into the functioning entities, or subsystems. One of the most important integrating subsystems is the highway or street network. Figure 3-1(a) shows a set of streets and rails connecting the various land uses.

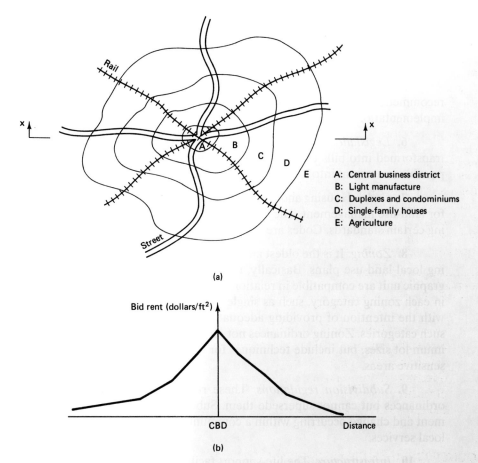

A: Central business district
B: Light manufacture
C: Duplexes and condominiums
D: Single-family houses
E: Agriculture

(a)

Bid rent (dollars/ft^2)

CBD   Distance

(b)

Figure 3-1   Components of Urban Spatial Structure.

3. *Urban spatial structure.* This structure formally combines the urban form through the urban interaction with a set of organizational rules into a city system. Figure 3-1(b) shows an example of a systemwide organizing mechanism, such as the competitive rent for different locations within the urban area. This mechanism produces an "ordering" of activities in terms of their location requirements and the rent that each person can afford to pay. This is merely one example; other organizing mechanisms are the functioning of government and public institutions and the accepted norms or social behavior.

4. *Comprehensive plan.* The basic overall plan is usually the comprehensive plan, sometimes referred to as the master plan or general plan. This plan, at the very minimum, is an official statement of a geographic unit's (city or region) policies and intentions pertaining to physical development in the years ahead. It establishes a framework for developing the geographic area under consideration and includes recommendations of what, where, when, and especially why certain developments in regard to housing, sewer, water, transportation, and so on have been made. The land-use plan, which is a basic part of the comprehensive plan, contains documentation of the analyses that were done leading to a determination of the "best" future physical development. A transportation plan is also a part of this comprehensive plan.

5. *Guidelines.* In some cases, a set of guidelines for development may serve as a recommended alternative in a land-use plan. Guidelines can be given importance as an implementation technique by putting them into legislation.

6. *Legislation.* Some of the recommendations from the land-use plan can be transformed into bills that can subsequently be submitted to the legislative body for possible enactment into law.

7. *Codes.* Housing and building codes are important implementation techniques for land-use management. They ensure the quality of community growth by establishing certain standards. Codes are most commonly used at the local municipal level.

8. *Zoning.* It is the oldest and most commonly used legal device for implementing local land-use plans. Basically, it is a means of assuring that land uses in a geographic unit are compatible in relation to one another. It allows for control of densities in each zoning category, such as single family, duplex, office, commercial, and so on, with the intention of providing adequate facilities (streets, water, sewer, school) for such categories. Zoning ordinances not only prescribe such items as setbacks and minimum lot sizes, but include techniques for preserving and protecting environmentally sensitive areas.

9. *Subdivision regulations.* These regulations complement the local zoning ordinances but cannot supersede them. Subdivision regulations control the development and change occurring within a community and encourage efficient and desirable local services.

10. *Infrastructure.* The life-support facilities of a geographic unit are collectively referred to as the infrastructure. It consists of all those basic elements that make an urban area function: transportation facilities, sewer and water facilities, highway, housing, harbors, pipelines, and so on.

4. CRITERIA FOR MEASURING AND COMPARING URBAN STRUCTURE (BOURNE, 1982)

It is useful to have a set of criteria that would help measure and compare urban form and spatial structure. This set of criteria brings together diverse concepts for evaluation purposes. Table 3-2 provides a list of 20 criteria grouped into four sections. Most of the criteria are self-explanatory in both their intent and application. The first section suggests that the internal structure of urban areas will differ if the building stock, the street system, and industrial plant are of different ages and constructed under differing policy, technology, and economic conditions. The second set of criteria consists of the more traditional and widely known indices. For example, cities of more or less similar

TABLE 3-2   CRITERIA FOR URBAN SPATIAL STRUCTURE

Level	Criteria	Description and examples
Context	1. Timing	Time and stage of development
	2. Functional character	Predominant mode and type of production (eg., service center, mining town)
	3. External environment	The socioeconomic and cultural environment in which the city is embedded
	4. Relative location	Position within the larger urban system (e.g., core–periphery contrasts)
Macro-form	5. Scale	Size: in area, population, economic base, income, etc.
	6. Shape	The geographic shape of the area
	7. Site and topographic base	The physical landscape on which the city is built
	8. Transport network	The type and configuration of transportation system
Internal form and function	9. Density	Average density of development; shape of density gradients (e.g., population)
	10. Homogeneity	The degree of mixing (or segregation) of uses, activities, and social groups
	11. Concentricity	The degree to which uses, activities, etc., are organized zonally about the city center
	12. Sectorality	The degree to which uses, activities, etc., are organized sectorally about the city center
	13. Connectivity	The degree to which nodes or subareas of the city are linked by networks of transportation, social interaction, etc.
	14. Directionality	The degree of elliptical orientation in interaction patterns (e.g., residential migration)
	15. Conformity	The degree of correspondence between function and form
	16. Substitutability	The degree to which different urban forms (e.g., buildings, areas, public bodies) developed for one function can be used (substituted) for another
Organization and behavior	17. Organizational principles	The underlying mechanism of spatial sorting and integration
	18. Cybernetic properties	The extent of feedback; the sensitivity of form to change
	19. Regulatory mechanisms	Internal means of monitoring and control (e.g., zoning, building controls, financial constraints)
	20. Goal orientation	The degree to which urban structure evolves toward a priori objectives

Source: Bourne, 1982. Reproduced by permission.

population located on different topographical sites will each produce a different urban form, and even their transportation system may differ.

The third set of criteria relates to indices of urban patterning that are easily measured. Combined, these indices provide a relatively comprehensive picture of the geometry of the city, but very little is known about the operation or behavior of the city itself. Figure 3-2 portrays the criteria of urban spatial structure for a typical city: density, homogeneity, concentricity, sectorality, connectivity, and directionality.

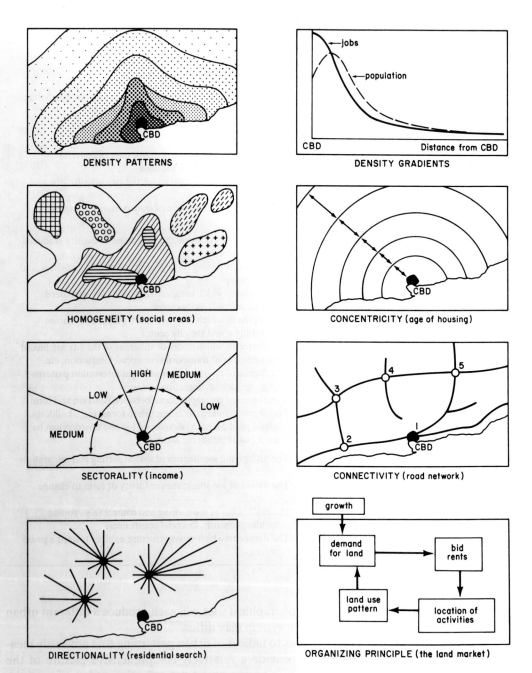

Figure 3-2   Selected Criteria of Urban Spatial Structure (Bourne, 1982).

This examination of the various criteria pulls together in a systematic fashion a variety of ideas from a wide field of knowledge. The interested reader may like to pursue this subject still further. A list of suggested readings is provided at the end of the chapter; see, particularly, Abler et al., 1971; Yeates, 1974; Deacon et al., 1976; Chapin et al., 1979; Meyer and Gómez-Ibáñez, 1981; and Bourne, 1982.

5. SOME SELECTED THEORIES AND TOPICS

The theory of urban planning is concerned with defining and understanding the contents, practices, and processes of planning. Theory is linked with practice and vice versa. For planners, this linkage is vital because planning, unlike the sciences, is a prescriptive (or normative), not a descriptive, activity. The planner's objective is not merely to describe the city and its components, but rather to propose ways in which they can be changed, hopefully for the better (Catanese and Snyder, 1979). In this section, the concept of accessibility is explained with the help of examples. Location theory and its applications to transportation are examined. A short account of zoning is also included.

5.1 Accessibility

The basic concept underlying the relationship between land use and transportation is accessibility. In its broadest context, accessibility refers to the ease of movement between places. Accessibility increases—either in terms of time or money—when movement becomes less costly. Also, the propensity for interaction increases as the cost of movement decreases (Blunden, 1971: Blunden and Black, 1984).

Example 1

From \ To	Node				Σ	Change
	A	B	C	D		
A	0(0)	6(4)	7(6)	9(8)	22(18)	−18%
B	6(4)	0(0)	6(5)	4(2)	16(11)	−31%
C	7(6)	6(5)	0(0)	7(5)	20(16)	−20%
D	9(8)	4(2)	7(5)	0(0)	20(15)	−25%

Note: Figures outside parentheses refer to original travel time; figures within parentheses refer to travel times after improvements.

Each node (*A*, *B*, *C*, *D*) represents an activity center, and each link (e.g., *AB*, *BC*) represents travel times in minutes (see Figure 3-E1). Transportation improvements are implemented on each link such that travel times are reduced. How do the transportation improvements affect the activity centers (land use)? (Revised travel times are indicated in parentheses.)

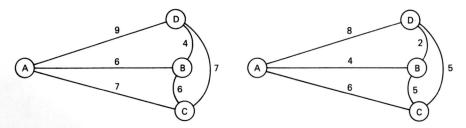

Figure 3-E1   Network for Example 1.

Solution

The matrix shows the existing and revised times of travel. The row sums are the accessibility measure for each node. Naturally, lower travel times mean greater accessibility. In all cases, there is a reduction in travel time: A, -18%; B, -31%; C, -20%; and D, -25%. It is apparent that activity center B has benefited the most, followed by D, C, and A.

Example 2

A downtown (D) is connected by arterials to activity/residential centers A, B, and C and to one another with travel times shown on the links. The arterials become more and more congested, with the result that travel times (in minutes) increase to new levels, as shown in Figure 3-E2, and most of the commercial and business centers located downtown establish branch centers in A, B, and C. Which activity center is likely to prosper most? What possible actions on the part of the city would improve the downtown?

	Activities Center			Downtown		
To From	A	B	C	D	Σ	Change
A	0(0)	8(7)	10(9)	4(16)	22(32)	45%
B	8(7)	0(0)	12(11)	3(15)	23(33)	43%
C	10(9)	12(11)	0(0)	5(20)	27(40)	48%
D	4(16)	3(15)	5(20)	0(0)	12(51)	325%

Note: Figures outside parentheses refer to original travel times; figures within parentheses refer to travel times after improvements.

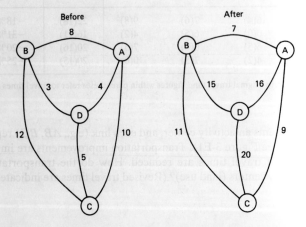

Figure 3-E2   Network for Example 2.

Solution

Activity centers A, B, and C are all likely to be equally benefited, as the difference between 43, 45, and 48 seems hardly significant. The downtown is certainly going to deteriorate rapidly. The possible ways by which it could be saved would be to reduce the travel time by improving the traffic flow on the arterials or to implement a bus system that would compete with the travel time to activity centers.

Personal accessibility is usually measured by counting the number of activity sites (also called *opportunities*) available at a given distance from the person's home and factoring that number by the intervening distance. Accessibility measures can be calculated for specific types of opportunities, such as shopping or working. One such measure is given by

$$A_i = \sum_j O_j d_{ij}^{-b} \tag{1}$$

where

A_i = accessibility of person i

O_i = number of opportunities at distance d from person i's home

d_{ij} = some measure of separation between i and j (such as travel time, travel cost, or just simply distance)

b = a constant

Such an accessibility index is a measure of the number of potential destinations available to a person and how easily he or she can reach them. The accessibility of a place with respect to other places in the city can be measured in a similar way, in which case A_i is the accessibility of zone i.

Example 3

A small city has three residential areas R_1, R_2, and R_3 with 1500, 2000, and 2500 workers, and two employment zones E_1 and E_2 with 2000 and 4000 job opportunities. The interzonal travel times in minutes are given in the following table. Find the actual and relative zonal accessibility of the residential areas assuming that $b = 1.0$.

o	d 1	2	R_0
1	10	12	1500
2	7	9	2000
3	6	8	2500
E_d	2000	4000	6000

Solution

$$A_o = \sum_d \frac{E_d}{t_{od}^b} \tag{2}$$

where

$$d = 1, 2; \quad o = 1, 2, 3$$

E_d = number of jobs in zone d

t_{od}^b = travel-time function

$$A_1 = \frac{2000}{10} + \frac{4000}{12} = 200 + 333 = 533$$

$$A_2 = \frac{2000}{7} + \frac{4000}{9} = 286 + 444 = 730$$

$$A_3 = \frac{2000}{6} + \frac{4000}{8} = 333 + 500 = 833$$

$$\text{Total} = 2096$$

and the relative accessibility is

$$A_1 = \frac{533}{2096} = 0.25$$

$$A_2 = \frac{730}{2096} = 0.35$$

$$A_3 = \frac{833}{2096} = 0.40$$

$$\text{Total} = 1.00$$

Discussion

Here the accessibility of everyone living in a particular zone is aggregated and therefore there is no way of distinguishing among different groups of people within a zone, such as those with cars and those without cars; for example, the 1500 workers living in zone R_1 are all assumed to own cars and reach zone E_1 in 10 minutes. A person's ability to reach different locations in a city depends only in part on the relative location of those places; it also depends on mobility (the ability to move to activity sites) and the transportation system that exists.

5.2 Location Theory

Theories of the location of activities, particularly residential, were developed in the 1960s by several regional scientists. By drawing examples from agricultural land economics, it is possible to conceive the specific use of a parcel of land as being a function of its distance from a market, assuming there is one market located at the center of a featureless field. The specific use of the land at any location will depend on its bid rent (L), as per the equation

$$L = E(p - a) - Efk \tag{3}$$

where

E = yield per unit land

p = market price per unit of commodity at site

a = production cost per unit of commodity at site

f = transportation cost per unit per unit of distance

k = distance to market

Note that for a particular commodity, k is the only variable. Efk, the total transportation cost, increases with distance, and $E(p - a)$ is a constant for a particular crop. At a certain distance $E(p - a)$ equals Efk, and this distance is known as the zero-rent margin; a particular crop grown at any farther distance would not be economically feasible.

Figure 3-3(a), shows the resulting bid-rent curves for three hypothetical commodities. Farmers will naturally select the crop (or commodity) whose profitability is

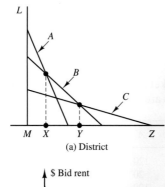

(a) District

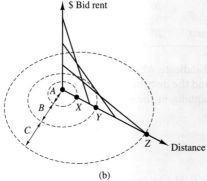

(b)

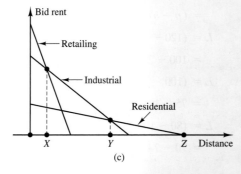

(c)

Figure 3-3 Bid-Rent vs. Distance from Market Center. (a) Distribution of Crops; (b) Three-Dimensional View of Crop Distribution; (c) Distribution of Land Use.

the highest for their location. The bid-rent curves for crops A and B intersect at a distance x from marketplace M, which means that it would be most profitable to grow crop A left of x. Similarly, crop B and crop C can best be grown between x and y and y and z, respectively. Figure 3-3(b) shows this in three dimensions.

Alonso has used von Thünen's theory in the urban context, by generating a series of land-use zones from the intersection of different bid-rent curves. Assuming that a city has one central business district (CBD) where all employment is located and that all transportation costs are linearly related to distance from the CBD, the bid-rent curves and corresponding land-use pattern can be sketched in terms of, say, three types of urban land use: retailing, industrial, and residential. See Figure 3-3(c). Notice that residential use has the shallowest of the three lines, with individual households making choices that maximize their individual satisfaction. For instance, the poor, with little disposable income, consume small amounts of space near the city center, where transportation costs are low. On the contrary, the rich, with much larger disposable income, pay about the same rent as the poor by living near the city's edge, but can afford to pay higher transportation costs for the longer commute to work.

Example 4

A concentric city with a single market at the center wants to produce four different crops, A through D, whose characteristics are given in dollars:

	A	B	C	D
Price at site per unit	120	100	80	50
Production cost per unit	20	25	10	10
Net price per unit at site	100	75	70	40
Transportation cost per unit	20	10	7	3.33

Sketch your results, and indicate which crop should be produced at what optimal distance from the city center, and the distribution. What implication does this theory have for transportation and city planning in the context of rent, housing costs, and distance from the city center?

Solution

Refer to Eq. 3.

$$L = E(p - a) - Efk \tag{3}$$

$$= (p - a) - fk$$

Crop A: $$L = (120 - 20) - 20k$$

$$= 100 - 20k$$

Crop B: $$L = (100 - 25) - 10k$$

$$= 75 - 10k$$

Crop C: $$L = (80 - 10) - 7k$$

$$= 70 - 7k$$

$$\text{Crop } D: \qquad L = (50 - 10) - 3.33\,k$$

$$= 40 - 3.33\,k$$

The four equations represent crops A, B, C, and D, and equating them pairwise gives distances 2.308, 8.182, and 12.12 miles representing the radii from the city center to which it is most profitable to grow crops A, C, and D, respectively. Note that crop B is obviously not profitable to grow. Figure 3-E4 illustrates the concentric rings where the crops are grown. Refer to Figure 3-E4(a), where the vertical axis represents the profit and the horizontal axis represents distance. Distance from market when profit = 0 is as shown below:

A	B	C	D
$\dfrac{100}{20} = 5$	$\dfrac{75}{10} = 7.5$	$\dfrac{70}{7} = 10$	$\dfrac{40}{3.33} = 12.12$

Figure 3-E4 shows that from an economic point of view:

1. Crop A should be grown from the center of the city to a distance of 2.308 distance units from the market.

2. Crop C should be grown from where A stops (2.308 distance units) to as far out as 8.182 distance units from the market.

3. Crop D should be grown from where C stops (8.182 distance units) to as far out as 12.12 distance units from the market.

4. Crop B should not be grown at all, because it would not be profitable to do so.

Example 4 illustrates that the transportation cost is a major factor of concern: the farther away from the market a crop is grown, the more it takes to get the product to the market. According to *rent theory*, the rent or cost of land is inversely proportional to the distance of the land from the CBD. Combining the two earlier phenomena, one can sketch the locations where the crops can be grown most economically.

5.3 Effects of Zoning

Activities, people, and locations all interact in such a way that everybody wishes to maximize their locations. People have reasons for living where they do. Businesses and industries also have locational preferences. Locational preferences result in patterns of concentration. Land-use planners regulate the compatibility of land-use patterns through zoning and other regulations.

An example will make these issues clear. Suppose that a small city has several parcels of land, each owned by a person who is free to sell the parcel to the highest bidder. People able to purchase these parcels of land on the open market pay a price.

This free-market sale, however, may possibly result in incompatible land use. For example, one of the parcels of land may be sold to a person who wants to establish a small factory that may be adjacent to a housing complex. To prevent any kind of nuisance, most communities enact zoning ordinances to regulate the use of land.

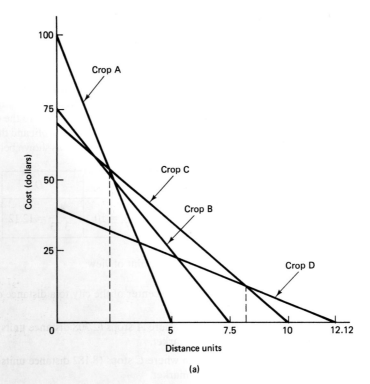

(a)

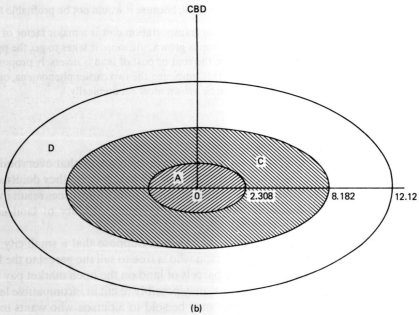

(b)

Figure 3-E4   Solution for Example 4: (a) Distance vs. Cost; (b) a Concentric City with One CBD

5.4 Land Values

A simple model of land value can be represented by a linear function, as follows:

$$LV_i = a - bD_i$$

where LV is land value, D is the distance from the CBD, and a and b are constants.

A somewhat more elaborate model can be represented by a power function, as follows:

$$LV_i = aD_i^{-b}$$

where land values decrease at a decreasing rate. In both these models, the unknowns can be determined by least-square (regression) analysis.

Models of land value for metropolitan areas can be quite complicated. For instance, a conceptual model for a large city located near the ocean could have an equation such as

$$LV_i = a - b_1C_i - b_2M_i - b_3E_i - b_4S_i$$

where LV_i is the land value at site i; C_i is the distance from the CBD; M_i is the distance from the ocean; E_i is the distance from the nearest elevated-subway station; and S_i is the distance from the nearest regional shopping center. Other models similar to the one described have used such variables as accessibility, amenities, and topographical features.

6. LAND USE AND TRANSPORTATION

The movement of people and goods in a city, referred to as *traffic flow*, is the joint consequence of land activity (demand) and the capability of the transportation system to handle this traffic flow (supply). Naturally, there is a direct interaction between the type and intensity of land use and the supply of transportation facilities provided. One of the primary objectives of planning any land-use and transportation system is to ensure that there is an efficient balance between land-use activity and transportation capability (Blunden and Black, 1984; ASCE, 1986).

The relationships between transportation and land development are viewed in three different contexts: (1) physical relationships at the macroscale, which are of long-term significance and generally considered as part of the planning process; (2) physical relationships at the microscale, which are both of short- and long-term significance and generally considered as urban design issues (often at the scale of particular sites or facilities); and (3) process relationships, which deal with the legal, administrative, financial, and institutional aspects of coordinating land and transportation development (Deacon et al., 1976).

Urban areas over the years have changed as a result of the dramatic shift from agricultural to urban employment. They are now centers of extensive financial influence and employment opportunities whose complexity and diversity undergo cycles of change related to numerous social and economic forces. Their physical characteristics

TABLE 3-3 EXAMPLES OF LAND-USE POTENTIAL

Type of land/activity	Measure
Residential	Population, dwelling units
Factories	Area, number of workers
Offices	Area, number of employees
Theaters	Seating capacity
Hotels	Number of rooms, floor area
Shopping center	Retail area, employees

are changing rapidly as a result of new time and distance relationships established by high-speed travel and low-cost electronic communication.

More people and higher percentages of the total population are living in urban areas but in less dense concentrations than ever before. Current dispersion of the population raises new issues as to the long-term impact on transportation effectiveness. As has been discussed before, land use and transportation form a closed loop. Like most other equilibrium systems, the land-use/transportation configuration eventually stabilizes.

Land-use potential is a measure of the scale of socioeconomic activity that takes place on a given area of land. A unique property of land use is its ability or potential to "generate" traffic. Hence, it is appropriate to relate the land-use potential of a parcel of land, having a particular activity, to generate a certain amount of traffic per day.

Table 3-3 provides typical examples of land-use potential. Note that traffic generation is a dynamic phenomenon and the intensity of traffic generation can be defined as a function of time and space. In a general sense, land use means the spatial distribution or geographical pattern of the city: residential areas, industry, commercial areas, retail business, and the space set aside for governmental, institutional, and recreational purposes. If the land uses by area for a city are known, it is possible to be able to estimate the traffic generated (Blunden and Black, 1984).

Trip generation provides the linkage between land use and travel. Land use for trip-generation purposes is usually described in terms of land-use intensity, character of the land-use activities, and the location within the urban environment.

Example 5

Data for shopping trips to shopping sites in various parts of a city are recorded. Calculate the shopping trip rates by location type, and discuss your results.

Solution

CBD:

$$\frac{7200 + 2500}{3000 + 1400} = 2.205 \text{ shopping trips/employee}$$

Shopping centers:

$$\frac{6000 + 12,000}{3000 + 1400} = 9.0 \text{ shopping trips/employee}$$

Zone	Location type	Number of employees	Number of shopping trips
1	CBD	3,000	7,200
2	CBD	1,400	2,500
3	Shopping center 1	600	6,000
4	Shopping center 2	1,400	12,000
5	Local center	15	50
6	Local center	50	140
7	Local center	85	300
8	Local center	105	380

Local centers:

$$\frac{50 + 140 + 300 + 380}{15 + 50 + 85 + 105} = 3.41 \text{ shopping trips/employee}$$

Discussion

The shopping trips per employee for shopping centers is the highest, followed by local centers and CBD. The analysis does not have to be tied to zones, but can be done on an individual basis. For example, the characteristics of shopping center 1 may be very different from those of shopping center 2. Here the location types have been aggregated, and this aggregation could mask the results.

A land-use/transportation system may be represented by a spatial array of land uses overlaid with a network representing the transportation system. Such a system is shown diagrammatically in Figure 3-4. Note that the land-use zones should ideally

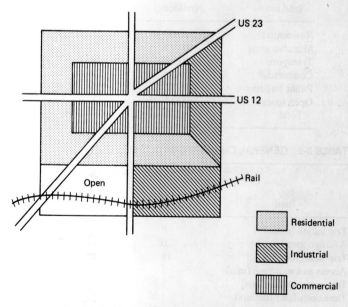

Figure 3-4   Typical Land-Use/Transportation Plan.

define an area of homogeneous land-use activity: for example, residential, commercial, industrial, and so on.

Transportation is a derived demand, that is, one does not generally make a trip just for the purpose of making a trip but for the specific purpose of going to work, or for shopping, or to get to school, and so on. Generally, trips are made for "profitable" purposes. Sometimes the "profits" are intangible. Between 10% and 25% of a person's income is spent on transportation, and probably the rewards derived from this activity justify this expenditure.

Land-use analysis is a convenient way to study the activities that provide the basis for trip generation because travel patterns (routes and traffic flows) are dictated by the transportation network and land-use arrangements. It must be remembered that a trip is an event linking an origin (say, one's home) and a destination (say, one's place of work). It is performed by traveling on a defined route, which has a certain length (in miles) and takes a certain time (in minutes) to travel.

An example of trip destinations and trip-generation rates by type of land use for a large metropolitan area in the midwest is provided in Table 3-4. Notice that residential, manufacturing, and public lands generate trips at about the same average rates, whereas commercial land generates trips at a rate nearly four times greater.

Table 3-5 shows the characteristics of typical work trips for two typical work locations in an urban area. It shows the relative importance of each component. Table 3-6 gives typical mean trip lengths of person-trips by land-use type.

TABLE 3-4   TRIP DESTINATIONS AND TRIP GENERATION

Type of land use	Person-trip destination (millions)	Area (mi^2)	Person-trip destination per square mile
Residential	5.606	180.6	31,000
Manufacturing	0.779	24.7	31,600
Transport	0.280	50.7	5,500
Commercial	2.449	21.1	116,000
Public building	0.782	23.1	33,800
Open spaces	0.315	114.9	2,700
Total	10.212	415.1	220,600

TABLE 3-5   GENERAL CHARACTERISTICS OF WORK TRIPS IN A CITY

Item	CBD trip Auto	CBD trip Transit	Suburban trip Auto	Suburban trip Transit
Trip length (miles)	6	7	8	9
Average speed (mph)	20	10	30	15
Trip time (min)	18	42	16	36
Access and wait time (min)	5	15	3	20
Trip costs/mile (dollars)	0.70	0.20	0.65	0.20
Time costs/hour (dollars)	1.00	1.00	1.00	1.00
Parking costs/hour (dollars)	1.00	—	0.25	—

TABLE 3-6   MEAN TRIP LENGTH OF PERSON
TRIPS BY LAND-USE TYPE

Land use	Mean trip length (miles)
Residential	4.3
Manufacturing	5.1
Commercial	
Retail	3.1
Service	4.8
Wholesale	5.9
Public buildings	3.6
Public open spaces	4.5
All land uses	4.3

7. URBAN GROWTH OR DECLINE

It is necessary to understand the possible causal relationships, feedbacks, and interactions between different sectors of the city (or urban area), particularly the land-use and transportation elements. Urban growth or decline is a result of this complex interaction.

It must be realized by now that transportation in terms of economic development is a derived demand and is therefore dependent on the development of other sectors of the economy. The principal objective of transportation planning consists in providing for the necessary movements of people and goods at a minimum overall cost to the economy.

A simple block diagram (Figure 3-5) portrays the interrelatedness or interdependence of the urban system. It shows that any financial allocation to improve the transportation facilities in an urban area eventually feeds back on itself. The figure also shows that urban land availability will eventually constrain urban growth.

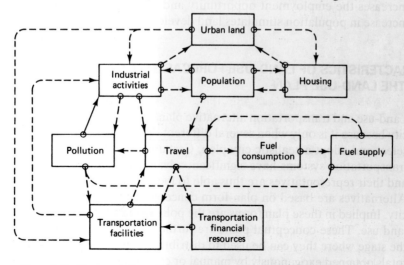

Figure 3-5   Transportation within the Urban System (Budhu and Grisson, 1985).

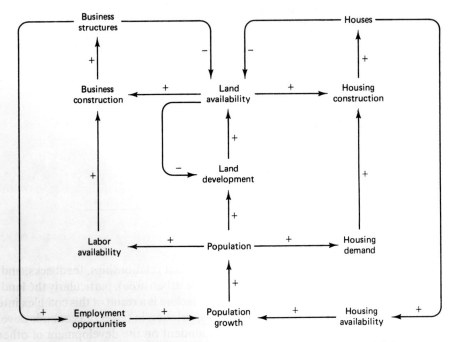

Figure 3-6 Transportation and the Socioeconomic Sector (Budhu and Grisson, 1985).

Figure 3-6 shows a conceptual framework of socioeconomic sector. Consider the following situation: More land is made available for downtown business expansion, business construction increases, and business structures increase, which in turn increases the employment opportunity and positively affects population. Finally, an increase in population stimulates land development.

8. CHARACTERISTICS OF LAND-USE FORECASTING AND THE LAND-USE PLAN

Land-use and transportation alternative plans are usually proposed and prepared for a city because it is only when several different options are examined that it is possible to select and adopt a realistic city plan. Certain combinations of land-use patterns and transportation systems have significant advantages in terms of efficiency. The public and their representatives are thus able to look at alternatives and identify advantages. Alternatives are based on plan-form concepts such as the radial, linear, or nucleated city. Implied in these plan forms are the policies concerning the location and density of land use. These conceptual plans are further developed and refined until they reach the stage where they can be used to distribute regional population and socioeconomic totals obtained exogenously by manual or computer procedures (Khisty, 1977).

8.1 Classification of Land-Use Models

Land-use models can be classified in numerous ways in terms of their sophistication. Level 1, the least sophisticated, involves the setting up of alternative physical patterns of land development. Such models do not get involved with the staging of transportation and other facilities. Generally, conventional manual methods are used to distribute growth in the study area. In level 2 models, the simple concept of the urban development process for spatially allocating households and employment is introduced along with the staging of transportation and other infrastructure construction. The main analytic components of this procedure are, for each land-use category, a set of location requirements and a set of space requirements. These judgments are made by the analyst, based on the given principles and standards, the special knowledge of local conditions, and what is considered to be in the best interest of the public. A further step is to introduce the use of some mathematical formulations, such as regression equations.

Level 3 models make more sophisticated use of concepts of the development process, including a wider range of policy specification. Models at this level are generally referred to as the *market simulation approach*. The archetype of this system of models was developed in the early 1960s by Ira Lowry of the Rand Corporation (Chapin and Kaiser, 1979). The general structure of the Lowry model has been used in large metropolitan studies in recent years (Khisty, 1977).

8.2 Land-Use Development Models

Land-use planning for a city is a complex task. In most democratic countries, land is allocated among alternative uses mainly in private markets with more or less public regulation. This results in cities developing mainly from locational decisions by a large number of private developers and buyers, each attempting to further personal and selfish interests. Land-use planning has been pursued by planners to predict wtth some degree of precision the spatial organization of population and economic activity in the region (Khisty, 1977; Smith, 1978).

Land-use models serve two distinct purposes: (1) forecasting the total activities of an urban area, and (2) allocating these activities among a predetermined set. Two simple land-use allocation models are described in the following sections: Hansen's accessibility model and the density saturation gradient method.

8.3 Hansen's Accessibility Model (Lee, 1973)

Hansen's model is designed to predict the location of population based on the premise that employment is the predominant factor in determining location. He suggested the use of an accessibility index, A_{ij}, where

$$A_{ij} = \frac{E_j}{d_{ij}^b}$$

where

A_{ij} = accessibility index of zone i with respect to zone j

E_j = total employment

d_{ij} = distance between i and j

b = an exponent

The overall accessibility index for zone i is therefore

$$A_{ij} = \sum_j \frac{E_j}{d_{ij}^b}$$

The amount of vacant land that is suitable and available for residential use is also an additional factor in attracting future population to the zone in question. This is referred to as *holding capacity* (H_i). The development potential of a zone D_i is, therefore,

$$D_i = A_i H_i$$

and population is distributed to zones on the basis of the relative development potential $A_i H_i / \Sigma A_i H_i$. If the total growth in population in a future year is G_t, the population allocated to zone i will be

$$G_i = G_t \frac{A_i H_i}{\Sigma A_i H_i} = G_t \frac{D_i}{\Sigma D_i}$$

Example 6

A small three-zone city has the following characteristics:

Zone	Total existing population	Holding capacity (acres)
1	2000	100
2	1000	200
3	3000	300
Total	6000	600

The travel times (in minutes) are given in the following table.

From i	To j 1	2	3
1	2	6	8
2	6	3	5
3	8	5	4

An exponent of 2 can be assumed based on work done with other cities of the same size. If the population of this city is expected to rise to 8000 persons in 20 years, how will the population be distributed by zone? Assume that the total employment in each zone is proportional to the total existing population in that zone.

Solution

Calculate A_{ij} and then A_i.

Zone	1	2	3	$\sum_i A_{ij}$
1	$2000/2^2 = 500$	$1000/6^2 = 28$	$3000/8^2 = 47$	575
2	$2000/6^2 = 56$	$1000/3^2 = 111$	$3000/5^2 = 120$	287
3	$2000/8^2 = 31$	$1000/5^2 = 40$	$3000/4^2 = 188$	259

Multiply A_i by H_i.

Zone	A_i	H_i	$D_i = A_i H_i$
1	575	100	57,500
2	287	200	57,400
3	259	300	77,700
Total			192,600

Calculate the relative development potential of each zone.

Zone	D_i	$\dfrac{D_i}{\Sigma D_i}$	G_t
1	57,500	0.299	2.392
2	57,400	0.298	2.384
3	77,700	0.403	3.224
Total	192,600	1.000	8.000

8.4 Density-Saturation Gradient Method (Khisty, 1977, 1979, 1981)

The density-saturation gradient (DSG) method was first used in the Chicago Area Transportation Study (CATS). Since then, many researchers have elaborated on this basic work. Three empirical rules are used in this method: (1) the intensity of land use declines as the distance or travel time to the CBD increases; (2) the ratio of the amount of land in use to the amount of available land decreases as distance from the CBD increases; and (3) the proportion of land devoted to each type of land use in an area remains stable.

Clark (1951) derived the basic equation for expressing this density–distance relationship (Yeates, 1974). The basic equation is

$$d_x = d_0 e^{-bx} \tag{3}$$

where

d_x = population density at distance x from the city center
d_0 = central density as extrapolated into the CBD of the city
b = density gradient or slope factor
e = base of natural logarithms

Clark made another assumption that is not dealt with by the density equation. He assumed that the higher downtown densities and the lower suburban densities will tend to equalize over time in most urban areas. This is supported by the findings of the most recent census, which indicate strong trends of population decline in the CBD and increased population movement toward peripheral and suburban areas. This observation implies that the density-saturation gradient is a function of age or regional location of the city and can be determined experimentally.

Holding capacity is given by the following expression:

$$HC_i = P_i + V_i d \tag{4}$$

where

HC_i = holding capacity of zone i
P_i = existing residential population of zone
V_i = vacant, available, and suitable land in zone i
d = anticipated average density at which all future residential development will occur

Also,

Percentage population saturation of zone i in a certain year

$$= \frac{\text{population of zone } i \text{ in a certain year}}{\text{holding capacity of zone } i} \times 100 \tag{5}$$

Vacant suitable land for residential development is estimated from data for planning areas, zoning ordinances, and zoning plans. The base-year percentage (population/residential) saturation is plotted against distance from the center of the city. The next step is the horizon-year projection of the percentage saturation curve. This is the most critical and subjective step. The only restraint on the projected curve is that the area under the new curve must account for the existing plus projected regional growth. Forecast population totals by analysis rings are determined by using appropriate ordinate values of the horizon-year curve. The distribution of ring totals to individual census tracts is completed with the help of residential development factors. Figure 3-7 shows a simple flowchart of the CATS estimating procedure.

The general procedure can be described as follows:

1. Establish the relationship between residential density and the distance from the CBD [Figure 3-8(a)].
2. Determine the percent population saturation for each zone and aggregate this percentage by ring and sector [Figure 3-8(b)]

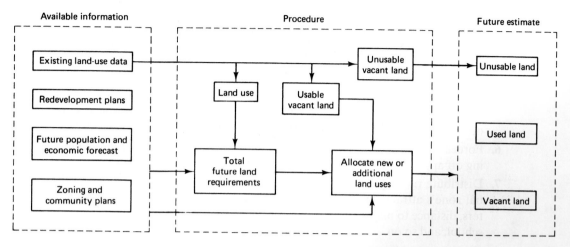

Figure 3-7   CATS Land-Use Estimating Procedure.

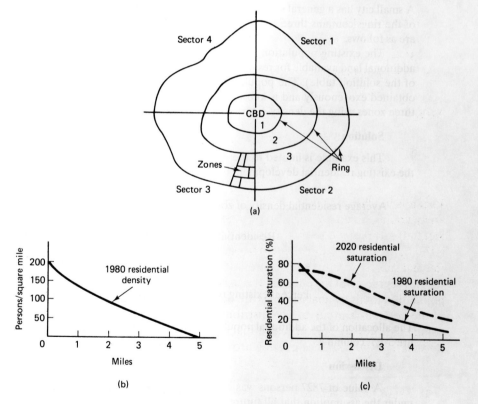

Figure 3-8   Typical City Showing Residential Density and Percent Residential Saturation as a Function of Distance from the City Center: (a) Small City Divided into Rings, Sectors, and Zones; (b) 1980 Residential Density; (c) 1980 and 2020 Residential Saturation by Distance from CBD.

3. Determine the percentage area of useful available land in each zone that has been earmarked for residential use. This percentage, known as the percentage residential saturation, is plotted against distance from the CBD [Figure 3-8(c)].

4. Obtain the total population for the forecast year for the city. This figure is determined exogenously.

5. Plot a curve representing the residential density, similar to the one plotted under step 2, such that the area under the curve is proportional to the total population obtained in step 4. This is the most critical and subjective step.

6. Forecast population totals by analysis rings. These totals are determined by scaling off appropriate ordinate values from the horizon-year curve.

7. Distribute ring totals to individual zones by subjectively weighting each individual zone's attractiveness according to such factors as distance to shopping centers, distance to major street systems or bus lines, residential capacity, nearness to school, and so on.

Example 7

A small city has a general structure consisting of three rings as shown in Figure 3-E7. One of the rings contains three typical zones (1, 2, and 3). The characteristics of these zones are as follows:

The existing population (for 1980) of each of the three zones is known, as is the additional land available for residential use (indicated in columns (5) and (2), respectively, of the solution table). The population of these three zones for the year 2000 has been obtained exogenously and is known to be 6100. Allocate this forecast population to the three zones using the density-saturation gradient method.

Solution

This example is limited to the analysis of residential land use only. The analysis of the existing residential development is as follows:

$$\text{Average residential density of zones (persons/acre)} = \frac{5006}{231} = 21.7 \text{ per acre}$$

$$\text{Residential capacity of zones} = 5006 + [21.7 \times 130]$$

$$= 7827 \text{ persons}$$

$$\text{Percent existing residential capacity} = \frac{5006}{7827} = 64\%$$

The allocation of the additional population (6100 − 5006 = 1094) is done as shown in Figure 3-E7, column 4.

Discussion

A value of 7827 persons was calculated as the holding capacity of the three zones under the assumption that all future residential development will take place at the same density as currently exists (i.e., 21.7 persons per acre). Exactly what value to adopt must result from an analysis of existing density patterns, the zoning policy, and of course, the judgment of the analyst?

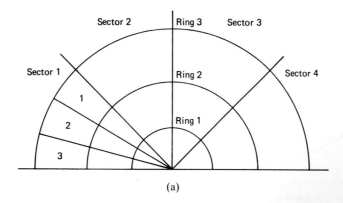

(a)

(1)	(2)	(3)	(4)	(5)
Zone	Total area (acres)	Existing resident population	Existing land in residential use	Maximum add'l. amt. of land available for residential use
1	287	1768	76	35
2	296	2008	103	40
3	144	1230	52	55
Totals (ring)	727	5006	231	130

(b)

(1)	(2)	(3)	(4)	(5)	(6)
Zone	Additional land available for residential use	Factor $\dfrac{Col.\ 2}{\Sigma\ Col.\ 2}$	Incremental growth in zone	Existing pop.	Forecast pop. (Col. 4 + Col. 5)
1	35	0.27	$0.27(6100 - 5006) = 295$	1768	2063
2	40	0.31	$0.31(6100 - 5006) = 339$	2008	2347
3	55	0.42	$0.42(6100 - 5006) = 460$	1230	1690
Total	130		1094	5006	6100

(c)

Figure 3-E7   (a) Rings, Sectors, and Zones; (b) Existing Population and Land Use; (c) Allocation of Additional Population.

Example 8

An almost circular city, 3 miles in radius, has a population of about 1600. Curves *a* and *c* in Figure 3-E8 indicate the existing density (persons per square mile) and percentage residential saturation, respectively. Each ring is about 1 mile wide. It is anticipated that this city will have a population of 2000 in the next 20 years. If the attractiveness of the rings is proportional to the distance from the CBD, how will you distribute the increase in population to the rings, keeping the following constraints in mind: (1) the average percent residential saturation should not exceed 60 and 45 in rings 2 and 3, respectively; and (2) hold the density at the center of the city at 120 persons/mi^2.

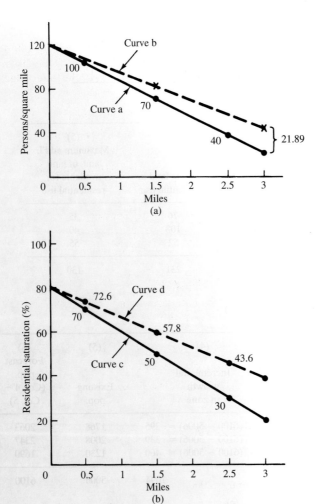

Figure 3-E8   (a) Density (Person/Square Mile) and (b) Percent Residential Saturation for a Three-Ring City.

Solution

Refer to Figure 3-E8.

Area of ring 1	$= 3.14 \times 1^2$	$= 3.14 \text{ mi}^2$
Area of ring 2	$= 3.14 \times (2^2 - 1^2)$	$= 9.42 \text{ mi}^2$
Area of ring 3	$= 3.14 \times (3^2 - 2^2)$	$= 15.70 \text{ mi}^2$

$$\text{Area of city} \qquad = 3.14 \times 3^2 \qquad = 28.26 \text{ mi}^2$$

$$\text{Population of ring 1} = 3.14 \times 100 \qquad = 314$$

$$\text{Population of ring 2} = 9.42 \times 70 \qquad = 659$$

$$\text{Population of ring 3} = 15.7 \times 40 \qquad = 628$$

$$\text{Total} = 1601$$

Allocate the increase in population $(2000 - 1601 = 399)$ and hold the density at 120 persons/mi^2 at the center, and increase the population density in proportion to the distance from the CBD. Therefore, if the increment in the density of the city at the outer edge of ring 3 is X persons/mi^2, then

At a distance of 3 miles, the increase in density will be X persons/mi^2.
At a distance of 0.5 mile, the increase in density will be $0.17X$ persons/mi^2.
At a distance of 1.5 miles, the increase in density will be $0.50X$ persons/mi^2.
At a distance of 2.5 miles, the increase in density will be $0.83X$ persons/mi^2.

The populations are

Ring 1: $\qquad 3.14\,(0.17X) + 314 \;=\; 0.53X + 314$ persons
Ring 2: $\qquad 9.42\,(0.50X) + 659 \;=\; 4.71X + 659$ persons
Ring 3: $\qquad 15.70\,(0.83X) + 628 = 13.03X + 628$ persons
Total population: $18.27X + 1601$ persons

$$18.27X + 1601 = 2000$$
$$X = 21.89 \text{ persons/mi}^2$$

Therefore, allot to

Ring 1: $\qquad (3.14)\,(0.17)\,(21.89) = \;\;12$ persons
Ring 2: $\qquad (9.42)\,(0.50)\,(21.89) = 103$ persons
Ring 3: $\qquad (15.70)\,(0.83)\,(21.89) = 285$ persons

Forecast population

Ring 1: $\qquad\qquad 314 + \;\;12 = 326$
Ring 2: $\qquad\qquad 659 + 103 = 762$
Ring 3: $\qquad\qquad 628 + 285 = 913$
$$\text{Total} \approx 2000$$

The percent residential saturation will be

Ring 1: $\qquad\qquad 70 \times \dfrac{326}{314} = 72.6\%$

Ring 2: $\qquad\qquad 50 \times \dfrac{762}{659} = 57.8\% < 60\% \text{ (average)}$

Ring 3: $\qquad\qquad 30 \times \dfrac{913}{628} = 43.6\% < 45\% \text{ (average)}$

The corresponding residential densities and percent residential saturation curves for the forecast year are shown in the figure as curves *b* and *d*, respectively.

Discussion

Note that, in general, the only mathematical constraint in allocating population to rings, sectors, and zones is that the city's area multiplied by the ordinates of the residential density curve must sum to the total projected population obtained exogenously. Also, care must be taken to verify if zoning ordinances and other city regulations are not being violated.

SUMMARY

Transportation can be visualized as the consequence of the fact that different types of land uses in the city are spatially separated. At the same time, enhanced mobility also can be seen as contributing to increased separation of land use.

This symbiotic relationship between transportation and land use produces the movement and traffic flow patterns seen in urban areas. The accessibility of places has a major impact on land values, and the location of a place within the transportation network determines its accessibility. Thus, in the long run, the transportation system, and the traffic flows on it, shapes the land-use pattern. Notice the large percentage of land occupied by streets and highways in any city (see Figure 3-9).

All movement in a city incurs a cost of some sort, measured in terms of time and/or money. There is a trade-off involved in the decision to make a trip. Because people generally value travel time and want to minimize it, they do not want to be too

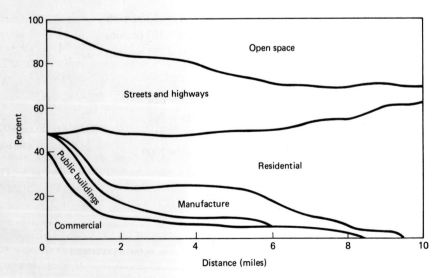

Figure 3-9   Percentage Distribution of Six Land Uses by Distance from the Center City.

far away from the places they need to visit regularly. This desire is an important determinant of land-use patterns in the city.

The basic concept underlying the relationship between land use and transportation is accessibility. In its broadest context, accessibility refers to the ease of movement between places.

Location theory provides a useful tool to understand where different urban activities are located and why. Zoning and its effects on land use show how the free-land market is partially controlled for the benefit of its citizens.

The land-use transportation relationship indicates that there are scores of variables affecting the links between land use and transportation. Some of the more important ones are financial resources, industrial activities, fuel cost, consumption and supply, business structure, employment opportunities, and population growth. All these variables and their connections confirm the fact that the urban fabric is one of the most complicated systems in the world.

Land-use development models help urban planners and transportation engineers to understand the complexities of the urban system and, above all, provide a means of allocating urban growth to various parts of the city using rational ideas.

REFERENCES

ABLER, R., J. S. ADAMS, and P. GOULD (1971). *Spatial Organization: The Geographer's View of the World*, Prentice-Hall, Englewood Cliffs, NJ.

AMERICAN SOCIETY OF CIVIL ENGINEERS (ASCE) (1986). *Urban Planning Guide*, Rev. Ed., ASCE manual 49, ASCE, New York.

BLUNDEN, W. R. (1971). *The Land Use/Transportation System*, Pergamon Press, Oxford, UK.

BLUNDEN. W. R., and J. A. BLACK (1984). *The Land Use/Transportation System*, 2nd Ed., Pergamon, Oxford, UK.

BOURNE, LARRY S. (1982). *Internal Structure of the City: Readings on Urban Form, Growth and Policy*, 2nd Ed., Oxford University Press, New York.

BUDHU, G., and D. GRISSON (1985). *Development of a Simulation Model to Study the Impact of Rapid Urban Growth on the Transportation Sector*, Transportation Research Record 1046, Transportation Research Board, Washington, DC.

CATANESE, A. J., and J. C. SNYDER (Eds.) (1979). *Introduction to Urban Planning*, McGraw-Hill, New York.

CHAPIN, F. STUART, and E. J. KAISER (1979). *Urban Land-Use Planning*, 3rd Ed., University of Illinois Press, Champaign, IL.

CLARK, COLIN (1951). Urban Population Densities, *Journal of the Royal Statistical Society, Series A*, Vol. 114.

DEACON, J. A., ET AL. (1976). *Urban Transportation and Land Use*, Final Report DOT-TST-76T-29, U.S. Department of Transportation, Washington, DC.

HANSEN, SUSAN (Ed.) (1986). *The Geography of Urban Transportation*, Guildford Press, New York.

KHISTY, C. J. (1977). *An Evaluation of Alternative Manual Land Use Forecasting Methods Used in Transportation Planning*, unpublished Ph.D. dissertation, Department of Civil Engineering, Ohio State University, Columbus, Ohio.

KHISTY C. J. (1979). *Land-Use Allocation Model for Small and Medium Sized Cities*, Transportation Research Record 730, Transportation Research Board, Washington, DC.

KHISTY. C. J. (1981). *Evaluation of Two Residential Models for Land Use Allocation*,

Transportation Research Record 820, Transportation Research Board, Washington, DC.

LEE, COLIN (1973). *Models in Planning*, Pergamon Press, Oxford, UK.

MEYER, JOHN R., and J. A. GÓMEZ-IBÁÑEZ (1981). *Autos, Transit and Cities*, Harvard University Press, Cambridge, MA.

MOORE, W. T. (1975). *An Introduction to Urban Development Models and Guidelines for*

Their Use in Urban Transportation Planning, U. S. Department of Transportation, Federal Highway Administration, Washington, DC.

SMITH, WALLACE F. (1978). *Urban Development—The Process and the Problems*, University of California Press, Berkeley, CA.

YEATES, MAURICE (1974). *An Introduction to Quantitative Analysis in Human Geography*, McGraw-Hill, New York.

EXERCISES

1. With the help of a diagram, sketch two possible density patterns for a city with a population of 360,000 people occupying 9000 acres. What is your rationale for the patterns you have chosen?

2. How would you go about measuring the incremental residential density of your city? How would you measure density for an apartment building, a single-family home, and a military base housing for 2000 soldiers? Could an aerial photograph(s) of the city help?

3. A land-use zone consists of 2000 single-family homes, 520 apartment units, and 3 hotels with 600 rooms each. Calculate the daily vehicular traffic generated by this zone, using the following equations:

$$t_s = 5 + 7.35\, U_s$$
$$t_a = 7.0 + 6.25\, U_a$$
$$t_h = 2 + 12\, U_h$$

where the subscripts s, a, and h refer to homes, apartments, and hotels, respectively, and t and U refer to trips and housing units, respectively. Why is there a significant difference in trip production among the units?

4. To what extent is the usefulness of zoning a function of city size? Examine the zoning ordinances in your city and list five important conclusions that you draw from your examination.

5. Select a city of your choice and obtain general plans (land-use plans would be preferred), showing residential, commercial, and industrial development for two or three points in time: for example, before World War II, after (say, the 1950s), and at present. What patterns of development can you discover taking place in this city vis-à-vis the transportation system? Write a short report on your findings.

6. In your opinion, what are the most important physical parameters for the efficient working of a city? List and prioritize your chosen parameters.

7. What would be the future outcome of your city if (a) zoning ordinances were abolished, and (b) the comprehensive plan for the city was revoked?

8. Three residential zones (1, 2, and 3) with 400, 500, and 700 resident workers and four employment zones (4, 5, 6, and 7) with 350, 450, 500, and 300 jobs are connected by a highway network having the following travel costs: $t_{14} = 10$, $t_{15} = 12$, $t_{16} = 14$, $t_{17} = 15$, $t_{24} = 8$, $t_{25} = 9$, $t_{26} = 10$, $t_{27} = 12$, $t_{34} = 4$, $t_{35} = 6$, $t_{36} = 8$, and $t_{37} = 15$. Assume that $b = 1.2$. Find the actual and relative zonal accessibility of the residential zones.

9. A concentric city with a single CBD has urban activities competing with one another for locations that will reduce their transportation costs. The three major activities are commercial, industrial, and residential. The gross profit per square foot, the transportation cost per mile, and the distance from the CBD for profitability for each activity are shown in the table. What is the likely pattern of the city?

Land use	Gross profit per square foot	Transport cost/mile	Limit in miles from CBD
Commercial (C)	100	50	2
Industrial (I)	80	20	4
Residential (R)	50	5	10

10. Refer to Problem 9. A commercial and industrial center is being planned 12 miles away from the city described in Problem 9, with negligible residential use. The characteristics of this center are as follows:

LAND USE	GROSS PROFIT	TRANSPORT COST/MILE
Commercial	70	35
Industrial	60	20

(a) How will this new center affect the distribution of land use?
(b) Sketch (plan and longitudinal section) the likely land use.

11. What will be the corresponding decisions if the transportation cost of crops A, B, C, and D in Example 4 has been changed to 35, 12, 8, and 4, respectively, as a result of urban growth?

12. A new city is likely to be developed in a remote dry desert of the southwest because of an oil field. The petroleum company currently has an employment of 2500 people, each person having an average of 3 people in the family, including the worker himself or herself. For every 10 people in town, 3 service jobs are generated a year. Estimate the population and the employment of this new city after 10 years. Will the growth of the city stabilize? What will be the saturated situation? The following assumptions may be made: (1) it is estimated that 2, 4, and 6 years from the date of settlement, the city will have an employment of 3500, 4200, and 4500, and that the maximum employment at the oil field will be 5000 and (2) a modified exponential growth applies.

13. A small city is divided into six residential zones and the zonal data derived from a survey are as follows:

Zone	Traffic generation	Number of homes
1	500	3250
2	630	4040
3	440	2835
4	760	4200
5	790	5700
6	1250	7250

Estimate the parameters of a linear regression model connecting traffic and homes and illustrate the data.

14. A four-zone city has the following characteristics:

Zone	Total existing population	Holding capacity (acres)
1	3000	300
2	2500	280
3	9000	500
4	4500	350

The travel times (in minutes) are as follows:

From i \ To j	1	2	3	4
1	5	10	12	15
2	10	4	9	20
3	12	9	3	14
4	15	20	14	6

An exponent of 2.2 can be used based on work done with other cities of the same size. If the city is likely to grow by 15% overall in 15 years, what would be the likely population located in each zone in the horizon year? What would be the percentage change in allocation of population to zones if the exponent were 1.8 and 2.0, respectively?

15. An expert on land-use planning challenges the figures calculated in Example 6 and has come up with the following distribution for the forecast year by using a more sophisticated method:

Zone	Population
1	1392
2	2280
3	4328

What should be the value of the exponent to match these figures? What conclusions do you draw from this exercise?

16. A large metropolitan area, semicircular in shape, has six land uses, as shown in Table 3-4. If the population of this metro area is 5 million people, answer the following questions.
   (a) What is the approximate percentage of the six land uses within the study area?
   (b) What is the ratio of the area of streets and highways to the area taken up by residential land?
   (c) What is the average residential density?
   (d) There are six major transportation corridors in this area. Sketch what you consider to be a reasonable distribution of residential, commercial, manufacture, public buildings, and open space.

17. An approximately 2-mile-square city has a central business district (CBD) of 1/2 mile square with a uniform population density of 100 persons/acre. The density from the edge of the CBD to the outer edge of the city on all four sides decreases uniformly to 5 persons/acre. Sketch the population density pattern and determine the total population. (1 mi^2 = 640 acres.)

18. A circular-shaped city has the following residential density function in relation to distance from the city center:

$$d_x = d_0 e^{-bx}$$

where

d = density (persons/acre)
x = distance from the city center (miles)
b = density gradient

At a distance of 2 miles from the city center, the average density is 21 persons/acre; at 3 miles, the density drops to 9 persons/acre. Establish the constants of the equation. What is the area of the city? (1 mi^2 = 640 acres.) Assume that the density at the outer edge of the city is 1 person/acre.

19. A land-use planner observed that in five zones of the city, the number of gas stations (Y) in relation to the population (X) in 1000's was

Y	2	7	3	5	8
X	1	5	3	2	4

   (a) Set up a linear regression equation connecting Y in terms of X.
   (b) Determine R^2 and apply the t-test.

20. (a) The population density at the center of a circular city is 200 persons/acre and the density decreases uniformly to the edge of the city, where it is 5 persons/acre. What is the approximate population of this city?
   (b) If the population density varies with distance from the city center as per the basic equation

$$d_x = d_0 e^{-bx}$$

what is the approximate population of this city?

21. The average number of trips (Y) made by households in 13 zones in an urbanized area is given along with the corresponding average family income (X):

Zone	1	2	3	4	5	6	7	8	9	10	11	12	13
Y	4.2	3.0	3.9	4.5	5.6	5.8	4.4	7.0	5.9	3.1	4.6	6.6	6.1
X	4.9	3.4	5.5	6.7	9.0	6.7	7.6	8.6	8.6	4.3	5.5	6.7	7.7

   (a) Plot the data to scale and draw a reasonable straight line connecting X and Y (by trial and error).
   (b) Set up a regression equation connecting Y in terms of X, determine R^2, and then apply the t-test.

22. Refer to Problem 21. The net residential density (X_1) (in dwelling units per acre) for the 13 zones is recorded as follows:

Zone	1	2	3	4	5	6	7	8	9	10	11	12	13
X_1	13.9	49.2	15.2	16.2	11.5	6.5	5.7	4.9	7.6	28.5	10.4	2.6	2.4

(a) Plot to scale trips/dwelling units/acre (Y) against X_1. What do you conclude from the figure? Do you feel that there is a linear relationship between X_1 and Y?

(b) Now calculate the logarithm (to base 10) of X_1, and plot to scale trips/dwelling units/acre (Y) against X_2 (log of X_1). What conclusions do you draw from this plot?

(c) Set up a regression equation connecting Y with X_2, determine R_2, and apply the t-test.

23. A medium-sized city in the Midwest is likely to increase its aggregate population in the next 10 years by 10%. Using Figures 3-5 and 3-6, what is your estimate of an increase/decrease in the various socioeconomic factors of this city, such as housing, labor, and so on? What additional information/data would be needed to make reasonably good estimates of these parameters?

24. A five-zone city having travel times (in minutes) as shown in the cells of the table is likely to add 1000 persons to its present population in the next 5 years.

(a) It is hypothesized that the distribution of this additional population will be in proportion to a zone's accessibility to employment. The additional employment in the next 5 years and its distribution is also shown. Assuming $b = 2$, use Hansen's model to distribute the growth of population.

Zone	1	2	3	4	5
1	1	3	8	5	12
2	3	1	6	2	9
3	8	6	1	4	7
4	5	2	4	1	7
5	12	9	7	7	1

ZONE	FUTURE EMPLOYMENT
1	150
2	30
3	200
4	100
5	25

(b) It is hypothesized that the additional population will be distributed on the basis of accessibility as well as on the available vacant land of each zone. What will be the zonal distribution?

ZONE	VACANT LAND (ACRES)
1	51
2	21
3	42
4	19
5	72

(c) If, in addition to conditions described in parts (a) and (b), it is hypothesized that the attractiveness of an individual zone plays a part in the location of the additional population, as per figures shown, how will the additional population be distributed?

ZONE	ATTRACTIVENESS
1	3
2	2
3	1
4	3
5	4

25. What are the various ways you can think of reducing commuting time and cost by using strategies involving land use? How would you prioritize these strategies?

26. A medium-sized city (population 100,000) is likely to attract a high-tech-oriented manufacturer to start production. It is a phased operation entailing a population growth of 5% per year for the next 3 years. What general effects will this rapid growth have on the city, particularly in terms of land development, land availability, construction, housing, employment opportunities, industrial activity, pollution, fuel consumption, transportation, financial resources, and so on? (*Hint:* Refer to Figures 3-5 and 3-6 as well as other literature cited.)

27. A small rural city of about 20,000 people is currently located 7 miles away from Interstate I-90 in Washington state. A two-lane highway runs through the city but is not connected directly to I-90. The state DOT plans to build an interchange (with on- and off-ramps) where this two-lane highway crosses I-90 and to improve the present 7-mile stretch of the two-lane highway to a four-lane highway. Answer the following questions from a land-use point of view.

(a) What is likely to happen to the land along the connecting highway near the interchange in the next decade?

(b) What is likely to happen to this city in the short run (5 to 10 years) and possibly in the long run (30 years)? Write a report on your fears and hopes for the future of this city.

Chapter 4

Vehicle and Human Characteristics

1. INTRODUCTION

Engineers and behavioral scientists have been working together for over 50 years to understand and design efficient person-machine systems. Their research has helped transportation engineers to understand how human beings (as vehicle operators, passengers, or pedestrians) interact with vehicles and the transportation facilities they use (Shinar, 1978).

It has been long recognized that the three main elements of the highway mode are the human element, the vehicle, and the environment. This recognition has provided safety engineers with a useful matrix (after Haddon, 1980; see Chapter 16), and details of this matrix are shown in Table 4-1, where these three elements are considered within the highway safety framework. If the management of the system is to be conducted efficiently, all applicable factors need to be considered. For instance, drivers need to have proper and adequate training and knowledge of the highway mode before they drive a vehicle; in the event of a crash, the vehicle needs to have the proper restraints, such as seat belts and air bags; and emergency medical service should be available to victims in case of a vehicle crash. Similarly, the vehicle and the human-made environment ought to have certain attributes at all three stages—precrash, crash, and postcrash—as countermeasures to cope with the dilemma posed by accidents (Lay, 1986; Homburger and Kell, 1988).

In this chapter, we draw together several topics that are treated in greater detail in other chapters: for example, traffic stream flow (Chapter 5), the geometric design of streets and highways (Chapter 6), highway capacity (Chapter 7), at-grade intersection design and traffic control devices (Chapters 8 and 9, respectively), and transportation safety (Chapter 16) are all intimately connected with this chapter.

First, a simplified framework of a model of the human–vehicle–environment system is described. In subsequent sections, the human and vehicle characteristics are dealt with. These characteristics are important in themselves, in the sense that the

TABLE 4-1   HIGHWAY SAFETY FACTORS

	Precrash	Crash	Postcrash
Human	Training Knowledge Skill Basic abilities Motives and attitudes	In-vehicle restraints worn by driver and fitted in vehicle	Emergency medical services Incident detection and assistance
Vehicle	Control system design Comfort system design Information systems design Laws and enforcement	Occupant protection system Control system design	Fire/fume control systems Design for ease of emergency access Repair capabilities
Environment	Geometrics and appurtenances Enforcement system Control system Weather and light conditions Road-surface conditions	Geometrics and appurtenances for energy absorption and forgiving highway	Geometrics for ease of emergency access Debris control and cleanup Restoration of road and traffic devices

Source: FHWA, 1980.

traffic engineer should realize that in the case of human beings, it is not enough to work for the "average" driver or the "average" pedestrian, because of the wide range of abilities of the "average" driver and pedestrian. A similar dilemma can be recognized when dealing with vehicle characteristics. The transportation system at any one moment is used by motor vehicles ranging from the smallest car to a large combination truck. Hence, the traffic engineer is looking more at "limitations" than at "averages."

2. A SIMPLIFIED FRAMEWORK

A simplified framework of a model that attempts to provide an understanding of the human–vehicle–environment system is shown in Figure 4-1. Proper driver education imparts knowledge about the human–vehicle–environment interaction, develops driving skills, and positively affects the attitude of the would-be driver. It has the potential for creating safer driving practice, resulting in reduced accidents. The laws and their enforcement provide guidance and motivation for safer and efficient driver behavior. Laws, therefore, should be realistic and comprehensible to be effective. It has been demonstrated that the mere existence of laws without effective enforcement is ineffective.

The roadway/roadside environments includes both the physical and ambient conditions, and the vehicle characteristics include the mechanical control system and information sources provided to the driver. The sensory field consists of many pieces of

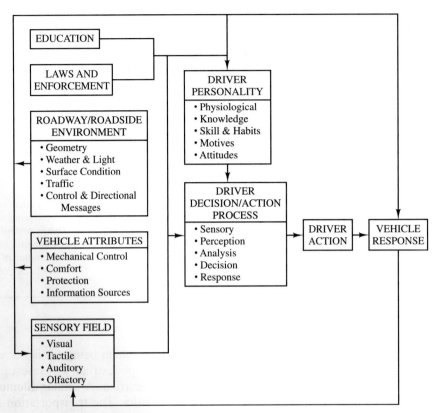

Figure 4-1　Human–Vehicle–Environment Operating System (FHWA, 1980).

information that the driver uses. For example, the visual field consists of objects, lines, and edges. Similar inputs received through the hearing and smelling abilities, together with those perceived by the sense of touch, make up the sensory field. A broad meaning is attached to the term "driver personality," encompassing the physiological attributes, knowledge, skills, and habits of the driver.

The driver decision process includes the classical chain of sensing, perceiving, analyzing, deciding, and responding. These activities are discussed later in this chapter. Finally, the vehicle response to the action taken by the driver is a function of vehicle characteristics and the roadway/roadside environment.

In summary, drivers have two functions in this system. First, they are using the system to move from one point to another in a certain period of time, taking into account safety, convenience, and comfort. They are also acting as the guidance and control system for the vehicle, which involves continuous "fine control" of the vehicle in terms of direction and speed. To do this, drivers must detect and select information from the general environment, including the highway geometry and translate the decisions into a set of actions on the vehicle. Through proper feedback, there is an intimate and continuous interaction among the highway geometrics, the vehicle, and the driver.

2.1 Driver Personality

A framework of a driver's personality is shown in Figure 4-2. As mentioned earlier, driver personality is a broad-based body of knowledge that deals with the driver's natural abilities, the learned capabilities, and his or her motives and attitudes. Good driving requires no exceptional natural abilities. Physical and psychological tests can reveal the need for mechanical and visual aids to correct most natural human deficiencies. On the other hand, the driver's learned capabilities have to be acquired by study and practice, and these acquisitions can be tested to indicate any shortcomings. To understand why drivers behave the way they do can be known from their motives and attitudes. Attitudes often determine how a driver reacts to a driving situation. Motives may be associated with fear of injury, fear of criticism, and feelings of social responsibility. Driving personality can be modified seriously and quickly by the use of alcohol, narcotics, and drugs. Illness, exhaustion, and discomfort can seriously impair driving efficiency, as shown in Figure 4-2.

2.2 Sensing

The driver can receive useful information regarding the safe control of the vehicle through feeling, seeing, hearing, and smelling. Thus, temperature and humidity, forces and rates of change, and vibrations and oscillations connected with the stability of the vehicle are some of the more common pieces of information that drivers pick up through their senses. Each of these senses is examined in this section.

Feeling.   A driver experiences forces acting on a vehicle, such as the force of gravity, acceleration, braking deceleration, and cornering acceleration. For example, the average driver may use 0.3 g of lateral acceleration while making a 90-degree turn on an urban street at 15 mph, but will generally not use more than 0.1 g on a long horizontal curve at 60 mph. Thus, drivers are strongly influenced, in accelerating or decelerating, by the speed and conditions of the highway.

Seeing.   Vision is the most important means of acquiring accurate information on the relationships among perceived objects and on traffic control messages. Certain characteristics of visual acuity are of special interest in transportation: static and dynamic visual acuity, depth perception, peripheral vision, "night" vision, and glare recovery. *Visual acuity* is the ability to see fine details clearly. The most acute vision is within a narrow cone of 3 to 5 degrees, although the limit of fairly clear sight is within 10 to 12 degrees. In view of this fact, it is necessary to place signs and symbols within this 10- to 12-degree cone of vision, and certainly within 20 degrees.

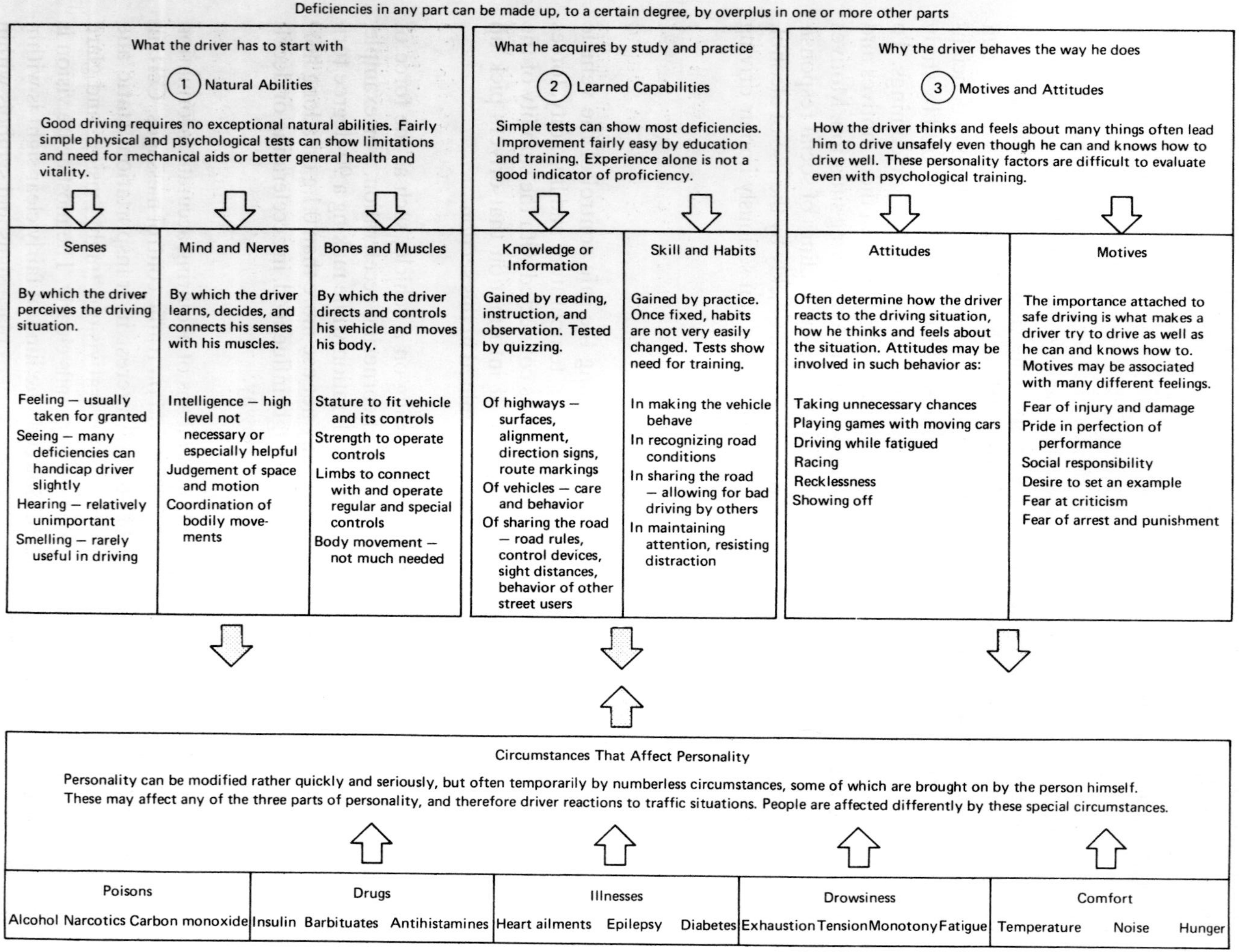

Deficiencies in any part can be made up, to a certain degree, by overplus in one or more other parts

① Natural Abilities	② Learned Capabilities	③ Motives and Attitudes
What the driver has to start with	What he acquires by study and practice	Why the driver behaves the way he does
Good driving requires no exceptional natural abilities. Fairly simple physical and psychological tests can show limitations and need for mechanical aids or better general health and vitality.	Simple tests can show most deficiencies. Improvement fairly easy by education and training. Experience alone is not a good indicator of proficiency.	How the driver thinks and feels about many things often lead him to drive unsafely even though he can and knows how to drive well. These personality factors are difficult to evaluate even with psychological training.

Natural Abilities

Senses	Mind and Nerves	Bones and Muscles
By which the driver perceives the driving situation.	By which the driver learns, decides, and connects his senses with his muscles.	By which the driver directs and controls his vehicle and moves his body.
Feeling – usually taken for granted Seeing – many deficiencies can handicap driver slightly Hearing – relatively unimportant Smelling – rarely useful in driving	Intelligence – high level not necessary or especially helpful Judgement of space and motion Coordination of bodily movements	Stature to fit vehicle and its controls Strength to operate controls Limbs to connect with and operate regular and special controls Body movement – not much needed

Learned Capabilities

Knowledge or Information	Skill and Habits
Gained by reading, instruction, and observation. Tested by quizzing.	Gained by practice. Once fixed, habits are not very easily changed. Tests show need for training.
Of highways – surfaces, alignment, direction signs, route markings Of vehicles – care and behavior Of sharing the road – road rules, control devices, sight distances, behavior of other street users	In making the vehicle behave In recognizing road conditions In sharing the road – allowing for bad driving by others In maintaining attention, resisting distraction

Motives and Attitudes

Attitudes	Motives
Often determine how the driver reacts to the driving situation, how he thinks and feels about the situation. Attitudes may be involved in such behavior as:	The importance attached to safe driving is what makes a driver try to drive as well as he can and knows how to. Motives may be associated with many different feelings.
Taking unnecessary chances Playing games with moving cars Driving while fatigued Racing Recklessness Showing off	Fear of injury and damage Pride in perfection of performance Social responsibility Desire to set an example Fear at criticism Fear of arrest and punishment

Circumstances That Affect Personality

Poisons	Drugs	Illnesses	Drowsiness	Comfort
Alcohol Narcotics Carbon monoxide	Insulin Barbituates Antihistamines	Heart ailments Epilepsy Diabetes	Exhaustion Tension Monotony Fatigue	Temperature Noise Hunger

Personality can be modified rather quickly and seriously, but often temporarily by numberless circumstances, some of which are brought on by the person himself. These may affect any of the three parts of personality, and therefore driver reactions to traffic situations. People are affected differently by these special circumstances.

Figure 4-2 Parts of Driver Personality (Baker, 1975).

Visual acuity depends on several factors for the same person, and the range of visual acuity is large for different age groups. For example, a person with 20/40 vision in broad daylight may have only 20/100 vision at night for the same length of time looking at an object. Although minimum standards of acuity have been prescribed for obtaining a driver's license, the ability of drivers to see varies with a host of variables.

Drivers need to have proper depth perception for judging distances and speeds. The primary mechanism that human beings utilize for depth perception is through binocular vision, although monocular parallax and other cues also assist in this process.

It is estimated that about 8% of all men and 4% of all women suffer from some degree of color blindness (reduced ability to distinguish between red and green). Visual deterioration increases with age and diseases of the eyes, such as cataract, and this deterioration is detected through periodic checks made by licensing authorities.

Peripheral vision relates to an individual's ability to see objects, not necessarily clearly. Such vision serves as a warning sign. The angle of peripheral vision normally varies between 120 and 180 degrees. It has been shown that the head can move 45 degrees to the right or left and 30 degrees up or down. When the head and eyes move from one object to another, an involuntary blink often occurs that blocks out what otherwise would be a streaming and blurred image. This fact is of importance for the placement of signs or signals.

The time required to see objects is also crucial. The fixational pause between eye movements ranges from about 0.1 to 0.3 second. While waiting to make a permitted left turn on green, the time required for a driver to scan an intersection from right to left and back amounts to about 1 second. Five propositions drawn from human factors literature apply directly to highway design:

1. As speed increases, visual concentration increases. Planes perpendicular to the road are prominent; parallel ones are not.

2. As speed increases, the point of visual concentration recedes. In other words, the eyes feel their way ahead of the wheels and try to allow the driver sufficient time for emergencies. At 25 mph, the driver is focusing about 600 ft ahead; at 45 mph, it is 1200 ft, and at 65 mph, it is 2000 ft.

3. As speed increases, peripheral vision diminishes. At 25 mph, the peripheral vision encompasses a horizontal angle of about 100 degrees; at 45 mph, this narrows down to 65 degrees; and above 60 mph, it is less than 40 degrees.

4. As speed increases, foreground details begin to fade. At 40 mph, the earliest point of clear vision is about 80 feet away. Foreground detail is greatly diminished at 50 mph and is negligible beyond 60 mph. Thus, only within an angle of 40 degrees and at a distance between 110 and 1400 feet is vision really adequate at 60 mph, an interval that is traversed in less than 15 seconds. Therefore, only large, simple shapes are meaningful at high speeds.

5. As speed increases, space perception becomes impaired.

Within the range of light levels, normally associated with night driving with headlights, it has been found that visual power decreases in acuity, contrast, form perception, and depth perception. The abilities to judge size, position, and motion of an object are also impaired. Glare from approaching headlights greatly reduces visibility. Night vision and the effects of glare have been shown to have greater negative impacts with increasing age.

Hearing and Smelling.   Hearing is important to the driver and pedestrian. Although usually not as important as vision in the act of driving, hearing can be helpful in preventing collision. Also, drivers can derive useful information through their hearing abilities about the vehicle engine; tires; warning sounds such as sirens, horns, bells, radio; and possibly other traffic sounds. It has been found that drivers having hearing problems can have 1.8 times more accidents than do drivers with normal hearing. The sense of smell is useful to the driver in detecting such emergencies as an overheated engine, burning brakes, smoking exhausts, and fires.

3. PERCEPTION REACTION

A person's process of extracting necessary information from the environment is called *perception.* As has been examined before, vision is the primary source. The driver's goal of moving from one point to another is achieved through three tasks: control, guidance, and navigation. Control relates to the physical manipulation of the vehicle, through lateral and longitudinal control by steering, accelerating, and braking. Information for controlling the vehicle is received by the driver through his or her natural sense mechanism. Guidance refers to the driver's task of selecting a safe speed and path on the highway, which is essentially a decision process. Thus, car following, overtaking, and passing are activities falling into this category. Information comes from the environment (the road), traffic control devices, and the surrounding traffic. Those activities that relate to the driver's ability to plan and execute a trip from origin to destination fall under the category of navigation, the information for which comes from maps, signs, and landmarks.

Sometimes the driver receives information at too fast a rate for proper absorption and this can lead to confusion and possible distress. When information is overloaded on a driver, he or she makes a selection on the basis of importance (or primacy). Usually, control information is more important than guidance information, and both of these are more important than navigation information.

With this background, it is useful to assess the time required from the point of perception to the point of reaction. This perception-reaction time is a key variable in many design considerations. Perception is divided into two parts: perception delay and apperception interval. *Perception delay* is the time between visibility and point of perception. *Apperception interval* is the time required to determine that there is a potential hazard. Reaction time is also defined as having two elements—reaction and total reaction—the latter includes the former. Reaction involves the analytical and decision-making portions of the driver's reaction process. Total reaction includes reaction plus the actual control response (e.g., bringing the foot to the brake). A value for perception-reaction time commonly used is 2.5 seconds.

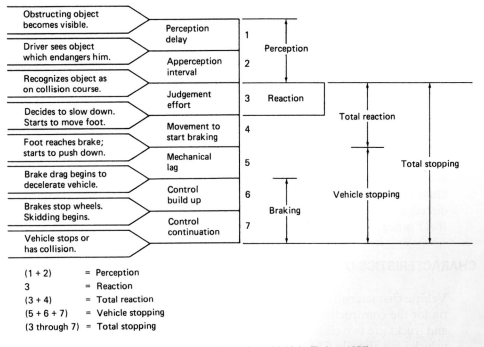

Figure 4-3   Example of Stopping a Vehicle (Baker, 1975).

An example of events that a typical driver would go through in case of an emergency stop on a local road are shown in Figure 4-3. A numerical example illustrating some of the concepts is given in Section 8.

4. DRIVER STRATEGY

Vanstrum and Caples (1971) describe a simple driving task model useful in understanding the relationship between driving behavior and the possibility of anticipating a hazard. The performance of a driver depends on the decision regarding the action needed and the moment at which it is taken, depending on the location of the obstacle or hazard, the relative speed of the driver, and the physical characteristics of the intervening space between the driver and the hazard.

Figure 4-4 shows a vehicle, located in its lane, moving at a speed that it will traverse distance 1 during the perception time, distance 2 during the time needed to reach a decision, and distance 3 during the reaction time. Distance 4 represents the minimum stopping distance. The letter T, the true point of no return, is the last point at which action can be taken to avoid a hazard. If the hazard is in motion, T is also in motion. M, the mental point of the driver, is the last point at which action must be taken (according to the driver's estimation), and A is the action point where the driver acts. The driver's margin of safety is distance AM and the driver's perception error of the situation

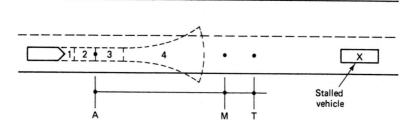

Figure 4-4 Driver Strategy (Vanstrum and Caples, 1971).

is the distance MT. The actual margin of safety is distance $AT(AT = AM + MT)$. Quantities AM and MT are usually plus going from M to A and T to M, taking the direction from the stalled vehicle (hazard) back to the driver as the positive direction. If AT is negative, the hazard cannot be avoided.

5. CHARACTERISTICS OF VEHICLES

Vehicle characteristics, which include such details as size and weight, dictate the criteria for the construction and improvement of transportation facilities. Passenger cars and trucks are two classes of vehicles considered in geometric design; details of these vehicles are given in Chapter 6 under the heading of design vehicles (as prescribed by AASHTO). This section deals with the kinematic and dynamic characteristics of vehicles. Many of the concepts introduced in this section are elaborated further in Chapter 6, including their practical applications.

6. KINEMATICS OF VEHICLES

The fundamental relationship connecting force and acceleration is given by the equation

$$F = m \times a \qquad (1)$$

where

F = force (lb)
m = mass
a = acceleration (ft/sec^2)

If acceleration is considered to be constant, then

$$\frac{dv}{dt} = a \qquad (2)$$

$$\int_{v_0}^{v} dv = \int_{0}^{t} a \, dt \qquad (3)$$

$$v = v_0 + at \qquad (4)$$

Also,

$$\frac{dx}{dt} = v \tag{5}$$

$$\int_0^x dx = \int_0^t (v_0 + at)\, dt \tag{6}$$

$$x = v_0 t + \tfrac{1}{2} at^2 \tag{7}$$

Distance as a function of speed can be obtained by substituting Eq. 4 into Eq. 7, giving

$$x = \frac{v^2 - v_0^2}{2a} \tag{8}$$

where

a = acceleration (ft/sec^2)
v = speed (ft/sec)
v_0 = initial speed (ft/sec)
x = distance (ft)
t = time (sec)

Nonuniform Acceleration.   In a real-world situation, it is often necessary to consider varying acceleration. One such condition is to relate acceleration in proportion to speed, and a specific case is to vary acceleration inversely to speed, as noted.

$$\frac{dv}{dt} = \alpha - \beta v \tag{9}$$

where α and β are constants. Inspection of this equation indicates that α is the maximum acceleration attainable, and α/β is the maximum possible speed. Integrating the equation between v and v_0, one gets

$$\left. \frac{-1}{\beta} \log (\alpha - \beta v) \right|_{v_0}^{v} = t$$

$$\frac{\alpha - \beta v}{\alpha - \beta v_0} = e^{-\beta t}$$

$$v = \frac{\alpha}{\beta} (1 - e^{-\beta t}) + v_0 e^{-\beta t} \tag{10}$$

The equation for distance as a function of time is

$$x = \frac{\alpha t}{\beta} - \frac{\alpha}{\beta^2} (1 - e^{-\beta t}) + \frac{v_0}{\beta} (1 - e^{-\beta t}) \tag{11}$$

and the acceleration–time relationship is

$$\frac{dv}{dt} = (\alpha - \beta v_0)e^{-\beta t} \tag{12}$$

Example 1

A truck traveling at 25 mph is approaching a stop sign. At time t_0 and at a distance of 60 ft, the truck begins to slow down by decelerating at 14 ft/sec². Will the truck be able to stop in time?

Solution

$$v = at + v_0$$

where

$$v = \text{final velocity} = 0 \text{ ft/sec}$$
$$v_0 = 25 \text{ mph} = 36.67 \text{ ft/sec}$$
$$a = 14 \text{ ft/sec}^2$$

Therefore,

$$0 = -14t + 36.67$$

giving

$$t = 2.62 \text{ sec}$$

The distance covered by the truck in these 2.62 seconds is

$$x = v_0 t + \tfrac{1}{2}at^2$$
$$= (36.67)(2.62) + \tfrac{1}{2}(-14)(2.62)^2$$
$$= 48.02 \text{ ft} < 60 \text{ ft}$$

which indicates that the truck will stop just in time.

Alternatively, a variation of Eq. 8 can be applied directly to determine the truck's stopping distance.

$$x = \frac{v_0^2}{2a}$$

The final speed, v, is zero as the vehicle comes to a complete stop. Here a is the deceleration rate. Thus,

$$x = \frac{(36.67)^2}{2 \times 14} = 48.02 \text{ ft}$$

As is discussed later in Chapter 6, $a = gf$ during braking when the motorist is attempting to stop the vehicle, where g is the acceleration due to gravity (32.2 ft/sec²), and f is the coefficient of friction between the tires and the road surface. Therefore, the braking distance for the vehicle can be stated as

$$x = \frac{v^2 - v_0^2}{2gf} \tag{13}$$

An effect for the grade (up or down) on the roadway can be included by modifying the equation as follows:

$$x = \frac{v^2 - v_0^2}{2g(f \pm s)} \tag{14}$$

where s represents the grade on the roadway. Use plus $(+)$ for an upgrade and minus $(-)$ for a downgrade. It is often convenient to use speed expressed in mph directly. A revised form of Eq. 14 can be expressed as

$$x = \frac{V^2 - V_0^2}{30(f \pm s)} \tag{15}$$

where V_0 and V are original and final speeds, respectively, in mph; g is substituted as 32.2 ft/sec^2.

Example 2

An impatient car driver stuck behind a slow-moving truck traveling at 20 mph decides to overtake the truck. The accelerating characteristic of the car is given by

$$\frac{dv}{dt} = 3 - 0.04v$$

where v is the speed (ft/sec), and t is the time (sec).

(a) What is the acceleration after 2, 3, 10, and 120 seconds?

(b) What is the maximum speed attainable by the car?

(c) When will the acceleration of the car approach zero?

(d) How far will the car travel in 120 seconds?

Solution

$$\frac{dv}{dt} = 3 - 0.04v = (\alpha - \beta v_0)e^{-\beta t}$$

Given $\alpha = 3$ ft/sec^2; $\beta = 0.04$ sec; also, 20 mph $= 29.33$ ft/sec.

(a) After 2 sec,

$$\frac{dv}{dt} = [3 - (0.04)(29.33)]e^{-0.04 \times 2}$$

$$= 1.686 \text{ ft/sec}^2$$

After 3 sec,

$$\frac{dv}{dt} = 1.618 \text{ ft/sec}^2$$

After 10 sec,

$$\frac{dv}{dt} = 1.223 \text{ ft/sec}^2$$

After 120 sec,

$$\frac{dv}{dt} = 0.015 \text{ ft/sec}^2$$

(b) Acceleration $= 3.0 - 0.04v$. Therefore, when acceleration $= 0, 3.0 - 0.04v = 0$ and $v = 75$ ft/sec $= 51.14$ mph.

(c) The acceleration approaches zero in approximately 400 sec.

(d)
$$x = \frac{\alpha t}{\beta} - \frac{\alpha}{\beta^2}(1 - e^{-\beta t}) + \frac{v_0}{\beta}(1 - e^{-\beta t})$$

$$= \frac{3(120)}{0.04} = \frac{3(1 - e^{-0.04 \times 120})}{(0.04)^2} + \frac{29.33(1 - e^{-0.04 \times 120})}{0.04}$$

$$= 7868 \text{ ft} = 1.49 \text{ mi}$$

Example 3

The impatient driver mentioned in Example 2 approaches an intersection controlled by a two-way stop sign. The through traffic is quite heavy, with an average gap of 5 sec. If this driver can achieve an acceleration of $dv/dt = (3 - 0.04v)$ ft/sec^2 and his perception-reaction time is 0.75 second, determine if he can clear the intersection. Assume that the width of the intersection is 24 ft and his car is 20 ft long.

Solution

The equation for calculating the distance as a function of time is

$$x = \frac{\alpha t}{\beta} - \frac{\alpha}{\beta^2}(1 - e^{-\beta t}) + \frac{v_0}{\beta}(1 - e^{-\beta t})$$

Here $\alpha = 3, \beta = 0.04, t = 5 - 0.75 = 4.25$ sec, and $v_0 = 0$:

$$x = \frac{3(4.25)}{0.04} - \frac{3(1 - e^{-0.04 \times 4.25})}{(0.04)^2} + 0$$

$$= 24.87 \text{ ft}$$

Thus, this vehicle is only able to cover 24.87 ft, whereas the intersection plus the length of the vehicle add up to 44 ft. Therefore, he is not able to clear the intersection safely.

Discussion

The average gap in the main stream of vehicles is 5 seconds. Hence, it would not be smart for this driver even to attempt to get through this intersection, particularly if the standard deviation of the gaps was large. It is assumed that the driver of the car begins to accelerate when the preceding vehicle clears the intersection.

7. DYNAMIC CHARACTERISTICS

A vehicle in motion has to overcome the resistance of the air, the resistance due to rolling, the resistance offered by the grade the vehicle is negotiating, and friction resistance. Air resistance is proportional to the cross-sectional area of the vehicle perpendicular to the direction of motion and the square of the speed of the vehicle. It has been shown that air resistance can be estimated by the expression (Harwood, 1992)

$$F_a = 0.5 \, \frac{2.15 \rho C_D A V^2}{g} \tag{16}$$

where

$\quad F_a$ = air resistance force (lb)
$\quad \rho$ = density of air (0.002385 lb/ft^3 at sea level)
$\quad C_D$ = aerodynamic drag coefficient
$\quad A$ = frontal cross-sectional area (ft^2)
$\quad V$ = vehicle speed (mph)
$\quad g$ = acceleration due to gravity (32.2 ft/sec^2)

The aerodynamic drag coefficient (C_D) is typically 0.4 for an average car and can vary from 0.15 to 0.5. Trucks have even higher drag coefficients. Rolling resistance represents a combination of internal friction (at wheel, axle, drive shaft bearings, and in transmission gears), the frictional slip between the tire and the road surface, flexing of tire rubber, rolling over rough particles, and climbing out of road surface depressions. Rolling resistance for passenger cars and trucks on a smooth pavement surface can be determined by Eqs. 17a and 17b, respectively.

$$R_r = (C_{rs} + 2.15 \, C_{rv} V^2) \, W \tag{17a}$$

$$R_r = (C_a + 1.47 \, C_b V) \, W \tag{17b}$$

where

$\quad R_r$ = rolling resistance force (lb)
$\quad C_{rs}$ = constant (typically, 0.012 for passenger cars)
$\quad C_{rv}$ = constant (typically, 0.65×10^{-6} sec^2/ft^2 for passenger cars)
$\quad C_a$ = constant (typically, 0.2445 for trucks)
$\quad C_b$ = constant (typically, 0.00044 sec/ft for trucks)
$\quad V$ = vehicle speed (mph)
$\quad W$ = gross vehicle weight (lb)

Forces acting on a vehicle in motion are shown in Figure 6-14; Figure 6-2 shows the acceleration and deceleration distances for passenger vehicles. Note that grade resistance is that component of the weight of the vehicle acting in the plane of the roadway. The friction between the tires of the vehicle and the road pavement creates a force that acts in opposition to forward movement.

Grade resistance force is represented by

$$R_g = \frac{WG}{100} \tag{18}$$

where:

R_g = grade resistance force (lb)
W = gross weight of vehicle (lb)
G = gradient (%)

Maximum power that a vehicle engine can deliver is a measure of its performance. Power can be expressed in units of horsepower (1 hp = 550 ft-lb/sec), based on the time rate of doing work. Power actually used by a motor vehicle for propulsion can be determined as follows:

$$P = 0.00267\,RV \qquad (19)$$

where:

P = power actually used (hp)
R = sum of resistance to motion $(R_r + R_a + R_g)$ (lb)
V = vehicle speed (mph)

Also negotiating a curve and acceleration would require additional power.

Example 4

Drivers with an average 20/40 vision travel at 55 mph in the curb lane of a freeway, where the exit ramps are designed for 25 mph. What should be the minimum distance of signs with 6-in. letters placed ahead of the exit? The following assumptions can be made: perception-reaction time = 2.5 seconds; $f = 0.17$; the pavement is on a 1% downgrade; drivers with 20/20 vision can read signs at 60 ft per inch of letter height.

Solution

Distance needed for braking from 55 to 25 mph:

$$\frac{V_2^1 - V_2^2}{30(f - s)} = \frac{(55)^2 - (25)^2}{30(0.17 - 0.01)} = 500 \text{ ft}$$

Drivers with 20/20 vision can read the signs at 60 × 6 = 360 ft; therefore, drivers with 20/40 vision can read the signs at 180 ft. Also, the perception-reaction time is 2.5 seconds and distance covered = 55 × 5280 × 2.5/60 × 60 = 202 ft. Therefore, the signs should be located 500 + 202 − 180 = 522 ft (minimum) ahead of the exit.

Discussion

Notice the uncertainty in prescribing the minimum distance because of the variation in the visual acuity of drivers and their perception-reaction time. The dilemma is to be able to choose a suitable average value that includes a large proportion of the driving population.

8. TIRE FRICTION

Although frictional forces are considered elsewhere in connection with vehicles, this is a good place to study the tire-friction mechanism. Two cases are examined here: forces acting on a standing tire and forces acting on a rolling tire.

In the case of the standing tire, vehicle loads are transmitted through the wheels to the tires and then on to the road. The load-carrying mechanism of the tire consists of two parts: the load-carrying capacity of the tire material and the load-carrying capacity of the compressed air in the tire. The inflation pressure multiplied by the contact patch area contributes approximately 85% to the load-carrying capacity of a tire, and the basic strength and stiffness of the side walls account for the rest.

In the case of a rolling tire, the situation is somewhat different. The rolling resistance of a tire is the force required to pull the tire over a surface. Rolling resistance is connected with the bending and straightening of the tread (and the tire carcass) as it meets and leaves the road surface. It is also connected with the nonuniform pressure distribution between the tire and the road. Radial tires, for example, exhibit a lower rolling resistance than conventional tires.

A tire that is subjected to a braking action develops a shear or traction force between its contact patch and the road surface. The braking torque acting between the brake shoe and brake drum causes the tire to decelerate, or, in other words, the velocity of the tire circumference is less than the forward velocity of the vehicle. This difference in velocity causes the tire to slip over the road surface. In fact, a locked sliding tire has 100% slip, and a free-rolling tire exhibits zero slip. Tire slip, s_T, is given by

$$s_T = \frac{v_v - v_T}{v_v} \tag{20}$$

where

s_T = tire slip (%)
v_v = vehicle velocity
v_T = circumferential velocity of tire (ft/sec)

The shear force is limited by the condition associated with a sliding tire.

The deceleration–time curve of the braking process is illustrated in Figure 4-5. The brake pedal displacement begins at time zero. After the application time, t_a, deceleration begins to increase and the maximum deceleration a_{max} is attained after

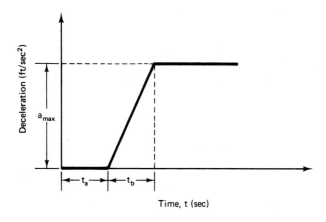

Figure 4-5   Idealized Deceleration Diagram.

the building time, t_b, has elapsed. When all wheels are locked, the braking distance, s_f, is given by

$$s_f = \frac{v^2}{2gf} = \frac{v^2}{64.4f} \qquad (21)$$

where v is the vehicle speed (ft/sec), and f is the tire–road friction coefficient. In Eq. 21, the deceleration is assumed to reach its maximum value at the instant of pedal application. However, the actual stopping distance is affected by time delays, such as t_a, and, therefore,

$$s = \frac{v^2}{2gf} + \left(\frac{t_a + t_b}{2}\right)v \quad \text{feet} \qquad (22)$$

where $gf = a =$ deceleration (ft/sec^2).

Example 5

A car is traveling along a road at a uniform velocity when at time zero the driver recognizes a hazard. At a moment 0.8 second later, the driver brakes her vehicle (locking the wheels), resulting in her vehicle sliding 90 ft in the same direction, at which time it strikes another stationary vehicle at 25 mph (Figure 4-E5). The road surface is dry and $f = 0.75$. The driver was traveling on a rural highway with a speed limit of 55 mph. Was she exceeding the speed limit?

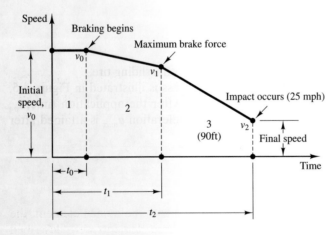

Figure 4-E5   Speed–Time Curve for Example 5.

Solution

There are three phases in this problem:

1. The driver lifts her foot from the accelerator and moves it to the brake pedal—the velocity is uniform.
2. Deceleration increases from zero to the maximum.
3. The braking system locks the wheels and deceleration is assumed to be constant until the vehicle strikes the stationary vehicle.

During the first phase, a driver reaction time of 0.5 second may be assumed. A linear increase in deceleration is assumed (from zero to locked wheels). Initial velocity (ft/sec) is given by

$$v_0 = v_1 + \frac{a}{2}(t_1 - t_0) \tag{23}$$

During the third phase, the deceleration is assumed to be uniform and, therefore,

$$v_1 = \sqrt{v_2^2 + 2as} = \sqrt{25(1.466)^2 + 2(0.75)(32.2)(90)}$$

$$= 75.43 \text{ ft/sec} = 51.46 \text{ mph}$$

As pointed out, the friction coefficient must be multiplied by 32.2 ft/sec^2 to yield the actual vehicle deceleration. Now

$$v_0 = v_1 + \frac{a}{2}(t_1 - t_0)$$

$$= 75.43 + (0.5)(0.75)(32.2)(0.8 - 0.5)$$

$$= 79.05 \text{ ft/sec} = 53.92 \text{ mph}$$

Obviously, she was not exceeding the speed limit of 55 mph.

Discussion

One might argue that if the driver had a slower reaction time, this fact would affect her adversely. Assume that her reaction time was 0.8 second; then her initial speed would be 51.46 mph.

SUMMARY

In this chapter, we synthesized several topics connected with the human element, the vehicle, and the environment. Characteristics of these three elements were described and put in an understandable format. The driver has two functions in the system: to move from one place to another in a certain period of time and as a guidance system for the vehicle. Information drawn from the general environment is translated into a set of meaningful decisions for safe passage of the vehicle. Naturally, the driver's personality comes into play at all times, including his or her perception and reaction abilities. The fundamental concepts of kinematics of vehicles were also described, taking into account uniform and nonuniform acceleration. Also, the dynamic characteristics of vehicles were discussed.

REFERENCES

BAKER, J. S. (1975). *Traffic Accidents Investigation Manual,* The Traffic Institute, Northwestern University, Evanston, IL.

FEDERAL HIGHWAY ADMINISTRATION (FHWA) (1980). *Highway Safety and Traffic Study Program*, prepared by Northwestern University, Evanston, IL.

HADDON, W. (1980). Advances in the Epidemiology of Injuries as a Basis of Public Policy, *Public Health Reports*, Vol. 95, No. 5, pp. 411–421.

HARWOOD, D. W. (1992). Vehicle Operating Characteristics, in *Traffic Engineering Handbook*, Institute of Transportation Engineers, Washington, DC.

HOMBURGER, W. S., and J. H. KELL (1988). *Fundamentals of Traffic Engineering*, University of California Press, Berkeley, CA.

LAY, M. G. (1986). *Handbook of Road Technology*, Vols. 1 and 2, Gordon and Breach, London.

SHINAR, D. S. (1978). *Psychology on the Road*, John Wiley, New York.

VANSTRUM, R. C., and B. G. CAPLES (1971). *Perception Model for Describing and Dealing with Driver Involvement in Highway Accidents*, Highway Research Record 365, Highway Research Board, Washington DC.

EXERCISES

1. The license plates of motor vehicles having letters 3 in. high can be read by a person with 20/20 vision at a maximum distance of 50 ft for every inch height of letter. A witness with 20/40 vision testifying in court regarding a speeding driver claims that he noted the license plate number accurately when he was 80 ft from this speeding car (going in the same direction as his own car). If he claims that he took 1.5 seconds to read the license plate, what is the reliability of his testimony?

2. A driver of a car traveling at 55 mph shifts her eyes from the center of the road to the left and tries to focus on a boy who is about to dart out across the highway. Estimate the distance traveled by the vehicle while the driver's eyes shift and fixate.

3. A driver with 30/40 vision needs 3 seconds to read a sign posted on a freeway. A person with 20/20 vision can read this sign at a distance of 225 ft. Do you feel that the driver has sufficient time to read this sign driving at a speed of 55 mph? At what speed would he have to travel just to be able to read the sign? Discuss this problem.

4. Refer to Example 4-1 in the text. A truck moving at 30 mph approaches this stop sign by decelerating at 15 ft/sec^2. Will it be able to stop in time, assuming all other conditions remain the same?

5. Refer to Example 4-3 in the text. If the average gap is observed to be 10 seconds, can this driver clear the intersection, all other conditions remaining the same? Discuss your answer noting that the gap is an average value.

6. A driver with 20/20 vision traveling at 55 mph in the median lane desires to exit the freeway at the exit ramp. If 5-in. letters are used, what is the minimum distance the sign should be placed ahead of the ramp exit? (Some assumptions that would be helpful are perception-reaction time = 2 sec; exit speed = 30 mph; $f = 0.15$; a driver with 20/20 vision can read a sign at 55 ft per inch of letter height.) Discuss your results.

7. A truck has been designed to maintain a constant speed of 55 mph on a +5% grade. The weight of the vehicle is 10,000 lb when fully loaded. What is the total resistance in still air

of this vehicle? What would be the resistance on a level road? What horsepower must be delivered to the wheels on the grade? If the efficiency is 80%, what must be the rated horsepower of the motor? Use

$$C_D = \text{air resistance parameter} = 0.40$$

$$A = \text{maximum cross-section} = 100 \text{ ft}^2$$

Discuss your results.

8. A truck that went out of control on an asphalt surface finally came to a halt on the gravel shoulder. Skid marks left by the truck were 230 ft on the asphalt surface and 40 ft on the gravel shoulder. Estimate the speed of the truck at the beginning of the skid. (Assume a range of values for the coefficient of friction for the two surfaces and discuss your results.)

9. The driver of a vehicle is following another vehicle in front of her on a two-lane highway at night. She keeps a clearance of one car length for every 10 mph of speed while she follows the lead car. However, while going downhill at 50 mph, the lead car crashes into the rear of an unlighted abandoned truck. At what speed will her car hit the wreckage? Assume a perception-reaction time between 0.5 and 2 seconds, a coefficient of friction of 0.35, the length of cars of 18 ft, and the road grade as 5%.

10. Deceleration is generally limited by either the condition of the brakes or the nature of the tire and roadway surface. A vehicle with extremely worn brakes will produce little deceleration no matter what the tire tread/road surface condition is. Discuss the mechanics of this statement. What would happen if the tires were brand new, the brakes were perfect, but there was ice on the pavement? Discuss.

11. A driver of a car applied the brakes and barely avoided hitting an obstacle on the roadway. The vehicle left 92-ft skid marks. Assuming that $f = 0.6$, determine whether the driver was in violation of the 45-mph speed limit at that location if she was traveling downhill on a 2.7-degree incline. Also compute the average deceleration developed.

12. Assume that a driver with normal vision can read a sign from a distance of 50 ft for each inch of letter height and that the roadway design is based on a driver with 20/40 vision. Determine how far away from an exit ramp a directional sign should be located to allow a safe reduction of speed from 60 to 25 mph, given a perception-reaction time of 1.5 sec, a coefficient of friction of 0.30, a letter size of 8 in., and a level freeway.

13. A car hits a tree at an estimated speed of 38 mph on a 4% downgrade. If skid marks are observed for a distance of 120 ft on a dry pavement ($f = 0.45$) followed by 250 ft on a grass-stabilized shoulder ($f = 0.20$), estimate the initial speed of the vehicle.

14. A car is being used in a test on level, but wet, concrete pavement to arrive at f values for stopping-distance equations. The minimum stopping distance achieved was 402 ft for an initial speed of 65 mph. What value of f was arrived at in this stop?

15. A driver with 20/40 vision needs 3 sec to read a sign posted on a freeway. A person with 20/20 vision can read this sign at a distance of 230 ft. Show through computation whether this driver has sufficient time to read the sign driving at a speed of 58 mph. At what speed would the driver have to travel just to be able to read the sign?

Chapter 5

Traffic Flow Characteristics

1. INTRODUCTION

In Chapter 4, we synthesized the fundamental characteristics of the driver, the vehicle, and the environment. In this chapter, we examine the flow of vehicles moving individually or in groups on a roadway or track, subject to constraints imposed by human behavior and vehicle dynamics.

It is vital to be able to design and operate transportation systems with the greatest possible efficiency and safety. Understanding the basic principles of traffic flow theory is one way of attaining this end. At the present time, there is no unified theory of traffic flow. Much of the knowledge currently available in this field is largely empirical.

In this chapter, we treat this topic at an elementary level, with the hope that the information will help engineers to understand problems occurring in the field, and provide the basic tools to solve these problems.

2. THE NATURE OF TRAFFIC FLOW

Traffic flow is a complex phenomenon. It requires little more than casual observation while driving on a freeway to discover that as traffic flow increases, there is generally a corresponding decrease in speed. Speed also decreases when vehicles tend to bunch together for one reason or another.

Traffic flow is a stochastic process, with random variations in vehicle and driver characteristics and their interactions. This statement needs some explanation. It is quite common to construct models of reality in which the effects of chance variation are ignored or averaged out, where any given input will produce an exactly predictable output. These models are deterministic. The alternative situation is to allow random variation in the model and then look at the probabilities of different outcomes. This method of stochastic modeling takes into consideration the variability among possible outcomes, not just average outcome (Lay, 1986a, 1986b).

3. APPROACHES TO UNDERSTANDING TRAFFIC FLOW

The interaction between vehicles and their drivers, and also among vehicles, is a highly complex process. There are three main approaches to the understanding and quantification of traffic flow. The first is a macroscopic approach that looks at the flow in an aggregate sense. Based on such physical analogies as heat flow and fluid flow, the macroscopic approach is most appropriate for studying steady-state phenomena of flow and hence best describes the overall operational efficiency of the system. The second is a microscopic approach that considers the response of each individual vehicle in a disaggregate manner. Here the individual driver–vehicle combination is examined, such as car maneuvering. This approach is used extensively in highway safety work. The third approach is the human-factor approach. Basically, it seeks to define the mechanism by which an individual driver (and his or her vehicle) locates himself or herself with reference to other vehicles and to the highway/guidance system. Notice that the microscopic and the human-factor approaches are closely related (Drew, 1968).

One way of combining all three approaches is to assume initially that a stream of traffic is composed of identical vehicles and identical drivers, thus permitting easy integration of the various approaches. The simplest combination also assumes that the traffic moves at uniform speed and that the vehicle spacing is dependent on speed. In other words, a vehicle's behavior is forced on it by other vehicles in the traffic stream. Indeed, speed is assumed to be the only variable that influences flow. Naturally, there is one particular vehicle flow associated with a speed adopted by the traffic stream (Lay, 1986a, 1986b).

4. PARAMETERS CONNECTED WITH TRAFFIC FLOW

There are at least eight basic variables or measures used in describing traffic flow, and several other stream characteristics are derived from these. The three primary variables are speed (v), volume (q), and density (k). Three other variables used in traffic flow analysis are headway (h), spacing (s), and occupancy (R). Also, corresponding to measures of spacing and headway are two parameters: clearance (c) and gap (g).

1. *Speed* is defined as a rate of motion, as distance per unit time, generally in miles per hour (mph) or kilometers per hour (km/hr). Because there is a broad distribution of individual speeds in a traffic stream, an average travel speed is considered. Thus, if travel times $t_1, t_2, t_3 \ldots, t_n$, are observed for n vehicles traversing a segment of length L, the average travel speed is

$$v_s = \frac{L}{\sum\limits_{i=1}^{n} \frac{t_i}{n}} = \frac{nL}{\sum\limits_{i=1}^{n} t_i} \tag{1}$$

where

v_s = average travel speed or space mean speed (mph)
L = length of the highway segment (miles)
t_i = travel time of the ith vehicle to traverse the section (hours)
n = number of travel times observed

Example 1

Three vehicles are traversing a 1-mile segment of a highway and the following observation is made:

Vehicle A: 1.2 min → 0.0200 hr/mi = 50 mph
Vehicle B: 1.5 min → 0.0250 hr/mi = 40 mph
Vehicle C: 1.7 min → 0.0283 hr/mi = 35.3 mph

What is the average travel speed of the three vehicles?

Solution

$$\text{Average travel time} = \frac{0.0200 + 0.0250 + 0.0283}{3}$$

$$= 0.0244 \text{ hr/mi}$$

The average travel speed = 1/0.0244 = 40.91 mph.

The average travel speed calculated is referred to as the *space mean speed*. It is called "space" mean speed because the use of average travel time essentially weights the average according to the length of time each vehicle spends in "space."

Another way of defining the "average speed" of a traffic stream is by finding the *time mean speed* (v_t). This is the arithmetic mean of the measured speeds of all vehicles passing, say, a fixed roadside point during a given interval of time, in which case, the individual speeds are known as "spot" speeds.

$$v_t = \frac{\sum\limits_{i=1}^{n} v_i}{n} \tag{2}$$

where v_i is the spot speed, and n is the number of vehicles observed.

Example 2

Three vehicles pass a mile post at 50, 40, and 35.3 mph, respectively. What is the time mean speed of the three vehicles?

Solution

$$\text{Average time mean speed} = v_t = \frac{\sum\limits_{i=1}^{n} v_i}{n} = \frac{50 + 40 + 35.3}{3} = 41.77 \text{ mph}$$

It can be shown that whereas the time mean speed is the arithmetic mean of the spot speeds, the space mean speed is their harmonic mean. Time mean speed is always greater than space mean speed except in the situation where all vehicles travel at the same speed. It can be shown that an approximate relationship between the two mean speeds is

$$v_t = v_s + \frac{\sigma_s^2}{v_s} \tag{3}$$

Also,

$$v_s = v_t - \frac{\sigma_t^2}{v_t} \tag{4}$$

where σ_s^2 is the variance of the space mean speeds. For example, if v_t is known to be 41.77 mph and the variance, $\sigma_t^2 = [\Sigma (v_i - v_t)^2]/n$, is equal to 37.58, then from Eq. 4, v_s can be found to be 40.91 mph, and this result checks with the figure indicated in Example 1.

2. Volume and rate of flow are two different measures. *Volume* is the actual number of vehicles observed or predicted to be passing a point during a given time interval. The *rate of flow* represents the number of vehicles passing a point during a time interval less than 1 hour, but expressed as an equivalent hourly rate. Thus, a volume of 200 vehicles observed in a 10-minute period implies a rate of flow of (200 × 60)/10 = 1200 veh/hr. Note that 1200 vehicles do not pass the point of observation during the study hour, but they do pass the point at that rate for 10 minutes.

Example 3

Calculate the rate of flow of vehicles from the following data:

Time period	Volume (vehicles)
4:00–4:15	700
4:16–4:30	812
4:31–5:00	1635
Total	3147

Solution

Although the volume = 3147 veh/hr, the individual rates of flow during the three time periods are 2800, 3248, and 3270 veh/hr, respectively.

3. *Density* or *concentration* is defined as the number of vehicles occupying a given length of lane or roadway, averaged over time, usually expressed as vehicles per mile (vpm). Direct measurement of density can be obtained through aerial photography, but more commonly it is calculated from Eq. 5 if speed and rate of flow are known.

$$q = v \times k \tag{5}$$

where

q = rate of flow (veh/hr)
v = average travel speed (mph)
k = average density (veh/mi)

Thus, a highway segment with a rate of flow of 1350 veh/hr and an average travel speed of 45 mph would have a density of k = (1350 veh/hr)/45 mph = 30 veh/mi. The proximity of vehicles in a traffic stream is given by density, which is a critical parameter in describing freedom of maneuverability.

4. Spacing and headway are two additional characteristics of traffic streams. *Spacing* (s) is defined as the distance between successive vehicles in a traffic stream as measured from front bumper to front bumper. *Headway* is the corresponding time between successive vehicles as they pass a point on a roadway. Both spacing and headway are related to speed, flow rate, and density.

$$\text{Avg. density } (k), \text{veh/mi} = \frac{5280, \text{ft/mi}}{\text{avg. spacing } (s), \text{ft/veh}} \tag{6}$$

$$\text{Avg. headway } (h), \text{sec/veh} = \frac{\text{avg. spacing } (s), \text{ft/veh}}{\text{avg. speed } (v), \text{ft/sec}} \tag{7}$$

$$\text{Avg. flow rate } (q), \text{veh/hr} = \frac{3600, \text{sec/hr}}{\text{avg. headway } (h), \text{sec/veh}} \tag{8}$$

Spacings of vehicles in a traffic lane can be generally observed from aerial photographs. Headways of vehicles can be measured using stopwatch observations as vehicles pass a point on a lane.

5. Lane occupancy is a measure used in freeway surveillance. If one could measure the lengths of vehicles on a given roadway section and compute the ratio:

$$R = \frac{\text{sum of lengths of vehicles}}{\text{length of roadway section}} = \frac{\Sigma L_i}{D} \tag{9}$$

then R could be divided by the average length of a vehicle to give an estimate of the density (k).

Example 4

Four vehicles, 18, 20, 21, and 22 ft long, are distributed over a length of a freeway lane 500 ft long. What is the lane occupancy and density?

Solution

$$R = \frac{18 + 20 + 21 + 22}{500} = 0.162$$

Average length of vehicle $= 20.25$ ft

$$k = 0.162 \times \frac{5280}{20.25} = 42.24 \text{ veh/mi}$$

Lane occupancy (LO) can also be described as the ratio of the time that vehicles are present at a detection station in a traffic lane compared to the time of sampling.

$$LO = \frac{\text{total time vehicle detector is occupied}}{\text{total observation time}} = \frac{\Sigma t_0}{T} \qquad (10)$$

Figure 5-1 illustrates the use of a detector in traffic engineering work. Here

$$t_0 = \frac{L + C}{v_s}$$

where L is the average length of vehicle, and C is the distance between the loop of the detector.

It is necessary to know the effective length of a vehicle as measured by the detector in use to calculate lane occupancy. Density can be calculated by using the expression

$$k = \frac{LO \times 5280}{L + C} \qquad (11)$$

In most cases, the detector is actuated as soon as the front bumper crosses the detector and remains on until the rear bumper leaves the detector.

Example 5

During a 60-sec period, a detector is occupied by vehicles for the following times: 0.34, 0.38, 0.40, 0.32, and 0.52 sec. Estimate the values of q, k, and v. (Assume that the loop-detector length $= 10$ feet and that the average length of vehicles $= 20$ ft.)

Solution

$$\Sigma t_0 = 0.34 + 0.38 + 0.40 + 0.32 + 0.52 = 1.96 \text{ sec}$$

$$n = 5$$

$$LO = 1.96 \times \frac{100\%}{60} = 3.27\%$$

The average effective length of a vehicle plus distance between loops is assumed $= 20 + 10 = 30$ ft; then

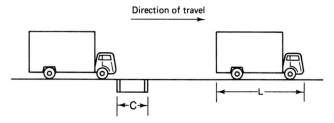

Direction of travel

L = length of vehicle
c = distance between loops of detector

Figure 5-1 Loop Detector.

$$k = \frac{3.27}{100} \left(\frac{5280}{30}\right) = 5.75 \text{ veh/mi}$$

$$v_s = \frac{n(L + C)}{\Sigma t_0} = \frac{5 \times 30}{1.96} = 76.53 \text{ ft/sec} = 52.18 \text{ mph}$$

$$q = k \times v_s = 5.75 \times 52.18 = 300 \text{ veh/hr}$$

It must be noted that in this example use is being made of an average effective length of a vehicle plus distance between detector loops totaling 30 ft. This assumption can lead to serious error, because the mixture of cars and trucks in a traffic stream can appreciably change the effective length of vehicles used in the calculations (Gerlough and Huber, 1975).

Lane occupancy applications are of much concern in on-line surveillance and control systems. For example, freeway surveillance is accomplished by monitoring lane occupancy from numerous stations such as entrance ramps. Proper records of lane occupancy can be useful in evaluating traffic stream performance.

6. *Clearance* and *gap* correspond to parameters of spacing (ft) and headway (sec). These four measurements are shown in Figure 5-2. The difference between spacing and clearance is obviously the average length of a vehicle in feet. Similarly, the difference between headway and gap is the time equivalence of the average length of a vehicle (L/v):

$$g = h - \left(\frac{L}{v}\right) \tag{12}$$

and

$$c = g \times v \tag{13}$$

where

$g =$ mean gap (sec)
$L =$ mean length of vehicles (ft)
$c =$ mean clearance (ft)
$h =$ mean headway (sec)
$v =$ mean speed (ft/sec)

Example 6

Figure 5-E6(a) shows a time–space plot of vehicles within a time–space domain. Based on the information contained in this figure, several parameters can be calculated.

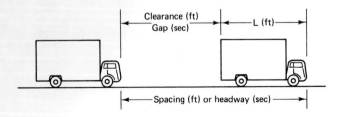

Figure 5-2   Clearance-Gap and Spacing-Headway Concept.

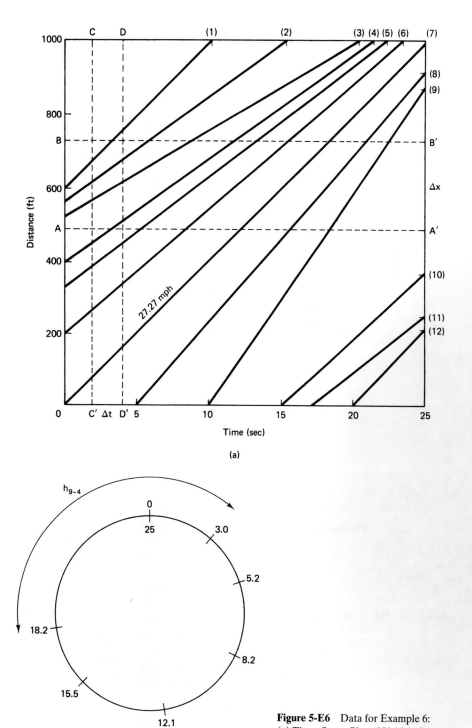

Figure 5-E6 Data for Example 6:
(a) Time–Space Plot of Vehicle Paths;
(b) Headways.

(a) An observer counts six vehicles crossing line A–A' in 25 sec. Calculate the flow of vehicles per hour (q).

$$q = \frac{n}{T} = \frac{6 \text{ veh}}{25 \text{ sec}} = 0.24 \text{ veh/sec} = 864 \text{ veh/hr}$$

(b) A timer is begun at time 0.0 second continuing for 25 seconds and the times at which the front end of vehicles pass the observation point A–A' are noted as follows [see Figure 5-E6(b)]:

Vehicle	Time of passing (sec)
4	3.0
5	5.2
6	8.2
7	12.1
8	15.5
9	18.2

What are the individual headways and the average headway? Because there are six vehicles, only the first five headways can be determined directly.

$$h_{4-5} = 2.2$$
$$h_{5-6} = 3.0$$
$$h_{6-7} = 3.9$$
$$h_{7-8} = 3.4$$
$$h_{8-9} = 2.7$$
$$h_{9-4} = 9.8$$
$$\text{Total} = 25.0$$

The final headway can be calculated as shown in Figure 5-E6(b):

$$\text{Average headway, } h = \frac{25}{6} = 4.17 \text{ sec}$$

$$\text{or } h = 3600/864 = 4.17 \text{ sec}$$

(c) A speed trap consisting of a record of times when vehicles passed two points, AA' and BB', was made as indicated on Figure 5-E6(a). Find the average speed of vehicles 4, 5, 6, 7, 8, and 9, assuming that the trap distance is 240 ft.

$$\text{Time-mean speed } v_t = \frac{161.03}{6} = 26.84 \text{ mph}$$

$$\text{Space-mean speed } v_s = \frac{6 \times 240}{38.8} = 37.11 \text{ ft/sec} = 25.30 \text{ mph}$$

Vehicle	Time of passing AA'	Time of passing BB'	Trap time (sec)	Speed (mph)
4	3.0	11.5	8.5	19.25
5	5.2	13.1	7.9	20.71
6	8.2	15.2	7.0	23.38
7	12.1	18.1	6.0	27.27
8	15.5	20.7	5.2	31.46
9	18.2	22.4	4.2	38.96
Total			38.8	161.03

Note that in calculating the space-mean speed, the numerator is the total distance traveled by the six vehicles within the roadway length and the denominator is the total time these six vehicles spent on this roadway section.

(d) At the time of observation, there were seven vehicles counted on the 1000-ft section of roadway lane. Calculate the density (k).

$$k = \frac{7 \text{ veh} \times 5280 \text{ ft/mi}}{1000 \text{ ft}} = 36.96 \text{ veh/mi}$$

(e) Two aerial photographs were taken, 2 seconds apart, and the positions of vehicles 1 through 7 as shown in Figure 5-E6(a) were noted. These seven vehicles are the same vehicles as those observed in section (d) of this example. Calculate the average speed and average flow.

Vehicle	Position 1	Position 2	Feet
1	680	760	80
2	610	670	60
3	560	615	55
4	450	510	60
5	380	440	60
6	270	350	80
7	60	160	100
Total			495

$$v_s = \frac{495}{7 \times 2} = 35.35 \text{ ft/sec} = 24.11 \text{ mph}$$

$$q = kv = 36.96 \times 24.11 = 891 \text{ veh/hr}$$

(f) Consider the time–space diagram [Figure 5-E6(a)] and the trajectories of the 12 vehicles. Within the time–space domain, calculate the flow, density, and speed.

$$q = \frac{\Sigma x_i}{A} \qquad k = \frac{\Sigma t_i}{A} \qquad v_s = \frac{\Sigma x_i}{\Sigma t_i}$$

where

x_i = distance traveled by the ith vehicle within the time–space domain

t_i = time taken by the ith vehicle to traverse the time–space domain

A = area of the time–space domain

A = 1000 × 25 = 25,000 ft-sec

$$q = \frac{7000 \text{ veh-ft}}{25,000 \text{ ft-sec}} = 0.28 \text{ veh/sec} = 1008 \text{ veh/hr}$$

$$k = \frac{194 \text{ veh-sec}}{25,000 \text{ ft-sec}} = 7.76 \times 10^{-3} \text{ veh/ft} = 40.97 \text{ veh/mi}$$

$$v_s = \frac{7000 \text{ veh-ft}}{194 \text{ veh-sec}} = 36.08 \text{ ft/sec} = 24.60 \text{ mph}$$

Vehicle, i	Distance traveled, x_i (veh-ft)	Time taken, t (sec)
1	400	10
2	440	15
3	480	20
4	600	21
5	680	22
6	800	23
7	1000	25
8	920	20
9	880	15
10	360	10
11	240	8
12	200	5
	$\Sigma x_i = 7000$	$\Sigma t = 194$

5. CATEGORIES OF TRAFFIC FLOW

Vehicle flow on transportation facilities may be generally classified into two categories.

1. *Uninterrupted flow* can occur on facilities that have no fixed elements, such as traffic signals, external to the traffic stream, that cause interruptions to traffic flow. Traffic flow conditions are thus the result of interactions among vehicles in the traffic stream and between vehicles and the geometric characteristics of the guideway/roadway system. Also, the driver of the vehicle does not expect to be required to stop by factors external to the traffic stream.

2. *Interrupted flow* occurs on transportation facilities that have fixed elements causing periodic interruptions to traffic flow. Such elements include traffic signals, stop signs, and other types of controls. These devices cause traffic to stop (or significantly slow down) periodically irrespective of how much traffic exists. Naturally, in this case, the driver expects to be required to stop as and when required by fixed elements that are part of the facility.

TABLE 5-1   TYPES OF TRANSPORTATION FACILITIES

Uninterrupted flow
   Freeways
   Multilane highways
   Two-lane highways
Interrupted flow
   Signalized streets
   Unsignalized streets with stop signs
   Arterials
   Transits
   Pedestrian walkways
   Bicycle paths

It should be noted that uninterrupted and interrupted flow are terms that describe the facility and not the quality of flow. A congested freeway where traffic is almost coming to a halt is still classified as an uninterrupted flow facility, because the reason for congestion is internal to the traffic stream. A well-timed signaling system on an arterial may result in almost uninterrupted traffic flow, but such a flow is likely to be interrupted for several reasons, primarily as part of the system, and is therefore classified as interrupted flow (TRB, 1985).

Table 5-1 provides a sample of the types of facilities under the categories of uninterrupted and interrupted flow facilities. Note that the categorization is approximate. Freeways, for instance, operate under the purest form of uninterrupted flow, and multilane and two-lane highways may also operate with almost uninterrupted flow, particularly in long segments between points of fixed interruptions, such as segments where signal spacing exceeds 2 miles. Pedestrian, bicycle, and transit flow are generally considered to be interrupted, although uninterrupted flow conditions can occur, say, for example, in a long busway, without stops.

6. THE UNINTERRUPTED TRAFFIC FLOW MODEL

This model can best be described by means of a typical curve of the form shown in Figure 5-3. Imagine several vehicles, driven by rational drivers along a section of freeway. As vehicles speed and spacing increases, the speeds approach the free speed, and drivers adopt their own speed when uninfluenced by other vehicles in the traffic stream (point C). It is useful to understand the situation at point A, which represents the maximum traffic density that occurs when traffic has virtually come to a complete stop, reminiscent of a linear parking lot. The dashed curve represents the normal flow behavior if all drivers were to have the same free speed (point D). It has been observed that drivers are uninfluenced by other vehicles in the traffic lane at flows of about 900 veh/hr or less, which is about half the capacity flow (point B). Note that in the forced-flow region, each vehicle adopts its minimum spacing and clearance distance.

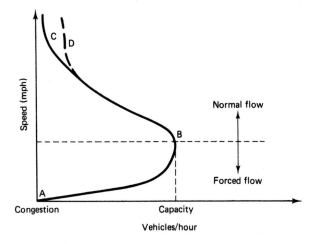

Figure 5-3   Speed–Flow Curve.

7. ANALYSIS OF SPEED, FLOW, AND DENSITY RELATIONSHIP

If it is hypothesized that a linear relationship exists between the speed of traffic on an uninterrupted traffic lane and the traffic density (veh/mi), as shown in Figure 5-4(a), then mathematically this relationship can be represented by

$$v = A - Bk$$

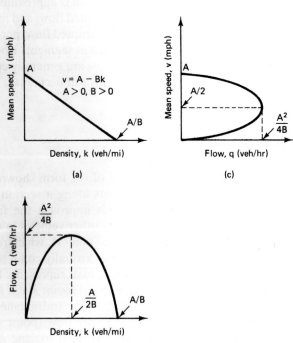

Figure 5-4   Speed–Flow–Density Curves.

where

v = mean speed of vehicles (mph)

k = average density of vehicles (veh/mi)

A, B = empirically determined parameters

Because the flow of an uninterrupted traffic stream is the product of the density and the speed, we have

$$q = kv = AK - Bk^2 \qquad (14)$$

$$q = kv = \frac{(v - A)v}{-B} = \frac{A}{B}v - \frac{v^2}{B}$$

At almost zero density, the mean free speed = A, and at almost zero speed, the jam density = A/B. The maximum flow occurs at about half the mean free speed and is equal to $A^2/4B$.

Figure 5-5 shows the theoretical relationship between flow (q) and density (k) on a highway lane, represented by a parabola. As the flow increases, so does the density, until the capacity of the highway lane is reached. The point of maximum flow (q_{max}) corresponds to "optimal" density (k_0). From this point onward to the right, the flow decreases as the density increases. At jam density (k_j), the flow is almost zero. On a freeway lane, this point may be likened to the traffic coming to a halt, where the lane appears to look like a parking lot.

If rays are drawn from zero through any point on the curve, the slope of the rays represents the corresponding space mean speed. The ray with a slope of v_f corresponds to the mean free speed and is tangential to the curve. This speed is possible when the density is near zero.

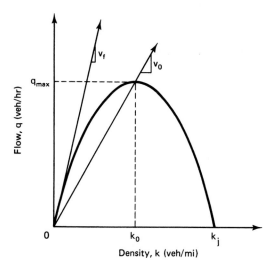

Figure 5-5   Flow-Density Curve.

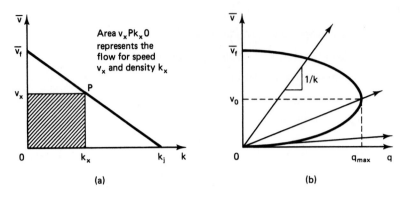

Figure 5-6   Speed–Density and Speed–Flow Curves.

Figure 5-6(a) shows a theoretical relationship between speed and density represented by a straight line. This relationship is not really true but is quite useful in practice. Flows can be calculated simply by multiplying coordinates of speeds and densities for any point on the straight line.

Finally, Figure 5-6(b) shows the theoretical relationship between speed and flow. Rays drawn from zero to any point on the curve have slopes whose inverse is equal to the density.

The hypothetical diagrams connecting mean speed, density, and flow shown in Figure 5-7 and the corresponding conditions on the road can best be discussed as follows. At point A, density is close to zero, and there are only a very few vehicles on the

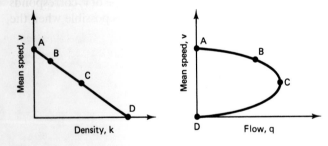

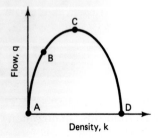

Figure 5-7 q–k–v Curves.

road; the volume is also close to zero and these few vehicles on the road can choose their own individual speeds, or change lanes with no restrictions. At point B, the number of vehicles has increased but the conditions are of "free flow" and there are hardly any restrictions, although such restrictions keep increasing steadily by the time point C is reached. From B to C, the flow conditions may be called "normal," but as density increases, drivers experience significant lack of freedom to maneuver their vehicles to the speed and lane of their choice. Around point C, traffic conditions begin to show signs of instability, and speeds and densities fluctuate with small changes in volume. Point C is the point of maximum volume, and further increases in density reduce speeds considerably. Such behavior is called *forced flow* and prevails all along from C to almost point D. Flow near point D is reduced almost to zero, with cars stacking up almost bumper to bumper. Point D is known as *jam density*. A driver would perceive excellent driving conditions from A to B, moderately good conditions from B to C, but increasingly deteriorating conditions from C to D.

8. EMPIRICAL STUDIES OF TRAFFIC STREAM CHARACTERISTICS

8.1 Macroscopic Models of Traffic Flow

(a) Greenshields' Model (Greenshields, 1935)

The general model connecting speed, flow, and density discussed so far is a linear model proposed by Greenshields (1935). As field measurements of speed, flow, and density became available, several researchers evolved traffic flow models based on actual curve fitting and statistical testing. The evaluation of models proceeded along two lines:

1. Relationships of q–k–v were tested in terms of goodness of fit to actual field data.
2. Relationships were supposed to satisfy certain boundary conditions (see Figure 5-4):
    (a) Flow is zero at zero density.
    (b) Flow is zero at maximum density.
    (c) Mean free speed occurs at zero density.
    (d) Flow–density curves are convex (i.e., there is a point of maximum flow).

$$v_s = v_f - \left(\frac{v_f}{k_j}\right) k \qquad (15)$$

Greenshields' model provides the slope and intercept by hand-fitting a straight line to plotted data or by using linear regression. This model satisfies all the four boundary conditions, although the statistical quality may be poor (e.g., low coefficients of determination and high standard errors).

(b) Greenberg's Model (Greenberg, 1959)

Greenberg developed a model taking speed, flow, and density measurements in the Lincoln Tunnel resulting in a speed–density model. He used a fluid-flow analogy concept, using the following form:

$$v_s = C\ln(k_j/k) \qquad \text{where } C \text{ is a constant} \tag{16}$$

Substituting q/k for v_s

$$q = Ck\ln(k_j/k) \tag{17}$$
$$= Ck(\ln k_j - \ln k)$$

Differentiating q with respect to k, we obtain

$$dq/dk = C[k(-1/k) + (\ln k_j - \ln k] = 0$$

And for maximum q,

$$C(-1 + \ln k_j - \ln k) = 0$$

Therefore,

$$\ln k_j - \ln k = 1 \tag{18}$$

and

$$\ln(k_j/k) = 1 \tag{19}$$

Substituting Eq. 19 in Eq. 16,

$$v_s(\text{maximum}) = C \tag{20}$$

so C is the speed at maximum flow.

Greenberg's model shows better goodness-of-fit as compared to Greenshield's model, although it violates the boundary conditions in that zero density can only be attained at an infinitely high speed.

Example 7

The speed–density relationship of traffic on a section of a freeway lane was estimated to be

$$v_x = 18.2\ln(220/k)$$

(a) What is the maximum flow, speed, and density at this flow?

(b) What is the jam density?

Solution

(a) $q = vk = 18.2k\ln(220k) = 18.2k(\ln 220 - \ln k)$

Maximum flow occurs at $dq/dk = 0$. Therefore,

$$\ln k_j - \ln k = 1$$
$$\ln 220 - \ln k = 1$$
$$5.394 - \ln k = 1$$
$$k = 80.93 \text{ veh/mile, when } q \text{ is maximum}$$
$$q_{max} = (18.2)(80.93)(1) = 1473 \text{ veh/hr and } v = 1473/80.93 = 18.2 \text{ mph}$$

(b) Jam density is obviously 220 veh/mi.

Example 8

Given k_j = 130 veh/mi; and k = 30 veh/mi, when v_s = 30 mph. Find q_{max}.

Solution

$$v_s = C \ln (k_j/k)$$

At k = 30 veh/mi, v_s = 30 mph. Therefore,

$$30 = C \ln (130/30)$$

$$C = 30/1.466 = 20.459 = v_s \text{ at maximum flow}$$

$$v_s = 20.459 \ln (k_j/k)$$

For maximum flow (q),

$$\ln k_j - \ln k = 1$$

$$\ln 130 - \ln k = 1$$

or

$$\ln k = 3.8675$$

$$k = 47.82 \text{ veh/mi}$$

$$q_{max} = (20.459)(47.82)(1)$$

$$= 978 \text{ veh/hr}$$

8.2 General and Linear Speed–Flow–Spacing–Density Curves

Figure 5-8 shows a series of plots connecting flow and speed, spacing and speed, density and flow. Note that the basic assumption is that the density–space relationship is not linear. Contours of equal spacing, equal flow, and equal speed are superimposed on the appropriate figures based on the fundamental equation, spacing = speed/flow. Also, the appropriate points of jam density and maximum (or optimum) flow are indicated.

If a linear relationship between density and speed is assumed, the corresponding results are shown in Figure 5-9. As mentioned before, this assumption was originally made by Greenshields on the basis of empirical data. Note again that in this case, the flow is a parabolic function of speed and that the maximum flow occurs at half the free speed.

Example 9

Assuming a linear speed–density relationship, the mean free speed is observed to be 60 mph near zero density, and the corresponding jam density is 140 veh/mi. Assume that the average length of vehicles is 20 ft.

(a) Write down the speed–density and flow–density equations.
(b) Draw the v–k, v–q, and q–k diagrams indicating critical values (Figure 5-E8).
(c) Compute speed and density corresponding to a flow of 1000 veh/hr.

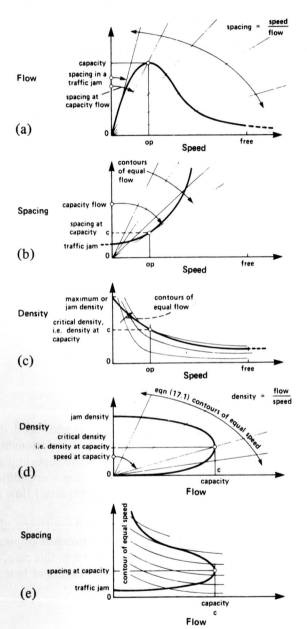

Figure 5-8   General Speed–Flow–Spacing Relationship (Lay, 1986a).

(d) Compute the average headways, spacings, clearances, and gaps when the flow is maximum.

Solution

(a) The equation for the linear relationship of v and k is

$$v = mk + C$$

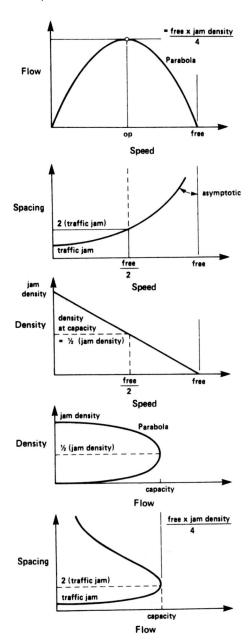

Figure 5-9   Speed–Flow–Spacing–Density Relationship Based on Free Speed and a Linear Speed–Density Curve (Lay, 1986a).

where m and C are constants. Substituting $k = 140$ when $v = 0$, and $k = 0$ when $v = 60$, we get $v = 60 - 0.43k$:

$$q = vk = (60 - 0.43k)k = 60k - 0.43k^2$$

(b) With $q = 60k - 0.43k^2$, to find the maximum value of q, differentiate q with respect to k and equate to zero.

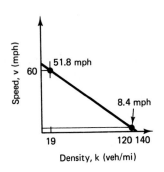

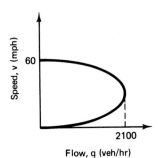

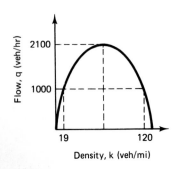

Figure 5-E8 q–k–v Curves for Example 8.

$$\frac{dq}{dk} = 60 - 0.86k = 0$$

Therefore,

$$k = 70 \text{ veh/mi}$$

and

$$q_{max} = (60 \times 70) - [0.43 \times (70)^2] = 2100 \text{ veh/hr. See Figure 5-E8.}$$

(c) If $q = 1000$ veh/hr, $1000 = 60k - 0.43k^2$. Therefore, $k = 120$ veh/mi or 19 veh/mi. Also, $v = 60 - 0.43k$, so substituting values of k results in

$$v = 8.4 \text{ mph} \quad \text{or} \quad 51.8 \text{ mph}$$

(d)

$$q = \frac{3600}{h}; \quad \text{therefore, } h = \frac{3600}{2100} = 1.714 \text{ sec}$$

$$h = \frac{s}{v}; \quad \text{therefore, } s = 1.714 \times 30 \times \frac{5280}{3600} = 75.42 \text{ ft}$$

$$g = h - \frac{L}{v} = 1.714 - \frac{20 \times 3600}{30 \times 5280} = 1.26 \text{ sec}$$

$$c = gv = 1.26 \times 30 \times \frac{5280}{3600} = 55.44 \text{ ft}$$

Refer to Figure 5-2 for definition of clearance, gap, spacing, and headway.

8.3 The Moving-Vehicle Estimation Method

Estimating the speed and flow of traffic from a moving vehicle is a useful field method. It consists of making a series of runs in a test vehicle, recording the number of vehicles that overtake the test vehicle, the number of vehicles passed by the test vehicle, and the travel time for the test vehicle.

Two cases are considered. In case 1, the test vehicle is stopped and the flow q, the number of cars overtaking the test vehicle, n_0, is recorded.

$$q = \frac{n_0}{T}$$

where T is the duration of the study.

In case 2, all vehicles are stopped except the test vehicle, and therefore the number of vehicles passed by the test vehicle $= n_s$.

$$k = \frac{n_s}{L}$$

where L is the length of the test section, and k is the density of traffic. Now $L = vT$, where v is the speed of the test vehicle. Therefore, $k = n_s/vT$, $n_0 = qT$, and $n_s = kL = kvT$. Therefore, $n_0 - n_s = qT - kvT$. If $n = n_0 - n_s$, then $n = qT - kvT$, or

$$\frac{n}{T} = q - kv \tag{21}$$

If several runs are made in this test vehicle, one can measure a whole set of values for n, T, and v and then solve for q and k. In practice, it may be convenient to drive the test vehicle both with traffic and against traffic. When the test vehicle is traveling with the traffic and against the traffic, the two equations corresponding to these two conditions are

$$\frac{n_w}{T_w} = q - kv_w \tag{22}$$

$$\frac{n_a}{T_a} = q + kv_a \tag{23}$$

Subscripts w and a correspond to "with traffic" and "against traffic," respectively. (The positive sign is placed because the test vehicle is traveling in the opposite direction.) Solving these two equations yields

$$q = \frac{n_w + n_a}{T_w + T_a} \tag{24}$$

To obtain the average travel time, T, of the traffic stream, substitute k for q/v and v_w for L/T_w:

$$\frac{n_w}{T_w} = q - \frac{q}{v}\frac{L}{T_w} = q - \frac{qT}{T_w}$$

Therefore,

$$T = T_w - \frac{n_w}{q} \qquad (25)$$

where q is given by

$$q = \frac{n_w + n_a}{T_w + T_a}$$

Example 10

A student riding his bicycle from campus on a one-way street takes 50 min to get home, of which 10 min was taken talking to the driver of a stalled vehicle. He counted 42 vehicles while he rode his bicycle, and 35 vehicles while he stopped. What are the travel time and flow of the vehicle stream?

Solution

Assuming that the vehicles counted while stopped represent the vehicle flow:

$$q = \frac{35}{10} = 3.5 \text{ veh/min} = 210 \text{ veh/hr}$$

$$T = T_w - \frac{n_w}{q} = 50 \text{ min} - \frac{42 \text{ veh}}{3.5 \text{ veh/min}} = 50 - 12$$

$$= 38 \text{ min}$$

Example 11

The data shown were noted in a travel time study on a 2-mile stretch of a highway using the moving-vehicle technique. Determine the travel time and flow in each direction on this section of the highway.

Direction	Travel time (min)	Number of vehicles		
		Traveling in opposite direction	Overtaking test vehicle	Overtaken by test vehicle
Northbound (NB)	T_N	N_N	n_N	
1	3.20	75	3	1
2	2.80	80	2	2
3	3.25	85	0	1
4	3.01	70	2	1
Average (NB)	3.07	77.5	1.75	1.25
Southbound (SB)	T_S	N_S	n_S	
1	3.30	78	4	0
2	3.25	74	2	2
3	3.40	79	0	2
4	3.35	82	3	3
Average (SB)	3.33	78.25	2.25	1.75

Solution

$$\text{Flow } NB = q_{NB} = (N_S + n_N)/(T_S + T_N)$$
$$= [78.25 + (1.75 - 1.25)]/(3.33 + 3.07) = 12.30 \text{ veh/min}$$

$$\text{Flow } SB = q_{SB} = (N_N + n_S)/(T_N + T_S)$$
$$= [77.5 + (2.25 - 1.75)]/(3.07 + 3.33) = 12.19 \text{ veh/min}$$

$$\text{Average travel time, northbound, } \overline{T}_N = T_N - n_N/q_{NB}$$
$$= 3.07 - (1.75 - 1.25)/12.30 = 3.029 \text{ min}$$

$$\text{Average travel time, southbound, } \overline{T}_S = T_S - n_S/q_{SB}$$
$$= 3.33 - (2.25 - 1.75)/12.19 = 3.289 \text{ min}$$

9. TRAJECTORY DIAGRAMS

The speed–flow–density relationships examined in previous sections of this chapter are considered with the aggregate behavior of vehicles in a traffic stream. If, however, we would like to examine the behavior of individual vehicles, a trajectory diagram can provide useful information. An example of a trajectory diagram is shown in Figure 5-10. Each trajectory represents the movement through space and time of a particular vehicle, and the combination of lines illustrates the interaction between vehicles. The slopes of the trajectories are the speeds of vehicles. Such diagrams have wide application in the study of platoon formation and dispersion, and the coordination of traffic.

9.1 Shock Waves and Bottlenecks

Consider a two-lane rural highway with a speed limit of 45 mph in rolling terrain where vehicles are prohibited from passing. Under this condition, it is not uncommon for a slow-moving truck to be seen crawling uphill at 15 mph, with a platoon of vehicles following the truck. Any additional vehicles joining this slow-moving platoon will be possibly having an approach speed of 45 mph, but may suddenly have to reduce their

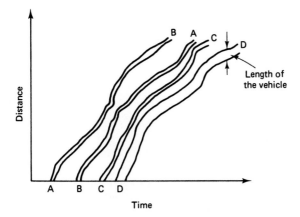

Figure 5-10   Time–Space Diagram.

speeds down to 15 mph, creating a "shock wave." This shock wave is indeed a sudden compression of vehicles that can lead to a rear-end collision if a driver is inattentive.

Now, suppose that the truck in question, with a platoon of impatient drivers behind it, negotiates the crest of the curve and is traveling downgrade with a speed of 45 mph. Here there is the possibility of another shock wave, this time of decompression.

A similar shock wave is likely to occur when, for example, a three-lane freeway is reduced to a two-lane freeway and a bottleneck situation arises. A reverse shock wave may occur after vehicles have negotiated the bottleneck, and the released conditions allow the compressed platoon to disperse, gain speed, and increase their headways.

Vehicular streams interrupted by traffic signals also represent a situation where vehicles approaching a stationary platoon are suddenly compressed (almost to jammed density), and then released when the signal turns green, representing decompression.

A generalized bottleneck formation is shown in Figure 5-11, in which the normal and bottleneck flow–density relationships are superimposed. A vehicle on the main road (flow q_A) with speed v_A (represented by the vector OA) approaches the bottleneck section of the main road. Vehicles in this section have a speed of v_B (flow q_B), represented by vector OB. The propagation of the shock wave (chord AB) in the platoon of vehicles following the lead car is represented by vector AB, with the result that vehicles approaching the bottleneck are forced to reduce their speeds to OB, which is called the *crawl speed*. This speed moves upstream.

When changes in flow are occurring, the changes through the stream of vehicles travel at a velocity, called the *wave velocity*, given by

$$u_w = \frac{dq}{dk} \tag{26}$$

and if

$$q_B = \text{platoon flow (veh/hr)}$$

$$k_B = \text{platoon density (veh/mi)}$$

$$q_A = \text{free flow}$$

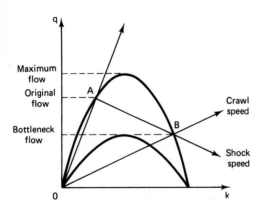

Figure 5-11   Shock Wave Measurements.

and

$$k_A = \text{free density}$$

then

$$q_B/k_B = v_B$$

and

$$q_A/k_A = v_A$$

Therefore,

$$u_w = \frac{q_B - q_A}{k_B - k_A} \tag{27}$$

where u_w is the speed of the shock wave.

If the sign of the shock wave is positive, the wave is proceeding in the direction of the stream flow (downstream); if the sign is negative, the shock wave is moving against the stream flow (upstream). A stationary shock wave exists if $u_w = 0$.

Example 12

A school zone (20 mph) of $\frac{1}{4}$-mile length is located on a 40-mph highway. Stream measurements at various sections upstream, middle of the school zone, and just downstream of the school zone, respectively, are as follows:

Approaching zone: $q_A = 1000$ veh/hr, $v_A = 40$ mph

Middle of zone: $q_B = 1100$ veh/hr, $v_B = 20$ mph

Downstream of zone: $q_C = 1200$ veh/hr, $v_C = 30$ mph

Sketch the q–k–v curves and indicate critical values. Calculate the intensity and direction of the shock waves created by this speed zone. What is the length of the platoon created by the speed zone and the time required to disperse it? (Assume that the speed-zone restriction operates for only 15 min in the morning and evening times.)

Solution

See Figure 5-E11.

$$q_A = 1000 \text{ veh/hr} \qquad v_A = 40 \text{ mph} \qquad k_A = 25 \text{ veh/mi}$$

$$q_B = 1100 \text{ veh/hr} \qquad v_B = 20 \text{ mph} \qquad k_B = 55 \text{ veh/mi}$$

$$q_C = 1200 \text{ veh/hr} \qquad v_C = 30 \text{ mph} \qquad k_C = 40 \text{ veh/mi}$$

It is obvious that there is a platoon working its way through the school zone at a speed of 20 mph. In other words, the front of the platoon moves at +20 mph forward relative to the highway. By the same token, the speed of the shock wave at the rear of the platoon is

$$u_{AB} = \frac{1100 - 1000}{55 - 25} = +\frac{100}{30} = 3.33 \text{ mph}$$

(moving downstream or in the direction of the stream flow).

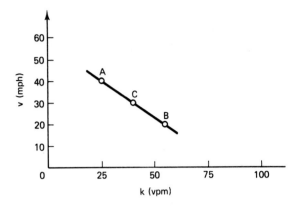

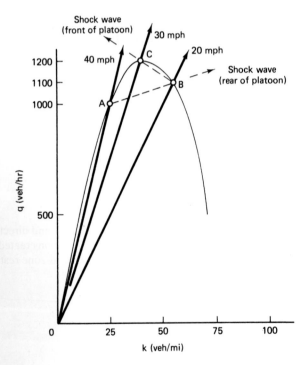

Figure 5-E12 q–k–v Curves for Example 12.

Therefore, the relative growth of the platoon is proportional to the relative speed between $20 - 3.33 = 16.67$ mph. The platoon grows at the rate of 16.67 mph as it moves forward. If the school zone operates for only 15 minutes every morning, the length of the platoon $= 16.67 \times \frac{1}{4}$ hour $= 4.17$ miles, and the number of vehicles in this platoon is $4.17 \times 55 = 230$.

A shock wave also develops where the school zone ends and the release (or decompression) zone begins. The speed of this shock wave is

$$u_{BC} \frac{1200 - 1100}{40 - 55} = -6.67 \text{ mph}$$

(moving upstream or opposite to the direction of the stream flow).

Note that the shock wave at the rear of the platoon moves downstream at +3.33 mph. Naturally, the relative speed of the two shock waves is $-6.67 - (+3.33) = 10$ mph. It will therefore take $4.17/10 = 0.417$ hour, or 25 minutes, to dissipate totally the platoon formed.

Example 13

Draw a time–distance sketch showing the front and rear of the platoon at any time from the start of the speed-zone enforcement until such time as the platoon has completely dispersed, based on Example 12.

Solution

Figure 5-E13 plots the front and rear of the platoon. The algebraic distance between the two represents the length of the platoon. Notice that the maximum length is $5.00 - 0.83 = 4.17$ miles. Also note that the platoon is finally dispersed 40 minutes (15 + 25) after the 15-minute speed enforcement, and 2.22 miles from the beginning of the $\frac{1}{4}$-mile-long school zone.

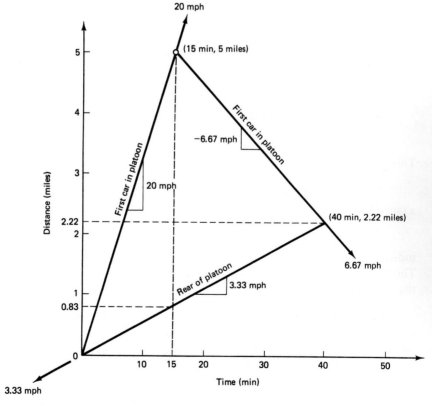

Figure 5-E13   Time–Distance Sketch for Example 13.

9.2 Shock Wave Propagation (Homburger, 1982)

Assuming Greenshields' traffic flow model, v_f represents the mean free speed and k_j the jam density.

$$v_i = v_f\left(1 - \frac{k_i}{k_j}\right)$$

and if

$$\frac{k_i}{k_j} = x$$

then

$$v_i = v_f(1 - x)$$

If there are two regions in a traffic stream flow having k_i/k_j values of x_1 and x_2, then

$$v_1 = v_f(1 - x_1) \qquad \text{and} \qquad v_2 = v_f(1 - x_2)$$

and

$$u_w = v_f[1 - (x_1 + x_2)]$$

where u_w is the velocity of the shock wave in terms of the two densities in the corresponding two regions. This test equation is useful in examining the shock wave in three common settings: small discontinuities in density, stopping situations, and starting situations.

Shock Wave Due to Nearly Equal Densities.   If x_1 and x_2 are nearly equal, then

$$u_w = v_f(1 - 2x_1)$$

This shock wave is referred to as a wave of discontinuity.

Shock Wave Caused by Stopping.   Here the density upstream x_1 is brought to a jam density condition and $x_2 = 1$. Therefore,

$$u_w = v_f[1 - (x_1 + 1)] = -(v_f)(x_1)$$

indicating that the shock wave of stopping moves upstream with a velocity of $v_f x_1$. Therefore, if a stream of cars stop at a signal at time $t = 0$, then at time t, the length of the platoon of stopped cars will be $(v_f)(x_1)(t)$.

Shock Wave Caused by Starting.   Assume that at time $t = 0$, a platoon of vehicles has accumulated at the stop bar at an intersection, and that the saturated (jam) density is such that $x_1 = 1$. At $t = 0$, the signal turns green permitting vehicles to be released at velocity v_2. Because

$$u_w = v_f[1 - (x_1 + x_2)], \quad \text{where } x_1 = 1$$

$$u_w = v_f[(1 - (1 + x_2))] = v_f(-x_2) = v_f x_2$$

Also,

$$v_2 = v_f(1 - x_2) \qquad x_2 = 1 - \frac{v_2}{v_f}$$

Therefore,

$$u_w = -v_f\left(1 - \frac{v_2}{v_f}\right) = -v_f + v_2 = -(v_f - v_2)$$

Assuming that vehicles depart at a velocity $v_2 = v_f/2$, the starting shock wave travels backwards with a speed of $u_f/2$.

Example 14

(a) A 2-lane highway traffic stream following Greenshields' model has the following characteristics: mean free speed, $v_f = 50$ mph, $k_j = 220$ veh/mile. What is the speed of the shock wave of discontinuity when $k_1 = 50$, $k_2 = 160$, and $k_3 = 110$ veh/mi?

(b) A traffic incident on this highway stops all traffic for 5 minutes, when the space mean speed is 45 mph and the density is 40 veh/mi. Calculate the shock-wave speed and the length of the stopped line of cars.

(c) Assuming the vehicles start moving at 25 mph after the incident is removed, calculate the speed of the starting wave.

Solution

(a) $$u_w = v_f(1 - 2x_1) \qquad \text{where } x_1 = k_i/k_j$$

When $k_1 = 50$, $\quad x_1 = 50/220 = 0.227, u_w = (50)[1 - 2(0.227)]$

$$= 27.27 \text{ mph, downstream}$$

When $k_2 = 160$, $\quad x_1 = 160/220 = 0.727, u_w = (50)[1 - 2(0.727)]$

$$= -22.73 \text{ mph, upstream}$$

When $k_2 = 110$, $\quad x_1 = 110/220 = 0.5 u_w = (50)[1 - 2(0.5)]$

$$= 0$$

(b) Shock wave caused by stopping:

$$u_w = -(v_f)(x_1) \qquad \text{where } x_1 = 40/220 = 0.1818$$

$$= -(50)(0.1818) = -9.09 \text{ (moving upstream)}$$

$$t = 5/60 = 1/12 \text{ hr}, \qquad \left(\tfrac{1}{12}\right)(9.09) = 0.7575 \text{ mi}$$

The number of vehicles in the line $= (0.7575)(220) = 167$

(c) $$u_w = -(v_f - v_2) = -(50 - 25) = -25 \text{ mph upstream}$$

This starting wave will overtake the stopping wave at a relative velocity of $-25.0 - (-9.09) = -15.91$ mph. The time to dissipate the line of 0.7575 miles $= 0.7575/15.91 = 0.0476$ hr $= 2.86$ minutes, and the point on the roadway will be 0.0476×25 mph $= 1.19$ miles upstream from the point of incident.

10. GENERAL MODEL OF VEHICLE STREAM FLOW

Consider two successive vehicles (or trains), called the *lead vehicle* and *following vehicle*, moving at a cruising speed, v (ft/sec), along a long stretch of highway or guideway (Figure 5-12). Using the fundamental equation of traffic flow, the minimum headway can be written as

$$\text{Minimum headway (sec)} = \frac{\text{minimum spacing (ft)}}{\text{cruise speed (ft/sec)}} \tag{28}$$

This minimum spacing consists of four components:

1. The distance covered during the perception-reaction time needed by the driver of the following vehicle after the brake lights of the lead vehicle come on
2. The difference between the braking distances of the following and lead vehicles
3. The distance between successive vehicles when stopped
4. The vehicle length

Expressing these components algebraically, the minimum spacing is given as

$$s_{\min} = vt_r + \left(\frac{v^2}{2b_2} - \frac{v^2}{2b_1}\right) + s_0 + L \tag{29}$$

where

$$v = \text{speed of vehicles (ft/sec)}$$
$$t_r = \text{perception-reaction time (sec)}$$
$$b_1, b_2 = \text{deceleration rate of leading and following vehicles, respectively (ft/sec}^2)$$
$$s_0 = \text{distance between two vehicles when stopped}$$
$$L = \text{length of the vehicle (ft)}$$

The choice of specific values for the six variables has important implication with respect to the level of safety provided by the system's operation. For instance, even though the maximum braking rate is vehicle-specific, the transit designer may choose to assume a rate lower than the maximum. The three braking or deceleration rates will be assumed here.

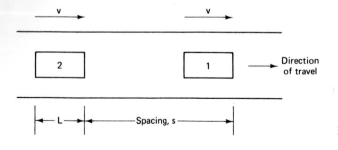

Figure 5-12   Vehicle Stream Flow.

b_n = normal braking rate, dictated by what passengers would feel comfortable with. Generally, the maximum value is limited by safety and comfort applicable to standing passengers (8 ft/sec^2)

b_e = emergency braking rate, which may cause discomfort to passengers and possible hazard (say, 24 ft/sec^2)

b_s = infinite braking rate (∞ ft/sec^2), "resulting in an instantaneous, stone-wall," or "brick-wall" stop, associated with a collision or sudden accident

There can be several safety policies adopted for controlling vehicles, assuming various combinations of the foregoing braking or deceleration rates for both the leading and following vehicles. Nine possible combinations of braking rates are listed in Table 5-2 (Vuchic, 1981). On inspection, all policies that assume that the following vehicle can brake faster than the lead vehicle as dictated by policies 6, 8 and 9 can be eliminated. Also, if the braking rates of both the lead and following vehicles are instantaneous (policy 3), it does not make sense to retain the policy. There are thus only five policies that can be considered valid—policies 1, 2, 4, 5, and 7. Table 5-2 shows the four regimes corresponding to the four valid policies, as well as regime 5, the hypothetical continuous train, operating at any constant speed without ever having to decelerate. The combinations of leading and following vehicle decelerations that depict safety regimes are shown in Figure 5-13 for a train 100 ft long. This figure plots spacing versus speed for these four regimes, as well as for a hypothetical continuous train (curve 5). Note that the higher the level of safety, the greater the required spacing; however, a larger spacing results in a lower capacity. Naturally, one may ask the question: How can one compromise between safety and capacity?

The relationship between mean (or average) spacing and mean density (or concentration) is given by the equation

$$s = \frac{5280}{k} \tag{30}$$

where s is the spacing (ft), and k is the density (veh/mi). Substituting Eq. 26 in Eq. 25 gives

TABLE 5-2   COMBINATIONS OF BRAKING RATES

Safety policy	Leading vehicle, b_1	Following vehicle, b_2	Case
1	∞	b_n	1
2	∞	b_e	2
3	∞	∞	—
4	b_e	b_n	4
5	b_e	b_e	3
6	b_e	∞	—
7	b_n	b_n	3
8	b_n	b_e	—
9	b_n	∞	—

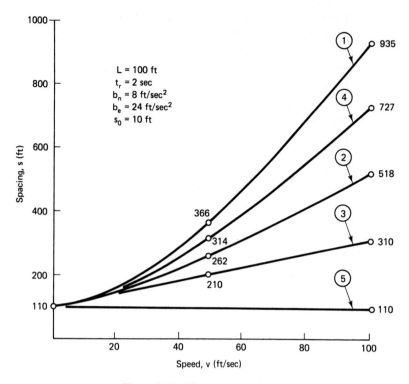

Figure 5-13 Distance versus Speed.

$$\frac{5280}{k} = vt_r + \left(\frac{v^2}{2b_2} - \frac{v^2}{2b_1}\right) + s_0 + L$$

Therefore,

$$k = \frac{5280}{vt_r + (v^2/2b_2 - v^2/2b_1) + s_0 + L} \tag{31}$$

This equation is plotted in Figure 5-14 with density (veh/mi) on the abscissa and speed v on the ordinate for the five regimes discussed earlier, including the limiting case of the hypothetical continuous train (curve 5). When speeds are very high and the corresponding density is very low, the condition is referred to as *free flow*.

If Eq. 31 is multiplied by the mean speed on both sides, this equation becomes

$$q = \frac{v}{vt_r + (v^2/2b_2 - v^2/2b_1) + s_0 + L} \tag{32}$$

Again, this equation is plotted using the characteristics shown in Figure 5-15 and for the five cases indicated.

Case 1: $b_1 = \infty, b_2 = b_n$ (∞ = instantaneous or brick-wall stop)
Case 2: $b_1 = \infty, b_2 = b_e$

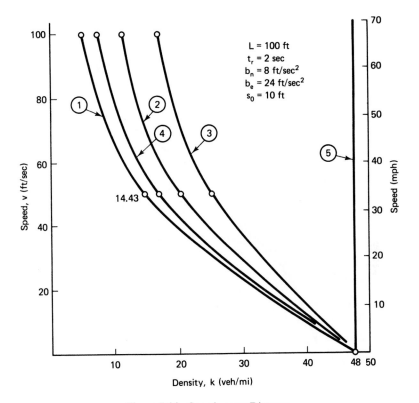

Figure 5-14   Speed versus Distance.

Case 3: $b_1 = b_e, b_2 = b_e$ or $b_1 = b_n, b_2 = b_n$

Case 4: $b_1 = b_e, b_2 = b_n$

Case 5:   Hypothetical continuous train

Figure 5-16 shows the relationship (for case 4, only) between flow and concentration using Eq. 31.

Each of the five valid regimes (represented by cases 1 through 5) are analyzed to establish the minimum headway (in seconds), the way capacity, and the optimum speed to obtain the maximum way capacity. Thus, for each policy condition, the way capacity is differentiated with respect to the cruise speed, and this is set to zero, giving the optimal cruise speed and the corresponding maximum way capacity (Vuchic, 1981).

Regime 1: $b_1 = \infty$ $b_2 = b_n$

$$s_{\min} = vt_r + \frac{v^2}{2b_n} - \frac{v^2}{\infty} + NL + s_0 \qquad \text{where } N = \text{vehicles per train}$$

$$h_{\min} = \frac{NL + s_0}{v} + t_r + \frac{v}{2b_n} \qquad \text{where } h_{\min} = \text{minimum headway (sec)}$$

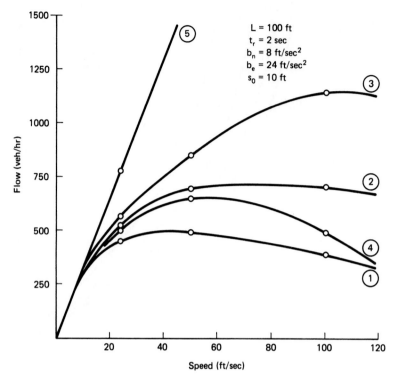

Figure 5-15   Flow versus Speed.

$$C_w = \frac{3600 N C_v}{(NL + s_0)/v + t_r + v/2b_n}$$ where C_v = passengers/veh

$$C_w = \frac{3600\, C_v}{h_{\min}}$$

$$v_{\text{opt}} = \sqrt{2(NL + s_0)\, b_n}$$ where C_w = way capacity

$$C_{w(\max)} = \frac{3600 N C_v}{[\sqrt{2(NL + s_0)/b_n}] + t_r}$$

This policy assumes that the following vehicle can be stopped at a normal braking rate in spite of the lead vehicle coming to a brickwall stop. Absolute safety is assured in this case. However, this policy results in longer minimum headways as compared to other policies, making it more expensive in terms of investment per passenger per hour.

Regime 2: $b_1 = \infty$ $b_2 = b_e$

$$s_{\min} = v t_r + \frac{v^2}{2b_e} - \frac{v^2}{\infty} + NL + s_0$$

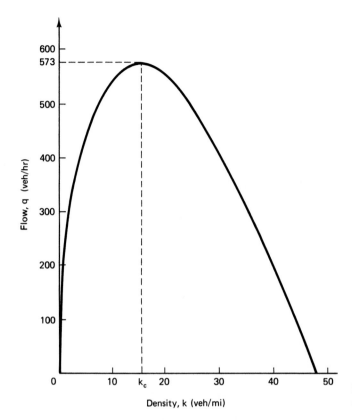

Figure 5-16 Flow versus Density (Regime 4).

$$h_{min} = \frac{NL + s_0}{v} + t_r + \frac{v}{2b_e}$$

$$C_w = \frac{3600NC_v}{(NL + s_0)/v + t_r + v/2b_e}$$

$$v_{opt} = \sqrt{2(NL + s_0)b_e}$$

$$C_{w(max)} = \frac{3600NC_v}{[\sqrt{2(NL + s_0)/b_e}] + t_r}$$

Regime 2 is somewhat less safe than regime 1, but it has a higher capacity at any cruise speed. This regime is not suitable for automated systems. Standees in such a system may be endangered.

Regime 3: $b_1 = b_e b_2 = b_e$ (or $b_1 = b_n b_2 = b_n$)

This regime is unacceptable because the lead and following vehicles are assumed to have the same braking rate. In this case:

$$s_{min} = vt_r + \frac{v^2}{2b_e} - \frac{v^2}{2b_e} + NL + s_0$$

$$= vt_r + NL + s_0$$

$$h_{min} = \frac{NL + s_0}{v} + t_r$$

$$C_w = \frac{3600NC_v}{(NL + s_0)/v + t_r}$$

This regime has no optimal speed nor a maximum C_w.

Regime 4: $b_1 = b_e$ $b_2 = b_n$

$$s_{min} = vt_r + \frac{v^2}{2b_n} - \frac{v^2}{2b_e} + NL + s_0$$

$$h_{min} = \frac{NL + s_0}{v} + t_r + \frac{v(b_e - b_n)}{2b_n b_e}$$

$$C_w = \frac{3600NC_v}{(NL + s_0)/v + t_r + v(b_e - b_n)/2b_n b_e}$$

$$v_{opt} = \sqrt{2(NL + s_0)b_e b_n/(b_e - b_n)}$$

$$C_{w(max)} = \frac{3600NC_v}{[\sqrt{2(NL + s_0)(b_e - b_n)/b_e b_n}] + t_r}$$

This regime allows for a comfortable braking rate of the following vehicle when the lead vehicle applies its emergency brakes. Naturally, the capacity is always higher than regime 1 and regime 2, provided $b_e < 2b_n$. However, the safety is lower than the other two.

Regime 5: $b_1 = b_2 = 0$

$$s = vt_r + \frac{v^2}{0} - \frac{v^2}{0} + NL + s_0$$

$$= vt_r + NL + 0$$

This is the case of a continuous train, and hence there is no maximum way capacity.

11. CENTRALLY VERSUS INDIVIDUALLY CONTROLLED MODES

This is an appropriate place to understand some fundamental differences between two principal modes. A mode of transportation is a particular configuration of right-of-way, vehicle technology, and rule of operation. For the purposes of this chapter all modes can be classified into two categories.

The first group consists of those modes whose rules of controlling the longitudinal spacing of vehicles, the stopping schedule of vehicles, and the time of operation are subject to their designer's decisions. This group can be referred to as *centrally controlled modes*. Most transit modes operating on exclusive or principally exclusive right-of-way fall into this group. It must be noted that vehicles or a group of connected vehicles forming a train of vehicles may be operated by a driver, or it may be operated automatically, but the headways (and other aspects of operation) between successive vehicles or trains are predecided (or predetermined) for the safe operation of these vehicles.

The second group consists of those modes where individual drivers make their own decisions regarding headways, speed, and so on, subject, of course, to state or local traffic laws. Private automobiles as well as transit vehicles operating in mixed traffic fall into this group. They are classified as *individually controlled modes*.

Centrally controlled systems (operating on an exclusive right-of-way) generally serve a series of stations along a line and their travel represents a discrete stop-and-go movement, with station-to-station travel as a basic element. Vehicle or train movement is quite deterministic, particularly when the system is fully automated. On the other hand, the movement of individually controlled vehicles, such as automobiles, buses, and streetcars, varies considerably, due to external factors such as other vehicles in the traffic stream, pedestrians crossing the streets, and signals. Naturally, travel times in this group are less precise, average speeds and headways being adopted for site-specific conditions.

As an example, Figure 5-17 shows typical plots of flow–density relationships for vehicles 20 ft long traveling under three different conditions.

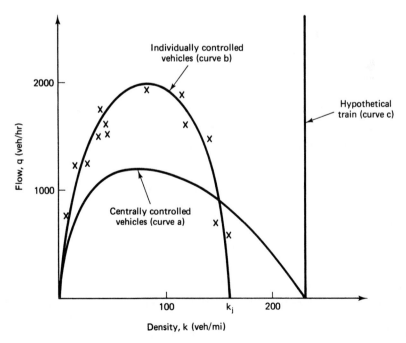

Figure 5-17   Centrally versus Individually Controlled Flow.

Curve *a* represents these 20-ft-long vehicles traveling on an exclusive right-of-way, automatically controlled, where the safety margin, after stop (s_0) = 3 ft, the normal and emergency braking rates (b_n and b_e) are 8 and 24 ft/sec², respectively, and the perception-reaction time of the following vehicle is 1 sec. Here the maximum flow is about 1200 veh/hr and the jam density is about 230 veh/mi. The flow–density relationship is represented by Eq. 32. This is a typical curve for a centrally controlled system.

Curve *b* represents a platoon of automobiles (20 ft long each) traveling in a freeway lane adopting their own speeds and headways. Each driver operates independently. This is a case of an individually controlled system. The data obtained by an observer reveal that the mean free speed is about 50 mph and the jam density is approximately 160 veh/mi. Fitting these data, one might find that they conform to Greenshields' equation.

Drawn on the same figure is curve *c*, which represents a chain of 20-ft vehicles, all linked mechanically to one another, operating on an exclusive right-of-way. This chain of vehicles, which forms a train, can be construed as vehicles at jam density, moving together at whatever speed they might choose, because the question of minimum headway (or safe braking distance) between individual vehicles does not apply.

Putting it all together, the three curves *a*, *b*, and *c* represent three different systems.

System a (centrally controlled vehicles):

$$L = 20 \text{ ft/veh}$$

$$s_0 = 3 \text{ ft}$$

$$t_r = 1 \text{ sec}$$

$$b_n = 8 \text{ ft/sec}^2$$

$$b_e = 24 \text{ ft/sec}^2$$

where

$$q = \frac{v}{vt_r + (v^2/2b_n - v^2/2b_e) + s_0 + L} \tag{33}$$

System b (individually controlled vehicles):

$$L = 20 \text{ ft/veh}$$

$$v_f = 50 \text{ mph}$$

$$k_j = 160 \text{ veh/mi}$$

where

$$q = kv$$

System c (hypothetical train):

$$L = \text{a continuous train (with cars all mechanically linked)}$$

$$v = \text{speed (can be whatever the system can sustain)}$$

Flow *q* depends on the speed and capacity of the train.

SUMMARY

In this chapter, we examined the nature of traffic flow. Because of the complexity posed in understanding the interaction between vehicles, several assumptions were made for establishing a general relationship between safe spacing and speed.

Parameters connected with traffic flow are defined and applied to everyday problems. The primary measurements are speed (v), flow (q), and density (k). Other measurements—spacing (s), headway (h), lane occupancy (LO), clearance (c), gap (g)—were also described. Two categories of traffic flow, uninterrupted and interrupted flow, were examined. The uninterrupted case of highway traffic was described in great detail, followed by a general model of vehicle stream flow under various safety policy constraints. Finally, a comparison of centrally controlled modes and individually controlled modes was made, showing how these two principal mode categories differ.

REFERENCES

DREW, D. R. (1968). *Traffic Flow Theory and Control*, McGraw-Hill, New York.

GERLOUGH, D. L., and M. J. HUBER (1975). *Traffic Flow Theory: A Monograph*, Special Report 165, Transportation Research Board–National Research Council, Washington, DC.

GREENBERG, H. (1959). An Analysis of Traffic Flow, *Operations Research*, Vol. 7, July, pp. 79–85.

GREENSHIELDS, B. D. (1935). *A Study of Traffic Capacity*, Highway Research Board–National Research Council, Washington, DC, pp. 448–477.

HOMBURGER, W. S. (Ed.) (1982). *Transportation and Traffic Engineering Handbook*, 2nd Ed., Prentice-Hall, Englewood Cliffs, NJ.

LAY, M. G. (1986a). *Handbook of Road Technology*, Vol. 2, *Traffic and Transport*, Gordon and Breach, New York.

LAY, M. G. (1986b). *Update for Source Book for Australian Roads*, Australian Road Research Board, Canberra, Chapter 17.

MITRIC, S. (1975). *Transportation Engineering Notes* (Mimeograph), Ohio State University, Columbus, Ohio.

TRANSPORTATION RESEARCH BOARD (TRB) (1985). *Highway Capacity Manual*, Special Report 209, National Research Council, Washington, DC.

VUCHIC, V. (1981). *Urban Public Transportation*, Prentice-Hall, Englewood Cliffs, NJ.

EXERCISES

1. A student recorded the following traffic counts of vehicles traveling on the curb lane of a freeway. Compute (a) the hourly flow, (b) the peak rate of flow for a 5-minute period, and (c) the peak rate of flow for a 15-minute period for this lane.

8:35–8:40 A.M.	104	9:05–9:10 A.M.	101
8:40–8:45 A.M.	109	9:10–9:15 A.M.	105
8:45–8:50 A.M.	116	9:15–9:20 A.M.	130
8:50–8:55 A.M.	122	9:20–9:25 A.M.	103
8:55–9:00 A.M.	130	9:25–9:30 A.M.	107
9:00–9:05 A.M.	121	9:30–9:35 A.M.	105

2. A helicopter pilot recorded the travel time of five vehicles on a 2-mile segment of a highway. Estimate the time-mean speed and space-mean speed of the vehicles.

VEHICLE	TRAVEL TIME (SEC)
1	161
2	173
3	145
4	159
5	182

3. A vehicle platoon was observed over a distance of 1000 ft on a single lane of Oak Street, entering at point *A* and departing at point *B*.

Vehicle	Time at *A* (sec)	Time at *B* (sec)
1	0	35
2	2	37
3	3	39
4	5	42
5	6	44
6	8	48

(a) Plot the trajectories of these six vehicles on graph paper and (b) calculate the average volume, density, and speed for these six vehicles observed, assuming the first 10 seconds at *A* as the period of observation.

4. Two aerial photographs were taken, 30 seconds apart, over one east-bound lane of I-90 near Spokane, Washington. The following results were recorded.

Vehicle	Positions	
	Photo 1	Photo 2
1	2000	2940
2	2100	3000
3	1900	2600
4	1500	2400
5	500	1630
6	0	1000

Plot these trajectories on graph paper and compute average flow, density, and speed over the 3000-ft length of the lane.

5. Four vehicles are traversing a 2-mile segment of a freeway and the following speed observations were made: vehicle 1, 56 mph; vehicle 2, 48 mph: vehicle 3, 45 mph; vehicle 4,

62 mph. What is the average time speed of the four vehicles? If the variance of the speeds is 48, what is the space mean speed of the vehicles?

6. The following speeds of cars and trucks were recorded on the median lane of I-90 in Spokane between 8 and 9 A.M. The length of the section was 1 mile. The speeds of the vehicles were 55, 54, 51, 55, 53, 53, 54, and 52 mph.
   (a) Compute the time mean speed and the variance about it.
   (b) Compute the space mean speed and the variance about it.

7. A car is driven from city A to city B and back. The outward journey is uphill and the car consumes gasoline at 20 miles/gallon. On the return journey mostly downhill, the consumption is 30 miles/gallon.
   (a) What is the harmonic mean of the car's gasoline consumption in miles/gallon?
   (b) If cities A and B are 50 miles apart, verify that the harmonic mean is the correct average, and not the arithmetic average. (Note that the harmonic mean is to space-mean speed as arithmetic mean is to time-mean speed.)

8. A loop detector having a length of 12 ft was observed to have six vehicles cross over it in a period of 148 sec, for the following durations: 0.44, 0.48, 0.50, 0.41, 0.49, and 0.55. Estimate the values of q, k, and v. The corresponding lengths of vehicles were 18, 21, 20, 23, 17, and 19 ft.

9. By assuming a linear speed–density relationship, the mean free speed on a highway facility lane equals 55 mph near zero density, and the jam density is observed to be about 170 veh/mi.
   (a) Plot the q–k–v curves in proper order.
   (b) Write down the speed–density and flow–density equations.
   (c) Comment on the maximum flow.
   (d) Compute speeds and densities corresponding to a flow of 900 veh/hr, describing traffic conditions from a driver's point of view.
   (e) Calculate average headways, gaps, clearance, and spacing at maximum flow and at jam density. Discuss all the results you obtain. Are they realistic?

10. (a) An inexperienced observer noted that a traffic stream on a freeway lane displayed average vehicle headways of 2.5 sec at a mean speed of 55 mph. Estimate the density, rate of flow, spacing, and gap of this traffic stream assuming the average length of vehicles is 20 ft.
    (b) During the same period, a traffic engineering student was the observer and was quite precise with his observation: he recorded that 55% of the vehicles were passenger cars 19 ft long, 40% were semitrailers 55 ft long, and 5% were articulated buses 60 ft long. How does this additional information affect your answers?

11. An engineer has the following data for a highway segment.

v (mph)	60	52	41	34	22
k (veh/mi)	11	43	62	80	103

   (a) Applying Greenshields' assumptions, estimate the mean free speed and jam density.
   (b) What are the maximum flow and the corresponding density?
   (c) What are the average headway and spacing when $k = 50$ veh/mi?
   (d) Compare your results obtained in (a), (b), and (c) with corresponding results applying Greenberg's assumptions.

12. Assuming that $v_s = C \ln(k_j/k)$, and given that $k_j = 100$ veh/mi and also that $k = 35$ veh/mi when $v = 40$ mph, find the maximum flow. Plot the q–k–v curves and indicate critical values.

13. An urban highway has the following flow characteristics:

$$q = 263v - 65v \ln v$$

Calculate q_{max}, v_{max}, and k_{max}. Plot q–k–v curves.

14. The AAA rule suggests that safe driving practice requires a clearance of one car length for every 10 mph of speed. Assuming that (1) the average car length is 18 ft, (2) the q–k–v curves follow Greenshields, (3) the mean free speed is 50 mph, and (4) the jam density is 150 veh/mi, test the hypothesis that densities in a continuous car stream can be explained by the AAA rule.

15. A travel-time study on a busy inner city route was conducted using the moving-vehicle method. Estimate the travel time and volume on this section of the highway. The data for this study are as follows:

| | | | Number of vehicles | | |
Run	Direction	Travel time (min)	Traveling in opposite direction	Overtaking test vehicle	Overtaken by test vehicle
1	NB	15.80	320	7	2
2	NB	16.20	332	10	3
3	NB	15.40	340	11	3
4	SB	15.93	305	12	4
5	SB	16.35	328	15	5
6	SB	16.33	345	5	0

Runs 1, 2, and 3 are northbound, and Runs 4, 5, and 6 are southbound.

16. A platoon of cars and trucks traveling on a two-lane rural highway have the following characteristics: $q = 1250$ veh/hr and $k = 100$ veh/mi. The lead truck of this platoon loses power and comes to a standstill at 10:44 A.M., but fortunately the driver restarts his engine and continues to move again at 10:47 A.M., releasing the stationary platoon at $q = 1800$ veh/hr with a speed of 20 mph. If $k_j = 235$ veh/mi, calculate the shock wave at the front and rear of the platoon, and sketch these values on a distance–time diagram.

17. A heavily traveled two-lane highway has a 2-mile section with steep grade. The highway has a speed limit of 55 mph. There have been complaints by users of this highway that trucks traveling upgrade cause undue delay to passenger cars during the morning peak hours. For instance, an observer on a Monday morning noted the following: space-mean-speed of vehicles was 48 mph with a density of 30 veh/mile. At 7:23 A.M., a truck leading the platoon traveling upgrade slowed down to 20 mph for a distance of 2 miles (7:29 A.M.), when it got off the highway. Vehicles behind this slow truck, having no chance to pass, formed a longer and longer platoon during this period, with a density of 90 veh/mi. Determine the length of the platoon. If delays of this nature occur daily, comment on what action should be taken to mitigate this problem.

18. Traffic on a two-lane highway is flowing at 1650 vph at a speed of 35 mph and a density of 50 veh/mi, when an overloaded truck enters this highway and travels for 2 miles before

exiting. A platoon forms behind the truck at a flow rate of 1200 vph and density of 180 veh/mi. The rear of the platoon is joined by other vehicles. Calculate the shock-wave velocities, and sketch the shock-wave measurements on a flow–density curve. Also, find the time the truck spends on the highway, the time needed to dissipate the platoon, and the time needed for the traffic flow to return to normal. Assume $k_j = 210$ veh/mi, and the mean free speed $v_f = 50$ mph. Assume that after the truck leaves the highway, the traffic stream picks up speed to 25 mph with a density of 105 veh/mile.

19. A four-lane arterial (with two lanes in each direction) has a capacity of 1000 vehicles/hour/lane. Due to some road repair work, the maximum capacity of two lanes in the northbound direction is reduced to 1100 veh/hr from 2000 veh/hr for a short length of 200 ft. If the total traffic upstream of the repair site in the northbound lanes is reasonably steady at 1500 veh/hr, find (a) the mean speed of traffic through the repair site, and (b) the rate at which the queue approaching the repair site grows. Assume the jam density is 200 vehicles/mile/lane.

20. A one-way street consisting of two lanes has a mean free speed of 40 mph and a jammed density of 130 vph/lane. The combined flow on this two-lane street is 1800 vph. If one of the lanes is temporarily closed for repairs for a short section, determine the following:
    (a) What is the mean speed of traffic through the bottleneck caused by the lane closure?
    (b) What is the mean speed of traffic immediately approaching the bottleneck?
    (c) What is the mean speed of traffic upstream of the bottleneck (and not influenced by the bottleneck)?
    (d) What is the rate at which a queue grows entering the bottleneck?

21. A traffic stream with a space-mean speed of 30 mph on a single lane of a two-lane street is brought to a halt when approaching a signalized intersection for a duration of 30 sec. The capacity of this lane is 1800 vph and the jam density is 180 vpm.
    (a) Determine the velocity of the stopping wave if the red phase lasts for 30 sec.
    (b) What distance upstream will vehicles be affected by the signal changing red?

22. Single light-rail vehicles 20 ft long run on an exclusive right-of-way, having the following characteristics: $L = 20$ ft/veh, $N = 1$ veh/train, $s_0 = 3$ ft, $t_r = 1$ sec, $b_n = 8$ ft/sec^2, $b_e = 24$ ft/sec^2. Calculate the optimal speed and the corresponding way capacity for regimes 1, 2, and 4.

23. In a centrally controlled mode, $L = 25$ ft, $s_0 = 5$ ft, $t_r = 1.5$ sec, $b_n = 10$ ft/sec^2, and $b_e = 20$ ft/sec^2. Draw to scale the following sketches, using all five regimes indicated in the text: (1) distance versus speed, (2) speed versus density, (3) flow versus speed, and (4) flow versus density.

24. A urban freeway runs parallel to a rail line. The rail line carries a line haul system of cars having the following characteristics: $L = 18$ ft/veh, $s_0 = 3$ ft, $t_r = 2$ sec, $b_n = 10$ ft/sec^2, and $b_e = 20$ ft/sec^2. The urban freeway has individual vehicles running on it having the following characteristics: the maximum flow of autos per lane is the same as the maximum flow of cars on the rail line, the length of the average automobile is 18 ft, the jam density = 170 veh/mi, and the flow is assumed to follow Greenshields' rules. Making suitable assumptions, plot the q–k–v curves for flows on the rail and the freeway lane. Discuss your results. If the railcars carry an average of 15 persons per car, while the average car occupancy for vehicles running on the freeway lane is 1.25, compare the person flows on the two systems. (Assume rail-line operates as per regime 2, and that the freeway has a total of four lanes).

25. The following linear relationship was estimated by an observer from data collected for a local freeway lane:

$$v = 30 - \frac{q}{120}$$

where v is the space mean speed (mph), and q is the flow (veh/hr). The average travel time (min) at flow q is given by $t(q) = 60d/v$, where d = length of the road (miles). The average travel time per hour for the flow q is $T(q) = q = 60d/v$, where d = length of the road (miles). The total travel time per hour for the flow q is $T(q) = qt(q)$. If the marginal travel time function $MT(q)$ is defined as the additional time imposed on the q vehicles by increasing the flow by one additional vehicle, then $MT(q) = dT(q)/dq$.

(a) Calculate the average, total, and marginal travel times per mile of trip length for a flow range of 200 to 2000 veh/hr.

(b) Plot the three time functions on the same graph.

Chapter 6

Geometric Design of Highways

1. INTRODUCTION

The geometric design of highways refers to the design of the visible dimensions of streets and highways. Its main purpose is to provide safe, efficient, and economical movement of traffic. The designer must also take into consideration the social and environmental impacts that are likely to occur because of the construction of new or reconstructed facilities. Among the many factors, driver behavior and traffic performance have the most influence on geometric design. The principles of geometric design have evolved from many years of experience and research. In the United States, the American Association of State Highway and Transportation Officials (AASHTO) has been responsible for developing the design standards. Of the several publications issued by AASHTO, *A Policy on Geometric Design of Highways and Streets* (AASHTO, 1994) is the principal reference.

The highway or street designer is concerned with at least four major areas of design: (1) locational design; (2) alignment design, which includes design controls and criteria; (3) cross-sectional design; and (4) access design. Each of these areas is described in this chapter.

2. LOCATIONAL DESIGN

One of the most crucial and important parts of the design process is the location of highways. The location procedure is an iterative process in which engineers, planners, economists, ecologists, and sociologists help to select several approximate locations based on information and data available, and then narrow their selection with the help of further information. Naturally, the location selection process has to be done with utmost care, taking into consideration the cost and operational efficiency of the proposed facility. The best route from the standpoint of user benefit and economy together with the socioeconomic and environmental impacts likely to be encountered are all taken into consideration.

The study begins with a survey to determine the traffic needs of the area and the controlling factors within the general route corridor area, such as terrain and existing and/or potential facilities. The following information will usually be of interest in locational planning and design:

- Land use, population distribution, and population density
- Geological structure of the region
- Potential for future industrial, farm, residential, or recreational development
- Existing roads, streets, and highways serving the area
- Existing utilities and facilities in the area
- Photographs of controlling features
- Photogrammetric maps of the area

A general outline of the steps necessary to choose an alignment follows:

1. Topographic maps are prepared (usually aerial and/or ground survey) for selecting a preliminary location.
2. A base map (scale 1 in. = 200 ft; contour intervals 2 to 5 ft) is used to study possible alternatives.
3. Future topographical details are added to select a final design and location, and this alignment is marked on a set of maps (1 in. = 100 ft; contour intervals to 2 ft).
4. By using a coordinate system for locating key features, horizontal and vertical controls are established to fix the final alignment.
5. In choosing the final alignment, several criteria are used. Some of the major criteria are the cost of the project, the cost to the users and nonusers, the environmental and social impacts created by the project, the short- and long-term impacts on various interest groups, and historical and archaeological impacts.

The availability of photogrammetric and computer techniques has revolutionized the process of alignment determination. Standard textbooks on surveying and highway engineering give details regarding the field work, data requirements, and plan preparation (JHK & Associates, 1980; Oglesby and Hicks, 1982; Wright, 1995).

3. DESIGN CONTROLS AND CRITERIA

Vehicle and pedestrian characteristics, traffic volume and its composition, assumed design speed, and the weight of the vehicles dictate the criteria for the optimization or improvement of various highway facilities.

3.1 Design Vehicles

AASHTO (1994) states the following:

> The physical characteristics of vehicles and the proportions of variously sized vehicles using the highways are positive controls in geometric design. Therefore, it is necessary to examine all vehicle types, select general class groupings, and establish representatively sized vehicles within each class for design use. Design vehicles are selected motor vehicles with the weight, dimensions, and operating characteristics used to establish highway design controls for accommodating vehicles of designated classes. For purposes of geometric design, each design vehicle has larger physical dimensions and larger minimum turning radius than those of almost all vehicles in its class. The largest of all the several design vehicles are usually accommodated in the design of freeways.

Passenger cars, buses, and trucks are three general classes of vehicles considered in geometric design. Passenger cars include all light vehicles and light delivery trucks, such as vans and pickups. The dimensions of twelve design vehicles are given in Table 6-1. The largest design vehicle to use a particular facility should be taken into consideration for design purposes. The minimum radii of the outside wheel paths are given in Table 6-2. Examples of the turning paths of passenger cars, trucks, and buses are shown in Figure 6-1.

Designers must take into consideration factors such as ground clearance, overhang, and vertical curvature of the highway when designing transportation facilities.

3.2 Vehicle Performance and Impacts

Highway design is greatly affected by the accelerating and decelerating characteristics of vehicles. Figure 6-2 gives the acceleration and deceleration distances for passenger cars on level grade.

Air pollution created by vehicle emission affects land uses adjacent to highways. Noise from vehicles also affects land use, particularly residential areas. Both these factors should be considered when considering alternatives. Vehicle fleet mix, vehicle speed, ambient air temperature, and vehicle fleet age are some of the factors that determine the severity of air pollution and noise levels.

Example 1

A passenger car is traveling at a speed of 40 mph on a dry level road.

   (a) What will be the speed if it accelerates for a distance of 600 ft?
   (b) From the final speed of part (a), what distance will it need to travel to decelerate comfortably to 30 mph?

   Solution

   (a) From Figure 6-2(a), reading along the curve for 40 mph toward the positive x direction, we can see where the curve and 600 ft meet; the y coordinate is 50 mph. Therefore, the new speed is 50 mph.

TABLE 6-1 DESIGN VEHICLE DIMENSIONS[a]

| Design vehicle type | Symbol | Overall | | | Overhang | | WB_1 | WB_2 | S | T | WB_3 | WB_4 |
		Height	Width	Length	Front	Rear						
Passenger car	P	4.25	7	19	3	5	11					
Single-unit truck	SU	13.5	8.5	30	4	6	20					
Single-unit bus	BUS	13.5	8.5	40	7	8	25					
Articulated bus	A-BUS	10.5	8.5	60	8.5	9.5	18		4[b]	20[b]		
Combination trucks												
Intermediate semitrailer	WB-40	13.5	8.5	50	4	6	13	27				
Large semitrailer	WB-50	13.5	8.5	55	3	2	20	30				
"Double-bottom" semitrailer—full trailer	WB-60	13.5	8.5	65	2	3	9.7	20	4[c]	5.4[c]	20.9	
Triple semitrailer	WB-96	13.5	8.5	102	2.5	3.3	13.5	20.7	3.3[d]	6[d]	21.7	21.7
Recreation vehicles												
Motor home	MH		8	30	4	6	20					
Car and camper trailer	P/T		8	49	3	10	11	18	5			
Car and boat trailer	P/B		8	42	3	8	11	15	5			
Motor home and boat trailer	MH/B		8	53	4	8	20	21	6			

[a] WB_1, WB_2, WB_3, WB_4, effective vehicle wheelbases; S, distance from the rear effective axle to the hitch point; T, distance from the hitch point to the lead effective axle of the following unit.
[b] Combine dimension 24; split is estimated.
[c] Combined dimension 9.4; split is estimated.
[d] Combined dimension 9.3; split is estimated.
Source: AASHTO, 1990.

TABLE 6-2 MINIMUM TURNING RADII OF DESIGN VEHICLES

Design vehicle type	Passenger car	Single-unit truck	Single-unit bus	Articulated bus	Semitrailer intermediate	Semitrailer combination large	Semitrailer-full trailer combination	Triple semi-trailer	Motor home	Passenger car with travel trailer	Passenger car with boat and trailer	Motorhome and boat trailer
Symbol	P	SU	BUS	A-BUS	WB-40	WB-50	WB-60	WB-96	MH	P/T	P/B	MH/B
Minimum design turning radius (ft)	24	42	42	38	40	45	45	50	40	24	24	50
Minimum inside radius (ft)	13.8	27.8	24.4	14.0	18.9	19.2	22.2	20.7	26.0	2.0	6.5	35

Source: AASHTO, 1990.

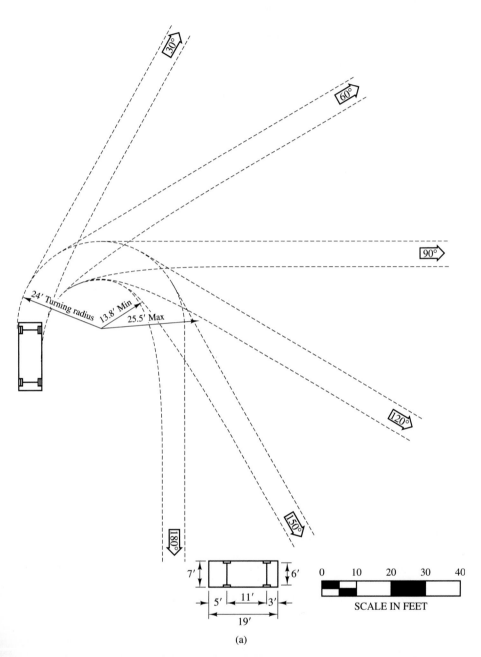

Figure 6-1   (a) Minimum Turning Path for P Design Vehicle (AASHTO, 1990).

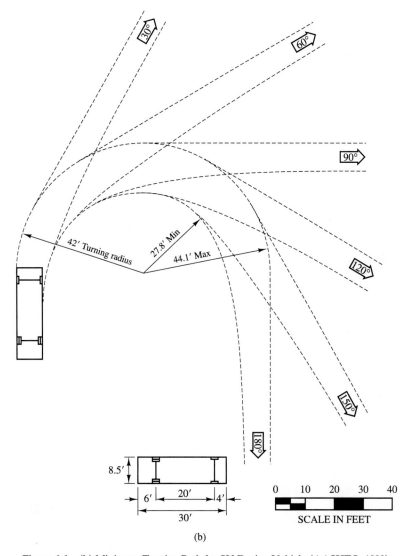

Figure 6-1   (b) Minimum Turning Path for SU Design Vehicle (AASHTO, 1990).

(b) From Figure 6-2(b), reading the initial speed, 50 mph, from the *y* axis, and project-
ing a horizontal line from that point until it hits the dashed line *C*, we can see that
the *x* coordinate is 240 ft. Therefore, the distance needed is 240 ft.

3.3 Driver Performance

The driving task is complex and demanding. This is particularly so when speeds are
high, time pressures bear on the driver, locations are unfamiliar, and when environmental conditions are adverse. Driver performance is one of the essential components
to be considered when designing highways.

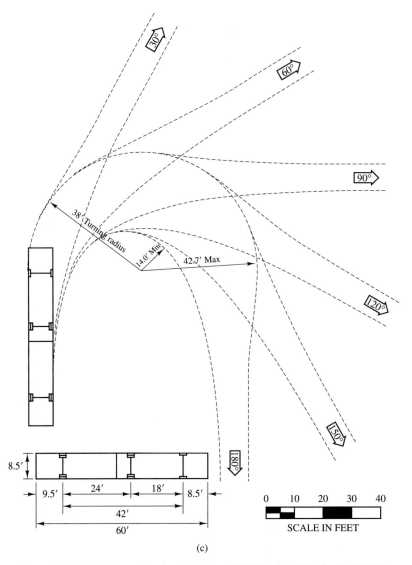

Figure 6-1 (c) Minimum Turning Path for A-BUS Design Vehicle (AASHTO, 1990).

The driving task depends on how information is received and used. The key to safe, efficient driving is error-free information handling. Driving performance activities fall into three levels: control, guidance, and navigation. Steering and speed control, road following, and trip planning and route following are examples of control, guidance, and navigation, respectively.

Information Handling. Whether they are aware of it or not, drivers perform several functions almost simultaneously. Several bits and sources of information compete for their attention. Through an attention-sharing process, drivers integrate the various

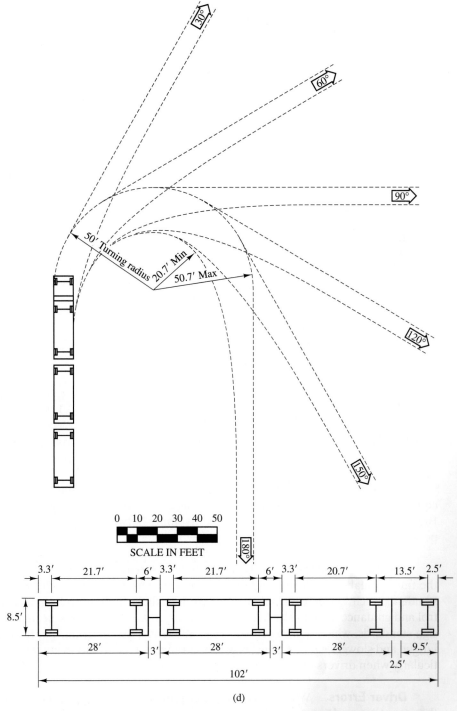

Figure 6-1 (d) Minimum Turning Path for WB-96 Design Vehicle (Triple Trailer) (AASHTO, 1990).

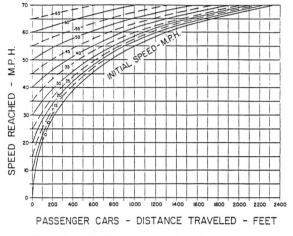

PASSENGER CARS - DISTANCE TRAVELED - FEET

Source: Michigan Report NCHRP 270 as developed by NY D.O.T. (33)

(a)

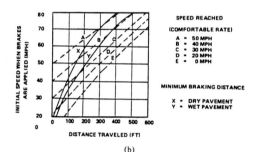

(b)

Figure 6-2 (a) Acceleration and (b) Deceleration Distances for Passenger Cars (AASHTO, 1990).

information inputs and maintain an awareness of the changing environment. Some decisions are made immediately; others are delayed, depending on the urgency demanded.

Reaction Time. Drivers' reaction times increase as a function of increased complexity. The chance of driver error increases as a function of longer reaction time. Brake reaction time for expected incidents averages between 2/3 and 2 seconds. Unexpected incidents clock 35% higher.

Speed. "Speed reduces the visual field, restricts peripheral vision, and limits the time available to receive and process information. Highways built to high design standards, such as freeways, help compensate for these limitations by simplifying control and guidance activities" (AASHTO, 1990, p. 51). At the same time, these high design standards, which aim to provide safe, efficient transportation, can lead to driver fatigue and slower reaction time, as well as a reduction in attention and vigilance, particularly when drivers overextend the customary length and duration of a trip.

Driver Errors. When drivers have to perform several highly complex tasks at the same time under extreme time pressure, they are prone to commit errors. Urban locations with closely spaced decision points, intensive land use, and heavy traffic are particularly sensitive areas where driver capabilities may be overloaded. In rural areas, drivers may face the opposite reaction of being mentally underloaded, which may lead to careless driving.

3.4 Traffic Characteristics

Traffic Volume.   Geometric features of design, such as width, alignments, and grades, are all directly affected by the traffic composition and volume. The general unit of measure for traffic is the average daily traffic (ADT), defined as the total volume during a given time period (in whole days), greater than 1 day and less than 1 year, divided by the number of days in that period. The ADT for a highway can be easily determined when continuous traffic counts are maintained. The ADT by itself is, however, not a practical measure of traffic because it does not adequately indicate the variation in the traffic in the year, the week, or different hours of the day. Naturally, we need an interval of time shorter than a day to more appropriately reflect the prevailing operating conditions. The peak-hour volume is the generally accepted criterion for use in geometric design. In most cases, a practical and adequate time period is 1 hour. It is the traffic volume expected to use the facility and is called the *design hourly volume* (DHV).

On rural highways, the 30th-highest hourly volume (30HV) of the year is generally considered as the DHV. Figure 6-3 shows the relationship between peak-hour and average daily traffic volume on rural arterials. The curves in the figure represent the hourly volumes of 1 year expressed as a percentage of ADT in a descending order of magnitude and the middle curve is the average for all locations studied and represents a highway with average fluctuation in traffic flow.

These curves lead to the conclusion that the hourly traffic used in design should be the 30th-highest hourly volume of the year (30HV). AASHTO policies provide a good explanation of both the 30HV and the DHV for urban highways. The DHV is generally expressed as a percentage (K) of the ADT. Observations indicate that the value of K ranges from 8 to 12% in urban areas and 12 to 18% in rural areas.

In urban areas especially, it is customary to know the peak flows within the peak hour. These peak-flow conditions vary according to the size of the city. For this reason, it is necessary to know not only the peak-hour flow, but also the peak 5-minute flows within the peak hour.

For two-lane highways, the DHV is the total traffic in both directions of travel. On multilane highways, knowledge of the hourly traffic volume in each direction of travel is essential for design. This becomes essential even on two-lane facilities with important intersections or where a widening of the roadway may be under consideration in future. The directional distribution (D) on highways typically defines the percentage in the heavier direction of travel. Traffic distribution by direction during peak hours is quite consistent from year to year and from day to day on a given highway. Typical values of D for rural and suburban highways range between 60 and 80%. In central business districts, it approaches 50%. Actual field measurements are recommended on multilane facilities for the determination of D.

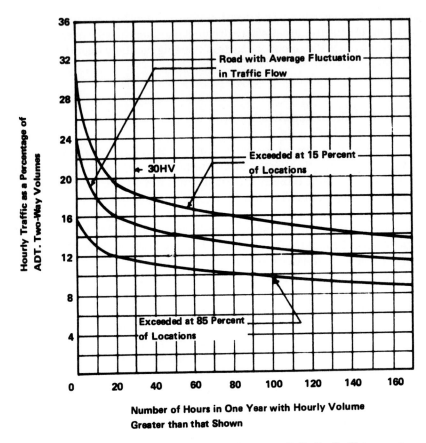

Figure 6-3 Relation between Peak-Hour and Average Daily Traffic Volumes on Rural Arterials (AASHTO, 1990).

The composition of traffic must be known for purposes of design. Passenger car, truck, and bus volumes are needed because each of these classes of vehicles have different operating characteristics, particularly in size and weight/horsepower ratio. If necessary, data on composition of traffic should be determined by traffic studies. Truck traffic is expressed as a percentage of total traffic during the design hour.

Operational Speed.   For the trip maker, speed is one of the most important factors in choosing a route or selecting a transportation mode. Economy, travel time, and convenience are directly related to speed. Highway speed is generally dependent on six conditions: the capability of drivers using the highway, the characteristics of the vehicle fleet using the highway, the physical characteristics of the highway and its road-sides, the weather, the presence of other vehicles (density), and the speed limitations (both legal or because of control devices). The aggregate effect of all of these conditions determines the speed on that stretch of highway.

The main objective in highway design is to satisfy the demands of the user in the safest and most economical manner. Provision should be made for a speed that satisfies nearly all drivers. AASHTO (1990) defines operating speed as "the highest overall speed at which a driver can travel on a given highway under favorable weather conditions and under prevailing traffic conditions without at anytime exceeding the safe speed as determined by the design speed on a section-by-section basis."

Design Speed.   "Design speed is the maximum safe speed that can be maintained over a specified section of a highway when conditions are so favorable that the design features of the highway govern" (AASHTO, 1990). Consistent with a desired degree of safety, mobility, and efficiency as well as constrained by environmental quality, economics, aesthetics, and sociopolitical impacts, every effort should be made to use as high a design speed as practicable.

A few pointers to keep in mind: The design speed should be consistent with the speed an average driver is likely to expect. Also, drivers do not adjust their speeds depending on the importance of the highway, but rather on the physical limitations and traffic thereon. A consistent design speed over a substantial length of roadway is an important factor. Any drastic speed changes should be avoided, and if such changes are necessary (due to topographic constraints), they should be gradual. Drivers should be warned well in advance of what to expect, by means of speed-zone signs and curve-speed signs. AASHTO (1990, pp. 63–68) provides a thorough discussion of design speeds. In summary, the maximum limit for low or lower design speed is 40 mph, and the minimum limit for high speed is 50 mph. The resultant intermediate design speed range of about 10 mph could have conditions approximating those of either high or low design speed, and such conditions would govern in selecting appropriate criteria.

Running Speed.   It is necessary to know actual vehicle speeds occurring on a highway section. This is called the running speed and is equal to the distance traveled by a vehicle divided by the time the vehicle is in motion. When vehicle flow is reasonably continuous, the spot speed at a section is the equivalent average running speed. The average spot speed is the arithmetic mean of all traffic speeds at a specified location. On longer sections, several spot speeds measured along the stretch of highway may be averaged to give the average running speed.

Free-Flow Speed.   Free-flow speed is the speed of traffic as density approaches zero. Practically, it is the speed at which drivers feel comfortable traveling under the physical, environmental, and traffic control conditions existing on an uncongested section of freeway or multilane highway. This definition is in a way similar to the operating speed when taken under low-volume conditions, and yet may imply speeds higher than the posted speed limit or even the design speed of the highway. Field determination of the free-flow speed is accomplished by performing travel-time studies during periods of low volume. Because free-flow speed is generally based on field measurements, it represents a practical value, and makes the operating speed, defined earlier, appear as a theoretical value. Figure 6-4 shows the speed-flow relationship for a typical uninterrupted-flow segment on a multilane highway under either ideal or nonideal conditions in which free-flow speed is known. Figure 6-4 indicates that the speed of traffic

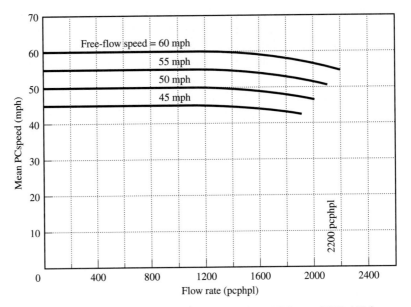

Figure 6-4   Speed–Flow Relationships on Multilane Highways (TRB, 1994).

on a multilane highway is insensitive to the traffic volume up to a flow rate of 1400 passenger cars per hour per lane (pcphpl).

3.5 Highway Capacity and Design Controls

Although Chapter 7 is devoted to highway capacity, the definitions of capacity and level of service are given here for convenience. The *Highway Capacity Manual* (TRB, 1994) defines capacity as "the maximum hourly rate at which persons or vehicles can be reasonably expected to traverse a point or uniform section of a lane or roadway during a given time period under prevailing roadway, traffic, and control conditions." Vehicle capacity represents the maximum number of vehicles, and person capacity represents the maximum number of persons. The latter is commonly used in evaluating public transit service, high-occupancy vehicle lanes, or pedestrian facilities. The concept of levels of service of a highway uses "qualitative measures that characterize operational conditions within a traffic stream, and their perception by motorists and passengers," with these conditions generally described by "such measures as speed and travel time, freedom to maneuver, traffic interruptions, comfort and convenience." Detailed information on these and other related concepts is given in Chapter 7.

Highway-capacity information serves three purposes: (1) to assess the adequacy or sufficiency of existing highway networks and to estimate when traffic growth is likely to exceed capacity; (2) to assist in the selection of the highway type and the dimensional needs of the network; and (3) to prepare estimates of operational improvements that are likely to be expected in the future.

Design volume is the volume of traffic, usually 10 to 20 years into the future, estimated to use a highway facility. The future year is the design year. The design hourly volume (DHV), mentioned earlier, is derived from the design volume. Design service volume, on the other hand, is the maximum hourly flow rate of traffic that a highway is likely to serve without the degree of congestion falling below a certain level of service (AASHTO, 1990, p. 76).

Acceptable Degrees of Congestion.   Highways are specifically designed for a particular level of service, so that they will serve the motorist without exceeding a certain degree of congestion. AASHTO (1994) suggests some principles to be kept in mind while deciding on the acceptable degree of congestion.

1. Traffic demand should not exceed the capacity of the highway even during short intervals of time.
2. The design traffic volumes per lane should not exceed the rate at which traffic can dissipate from a standing queue.
3. Motorists should be given some latitude in the choice of speeds (speeds could be related to the length of the trip).
4. Operating conditions should provide a degree of freedom from driver tension that is consistent with trip length and duration.
5. There are practical limitations to having an ideal freeway.
6. The attitude of motorists toward adverse operating conditions is influenced by their awareness of the construction and right-of-way costs.

Once a level of service has been selected, it is essential that all elements of the roadway are consistently designed to this level. Also, each level of service is matched with an accompanying service volume, and this is the design service volume. Although designers may possibly select the most appropriate level of service for highway facilities, Table 6-3 can serve as a useful guide.

Access Control.   Access control implies regulated limitation of public access to and from properties abutting the highway facility. These regulations are characterized

TABLE 6-3   GUIDE FOR SELECTION OF DESIGN LEVELS OF SERVICE

Highway type	Type of area and appropriate level of service[a]			
	Rural level	Rural rolling	Rural mountainous	Urban and suburban
Freeway	B	B	C	C
Arterial	B	B	C	C
Collector	C	C	D	D
Local	D	D	D	D

[a] General operating conditions for levels of service: B, reasonably free flow, but speeds beginning to be restricted by traffic conditions; C, in stable flow zone, but most drivers restricted in freedom to select their own speed; D, approaching unstable flow, drivers have little freedom to maneuver; E, unstable flow, may be short stoppages.
Source: AASHTO, 1994.

as full control of access, partial control of access, and driveway and approach regulations. Freeways require full control of access, whereas other classes of highways may have varying degrees of access control. For highways without full control of access, additional safety can be achieved by limiting the number of driveways and intersections. Other techniques aim to (1) limit the number of conflict points, (2) separate basic conflict areas, (3) reduce maximum deceleration requirements, and (4) remove turning vehicles or queues from certain portions of through lanes. The type of street or highway to be built should be coordinated with the local land-use plan to ensure that the desired degree of access control can be maintained.

Design Designation. In summary, it is evident that there are several major controls governing the design of highways. These are

> ADT (current and future years)
> DHV (future year)
> D (directional distribution)
> T (percentage of trucks)
> v (design speed)
> LOS (design level of service)
> AC (access control)
> K (percentage of ADT)

4. ELEMENTS OF DESIGN

4.1 Sight Distance

For operating a motor vehicle safely and efficiently, it is of utmost importance that the driver have the capability of seeing clearly ahead. Therefore, sight distance of sufficient length must be provided so that drivers can operate and control their vehicles safely. Sight distance is the length of roadway ahead visible to the driver. Sight distance is discussed in four important cases:

1. The distances required by motor vehicles to stop
2. The distances needed at complex locations
3. The distances required for passing and overtaking vehicles, applicable on two-lane highways
4. The criteria for measuring these distances for use in design

Stopping Sight Distance. At every point on the roadway, the minimum sight distance provided should be sufficient to enable a vehicle traveling at the design speed to stop before reaching a stationary object in its path. Stopping sight distance is the aggregate of two distances: brake reaction distance and braking distance.

Brake reaction time is the interval between the instant that the driver recognizes the existence of an object or hazard ahead and the instant that the brakes are actually applied. Extensive studies have been conducted to ascertain brake reaction time. Minimum reaction times can be as little as 1.64 seconds: 0.64 for alerted drivers plus 1 second for the unexpected signal. Some drivers may take over 3.5 seconds to respond under similar circumstances. For approximately 90% of drivers, a reaction time of 2.5 seconds is considered adequate. This value is therefore used in Table 6-4.

The braking distance of a vehicle on a roadway may be determined by the formula

$$d = \frac{v^2}{2g(f \pm s)} \tag{1}$$

where

d = braking distance (ft)

s = slope of the road (in decimal form)

v = initial speed (ft/sec)

g = 32.2 ft/sec^2

f = coefficient of friction between tires and roadway

The value of f depends on the pressure; composition and tread of tires; type and condition of pavement surface; and the presence of moisture, mud, snow, or ice. Figure 6-5 shows curves derived from the results of many tests in both wet and dry pavement conditions. Because of lower coefficients of friction on wet pavements as compared with dry, the wet condition governs in determining stopping distances for use in design.

If we add driver reaction time to Eq. 1, we get

TABLE 6-4   STOPPING SIGHT DISTANCE (WET PAVEMENTS)

Design speed (mph)	Assumed speed for condition (mph)	Brake reaction Time (sec)	Brake reaction Distance (ft)	Coefficient of friction, f	Braking distance on level (ft)	Stopping sight distance Computed (ft)	Stopping sight distance Rounded for design (ft)
20	20–20	2.5	73.3–73.3	0.40	33.3–33.3	106.7–106.7	125–125
25	24–25	2.5	88.0–91.7	0.38	50.5–54.8	138.5–146.5	150–150
30	28–30	2.5	102.7–110.0	0.35	74.7–85.7	177.3–195.7	200–200
35	32–35	2.5	117.3–128.3	0.34	100.4–120.1	217.7–248.4	225–250
40	36–40	2.5	132.0–146.7	0.32	135.0–166.7	267.0–313.3	275–325
45	40–45	2.5	146.7–165.0	0.31	172.0–217.7	318.7–382.7	325–400
50	44–50	2.5	161.3–183.3	0.30	215.1–277.8	376.4–461.1	400–475
55	48–55	2.5	176.0–201.7	0.30	256.0–336.1	432.0–537.8	450–550
60	52–60	2.5	190.7–220.0	0.29	310.8–413.8	501.5–633.8	525–650
65	55–65	2.5	201.7–238.3	0.29	347.7–485.6	549.4–724.0	550–725
70	58–70	2.5	212.7–256.7	0.28	400.5–583.3	613.1–840.0	625–850

Source: AASHTO, 1990.

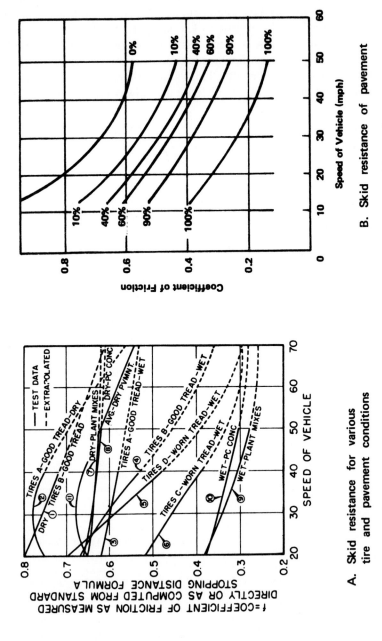

A. Skid resistance for various tire and pavement conditions

B. Skid resistance of pavement

Figure 6-5 Variation in Coefficient of Friction with Vehicular Speed (AASHTO, 1990).

$$d = \frac{v^2}{2g(f \pm s)} + t_r v \tag{2}$$

where t_r is the driver reaction time (sec). If speed is given in mph, Eq. 2 can be rewritten as

$$d = \frac{V^2}{30(f \pm s)} + 1.47 t_r V \tag{3}$$

Note that the units for d are in feet and V is in mph, assuming that 1 ft/sec = 0.68 mph (or 1.47 ft/sec = 1 mph).

Example 2

An alert driver (with a reaction time 0.5 second) is driving downhill on a 4% grade at 35 mph on a dry pavement when suddenly a person steps from behind a parked car in the path of the driver, at a distance of 100 ft.

(a) Can the driver stop in time (locked wheels)? $f = 0.7$.
(b) Can the driver stop in time on a rainy day? $f = 0.4$.

Solution

(a) $$d = \frac{(35)^2}{30(0.7 - 0.04)} + (1.47)(35)(0.5) = 87.6 \text{ ft}$$

Hence, the driver will be able to stop in time.

(b) $$d = \frac{(35)^2}{30(0.4 - 0.04)} + (1.47)(35)(0.5) = 139.1 \text{ ft}$$

In this case, the driver will not be able to stop in time.

Discussion

The driver's reaction time, the condition of the road pavement, and the prevailing weather all play a significant role in this problem.

Decision Sight Distance.   Although stopping sight distances are generally sufficient to allow competent and alert drivers to stop their vehicles under ordinary circumstances, these distances are insufficient when information is difficult to perceive. When a driver is required to detect an unexpected or otherwise difficult-to-perceive information source, a decision sight distance should be provided. Interchanges and intersections, changes in cross-section such as toll plazas and lane drops, and areas with "visual noise" are examples where drivers need decision sight distances. Table 6-5 provides values used by designers for appropriate sight distances. These values are applicable to most situations and have been developed from empirical data. The following avoidance maneuvers are covered in Table 6-5:

TABLE 6-5 DECISION SIGHT DISTANCE

Design speed (mph)	Decision sight distance for avoidance maneuver (ft)				
	A	*B*	*C*	*D*	*E*
30	220	500	450	500	625
40	345	725	600	725	825
50	500	975	750	900	1025
60	680	1300	1000	1150	1275
70	900	1525	1100	1300	1450

Source: AASHTO, 1990.

- Avoidance Maneuver A: Stop on rural road
- Avoidance Maneuver B: Stop on urban road
- Avoidance Maneuver C: Speed/path/direction change on rural road
- Avoidance Maneuver D: Speed/path/direction change on suburban road
- Avoidance Maneuver E: Speed/path/direction change on urban road

In computing and measuring decision sight distances, the 3.5-ft eye height and 6-in. object height criteria for stopping sight distance should be adopted.

Passing Sight Distance for Two-Lane Highways. On most two-lane, two-way highways, vehicles frequently overtake slower-moving vehicles by using the lane meant for the opposing traffic. To complete the passing maneuver safely, the driver should be able to see a sufficient distance ahead. Passing sight distance is determined on the basis that a driver wishes to pass a single vehicle, although multiple-vehicle passing is permissible.

Based on observed traffic behavior, the following assumptions are made:

1. The overtaken vehicle travels at a uniform speed.
2. The passing vehicle trails the overtaken vehicle as it enters a passing section.
3. The driver requires a short period of time to perceive whether a clear passing section is available and to start maneuvering.
4. The passing vehicle accelerates during the maneuver, during the occupancy of the left lane, at about 10 mph higher than the overtaken vehicle.
5. There is a suitable clearance length between the passing vehicle and the oncoming vehicle.

The minimum passing sight distance for two-lane highways is determined as the sum of the four distances shown in Figure 6-6.

d_1 = distance traveled during perception and reaction time and during the initial acceleration to the point of encroachment on the left lane

d_2 = distance traveled while the passing vehicle occupies the left lane

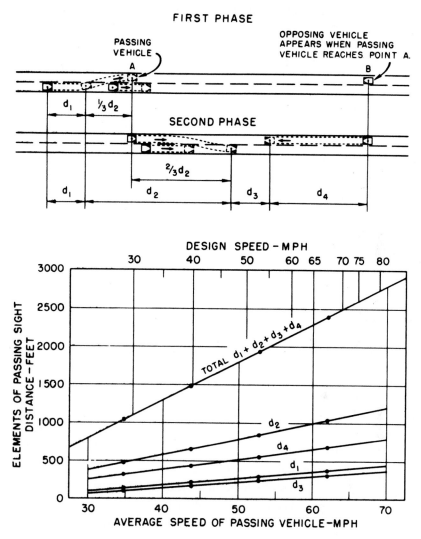

Figure 6-6   Elements of and Total Passing Sight Distance for Two-Lane Highways (AASHTO, 1990).

d_3 = distance between the passing vehicle at the end of its maneuver and the opposing vehicle

d_4 = distance traveled by an opposing vehicle for two-thirds of the time the passing vehicle occupies the left lane, or two-thirds of d_2

Safe passing sight distances for various speed ranges determined from distance and time values observed in the field are summarized in Table 6-6.

In general,

TABLE 6-6 ELEMENTS OF SAFE PASSING SIGHT DISTANCE FOR TWO-LANE HIGHWAYS

Speed group (mph):	30–40	40–50	50–60	60–70
Average passing speed (mph):	34.9	43.8	52.6	62.0
Initial maneuver:				
a = average acceleration[a] (mph/sec)	1.40	1.43	1.47	1.50
t_1 = time[a] (sec)	3.6	4.0	4.3	4.5
d_1 = distance traveled (ft)	145	215	290	370
Occupation of left lane:				
t_2 = time[a] (sec)	9.3	10.0	10.7	11.3
d_2 = distance traveled (ft)	475	640	825	1030
Clearance length:				
d_3 = distance traveled[a] (ft)	100	180	250	300
Opposing vehicle:				
d_4 = distance traveled (ft)	315	425	550	680
Total distance, $d_1 + d_2 + d_3 + d_4$ (ft)	1035	1460	1915	2380

[a] For consistent speed relation, observed values are adjusted slightly.
Source: AASHTO, 1990.

$$d_1 = 1.47t_1\left(V - m + \frac{at_1}{2}\right) \tag{4}$$

$$d_2 = 1.47Vt_2 \tag{5}$$

$$d_3 = \text{varies from 110 to 300 ft}$$

$$d_4 = \frac{2d_2}{3} \tag{6}$$

where

t_1 = time of initial maneuver (sec)
t_2 = time passing vehicle occupies the left lane (sec)
V = average speed of passing vehicle (mph)
m = difference in speed of passed vehicle and passing vehicle (mph)

Table 6-7 represents the likely passing speeds on two-lane highways.

Height of Driver's Eye and Height of Object. The distance along a roadway such that an object of specified height is continuously visible to the driver is the sight distance. This distance is dependent on the height of the driver's eye above the road surface, the specified object height above the road surface, and the height of sight obstructions within the line of sight.

For all sight distance calculations, the height of the driver's eye is considered to be 3.5 ft above the road surface, both for stopping and passing sight distances. The height of the object is considered to be 6 in. above the road surface for stopping sight distance calculations and 4.25 ft for passing sight distance calculations.

TABLE 6-7 MINIMUM PASSING SIGHT DISTANCE FOR DESIGN
OF TWO-LANE HIGHWAYS

Design speed (mph)	Assumed speeds (mph)		Minimum passing sight distance (ft)	
	Passed vehicle	Passing vehicle	Figure 6-6	Rounded
20	20	30	810	800
30	26	36	1090	1100
40	34	44	1480	1500
50	41	51	1840	1800
60	47	57	2140	2100
65	50	60	2310	2300
70	54	64	2490	2500

Source: AASHTO, 1990.

Obstructions that commonly occur on roadways between the driver's eye and the object are points on a crest vertical curve. On horizontal curves, the obstruction may be a crest vertical curve or else some physical feature outside the traveled way, such as a longitudinal barrier or the backslope of a cut section. All highway construction plans should be checked in both the vertical and horizontal planes for sight distance obstructions. Methods for scaling sight distances are demonstrated in Figure 6-7. A typical sight distance record is shown at the bottom of this figure.

4.2 Horizontal Alignment

Horizontal alignment consists of a series of straight sections of highway joined by suitable curves. It is necessary to establish the proper relation between design speed and curvature and also the joint relationships with superelevation and side friction.

Horizontal curves are described by either their radius or the degree of the curve. The degree of a curve is the central angle subtended by an arc of 100 ft measured along the center of the road.

Superelevation.   A vehicle is forced radially outward by centrifugal force when it moves in a circular path. The vehicle weight component creates side friction between the road surface and tires to counterbalance the centrifugal force. In addition, the superelevated section of a highway offsets the tendency of the vehicle to slide outward. See Figure 6-8, where

$$W = \text{weight of vehicle}$$

$$\beta = \text{angle of pavement cross-slope}$$

$$e = \text{rate of superelevation} = \tan \beta$$

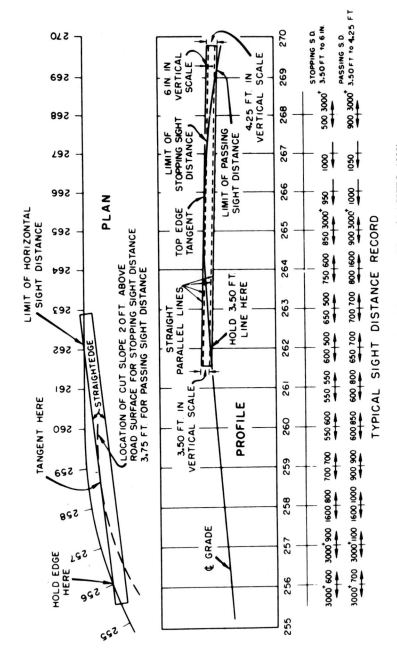

Figure 6-7 Scaling and Recording Sight Distances on Plans (AASHTO, 1990).

186

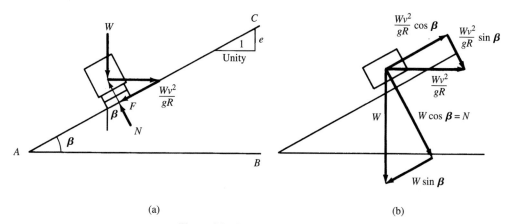

(a)   (b)

Figure 6-8   Superelevation Theory.

$$F = \text{side friction}$$

$$= \mu N = \mu W \cos \beta + \mu \frac{Wv^2}{gR} \sin \beta,$$

where μ is the side friction factor, but because $\sin \beta$ is small, the second term can be neglected

$$g = 32.2 \text{ ft/sec, or } 9.81 \text{ m/sec}^2$$

$$v = \text{longitudinal velocity (ft/sec)}$$

When all the forces on the vehicle are in equilibrium,

$$W \sin \beta + F = \frac{Wv^2}{gR} \cos \beta$$

$$\therefore W \sin \beta + \mu W \cos \beta = \frac{Wv^2}{gR} \cos \beta \tag{7}$$

Dividing by $W \cos \beta$, we get

$$\tan \beta + \mu = \frac{v^2}{gR} \qquad \text{or} \qquad e + \mu = \frac{v^2}{gR} \tag{8}$$

Changing v ft/sec to V mph, and considering that F can act to the right as well as to the left, we have

$$e \pm \mu = \frac{V^2}{15R} \tag{9}$$

where

$$R = \text{radius of curve (ft)}$$
$$V = \text{vehicle speed (mph)}$$

There is a practical limit to the rate of superelevation, depending on driver comfort and physical limitations such as the presence of ice and snow. If the rate of superelevation is in excess of that required for normal speeds, steering of vehicles will be somewhat difficult.

The coefficient of side friction is expressed as

$$\mu = \frac{V^2}{15R} - e \qquad (10)$$

Values of μ should be substantially less than the coefficient of friction of impending skid, because a margin of safety is necessary. Studies show that the maximum side friction between new tires and wet concrete pavements ranges from about 0.5 at 25 mph to approximately 0.35 at 60 mph. For normal, wet, concrete pavement and smooth tires, the value is about 0.35 at 45 mph. However, curve design cannot be based entirely on available side friction factor. An important criterion is to consider the point at which the driver feels a sense of discomfort from the effect of centrifugal force. For speeds up to 60 mph, a maximum value of 0.16 is recommended. The side friction factor should not exceed 0.10 for speeds of 70 mph or higher. At lower speeds, drivers are more tolerant of discomfort and higher values may be used in design.

Several factors dictate the maximum rates of superelevation: climate conditions, terrain, location (urban or rural), and frequency of very slow-moving vehicles. No single maximum superelevation rate is universally applicable. The common superelevation rate is 0.10, where snow and ice are not prevalent. Where snow and ice are factors, a superelevation rate of 0.08 is a logical maximum to minimize slipping across the highway. A rate of 0.120 may be used on low-volume, gravel-surfaced roads to facilitate cross-drainage. A low maximum rate, usually 0.04 to 0.06, is common practice where traffic congestion or extensive marginal development acts to curb top posted speeds. Table 6-8 gives the maximum degree of curves and the minimum radius for each of the four maximum superelevation rates for design speeds from 20 to 70 mph.

Example 3

Determine a proper superelevation rate for a low-volume, gravel-surfaced road with a design speed of 50 mph and a degree of curvature of 8 degrees.

Solution

Assumption: Side friction coefficient, $\mu = 0.14$ (at 50 mph speed).

Calculation: Because

$$\frac{2\pi R}{360} = \frac{100}{D}$$

then

$$R = \frac{36,000}{2\pi D} = 716.2 \text{ ft}$$

TABLE 6-8 MAXIMUM DEGREE OF CURVE AND MINIMUM RADIUS DETERMINED
FOR LIMITING VALUE OF *e* AND *f*: RURAL HIGHWAYS AND HIGH-SPEED URBAN STREETS

Design speed (mph)	Maximum[a] *e*	Maximum *f*	Total *(e + f)*	Maximum degree of curve	Rounded maximum degree of curve	Maximum radius (ft)
20	0.04	0.17	0.21	44.97	45.0	127
30	0.04	0.16	0.20	19.04	19.0	302
40	0.04	0.15	0.19	10.17	10.0	573
50	0.04	0.14	0.18	6.17	6.0	955
60	0.04	0.12	0.16	3.81	3.75	1528
20	0.06	0.17	0.23	49.25	49.25	116
30	0.06	0.16	0.22	20.94	21.0	273
40	0.06	0.15	0.21	11.24	11.25	509
50	0.06	0.14	0.20	6.85	6.75	849
60	0.06	0.12	0.18	4.28	4.25	1348
65	0.06	0.11	0.17	3.45	3.5	1637
70	0.06	0.10	0.16	2.80	2.75	2083
20	0.08	0.17	0.25	53.54	53.5	107
30	0.08	0.16	0.24	22.84	22.75	252
40	0.08	0.15	0.23	12.31	12.25	468
50	0.08	0.14	0.22	7.54	7.5	764
60	0.08	0.12	0.20	4.76	4.75	1206
65	0.08	0.11	0.19	3.85	3.75	1528
70	0.08	0.10	0.18	3.15	3.0	1910
20	0.10	0.17	0.27	57.82	58.0	99
30	0.10	0.16	0.26	24.75	24.75	231
40	0.10	0.15	0.25	13.38	13.25	432
50	0.10	0.14	0.24	8.22	8.25	694
60	0.10	0.12	0.22	5.23	5.25	1091
65	0.10	0.11	0.21	4.26	4.25	1348
70	0.10	0.10	0.20	3.50	3.5	1637

[a] In recognition of safety considerations, use of $e_{max} = 0.04$ should be limited to urban conditions.
Source: AASHTO, 1990.

Also,

$$e + \mu = \frac{V^2}{15R}$$

Therefore,

$$e = \frac{V^2}{15R} - \mu = \frac{50^2}{15(716.2)} - 0.14 = 0.09$$

Discussion

To facilitate cross-drainage, a commonly used superelevation rate is 0.12. Therefore, $e = 0.12$ is used.

Transition (Spiral) Curves. When vehicles enter or leave a circular horizontal curve, the gain or loss of centrifugal force cannot be effected instantaneously, considering safety and comfort. In such cases, the insertion of transition curves between tangents and circular curves warrants consideration. A properly designed transition curve provides the following advantages:

- A natural, easy-to-follow path for drivers such that the centrifugal force increases and decreases gradually as a vehicle enters and leaves a circular curve
- A convenient desirable arrangement for superelevation runoff
- Flexibility in the widening of sharp curves
- Enhancement in the appearance of the highway

A basic expression used for computing the minimum length of a spiral is

$$L = \frac{3.15V^3}{RC} \qquad (11)$$

where

$L =$ minimum length of spiral (ft)

$V =$ speed (mph)

$R =$ curve radius (ft)

$C =$ rate of increase of centrifugal acceleration (ft/sec^3)

Values for C that range between 1 and 3 are used in highway design. A practical control for the length of a spiral is when it equals the length required for superelevation runoff.

Change in cross-slope may be accomplished by rotating the pavement (1) about the center line and (2) about the inside or outside edge. Figure 6-9 illustrates these two cases. In the design of divided highways, streets, and parkways, the inclusion of a median in the cross-section alters somewhat the superelevation runoff treatment.

Sight Distance on Horizontal Curves. Where there are sight obstructions, such as walls, cut slopes, buildings, and trees, on the inside of horizontal curves, adequate sight distance must be provided. This provision may need adjustment in alignment of the road. Specific studies are needed in each case because of the many variables involved. It is recommended that the designers check the actual condition by using a design speed and selected sight distance as a control.

For design purposes, the line of sight is a chord of the curve, and the applicable stopping sight distance is measured along the center line of the inside lane around the curve. The design charts in Figure 6-10 show the required middle ordinates for clear sight areas to satisfy the lower and upper values, respectively, of stopping sight distance required for curves of various degrees.

The relationship between the radius of the curve, R (ft); the degree of curve, D; the stopping sight distance, S (ft); and the midordinate, M (ft), is as follows:

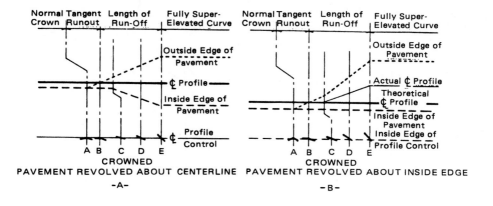

PAVEMENT REVOLVED ABOUT CENTERLINE
-A-

PAVEMENT REVOLVED ABOUT INSIDE EDGE
-B-

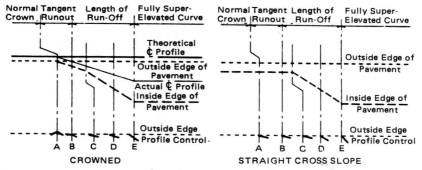

PAVEMENT REVOLVED ABOUT OUTSIDE EDGE
-C-

PAVEMENT REVOLVED ABOUT OUTSIDE EDGE
-D-

Note Angular breaks to be appropriately
      rounded as shown by dotted line.
      (See text)

Figure 6-9   Methods of Attaining Superelevation (AASHTO, 1990).

$$M = \frac{5730}{D}\left(1 - \cos\frac{SD}{200}\right) \tag{12}$$

Because

$$R = \frac{5730}{D} \quad \text{and} \quad \theta = \frac{SD}{200} \tag{13}$$

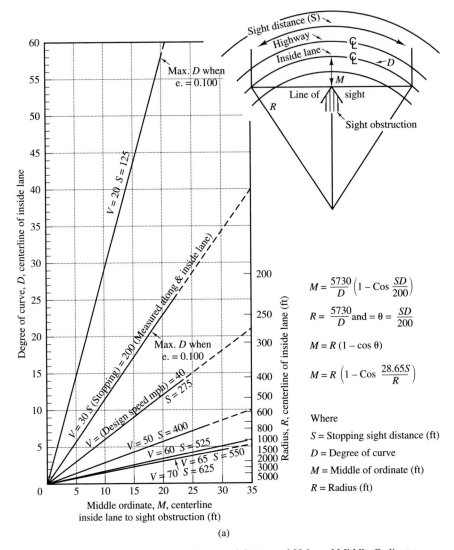

$$M = \frac{5730}{D}\left(1 - \text{Cos}\ \frac{SD}{200}\right)$$

$$R = \frac{5730}{D}\ \text{and} = \theta = \frac{SD}{200}$$

$$M = R\ (1 - \cos\theta)$$

$$M = R\left(1 - \text{Cos}\ \frac{28.65S}{R}\right)$$

Where

S = Stopping sight distance (ft)

D = Degree of curve

M = Middle of ordinate (ft)

R = Radius (ft)

Figure 6-10 Relation between Degree of Curve and Value of Middle Ordinate Necessary to Provide Stopping Sight Distance on Horizontal Curves under Open Road Conditions: (a) Range of Lower Values; (b) Range of Upper Values (AASHTO, 1990).

therefore,

$$M = R(1 - \cos\theta) \tag{14}$$

$$M = R\left(1 - \cos\frac{28.65S}{R}\right) \tag{15}$$

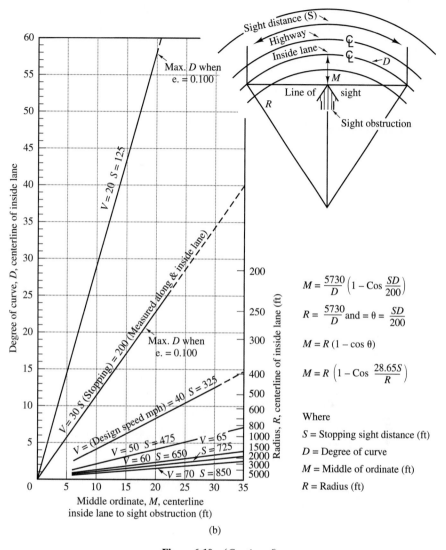

Figure 6-10   (*Continued*)

The relationship between the radius of the curve, R; the long chord, L; and the midordinate, M, is as follows (Figure 6-11):

$$R^2 = (R - M)^2 + \left(\frac{L}{2}\right)^2 \tag{16}$$

or

$$R = \frac{M}{2} + \frac{L^2}{8M} \tag{17}$$

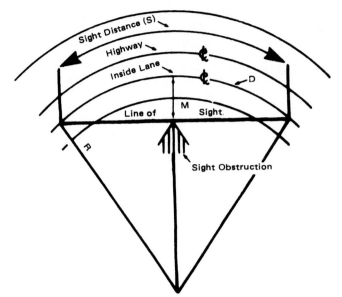

Figure 6-11 Line of Sight and Sight Obstruction.

Example 4

The corner of a building is situated next to a horizontal curve with a radius of 132 ft on a rural highway. The inside lane is 10 ft wide and the inside edge of the road is 6 ft from the corner of the building. Determine what speed limit should be imposed on this section of the highway.

Solution

Assumption: The shoulder width (the distance from the edge of the inside lane to the edge of the road) is 4 ft, the coefficient of friction is 0.4, and the reaction time of the driver, t_r, is 2.5 sec.

Calculation:

$$\text{Radius of the inside lane} = 132 - \frac{10}{2} = 127 \text{ ft}$$

$$\text{Midordinate, } M = 6 + 4 + 5 = 15 \text{ ft}$$

$$D = \frac{5729.58}{R} = \frac{5729.58}{127} = 45°$$

Also,

$$M = R\left(1 - \cos\frac{SD}{200}\right)$$

where S is the stopping sight distance. Therefore,

$$S = \frac{200}{D}\cos^{-1}\left(1 - \frac{M}{R}\right) = \frac{200}{45}\cos^{-1}\left(1 - \frac{15}{127}\right)$$

$$= 125 \text{ ft}$$

Also, the stopping distance on a level road is given by

$$S = \frac{V^2}{30f} + 1.47t_r V$$

giving

$$0 = V^2 + 30f(1.47t_r V - S) = V^2 + 30(0.4)[1.47(2.5)V - 125]$$

Therefore, $V = 22.5$ mph, and the speed limit should be 20 mph. This results also can be checked from Figure 6-10(b): Using $M = 15$ ft and $D = 45°$, we get $V = 20$ mph and $S = 125$ ft.

Discussion

Design graphs could be used for varying site conditions. The speed obtained from the graphs should be checked for the adequate stopping sight distance.

The minimum passing sight distance for a two-lane road or street is about four times as great as the minimum stopping sight distance at the same speed. Passing sight distance is measured between a height of eye of 3.50 ft and height of object of 4.25 ft. The sight line near the center of the area inside a curve is about 1.75 ft higher than for stopping sight distance. The resultant lateral dimension for normal highway cross-sections is cut between the center line of the inside lane and the midpoint of the sight line is from 3.75 to 10.5 ft greater than that for stopping sight distance. Naturally, for practical purposes, passing sight distance sections must be confined to tangents and very flat alignment conditions. Graphical methods are most useful for checking these conditions.

4.3 Vertical Alignment

Grades.   Road and street alignment is influenced by the topography of the traversed land. Because the objective is to encourage uniform vehicle operation, it is necessary to provide proper grades and vertical curves. Passenger cars can readily negotiate grades as high as 4 to 5% without appreciable loss in speed. Speeds decrease progressively with an increase in the ascending grade. Passenger car speeds are generally higher on downgrades than on level sections. Truck speeds, on the other hand, are influenced greatly by grades. On level sections, truck speeds are about the same as those for passenger cars.

On downgrades, truck speeds increase by about 5% and decrease about 7% or more on upgrades as compared with operation on level sections. Tables from AASHTO (1990) provide the maximum grade controls employed in terms of design speed for the following:

Local rural roads: Table 6-9
Recreational roads: Table 6-10
Rural and urban collectors: Table 6-11
Rural arterials: Table 6-12
Urban arterials: Table 6-13
Urban and rural freeways: Table 6-14

TABLE 6-9   MAXIMUM GRADES (%) FOR LOCAL RURAL ROADS

Type of terrain	Design speed (mph)				
	20	30	40	50	60
Level	—	7	7	6	5
Rolling	11	10	9	8	6
Mountainous	16	14	12	10	—

Source: AASHTO, 1990.

TABLE 6-10   MAXIMUM GRADES (%) FOR RECREATIONAL ROADS

Type of terrain	Design speed (mph)			
	10	20	30	40
Level	8	8	7	7
Rolling	12	11	10	9
Mountainous	18	16	14	12

Source: AASHTO, 1990.

TABLE 6-11   MAXIMUM GRADES (%) FOR RURAL AND URBAN COLLECTORS[a]

Type of terrain	Design speed (mph)					
	20	30	40	50	60	70
	Rural collectors					
Level	7	7	7	6	5	4
Rolling	10	9	8	7	6	5
Mountainous	12	10	10	9	8	6
	Urban collectors					
Level	9	9	9	7	6	5
Rolling	12	11	10	8	7	6
Mountainous	14	12	12	10	9	7

[a] Maximum grades shown for rural and urban conditions of short lengths (less than 500 ft) and on one-way downgrades and on low-volume roads may be 2% steeper.
Source: AASHTO, 1990.

TABLE 6-12   MAXIMUM GRADES (%) FOR RURAL ARTERIALS

	Design speed (mph)			
Type of terrain	40	50	60	70
Level	5	4	3	3
Rolling	6	5	4	4
Mountainous	8	7	6	5

Source: AASHTO, 1990.

TABLE 6-13   MAXIMUM GRADES (%) FOR URBAN ARTERIALS

	Design speed (mph)			
Type of terrain	30	40	50	60
Level	8	7	6	5
Rolling	9	8	7	6
Mountainous	11	10	9	8

Source: AASHTO, 1990.

TABLE 6-14   MAXIMUM GRADES (%) FOR URBAN AND RURAL FREEWAYS[a]

	Design speed (mph)		
Type of terrain	50	60	70
Level	4	3	3
Rolling	5	4	4
Mountainous	6	6	5

[a] Grades 1% steeper than the value shown may be used for extreme cases in urban areas where development precludes the use of flatter grades and for one-way downgrades except in mountainous terrain.
Source: AASHTO, 1990.

Maximum grades of about 5% are considered appropriate for a design speed of 70 mph. For a design speed of 30 mph, maximum grades generally are in the range of 7 to 12% depending on topography, with an average of about 8%. Control grades for 40-, 50-, and 60-mph design speeds are intermediate between the extremes noted before. It is recommended that the maximum design grade should be used infrequently rather than as a value to be used in most cases.

Maximum grade in itself is not a complete design control. The length of a particular grade should also be considered. The maximum length of a designated upgrade, called the critical length, is one on which a loaded truck can operate without an unreasonable reduction in speed. If efficient operation is to be maintained, design adjustments, such as changes in location to reduce grades or provision of additional lanes, should be made.

It has been demonstrated that regardless of the average speed on the highway, the greater a vehicle deviates from the average speed, the greater its chances of becoming involved in an accident. Figure 6-12 shows the results of this research.

The common basis for determining the critical length of grade is a reduction in speed of trucks below the average running speed. Of course, the ideal case would be when all traffic operates at this speed, but this is not practical. An examination of Figure 6-12 shows that the accident involvement rate increases significantly when the truck speed reduction exceeds 10 mph. Based on this observation, it is recommended that a 10-mph reduction criterion be used as the general design guide for determining critical lengths of grade.

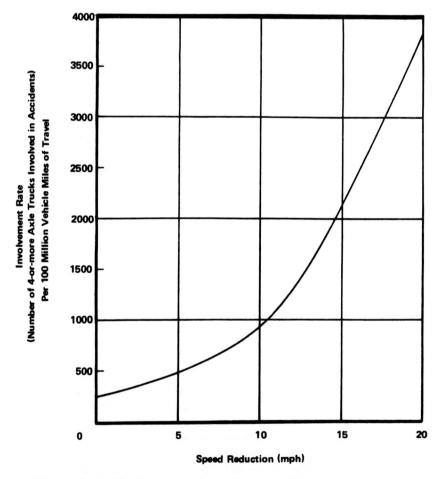

Figure 6-12 Accident Involvement Rate of Trucks for Which Running Speeds Are Reduced below Average Running Speed of All Traffic (AASHTO, 1990).

The length of any given grade that will cause the speed of a representative truck (300 lb/hp) entering a grade at 55 mph to be reduced below the average running speed of all traffic is shown in Figure 6-13. The curve showing a 10-mph speed reduction is the general design grade for determining critical grade lengths.

Example 5

What is the critical length of grade for a highway with a design speed of 60 mph with an upgrade of 4%? If the design speed were 40 mph, what would be the change in critical length?

Solution

If one follows the 10-mph, speed-reduction curve shown in Figure 6-13, it intersects the 4% upgrade line at 1000 ft. Provide a critical grade length of 1000 ft or less. If the design speed were 40 mph, the initial and minimum tolerable speeds on the grade would be different, but for the same permissible speed reduction, the critical length would be 1000 ft.

Discussion

This is so because when a truck approaches an upgrade, the driver often increases speed to make a climb on the upgrade at as high a speed as possible.

Climbing Lanes.   An extra lane, called a climbing lane on the upgrade side of a two-lane highway, is desirable where the length of the grade causes a reduction of 10 mph or more in the speed of loaded vehicles. Such provision is justified when the

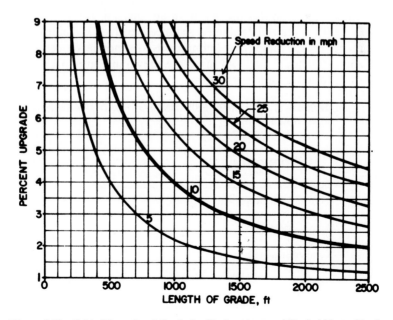

Figure 6-13   Critical Lengths of Grade for Design, Assumed Typical Heavy Truck of 300 lb/hp, Entering Speed = 55 mph (AASHTO, 1990).

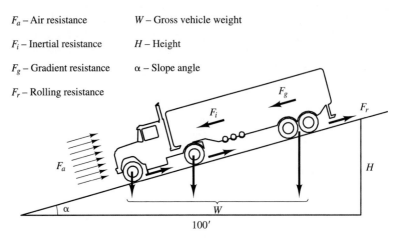

F_a – Air resistance

W – Gross vehicle weight

F_i – Inertial resistance

H – Height

F_g – Gradient resistance

α – Slope angle

F_r – Rolling resistance

Figure 6-14   Forces Acting on Vehicle in Motion (AASHTO, 1990).

traffic volumes are higher and the percentage of vehicles with high weight/horsepower ratios is high. The *Highway Capacity Manual* (TRB, 1994) and AASHTO (1990) should be consulted for determining the level of service desired on a particular two-lane highway with regard to grade and the need for a climbing lane.

Emergency Escape Ramps.   The provision of an emergency escape ramp on long, descending grades is appropriate for slowing and stopping out-of-control vehicles away from the main stream of traffic. Loss of braking ability through overheating or mechanical failure results in the driver losing control of the vehicle.

Figure 6-14 illustrates the action of air, inertial, gradient, and rolling resistance on a vehicle. The values for rolling resistance are given in Table 6-15. Four basic types of emergency escape ramps are commonly used: sandpile, descending grade, horizontal grade, and ascending grade, and these are shown in Figure 6-15. Each type is suitable for a particular topographic condition.

TABLE 6-15   ROLLING RESISTANCE OF ROADWAY SURFACING MATERIALS

Surfacing material	Rolling resistance (lb/1000 lb GVW)[a]	Equivalent grade[b] (%)
Portland cement concrete	10	1.0
Asphalt concrete	12	1.2
Gravel, compacted	15	1.5
Earth, sandy, loose	37	3.7
Crushed aggregate, loose	50	5.0
Gravel, loose	100	10.0
Sand	150	15.0
Pea gravel	250	25.0

[a] Gross Vehicle Weight.
[b] Rolling resistance expressed as equivalent gradient.
Source: AASHTO, 1990.

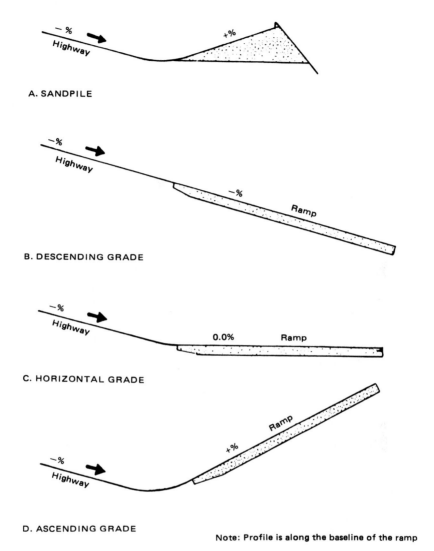

A. SANDPILE

B. DESCENDING GRADE

C. HORIZONTAL GRADE

D. ASCENDING GRADE

Note: Profile is along the baseline of the ramp

Figure 6-15   Basic Types of Emergency Escape Ramps (AASHTO, 1990).

Vertical Curves.   The gradual change between tangent grades is accomplished by means of any one of the sag or crest curves depicted in Figure 6-16. Vertical curves should result in a design that is safe, comfortable in operating, pleasing in appearance, and adequate for drainage. The provision of ample sight distances for the design speed adopted is the major control factor for safe operation on crest vertical curves. In all cases, the minimum stopping sight distance should be provided.

Parabolic curves are usually used in highway design. The vertical offset from the tangent varies as the square of the horizontal distance from the point of vertical curvature or tangency (PVC or PVT). The geometry of a vertical curve is shown in Figure 6-17. The general equation of the parabola is

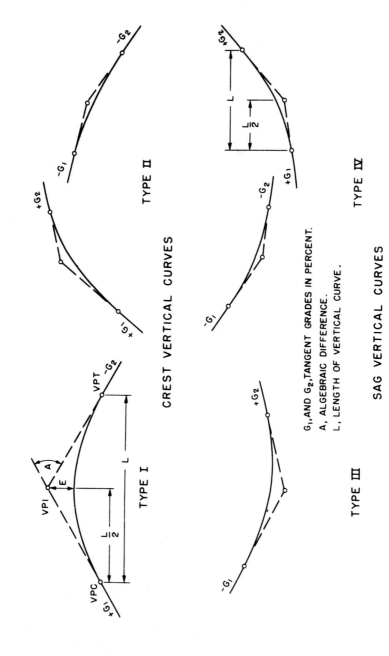

Figure 6-16 Types of Vertical Curves (AASHTO, 1990).

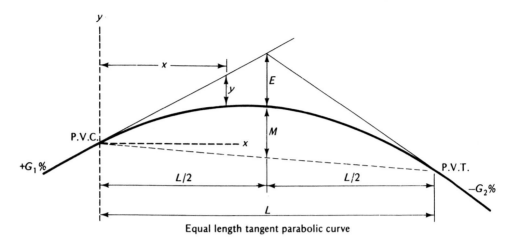

Figure 6-17   Geometry of the Vertical Curve (Carter and Homburger, 1978).

$$y = ax^2$$

The rate of change in slope of the tangent to parabola is

$$\frac{dy}{dx} = 2ax$$

The rate of change in grade per station is

$$\frac{d^2y}{dx^2} = 2a = \frac{100A}{L}$$

Where $A\%$ is the algebraic difference in grades, $(G_2 - G_1)$, and L, the length of the curve in feet or meters is horizontal projection of the curve on x-axis.

$$\text{Midcurve offset } E = \frac{AL}{800} = M \tag{18}$$

Other offsets vary as the square of the distances from PVC (or PVT). For example,

$$\frac{y}{x^2} = \frac{E}{(L/2)^2} \quad \text{or} \quad y = \frac{Ax^2}{200L} \tag{19}$$

Crest Vertical Curves.   The minimum lengths of crest vertical curves are dictated by sight distance requirements, that is, the ability of the driver to see an obstacle over the crest of the curve, within safe stopping distance. The basic equations for the length of a parabolic vertical curve (L) are given in terms of the algebraic difference in grades A and sight distance S.

Figure 6-18 illustrates two cases of a vertical curve, one in which $S > L$, and the other where $S < L$. The height of the driver's eye, H_1, and the height of the object, H_2,

A: Case 1 Sight distance greater than
length of vertical curve $(S > L)$

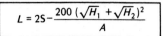

$$L = 2S - \frac{200 \left(\sqrt{H_1} + \sqrt{H_2}\right)^2}{A}$$

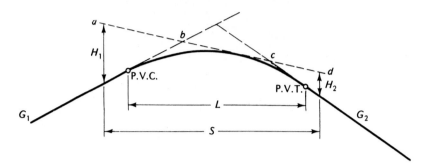

B: Case 2 Sight distance less than
length of vertical curve $(S < L)$

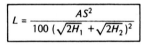

$$L = \frac{AS^2}{100 \left(\sqrt{2H_1} + \sqrt{2H_2}\right)^2}$$

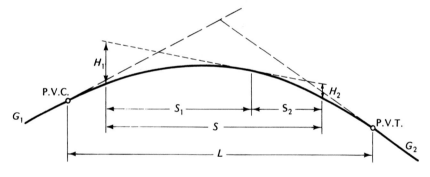

where: L = Length of vertical curve, feet or meters
S = Sight distance, feet or meters
A = Algebraic difference in grads, percent
H_1 = Height of eye above roadway surface, feet or meters
H_2 = Height of object above roadway surface, feet or meters

Figure 6-18   Sight Distance of Crest Vertical Curves (Carter and Homburger, 1978).

are vertical offsets to the tangent line of sight. The problem is to find the slope of the sight line that will make the distance ad, in Figure 6-18(a), a minimum. If g represents the difference between the gradient of the sight line and the gradient G_1, then $A - g$ will be the difference between the gradient of the sight line and gradient G_2. In case 1, $S > L$, use is also made of the property of the parabola that the horizontal projection of the intercept formed by a tangent is equal to one-half the projection of the long chord of the parabola. The sight distance may be expressed as the sum of the horizontal projections:

$$S = \frac{100H_1}{g} + \frac{L}{2} + \frac{100H_2}{A - g} \tag{20}$$

To find the slope of the sight line that will make S a minimum, set $dS/dG = 0$; hence,

$$\frac{dS}{dG} = -\frac{100H_1}{g^2} - \frac{100H_2}{(A - g)^2} = 0$$

$$g = \frac{A\sqrt{H_1 H_2} - H_1 A}{H_2 - H_1} \tag{21}$$

Substituting for g in equation 20 and solving for L gives the minimum length of vertical curve needed to provide the necessary sight distance, S:

$$L = 2S - \frac{200\ (\sqrt{H_1} + \sqrt{H_2})^2}{A} \tag{22}$$

In case 2, $S < L$, use is made of the basic offset property of the parabolic curve, giving

$$\frac{H_1}{AL/800} = \frac{S_1^2}{(L/2)^2} \quad \text{and} \quad \frac{H_2}{AL/800} = \frac{S_2^2}{(L/2)^2} \tag{23}$$

Solving for S_1 and S_2 and summing to get S, we get

$$L = \frac{AS^2}{100(\sqrt{2H_1} + \sqrt{2H_2})^2} \tag{24}$$

The criteria for height of eye and object are given in Table 6-16. If H_1 is assumed to be 3.50 ft and H_2 0.5 ft, then Eqs. 22 and 24 can be written as

TABLE 6-16   CRITERIA IN DESIGN OF CREST CURVES

	Stopping sight distance (ft)	Passing sight parameter distance (ft)
Height of eye, H_1	3.50	3.50
Height of object, H_2	0.50	4.25
Minimum length of curve		
$\quad S > L$	$2S - 1329/A$	$2S - 3093/A$
$\quad S < L$	$AS^2/1329$	$AS^2/3093$

Source: AASHTO, 1990.

$$L = 2S - \frac{1329}{A} \qquad \text{for } S > L \qquad\qquad (25)$$

$$L = \frac{AS^2}{1329} \qquad \text{for } S < L \qquad\qquad (26)$$

which apply to stopping sight distances.

Similarly, if the values of H_1 and H_2 are assumed to be 3.50 ft and 4.25 ft, respectively, Eqs. 22 and 24 can be written as

$$L = 2S - \frac{3093}{A} \qquad \text{for } S > L \qquad\qquad (27)$$

$$L = \frac{AS^2}{3093} \qquad \text{for } S < L \qquad\qquad (28)$$

which apply to passing sight distances.

Example 6

A $+3.9\%$ grade intersects a $+1.1\%$ grade at station $20 + 50.00$ and elevation $1005 + 00$ ft.

(a) Determine the minimum length of the crest vertical curve for a design speed of 50 mph.

(b) Calculate the location of the PVC and the elevation of the middle point of the curve.

Solution

Assumption: The height of the driver's eye, $H_1 = 3.5$ ft; the height of the object for stopping $H_2 = 0.5$ ft. This crest vertical curve is a parabolic curve.

Calculation:

(a) $A = 1.1 - 3.9 = -2.8\%$. From Table 6-4, the minimum stopping distance, $S = 475$ ft for $V = 50$ mph. If $S \leq L$, then

$$L = \frac{|A|S^2}{100(\sqrt{2H_1} + \sqrt{2H_2})^2} = \frac{|-2.8|(475)^2}{100(\sqrt{2 \times 3.5} + \sqrt{2 \times 0.5})^2}$$

$$= 475.30 \text{ ft}$$

This is true because $475 < 475.30$, so $L = 475.30$ ft.

(b) The middle point of the curve is at $x = L/2 = 237.65$ ft, so the location of PVC $= 2050.00 - 237.65 = 1812.35$ or station $18 + 12.35$. The midcurve offset is given by

$$E = \frac{AL}{800} = \frac{-2.8 \times 475.30}{800} = -1.66 \text{ ft}$$

Therefore, the elevation of the middle of the curve $= 1005.00 - 1.66 = 1003.34$ ft.

Discussion

If the calculated L is less than S in part (a), the equation for L with $L < S$ should be used. The length of the crest curve should provide a safe stopping distance and at the same time meet the specific design requirement of the road geometry. Once the design equation for a curve is known, the location and elevation of any point on the curve can be obtained.

Sag Vertical Curves.   No single criterion is used for determining lengths of sag vertical curves. The criteria generally used are (1) headlight sight distance, (2) rider comfort, (3) drainage control, and (4) a rule of thumb for general appearance. The most widely used criterion is headlight sight distance. When a vehicle traverses a sag vertical curve at night, the portion of highway lighted ahead is dependent on the position of the headlights and the direction of the light beam. A headlight height of 2 ft and a 1-degree upward divergence of the light beam from the longitudinal axis of the vehicle is the usual configuration adopted.

For case 1, where $S > L$, the intersection of the light beam with the pavement requires a curve length of

$$L = 2S - \frac{200(H + S \tan B)}{A} \tag{29}$$

and when $H = 2$ ft and $B = 1$ degree,

$$L = 2S - \frac{400 + 3.55S}{A} \tag{30}$$

For case 2, where $S < L$, the length of vertical curve is

$$L = \frac{AS^2}{200(H + S \tan B)} \tag{31}$$

and when $H = 2$ ft and $B = 1$ degree,

$$L = \frac{AS^2}{400 + 3.55S} \tag{32}$$

where

L = length of sag curve (ft)
S = light beam distance (ft)
A = algebraic difference in grade (%)

It is generally desirable to relate S, light beam distance, to safe stopping sight distance as the vertical curve should be long enough so that the two distances are nearly the same. Accordingly, it is appropriate to use stopping sight distance for various speeds as the S value in previous equations.

The discomfort experienced by users of sag curves is greater than that experienced with crest curves because gravitational and centrifugal forces are combining rather than opposing. Riding comfort on sag vertical curves is achieved where the centrifugal acceleration does not exceed 1 ft/sec^2. The general expression for a criterion is

$$L = \frac{AV^2}{46.5} \tag{33}$$

where V is the design speed (mph).

The length of vertical curve required to satisfy this comfort criterion at various speeds is about 50% of that required to satisfy the headlight sight distance requirement. When a sag vertical curve occurs at an underpass, the overhead structure may create a problem by shortening the sight distance. Under such circumstances, the length of vertical curve required for both adequate sight distance and clearance may be determined, as shown in Figure 6-19, using the following equations:

Case 1: Sight distance greater than length of vertical curve ($S > L$):

$$L = 2S - \frac{800}{A}\left(C - \frac{H_1 + H_2}{2}\right)$$

Case 2: Sight distance less than length of vertical curve ($S < L$):

$$L = \frac{AS^2}{800}\left(C - \frac{H_1 + H_2}{2}\right)^{-1}$$

5. CROSS-SECTION ELEMENTS

As mentioned before, highways are classified by function, and each class of highway, such as local roads and streets, collector roads and streets, rural and urban arterials, and freeways, has unique design features. Thus, the elements of the cross-section depend on the usage of the facility. However, almost all highway cross-sections are composed of four basic elements:

1. The paved road, consisting of the traffic-bearing lanes
2. The road margins, consisting of the shoulders, curbs, drainage channels, and medians
3. The traffic separation devices, which include traffic barriers, median barriers, and crash cushions
4. The provision of bikeways and sidewalks

Traffic characteristics and the desired level of service will dictate the configuration and dimension of each of these four elements. The extent of the ultimate expansion of the facility may also have an effect on the choice, configuration, and dimension of the elements. For example, the likely range of dimensions for four- to eight-lane arterial streets is shown in Figure 6-20. These cross-sections, both with and without frontage roads, are grouped in three categories, representing desirable, intermediate,

Case I Sight distance greater than
length of vertical curve ($S > L$)

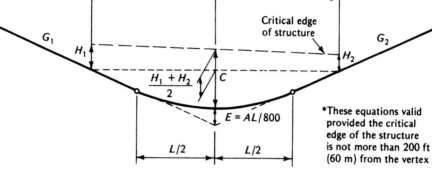

$$L = 2S - \frac{800}{A}\left(C - \frac{H_1 + H_2}{2}\right) \quad *$$

*These equations valid
provided the critical
edge of the structure
is not more than 200 ft
(60 m) from the vertex

Case 2 Sight distance less than
length of vertical curve ($S < L$)

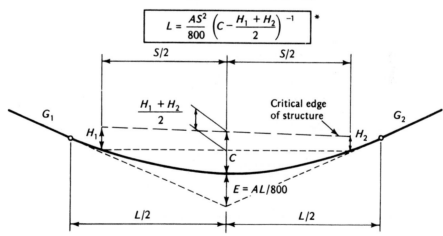

$$L = \frac{AS^2}{800}\left(C - \frac{H_1 + H_2}{2}\right)^{-1} \quad *$$

where L = Length of vertical curve, feet or meters
A   = Algebraic difference in grades, percent
S   = Sight distance, feet or meters
C   = Vertical clearance of underpass, feet or meters
H_1 = Vertical height or eye above roadway surface, feet or meters
H_2 = Vertical height of object above roadway surface, feet or meters

Figure 6-19   Sight Distance at Underpasses (Carter and Homburger, 1978).

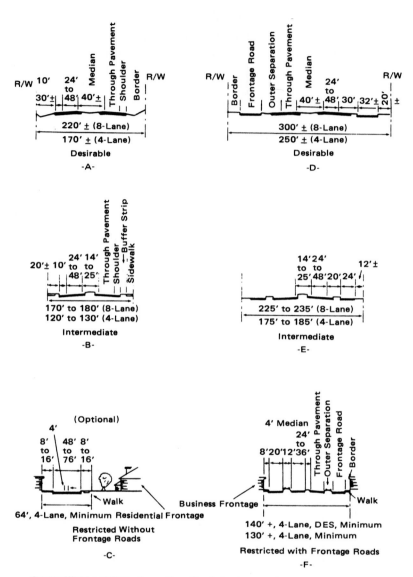

Figure 6-20 Cross-Section of Arterial Streets with and without Frontage Roads (AASHTO, 1990).

and restricted widths. Details of other classes of highways and streets are given in AASHTO (1990).

5.1 Pavement Characteristics

Pavement selection is generally determined by the volume and composition of traffic, soil characteristics, weather, performance of pavements in the region, availability of materials, the initial cost, and the overall annual maintenance and service life cost. The structural design of pavements is given in several textbooks (AASHTO, 1989, 1990; Huang, 1993), and reference should also be made to AASHTO (1993).

Some of the most important pavement characteristics in relation to geometric design are the ability of the pavement surface to retain its shape and dimensions over its life cycle, the ability to drain surface and subsurface rainwater, and the ability to offer adequate skid resistance to vehicles using it.

5.2 Pavement Cross-Slope

Two-lane and wider undivided pavements on tangents are sloped from the center to each edge to prevent water from collecting on the pavement surface. The cross-slope may be plane or curved or a combination of the two. Curved cross-sections are usually parabolic. On divided highways, each one-way pavement may be crowned separately. Roadway sections for divided highways are shown in Figure 6-21. Cross-slopes up to 2% are barely perceptible by drivers, but slopes greater than 2% can create steering and skidding problems. On high-type, two-lane pavements, the acceptable rate of cross-slope is from 1.5 to 2%. Intermediate- and low-surface types can have cross-slopes ranging from 1.5 to 3% and from 2 to 6%, respectively.

5.3 Lane Widths

Ten- to 13-ft lane widths are generally adopted on most U.S. highways, although the 12-ft width predominates. The capacity of highways is markedly affected by the lane width, and this effect is explained in Chapter 7. The use of 11-ft lanes is acceptable in urban areas, where the right-of-way and costs can be a constraint. Ten-foot lanes are acceptable only on low-speed facilities. Auxiliary lanes at intersections and interchanges often help to facilitate traffic movements. Such lanes should not be less than 10 ft wide.

5.4 Shoulders

Right shoulders should be provided along all highways. They are used for accommodating stopping vehicles in emergencies and for lateral support of the subbase, base, and surface courses. They vary in width from 2 to 12 ft. Graded and usable shoulders are shown in Figure 6-22. Shoulders may be surfaced with gravel, shell, crushed rock, mineral or chemical additives, bituminous surface treatments, and various forms of asphaltic or concrete pavements. Some important advantages of shoulders are as follows:

- Space is available for the motorist who is in an emergency situation; pedestrians and bicyclists can also use the shoulder.
- Sight distance is improved in cut sections.
- Capacity is improved.

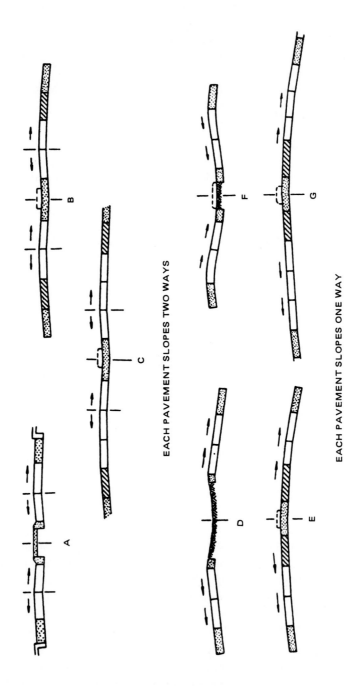

EACH PAVEMENT SLOPES TWO WAYS

EACH PAVEMENT SLOPES ONE WAY

Figure 6-21 Road Sections for Divided Highways (AASHTO, 1990).

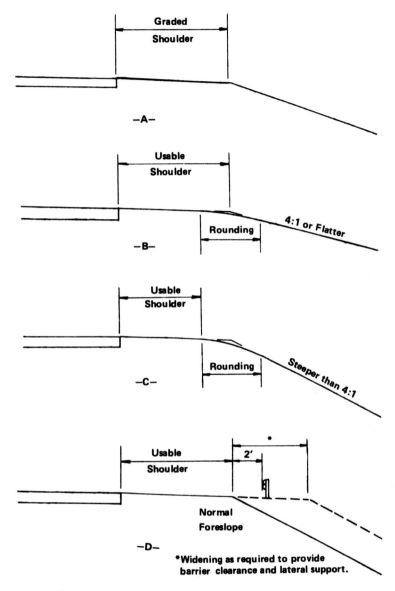

Figure 6-22   Graded and Usable Shoulders (AASHTO, 1990).

- Lateral clearance is provided for signs and guardrails.
- Structural support is provided for the pavement.

A shoulder should be continuous, regardless of its width. Shoulders on structures such as bridges should have the same width as the shoulders on the regular roadway. Any narrowing of shoulders may cause serious operating and safety problems.

5.5 Curbs

Curbs are used for the following reasons: drainage control, pavement-edge delineation, right-of-way reduction, aesthetics, delineation of pedestrian walkways, and reduction of maintenance operations. Urban highways benefit from the use of curbs in traffic safety and reduction of accidents. Rural highways are seldom provided with curbs, as they do not serve any useful purpose.

Barrier and mountable curbs are the two general classes of curbs provided for urban streets. Figure 6-23 shows typical curbs. Barrier curbs may be provided along the faces of long walks and tunnels. Mountable curbs allow vehicles to cross over if needed (Carter and Homburger, 1978).

5.6 Drainage Channels

The collection and conveyance of surface water from highway rights-of-way is performed by drainage channels. They include (1) roadside channels in cut sections,

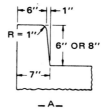

BARRIER CURBS

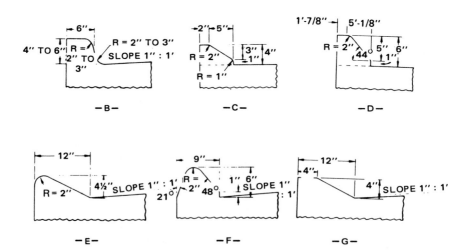

MOUNTABLE CURBS

Figure 6-23   Typical Highway Curbs (AASHTO, 1990).

(2) toe-of-slope channels to remove the water to natural water courses, (3) intercepting channels placed back of the top of cut slopes to intercept surface water from settling into the cut section, and (4) chutes to carry collected water down steep cut or fill slopes.

Control of surface drainage is the primary purpose for construction of roadside channels, and open-channel ditches cut into the roadside are one of the most economical methods. The depth of channel should be sufficient to remove the water without saturation of the pavement subgrade. Standard textbooks on highway engineering and drainage may be consulted for details on highway drainage (Oglesby and Hicks, 1982; Wright, 1995).

5.7 Traffic Barriers

The *Roadside Design Guide* (AASHTO, 1989) contains state-of-the-art information on this subject. Only a brief outline is provided here.

Traffic barriers are used to minimize the severity of potential accidents that may occur when vehicles leave the traveled way. It must be pointed out that barriers are a hazard in themselves, and therefore minimizing the use of such barriers is justified. Traffic barriers include longitudinal barriers, such as roadside barriers and median barriers, and crash cushions. The latter function primarily by decelerating errant vehicles to a stop or by redirection.

Three types of longitudinal barriers are generally in use: flexible, semirigid, and rigid. They differ in the amount of barrier deflection that takes place when the barrier is struck. Flexible systems, usually the cable type, depend on their tensile strength and are more forgiving than the other types. The semirigid system consists of rails and posts, resistance being obtained through the combined flexural and tensile strength of the rail. Rigid systems such as concrete barriers do not deflect on impact. Energy is dissipated by raising the vehicle and the deformation of the vehicle itself on collision. Because lateral deflection of longitudinal barriers is a major concern, the selection of a longitudinal barrier system depends on the space available at the highway site for accommodating this deflection.

5.8 Medians

A median is a portion of a divided highway (with four or more lanes) separating the traveled way for traffic in the opposing direction. The median width is the dimension between the through-lane edges and includes the left shoulders, if any. A median provides freedom from the interference of opposing traffic, a recovery area for out-of-control vehicles, a stopping area in case of emergencies, room for speed changes and storage of left- and U-turning vehicles and for less headlight glare, and space for additional future lanes.

Although medians should be as wide as possible, economic factors often limit this width because of cost of land and general maintenance. Medians are classified as traversable, deterring, or barrier. Traversable medians are merely paint stripes or buttons and are easily traversable. If a mountable curb or corrugation is provided, it is known as a deterring median. A barrier median is usually in the form of a guardrail or a concrete wall that prevents traffic from crossing over. AASHTO (1989) should be referred to for details.

5.9 Bike Paths and Sidewalks

Details of these two important provisions are given in Chapter 12. Provision for bicycle facilities should be in accordance with the *Guide for Development of New Bicycle Facilities* (AASHTO, 1981).

5.10 Right-of-Way

Right-of-way is determined by traffic requirements, topography, land use, cost of acquisition, intersection design, and the possibility of ultimate expansion. The required width of the right-of-way is the summation of the various cross-sectional elements, such as the number of lane widths, median width, auxiliary lanes, shoulders, frontage roads (if any), side slopes, and retaining walls. In urban areas, where land is expensive, the designer has to make a decision regarding a satisfactory width of land that will give the maximum service within a limited width of right-of-way. Economic considerations, physical constraints, and environmental reasons may decide right-of-way widths.

Example 7

Estimate the total width of right-of-way (R/W) of a depressed divided rural highway with three lanes in each direction. Show the median, shoulders, lanes, side slopes, and so on, with recommended dimensions on a cross-section of the highway.

Solution

We use

$$\text{Lane width} = 12 \text{ ft}$$

$$\text{Cross-slope} = 2\%$$

$$\text{Shoulder width} = 6 \text{ ft}$$

$$\text{Median width} = 4 \text{ ft}$$

$$\text{Side slope} = 1:4 \text{ and width} = 20 \text{ ft}$$

Then

$$\text{R/W} = 2[20 + 6 + (3 \times 12) + 6] + 4 = 140 \text{ ft. Refer to Figure 6-E7.}$$

Discussion

The side slope value depends on the depth of cut. In practice, if the depth of cut is small, the slope can be steeper. But if the depth of cut is large, the slope will be flatter, to avoid land and rock slides.

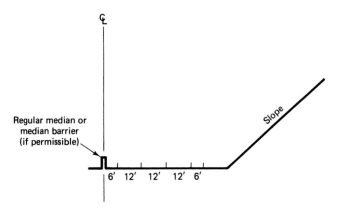

Figure 6-E7   Right-of-Way.

REFERENCES

AMERICAN ASSOCIATION OF STATE HIGHWAY AND TRANSPORTATION OFFICIALS (AASHTO) (1981). *Guide for Development of New Bicycle Facilities*, AASHTO, Washington, DC.

AMERICAN ASSOCIATION OF STATE HIGHWAY AND TRANSPORTATION OFFICIALS (AASHTO) (1989). *Roadside Design Guide*, AASHTO, Washington, DC.

AMERICAN ASSOCIATION OF STATE HIGHWAY AND TRANSPORTATION OFFICIALS (AASHTO) (1990). *A Policy on Geometric Design of Highways and Streets*, AASHTO, Washington, DC.

AMERICAN ASSOCIATION OF STATE HIGHWAY AND TRANSPORTATION OFFICIALS (AASHTO) (1993). *Guide for Design of Pavement Structures*, AASHTO, Washington, DC.

AMERICAN ASSOCIATION OF STATE HIGHWAY AND TRANSPORTATION OFFICIALS (AASHTO) (1994). *A Policy on Geometric Design of Highways and Streets*, metric, AASHTO, Washington, DC.

CARTER, EVERETT C., and WOLFGANG S. HOMBURGER (1978). *Introduction to Transportation Engineering*, Reston Publishing, Reston, VA.

HUANG, YANG H. (1993). *Pavement Analysis and Design*, Prentice Hall, Englewood Cliffs, NJ.

JHK & ASSOCIATES (1980). *Design of Urban Streets*, Technology Sharing Report 80-204. U.S. Department of Transportation. Washington, DC.

OGLESBY, CLARKSON H., and R. GARY HICKS (1982). *Highway Engineering*, 4th Ed., John Wiley, New York.

TRANSPORTATION RESEARCH BOARD (TRB) (1994). *Highway Capacity Manual*, 3rd Ed., Special Report 209, National Research Council, Washington, DC.

WRIGHT, PAUL H. (1995). *Highway Engineering*, John Wiley, New York.

EXERCISES

1. The traffic counts in the accompanying table were obtained on a section of a highway for 10 weekdays.
   Calculate the average daily traffic (ADT) on the highway.

2. A rural multi-lane highway carries 750 passenger cars per hour per lane in one direction. Using Figure 6-4, determine the average running speed on this highway.

TRAFFIC COUNTS

Day	Vehicle count
1	10,000
2	9,800
3	10,500
4	10,030
5	9,950
6	11,000
7	9,700
8	9,900
9	10,010
10	10,400

3. A rural highway has a +6% grade on a mountainous region. If the maximum and minimum speed limit on this highway for trucks is 55 mph and 45 mph, respectively, determine the "critical length" of this grade.

4. A transition curve is needed to connect a circular section with a straight section of a rural highway. If the design speed on the highway is 65 mph and the radius of curvature of the circular section is 1000 ft, determine the length of the transition curve. (Assume $C = 2$.)

5. A level, rural, two-lane highway has a 10-degree curved section as it passes by a large building. The obstructing corner of the building is 24.5 ft away from the center line of the highway.
   (a) If the lanes of the highway are 11 ft wide, what is the stopping sight distance on this section?
   (b) What is the safe operating speed on the section in case the corner is only 17 ft away from the center line of the highway?

6. A section of a two-lane rural highway consists of a −3% grade intersecting a +2% grade. The speed limit on the highway is 55 mph. The height of headlights of an average car is 2 ft with a beam angle of 1 degree. Determine the length of sag vertical curve that joins the two grades, considering a safe passing maneuver. Make any necessary assumptions.

7. A +7.5% grade meets a horizontal grade on a section of a rural mountainous highway. If the length of the crest vertical curve formed in that section is 300 ft long, determine the safe operating speed on the highway.

8. A parabolic sag vertical curve is formed when a rural highway passes under a bridge joining a −8% grade to the horizontal grade. The vertical point of intersection of these two grades lies 3 ft below the curve. If the vertical clearance of the underpass is only 10 ft, what speed limit would you suggest should be imposed at this spot, considering a safe stopping sight distance?

9. **(a)** A rural highway section consists of a 1000-ft length of 4% downgrade followed immediately by a 6% upgrade. If a speed reduction of 15 mph is permissible for trucks, what length of the 6% upgrade would you recommend?
   (b) This highway (farther down the road) has a 2% upgrade for about 1500 ft followed by another 5% upgrade. What critical length would you recommend for the 5% upgrade if the reduction in speed for trucks is limited to 10 mph?

10. A parabolic vertical curve connects a +4.5% grade with a −4.0% grade on a rural highway at station 875 + 25.4 and elevation 512.6 ft above the sea level. The length of the curve is 550 ft. Calculate the location and elevation of the PVC and the PVT, and elevations at stations 870 + 50 and 876 + 50.

11. Determine the maximum degree of a horizontal curve for a roadway where the design speed is 65 mph. The side friction factor, f, and the superelevation rate, e, are in the range of 0.10 to 0.15 and 0.08 to 0.10, respectively.

12. **(a)** A two-lane rural highway passing through a forested area has sight obstructions located 20 ft from the center line of the inside lane. Inadequate sight distance has been blamed for some accidents that have occurred on this section of highway where it negotiates a 15-degree horizontal curve, with a posted speed of 45 mph. Analyze the problem and suggest actions to mitigate it.

    (b) What speed limit would you recommend for this problem section of highway? The coefficient of friction for the pavement varies between 0.40 and 0.31 for speeds between 20 and 45 mph, respectively.

13. A level, rural, two-lane highway has a 6-degree horizontal curve. What should be the rate of superelevation to develop the maximum permissible side friction on this section at a design speed of 65 mph? Schematically show the method of attaining the superelevation by revolving the pavement about the outside edge. (Assume $\mu = 0.11$.)

14. Develop an expression for stopping sight distance, S, in terms of radius of curvature, R, and middle ordinate, M, from the following formula:

$$M = \frac{5730}{D} \text{vers} \frac{SD}{200}$$

    where D is the degree of curvature, and vers $\theta = 1 - \cos \theta$.

15. Calculate the minimum length of a crest vertical curve that connects a -3% grade with a $+2\%$ grade on a highway. Determine whether a speed of 65 mph can be maintained on this section. Make any necessary assumption.

16. **(a)** A highway design includes the intersection of a $+4.8\%$ grade with a -4.2% grade at station $1052 + 75$ at elevation 851.50 feet above sea level. Calculate the center line elevation along this highway for every 50-ft station on a parabolic vertical curve of 600-ft length.

    (b) What is the minimum stopping sight distance on this vertical curve and the maximum safe speed on this section for both wet and dry pavement conditions? Use appropriate values of f.

17. A Seattle, Washington, highway spans across a lake and experiences average traffic flow fluctuations. Find the peak unidirectional flow if the projected ADT = 60,000 veh/day and the directional distribution is 60%.

18. On a curved segment of a rural two-lane highway in Willamette Valley wine country, the sight distance is 1000 ft and the curve is signed and marked as a passing zone. Should the passing zone be used in combination with a 50-mph, posted speed limit? If not, what maximum speed does the 1000-ft, passing sight distance allow for?

Chapter 7

Highway Capacity

1. INTRODUCTION

Chapter 5 dealt mainly with the relationship between speed, density, and rate of flow for centrally and individually controlled vehicles. Also, two categories of traffic flow were examined: interrupted and uninterrupted. One of the conclusions drawn from studying the characteristics of traffic flow is that when flow rates approach optimum (or capacity), there is the likelihood of congestion setting in, causing excessive delay to vehicles. In this chapter, we apply these concepts in a practical way.

Imagine a single vehicle traversing a freeway section at 3 A.M. It happens to be the sole vehicle using the facility. If speed limits were ignored by the driver of this vehicle, she could drive at whatever speed liked consistent with the condition and characteristics of the vehicle, the driver's ability, and the geometrics of the freeway section. The driver could possibly choose to travel at 90 mph. On the other hand, this very section of freeway could be highly congested at 5 P.M., when the driver of our vehicle is stuck in a lane, crawling at near-zero speed, without even a chance to change her lane and get off the freeway. In short, the same freeway section offers differing operational "quality of service" to drivers (and passengers) depending on the amount of congestion prevailing on a particular section of freeway. These concepts—capacity and "quality" or level of service—are most useful and important to the transportation engineer. They have broadened the use of procedures in operational analysis and created a quality scale that has become an integral part of the traffic engineering profession. Even decision makers and public groups use these concepts as major descriptors when communicating with traffic engineers, and vice versa.

In this chapter, we describe the basic definitions and concepts relating to capacity and level of service. It also presents procedures for determining the capacity and level of service of uninterrupted flow transportation facilities: freeways, multilane highways, and two-lane highways. Interrupted traffic flow facilities, such as signalized intersections and arterials, are dealt with in a separate chapter.

2. HIGHWAY CAPACITY AND LEVEL OF SERVICE

Transportation modes that make use of highways and are controlled by individual drivers are referred to as individually controlled modes. The *Highway Capacity Manual* (HCM) (TRB, 1994) is the standard reference work used in this area. Over the years, the HCM has emerged as a collection of the latest proven techniques for estimating highway capacity.

Two major types of transportation facilities are described in the HCM:

1. Uninterrupted flow facilities
    (a) Freeways
    (b) Multilane highways
    (c) Two-lane highways
2. Interrupted flow facilities
    (a) Signalized intersections
    (b) Unsignalized intersections
    (c) Arterials
    (d) Transit
    (e) Pedestrian ways
    (f) Bikeways

The analysis of these facilities varies considerably. Note, however, that the details presented in the HCM on transit, bicycles, and pedestrians focus on those aspects that interact with traffic using the street/highway. "In general, the capacity of a facility is the maximum hourly rate at which persons or vehicles can reasonably be expected to traverse a point or uniform section of a lane or roadway during a given time period under prevailing roadway, traffic, and control conditions" (TRB, 1994). A 15-minute period is generally used. Roadway conditions refer to the type of facility, its geometric characteristics, the number of lanes (by direction), lane and shoulder widths, lateral clearances, design speed, horizontal and vertical alignments, and availability of queuing space at intersections. Traffic conditions refer to the distribution of vehicle types using the facility, the amount and distribution of traffic in available lanes of a facility, and the directional distribution. The types and specific design of control devices (such as traffic signals and their timing) and traffic regulations on the facility constitute control conditions (TRB, 1994).

Level of service (LOS) is a qualitative measure describing operational conditions within a traffic stream and their perception by motorists and/or passengers. Factors such as speed and travel time, freedom to maneuver, traffic interruptions, and comfort and convenience are generally included as conditions affecting LOS. Each facility can be measured on the basis of six levels of service, A through F, A representing the best operating conditions and F the worst (TRB, 1994).

The maximum rate of flow that can be accommodated by a facility at each LOS (except LOS F) is described as the service *flow rate*. Thus, every facility has five service flow rates, corresponding to each LOS (A through E), and the service flow rate for a designated LOS is the maximum hourly rate at which persons or vehicles can reasonably be expected to traverse a point or uniform section of a lane or roadway during a given time period under prevailing roadway, traffic, and control conditions.

Note that each LOS represents a range of conditions defined by a range of one or more operational parameters. Although the concept of LOS attempts to address a wide variety of operating conditions, limitations on data collection and their availability make it impractical to consider the full range of operational parameters for every type of transportation facility. The parameters selected to define LOS for each facility type are called *measures of effectiveness* (MOEs) and represent those measures that best describe the quality of operation on the facility. For example, density [passenger cars per mile per lane (pc/mi/ln)] is the MOE for basic freeway segments, and time delay (%) is the MOE considered for two-lane highways (TRB, 1994).

3. BASIC FREEWAY CAPACITY STUDIES

3.1 Definitions

A freeway is a divided highway facility having two or more lanes in each direction for the exclusive use of traffic, with full control of access and egress. In the highway hierarchy, the freeway is the only facility that provides completely uninterrupted flow. A freeway is composed of three subcomponents: the basic freeway segments, weaving areas, and ramp junctions. Figure 7-1 shows these subcomponents. Only basic freeway sections are dealt with.

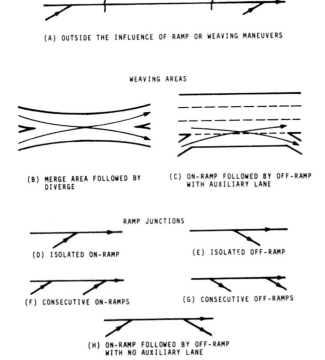

Figure 7-1 Freeway Components (TRB, 1994).

Freeway capacity is the maximum sustained (15 minutes) rate of flow in vehicles per hour (veh/hr) at which traffic can pass a point or uniform segment of freeway under prevailing roadway and traffic conditions. Roadway characteristics include the number and width of lanes, right-shoulder lateral clearance, interchange spacing, grades, and lanes configurations. Traffic conditions include the percentage composition of the traffic stream by vehicle type, lane distribution characteristics, and driver characteristics (such as weekday commuters and recreational drivers).

3.2 Freeway Flow Characteristics

Traffic flow within a basic freeway segment can be generally described in three flow types: free flow, queue discharge flow, and congested flow. Each flow type represents different conditions on the freeway and can be defined by speed–flow–density ranges. Free flow represents traffic flow that is unaffected by upstream or downstream conditions. At low to moderate flows, this flow type is generally defined within a speed range of 55 to 70 mph and, at high flow rates, between 45 to 65 mph. Queue discharge flow represents traffic flow after passage through a bottleneck and in the process of acceleration back to the free-flow speed of the freeway. This flow type is generally defined within a narrow range of flows, 2000–2300 passenger cars per hour per lane (pc/hr/ln), with speeds ranging from 35 mph up to the free-flow speed. This acceleration to free flow may occur within 1/2 to 1 mile from the bottleneck. Queue discharge rate is approximately 5% lower than the free-flow rate. Congested flow represents traffic flow that is influenced by the effects of a downstream bottleneck. Flow can vary widely in range of flows and speeds. Queues, within which vehicles stop and move, can stretch to thousands of feet. The basic characteristics are established for a set of ideal conditions:

- 12-ft minimum lane widths
- 6-ft minimum right-shoulder lateral clearance between the edge of the travel lane and the nearest obstacle or object that influences travel behavior. The minimum median lateral clearance is 2 ft
- all passenger cars in the traffic stream
- 10 or more lanes
- interchanges spaces every 2 miles or more
- level terrain, with grades no greater than 2%

These conditions represent the highest type of basic freeway section, one with a free-flow speed of 70 mph. Prevailing roadway and traffic conditions, as they vary from ideal conditions, will in turn affect the free-flow speed.

Figure 7-2 shows speed-flow relationships for the free-flow regime on basic freeway segments under ideal or non-ideal conditions.

3.3 Free-Flow Speed

Free-flow speed is defined as the speed that occurs when density and flow are zero. All recent studies suggest that speed on freeways is insensitive to flow over a broad range of flow rate. Figure 7-2 shows speed to be constant for flow rates up to 1300 veh/hr/ln

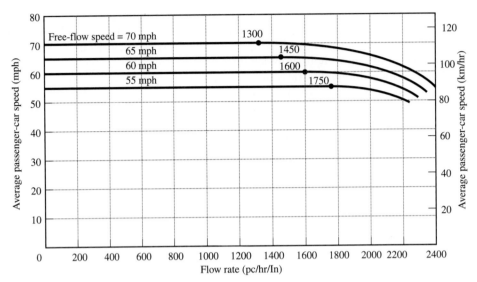

Figure 7-2   Speed–Flow Relationship–Free Flow (NCHRP 3-45, 1995).

for 70 mph free-flow speeds. For freeways with lower free-flow speeds, the speed remains constant even at higher rate of flow. This effect became very noticeable with the introduction of the 55 mph speed restriction on urban freeways. In practice, the free-flow speed can be easily measured in the field as the average speed of vehicles when the flow rate is less than 1300 pc/hr/ln. The 70-mph curve of Figure 7-2 represents ideal conditions.

Capacity.   The curves of Figure 7-2 show two regions of flow: (1) A region over which the speed is insensitive to the increasing flow rate. This range extends to 1300 pc/hr/ln for 70-mph freeways, and to 1750 pc/hr/ln for 55-mph freeways. (2) A region over which increasing flow rate causes a drop in speed. The value of capacity for freeways with ideal traffic and roadway conditions is 2400 pc/hr/ln. The value of capacity varies with free-flow speed of the freeway. The minimum capacity is 2250 pc/hr/ln, representing a free-flow speed of 55 mph. The capacities for free-flow speeds of 65 and 60 mph are 2350 and 2300 pc/hr/ln, respectively. Figure 7-2 shows that capacity flow may occur at speeds ranging from 49 to 55 mph, depending on the free-flow speed of the facility. Speed drop, with increased flow, is smaller on a freeway section with low free-flow speed compared to a high free-flow speed. Density at capacity varies from 43.6 pc/mi/ln at a free-flow speed of 70 mph to 46.0 pc/mi/ln at 55 mph.

3.4 Level of Service

Although speed is a major concern of the motorists using a freeway facility, it remains nearly constant over a wide range of flows. Freedom to maneuver within the traffic stream and proximity to other vehicles are equally important and are used, in preference over speed, in describing the level of service. Besides, density increases throughout the range of flows up to capacity, and therefore provides a better measure of effectiveness. The densities used to define levels of service for basic freeway sections are as follows:

LEVEL OF SERVICE	MAXIMUM DENSITY
A	$\leq$ 10 pc/mi/ln
B	$\leq$ 16 pc/mi/ln
C	$\leq$ 24 pc/mi/ln
D	$\leq$ 32 pc/mi/ln
E	$\leq$ 43.6 pc/mi/ln (70-mph free-flow speed)
	$\leq$ 44.3 pc/mi/ln (65-mph free-flow speed)
	$\leq$ 45.1 pc/mi/ln (60-mph free-flow speed)
	$\leq$ 46.0 pc/mi/ln (55-mph free-flow speed)
F	Exceeds corresponding limit for LOS E

Level-of-service criteria for basic freeway sections are shown in Table 7-1. To be within a given level of service, the density criterion must be met. Figure 7-3 illustrates the relationship among speed, flow, and density for basic freeway sections operating under ideal conditions. Various levels of service are shown using density boundary values. Either Table 7-1 or Figure 7-3 may be used in making level-of-service determinations.

LOS A: Free-flow operation; free-flow speed generally prevails; vehicles completely unimpeded in their ability to maneuver within the traffic stream; average spacing of 528 ft. The effects of incidents are local and minimal.

LOS B: Reasonably free flow; generally free-flow speed; ability to maneuver within the traffic stream slightly restricted; average spacing 330 ft. Local deterioration of service is more severe with incidents.

LOS C: Flow with speeds still at or near free-flow speed; freedom to maneuver within the traffic stream noticeably restricted and lane changes require more care and vigilance by the driver; average spacing 220 ft. Local deterioration due to incidents is substantial and queues may be expected to form behind any significant blockage.

LOS D: Speeds begin to decline slightly with increasing flow; density deteriorates more rapidly; freedom to maneuver is more noticeably limited; average spacing 165 ft. Minor incidents can be expected to cause queuing.

LOS E: At its lower boundary describes capacity; operations are volatile and virtually no useable gaps exist in the traffic stream; maneuverability within the traffic stream is extremely limited; average spacing 110 ft at speeds still over 50 mph. Any incident can be expected to produce a serious breakdown with extensive queuing.

LOS F: Describes breakdown in vehicular flow at points of recurring congestion such as merge, weave, or lane drop locations. It can also be caused by traffic incidents. In all cases, breakdowns occur when the ratio of arrival flow rate to actual capacity exceeds 1.0.

TABLE 7-1   LEVEL OF SERVICE CRITERIA FOR BASIC FREEWAY SECTIONS

Level of service	Maximum density (pc/mi/ln)	Minimum speed (mph)	Maximum service flow rate (pc/ph/pl)	Maximum volume/capacity (v/c) ratio
Free-Flow Speed = 70 mph				
A	10.0	70.0	700	0.29
B	16.0	70.0	1120	0.47
C	24.0	68.0	1632	0.68
D	32.0	64.0	2048	0.85
E	43.6	55.0	2400	1.00
F	var	var	var	var
Free-Flow Speed = 65 mph				
A	10.0	65.0	650	0.28
B	16.0	65.0	1040	0.44
C	24.0	64.5	1548	0.66
D	32.0	62.0	1984	0.84
E	44.3	53.0	2350	1.00
F	var	var	var	var
Free-Flow Speed = 60 mph				
A	10.0	60.0	600	0.26
B	16.0	60.0	960	0.42
C	24.0	60.0	1440	0.63
D	32.0	58.0	1856	0.81
E	45.1	51.0	2300	1.00
F	var	var	var	var
Free-Flow Speed = 55 mph				
A	10.0	55.0	550	0.24
B	16.0	55.0	880	0.39
C	24.0	55.0	1320	0.59
D	32.0	54.5	1744	0.78
E	46.0	49.0	2250	1.00
F	var	var	var	var

Source: NCHRP 3-45, 1995.

3.5 Use of the *Highway Capacity Manual*

The determination of level of service for a basic freeway section generally involves three components, namely, flow rate, free-flow speed, and level of service. An equivalent passenger-car flow rate is calculated using Eq. 1 to allow for the effect of heavy vehicles and variation of traffic flow during the hour in the traffic stream.

$$v_p = \frac{V}{\text{PHF} \times N \times f_{HV}} \tag{1}$$

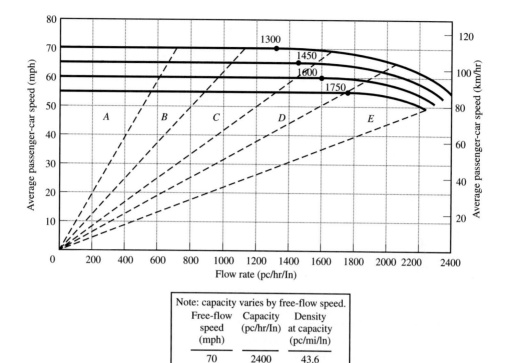

Note: capacity varies by free-flow speed.

Free-flow speed (mph)	Capacity (pc/hr/ln)	Density at capacity (pc/mi/ln)
70	2400	43.6
65	2350	44.3
60	2300	45.1
55	2250	46.0

Figure 7-3   Level-of-Service Criteria Illustrated (NCHRP 3-45, 1995).

where

v_p = passenger-car equivalent flow rate, pc/hr/ln
V = hourly volume, vph
PHF = peak-hour factor
N = number of lanes
f_{HV} = heavy vehicle adjustment factor

Observations of traffic flow consistently indicate that the flow rates found in the peak 15-minute period within an hour are not sustained over the entire hour. The peak-hour factor in Eq. 1 accounts for this phenomenon. On freeways, typical peak-hour factors range from 0.80 to 0.95. Lower peak-hour factors are characteristic of rural freeways or off-peak conditions. Higher factors are typical of urban and suburban peak-hour conditions. If local data are unavailable, 0.85 and 0.90 may be used for rural and urban/suburban peak-hour conditions, respectively.

Adjustments for the presence of heavy vehicles in the traffic stream apply for trucks, buses, and recreational vehicles (RVs). Trucks and buses are treated identically. f_{HV} is determined using a two-step process. The passenger-car equivalent of each truck/bus (E_T) and recreational vehicle (E_R) is found for the traffic and roadway conditions under study. Using the value of E_T and E_R and the percentage of each type of vehicle in the traffic stream (P_T and P_R), the adjustment factor f_{HV} may be computed. The impact of heavy vehicles on traffic flow depends on grade conditions as well as the traffic composition.

Passenger-car equivalents can be selected for two conditions:

1. *Extended general freeway segments.* It is often possible to consider an extended length of freeway containing a number of upgrades, downgrades, and level segments as a single uniform segment. As a rule, extended general segments may be used where no one grade of 3% or greater is longer than 1/4 mile or longer than 1/2 mile for grades less than 3%. Extended general freeway segments are further classified into level, rolling, and mountainous terrain. Level terrain includes short grades of less than 2%. Rolling terrain causes heavy vehicles to reduce their speeds substantially below those of passenger cars, but not to crawl speeds for any significant length of time, which typifies mountainous terrain. Refer to Table 7-2 for values of E_T and E_R for extended general freeway segments.

2. *Specific grades.* Any grade less than 3% and longer than 1/2 mile, or any grade of 3% or more and longer than 1/4 mile is usually analyzed as a separate segment. Tables 7-3 and 7-4 give values of E_T and E_R for specific upgrade sections. There is little specific data on the impact of heavy vehicles on traffic flow on downgrades. In general, if the downgrade is not so severe as to cause trucks to shift into low gear, they may be treated as level-terrain segments. Where more severe downgrades occur, Table 7-5 gives guidelines for use in selecting downgrade values of E_T. For recreational vehicles, downgrades may be treated as level terrain. If there is a continuous series of grades in the freeway alignment, an average grade to the point in question can be used. This average grade is computed from dividing the total rise (ft) from the beginning of the composite grade by the length of the grade (ft). The average grade technique is acceptable for grades in which all subsections are less than 4%, or where the total length of the composite grade is less than 4000 ft. For more severe composite grades, HCM provides a detailed technique, which is based on the vehicle performance curves and equivalent speeds.

The adjustment factor, f_{HV}, for heavy vehicles is computed using the following equation:

TABLE 7-2 PASSENGER-CAR EQUIVALENTS ON EXTENDED GENERAL FREEWAY SEGMENTS

Category	Type of terrain		
	Level	Rolling	Mountainous
E_T for trucks and buses	1.5	3.0	6.0
E_R for recreational vehicles	1.2	2.0	4.0

Source: NCHRP 3-45, 1995.

TABLE 7-3 PASSENGER-CAR EQUIVALENTS FOR TRUCKS AND BUSES ON SPECIFIC UPGRADES

Grade (%)	Length (miles)	Passenger-car equivalent, E_T								
Percent trucks and buses		2	4	5	6	8	10	15	20	25
<2	All	1.5	1.5	1.5	1.5	1.5	1.5	1.5	1.5	1.5
2	$0-\frac{1}{4}$	1.5	1.5	1.5	1.5	1.5	1.5	1.5	1.5	1.5
	$\frac{1}{4}-\frac{1}{2}$	1.5	1.5	1.5	1.5	1.5	1.5	1.5	1.5	1.5
	$\frac{1}{2}-\frac{3}{4}$	1.5	1.5	1.5	1.5	1.5	1.5	1.5	1.5	1.5
	$\frac{3}{4}-1$	2.5	2.0	2.0	2.0	1.5	1.5	1.5	1.5	1.5
	$1-1\frac{1}{2}$	4.0	3.0	3.0	3.0	2.5	2.5	2.0	2.0	2.0
	$>1\frac{1}{2}$	4.5	3.5	3.0	3.0	2.5	2.5	2.0	2.0	2.0
3	$0-\frac{1}{4}$	1.5	1.5	1.5	1.5	1.5	1.5	1.5	1.5	1.5
	$\frac{1}{4}-\frac{1}{2}$	3.0	2.5	2.5	2.0	2.0	2.0	2.0	1.5	1.5
	$\frac{1}{2}-\frac{3}{4}$	6.0	4.0	4.0	3.5	3.5	3.0	2.5	2.5	2.0
	$\frac{3}{4}-1$	7.5	5.5	5.0	4.5	4.0	4.0	3.5	3.0	3.0
	$1-1\frac{1}{2}$	8.0	6.0	5.5	5.0	4.5	4.0	4.0	3.5	3.0
	$>1\frac{1}{2}$	8.5	6.0	5.5	5.0	4.5	4.5	4.0	3.5	3.0
4	$0-\frac{1}{4}$	1.5	1.5	1.5	1.5	1.5	1.5	1.5	1.5	1.5
	$\frac{1}{4}-\frac{1}{2}$	5.5	4.0	4.0	3.5	3.0	3.0	3.0	2.5	2.5
	$\frac{1}{2}-\frac{3}{4}$	9.5	7.0	6.5	6.0	5.5	5.0	4.5	4.0	3.5
	$\frac{3}{4}-1$	10.5	8.0	7.0	6.5	6.0	5.5	5.0	4.5	4.0
	>1	11.0	8.0	7.5	7.0	6.0	6.0	5.0	5.0	4.5
5	$0-\frac{1}{4}$	2.0	2.0	1.5	1.5	1.5	1.5	1.5	1.5	1.5
	$\frac{1}{4}-\frac{1}{3}$	6.0	4.5	4.0	4.0	3.5	3.0	3.0	2.5	2.0
	$\frac{1}{3}-\frac{1}{2}$	9.0	7.0	6.0	6.0	5.5	5.0	4.5	4.0	3.5
	$\frac{1}{2}-\frac{3}{4}$	12.5	9.0	8.5	8.0	7.0	7.0	6.0	6.0	5.0
	$\frac{3}{4}-1$	13.0	9.5	9.0	8.0	7.5	7.0	6.5	6.0	5.5
	>1	13.0	9.5	9.0	8.0	7.5	7.0	6.5	6.0	5.5
6	$0-\frac{1}{4}$	4.5	3.5	3.0	3.0	3.0	2.5	2.5	2.0	2.0
	$\frac{1}{4}-\frac{1}{3}$	9.0	6.5	6.0	6.0	5.0	5.0	4.0	3.5	3.0
	$\frac{1}{3}-\frac{1}{2}$	12.5	9.5	8.5	8.0	7.0	6.5	6.0	6.0	5.5
	$\frac{1}{2}-\frac{3}{4}$	15.0	11.0	10.0	9.5	9.0	8.0	8.0	7.5	6.5
	$\frac{3}{4}-1$	15.0	11.0	10.0	9.5	9.0	8.5	8.0	7.5	6.5
	>1	15.0	11.0	10.0	9.5	9.0	8.5	8.0	7.5	6.5

Note: If the length of grade falls on a boundary, apply the longer category; interpolation may be used to find equivalents for intermediate percent grades.
Source: NCHRP, 1995.

TABLE 7-4 PASSENGER-CAR EQUIVALENT FOR RECREATIONAL VEHICLES ON SPECIFIC UPGRADES

Grade (%)	Length (miles)	Passenger-car equivalent, E_R								
Percent RVs		2	4	5	6	8	10	15	20	25
< 2	All	1.2	1.2	1.2	1.2	1.2	1.2	1.2	1.2	1.2
3	$0-\frac{1}{2}$	1.2	1.2	1.2	1.2	1.2	1.2	1.2	1.2	1.2
	$>\frac{1}{2}$	2.0	1.5	1.5	1.5	1.5	1.5	1.2	1.2	1.2
4	$0-\frac{1}{4}$	1.2	1.2	1.2	1.2	1.2	1.2	1.2	1.2	1.2
	$\frac{1}{4}-\frac{1}{2}$	2.5	2.5	2.0	2.0	2.0	2.0	1.5	1.5	1.5
	$>\frac{1}{2}$	3.0	2.5	2.5	2.0	2.0	2.0	2.0	1.5	1.5
5	$0-\frac{1}{4}$	2.5	2.0	2.0	2.0	1.5	1.5	1.5	1.5	1.5
	$\frac{1}{4}-\frac{1}{2}$	4.0	3.0	3.0	3.0	2.5	2.5	2.0	2.0	2.0
	$>\frac{1}{2}$	4.5	3.5	3.0	3.0	3.0	2.5	2.5	2.0	2.0
6	$0-\frac{1}{4}$	4.0	3.0	2.5	2.5	2.5	2.0	2.0	2.0	1.5
	$\frac{1}{4}-\frac{1}{2}$	6.0	4.0	4.0	3.5	3.0	3.0	2.5	2.5	2.0
	$>\frac{1}{2}$	6.0	4.5	4.0	4.0	3.5	3.0	3.0	2.5	2.0

Note: If the length of grade falls on a boundary, apply the longer category; interpolation may be used to find equivalents for intermediate percent grades.
Source: NCHRP, 1995.

TABLE 7-5 PASSENGER-CAR EQUIVALENTS FOR TRUCKS AND BUSES ON SPECIFIC DOWNGRADES

Percent downgrade (%)	Length of grade (miles)	Passenger-car equivalent, E_T			
		Percent trucks/buses			
		5	10	15	20
< 4	All	1.5	1.5	1.5	1.5
4	≤ 4	1.5	1.5	1.5	1.5
4	> 4	2.0	2.0	2.0	1.5
5	≤ 4	1.5	1.5	1.5	1.5
5	> 4	5.5	4.0	4.0	3.0
≥ 6	≤ 4	1.5	1.5	1.5	1.5
> 6	> 4	7.5	6.0	5.5	4.5

Source: NCHRP, 1995.

$$f_{HV} = \frac{1}{1 + P_T(E_T - 1) + P_R(E_R - 1)} \tag{2}$$

where

f_{HV} = heavy-vehicle adjustment factor

P_T, P_R = proportion of trucks/buses and recreational vehicles, respectively, in the traffic stream

E_T, E_R = passenger car equivalent for trucks/buses and recreational vehicles, respectively, in the traffic stream

In many cases where recreational vehicles represent a small percentage, it may be convenient to consider all heavy vehicles as typical trucks. Thus, a traffic stream consisting of 14% trucks, 3% recreational vehicles, and 1% buses might be analyzed as having 18% trucks. It is generally acceptable to do this where the percentage of trucks and buses in the traffic stream is at least five times the percentage of RVs present.

Determination of Free-Flow Speed.   The mean speed of passenger cars measured under low to moderate flows, up to 1300 pc/hr/ln, is the free-flow speed. It can be measured directly in the field or estimated using guidelines. For a field measurement, the study should be conducted at a representative location within the section being evaluated. A systematic sample of at least 100 cars across all lanes during off-peak hours could provide a satisfactory measurement. If field-measured data are used, no subsequent adjustments are made to the free-flow speed as it reflects the net effect of all conditions at the study site. If field measurement of free-flow speed is not possible, it can be estimated using the following equation:

$$FFS = 70 - f_{LW} - f_{LC} - f_N - F_{ID} \tag{3}$$

where

FFS = estimated free-flow speed (mph)

f_{LW} = adjustment for lane width from Table 7-6 (mph)

f_{LC} = adjustment for right-shoulder lateral clearance from Table 7-7 (mph)

f_N = adjustment for number of lanes from Table 7-8 (mph)

f_{ID} = adjustment for interchange density for Table 7-9 (mph).

The ideal free-flow speed (70 mph) is reduced by the adjustment factors in Tables 7-6 to 7-9 when roadway conditions differ from ideal. No adjustment factors are recommended for left-shoulder clearance of less than 2 ft. It is considered rare and professional judgment is recommended. Only mainline lanes, both basic and auxiliary, are included in counting the number of lanes. High-occupancy-vehicle (HOV) lanes should not be included. Interchange density is determined over a 6-mile section of freeway (3 miles upstream and 3 miles downstream) in which the freeway section being studied is located. An interchange is defined as having at least one on-ramp.

TABLE 7-6   ADJUSTMENT FACTORS FOR
LANE WIDTH

Lane width (feet)	Reduction in free-flow speed, f_{LW} (mph)
≥ 12	0.0
11	2.0
10	6.5

Source: NCHRP 3-45, 1995.

TABLE 7-7   ADJUSTMENT FACTORS FOR RIGHT-SHOULDER
LATERAL CLEARANCE

Right-shoulder lateral clearance (feet)	Reduction in free-flow speed, f_{LC} (mph)		
	Lanes (one direction)		
	2	3	4 or more
≥ 6	0.0	0.0	0.0
5	0.6	0.4	0.2
4	1.2	0.8	0.4
3	1.8	1.2	0.6
2	2.4	1.6	0.8
1	3.0	2.0	1.0
0	3.6	2.4	1.2

Source: NCHRP 3-45, 1995.

TABLE 7-8   ADJUSTMENT FACTORS FOR
NUMBER OF LANES

Number of lanes (one direction)	Reduction in free-flow speed, f_N (mph)
≥ 5	0.0
4	1.5
3	3.0
2	4.5

Source: NCHRP 3-45, 1995.

TABLE 7-9 ADJUSTMENT FACTORS FOR INTERCHANGE DENSITY

Interchanges per mile	Reduction in free-flow speed, f_{ID} (mph)
≤ 0.50	0.0
0.75	1.3
1.00	2.5
1.25	3.7
1.50	5.0
1.75	6.3
2.00	7.5

Source: NCHRP 3-45, 1995.

Determination of Level of Service.   The level of service on a basic freeway section can be determined directly from Figure 7-3 using free-flow speed and the flow rate. The following steps illustrate the procedure:

1. Define and segment the freeway section as appropriate.
2. On the basis of measured or estimated free-flow speed, construct a typical curve as shown in Figure 7-3. The curve should intersect the y axis at the free-flow speed.
3. Enter the flow rate, v_p, on the horizontal axis and read up to the curve identified in step 2 and determine the average passenger-car speed and level of service corresponding to that point.
4. Determine the density of flow as follows:

$$D = v_p / S_{pc} \tag{4}$$

where

D = density (pc/mi/ln)
v_p = flow rate (pc/hr/ln)
S_{pc} = average passenger-car speed (mph)

The level of service can also be determined using the density ranges provided in Table 7-1.

3.6 Applications

Procedures for the application of the HCM (TRB, 1994; NCHRP, 1995) to freeway problems fall under three categories: (1) operational analysis, (2) designing, and (3) planning.

Operational Analysis.   Given known or projected geometric and traffic conditions to estimate the LOS, speed, and density of the traffic stream. This analysis is useful in evaluating impacts of alternative designs.

Steps for Operational Analysis

1. Calculate the hourly flow rate in pc/hr/ln for the direction of flow being analyzed using Eq. 1. Compute the heavy-vehicle adjustment factor using Eq. 2 and Tables 7-2 through 7-5.
2. Obtain the free-flow speed through field measurement or estimate using Eq. 3. Use adjustment factors for lane width, right-shoulder lateral clearance, number of lanes, and interchange density from Tables 7-6 through 7-9.
3. Select an appropriate curve in Figure 7-3 or construct an interpolated speed-flow speed curve, if necessary.
4. Reading up from the flow rate as calculated in step 1, determine the average passenger-car speed and level of service.
5. Determine the density using Eq. 4.
6. Determine the maximum service flow rate, maximum volume/capacity (v/c) ratio, and maximum density for a given level of service using Table 7-1.

Designing.   Given a forecast demand volume with known design standards and a desired LOS to determine the number of lanes.

Steps for Designing

1. Assume a four-lane freeway.
2. Convert anticipated traffic characteristics (PHF, percent trucks/buses and RVs) and directional design hour volume to flow rate in pc/hr/ln using Eqs. 1 and 2 and appropriate tables.
3. Establish anticipated roadway conditions from design guidelines, including lane width, right-shoulder clearance, and terrain or specific grade information.
4. Estimate free-flow speed using Eq. 3 and Tables 7-6 through 7-9.
5. Using Figure 7-3, select or interpolate an appropriate speed-flow curve as necessary. Determine the level of service.
6. If the calculated level of service is worse than the desired level of service, repeat steps 2 through 5 for 6, 8, and 10 lanes as necessary.

Planning.   Given a forecast demand volume to determine the number of lanes. This determination is generally a preliminary estimate based on approximate data. At the planning stage, it is quite possible that details of specific grades and geometric features are not available. Capacity analysis is therefore approximate. Only the following data are needed to conduct a planning analysis: a forecast of AADT in the anticipated design year, a forecast of the likely heavy-vehicle percentage, and a general classification of terrain type.

Steps for Planning

1. Convert AADT to DDHV using the relationship

$$DDHV = AADT \times K \times D \tag{5}$$

where

AADT = forecast average annual daily traffic (veh/day)

DDHV = directional design hour volume (veh/hr)

K = percent of AADT occurring in peak hour

D = percent of peak-hour traffic in the heavier direction

Values of K and D are based on local conditions, but in the absence of values the following figures could be used. For K:

Urban freeways: 0.08 to 0.09
Rural freeways: 0.15 to 0.20

As land-use densities increase, the K factor decreases because traffic demand is distributed more smoothly throughout the day. For D:

Urban circumferential freeways: 0.52
Urban radial freeways: 0.55
Rural freeways: 0.75

2. Use either operational or design analysis procedures specified before.

3.7 Examples

Example 1

At a rural section of a freeway, free-flow speed is observed as 66 mph through field measurement. Determine the level of service of this section when the flow rate is 2350 pc/hr/ln.

Solution

A free-flow speed curve for 66 mph is shown in Figure 7-4 with a dashed line. At this speed, the capacity can be observed as 2375 pc/hr/ln. Locating a flow rate of 2350 pc/hr/ln

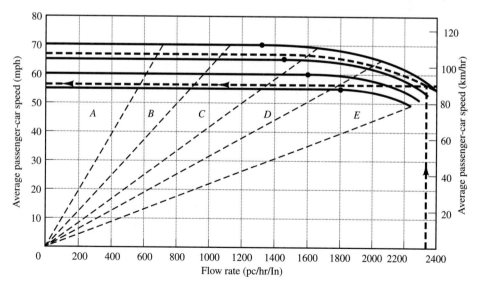

Figure 7-4 Example of Graphic Solution Using Speed-Flow Curves (NCHRP 3-45, 1995).

on the horizontal axis and reading up to the dashed curve, one finds the average passenger-car speed is approximately 56 mph and the level of service is E.

Example 2

Given:

> Four-lane urban freeway
> Interchange density 2.0/mile
> Directional peak flow 1950 veh/hr
> 5% trucks
> PHF 0.90
> 11-ft lanes; obstructions 4 ft from pavement edge
> Rolling terrain

Determine: (a) LOS; and (b) how much additional traffic could be accommodated before reaching capacity.

Solution

(a) Find the service-flow rate:

$$v_p = \frac{V}{\text{PHF} \times N \times f_{HV}} \qquad\qquad f_{HV} : E_T = 3.0 \text{ (rolling terrain)}$$

$$P_T = 0.05$$

$$v_p = \frac{1950 \text{ vph}}{0.90 \times 2 \times 0.90} = 1203 \text{ pc/hr/ln} \qquad f_{HV} = \frac{1}{1 + 0.05(3-1)} = 0.90$$

Find the free-flow speed:

$$\text{FFS} = 70 - f_{LW} - f_{LC} - f_N - F_{ID}$$

$f_{LW} = 2.0$ mph (11 ft lanes; Table 7-6)
$f_{LC} = 1.2$ mph (4 ft clearance; Table 7-7)
$f_N = 4.5$ mph (2 lanes/direction; Table 7-8)
$f_{ID} = 7.5$ mph (2 interchange/mi; Table 7-9)

$$\text{FFS} = 70 - 2.0 - 1.2 - 4.5 - 7.5 = 54.8 \text{ mph}$$

For $v_p = 1203$, FFS = 54.8: LOS reads "C" from the speed-flow curve of Figure 7-5 or Table 7-1.

(b) Find the additional traffic to reach capacity:
Follow the speed-flow curve to terminus of LOS E → 2250 pc/hr/ln or find MSF for LOS E and FFS = 54.8 mph using Table 7-1 → 2250.

Additional traffic (peak 15-minute passenger-car flow rate)

$$= (2 \text{ lanes}) (2250 \text{ pc/hr/ln}) - (2 \text{ lanes}) (1203 \text{ pc/hr/ln})$$

$$= 2094 \text{ pc/hr (directional flow)}$$

To convert to hourly flow:

$$2094 \times \text{PHF} \times f_{HV} = 2094 \times 0.90 \times 0.90 = 1696 \text{ veh/hr}$$

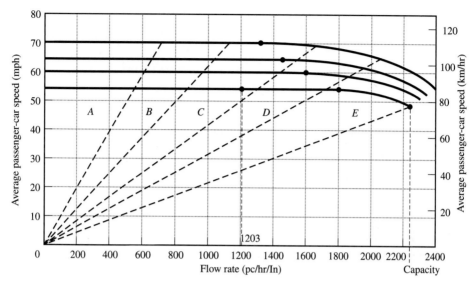

Figure 7-5   Example of Graphic Solution Using Speed-Flow Curves (NCHRP, 1995).

Example 3

Given:

> Level terrain—extended section
> Urban area with 1.5 interchanges/mile
> DDHV 4050 veh/hr
> 10% trucks, no buses, no RVs
> PHF 0.95
> Full shoulder and 12-ft-wide lanes expected to be provided

Determine: Number of lanes to operate at LOS C.

Solution

Determine number of lanes to operate at LOS C.
Find FFS:

$$FFS = 70 - f_{LW} - f_{LC} - f_N - F_{ID}$$

$f_{LW} = 0$ (expected 12-ft lane width)
$f_{LC} = 0$ (expected full clearance)

Make initial assumption for number of lanes:

$f_N = 3.0$ (initially, assume three lanes; Table 7-8)
$f_{ID} = 5.0$ mph (1.5 interchange/mi; Table 7-9)

$$FFS = 70 - 3.0 - 5.0 = 62 \text{ mph}$$

Find N for given DDHV:

$$v_p = \frac{V}{PHF \times N \times f_{HV}} \qquad\qquad f_{HV} = \frac{1}{1 + 0.10(1.5 - 1)} = 0.95$$

$$N = \frac{V}{v_p \times PHF \times f_{HV}} \qquad\qquad P_T = 0.10, E_T = 1.5 \text{ (level; Table 7-2)}$$

$$= \frac{4050 \text{ vph}}{1485 \times 0.95 \times 0.95} = 3.0 \text{ lanes}$$

(v_p = MSF for LOS C and FFS of 62 mph = 1485. Interpolate from Table 7-1 or estimate from speed–flow curve in Figure 7-6.)

Use N = 3 lanes.

Example 4

Given:

> Radial freeway being planned in urban area
> Freeway access expected to be provided at 1-mile spacing
> Expected AADT = 70,000 veh/hr
> 15% trucks
> PHF = 0.90
> Rolling terrain

Determine: Number of lanes for LOS C.

Solution

Choose K = 0.09, D = 0.55.

$$DDHV = AADT \times K \times D$$

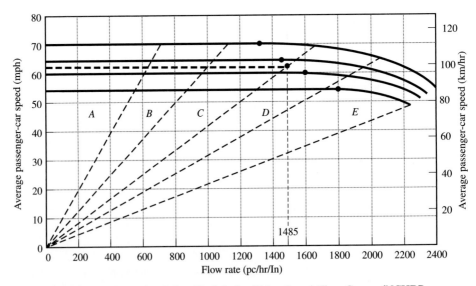

Figure 7-6 Example of Graphic Solution Using Speed–Flow Curves (NCHRP, 1995).

$$\text{DDHV} = 70,000 \times 0.09 \times 0.55 = 3465 \text{ veh/hr}$$

$$\text{FFS} = 70 - f_{LW} - f_{LC} - f_N - F_{ID}$$

$$f_{LW} = 0$$

$$f_{LC} = 0$$

$$f_N = 4.5\,(N = 2), 3.0\,(N = 3), 1.5\,(N = 4)$$

$$f_{ID} = 2.5\,(\text{interchange/mi; Table 7.9})$$

$$\text{FFS} = 70 - 0 - 0 - 4.5 - 2.5 = 63\,(N = 2)$$

$$\text{FFS} = 70 - 0 - 0 - 3.0 - 2.5 = 64.5\,(N = 2)$$

$$\text{FFS} = 70 - 0 - 0 - 1.5 - 2.5 = 66\,(N = 2)$$

Find the service flow for 2, 3, and 4 lanes:

$$f_{HV} = \frac{1}{1 + 0.15(3 - 1)} = 0.77 \qquad E_T = 3.0 \text{ (Table 7-2)}$$

$$N = 2: v_p = \frac{V}{\text{PHF} \times N \times f_{HV}} = \frac{3465}{0.95 \times 2 \times 0.77} = 2368 \text{ pc/hr/ln}$$

$$N = 3: v_p = \frac{V}{\text{PHF} \times N \times f_{HV}} = \frac{3465}{0.95 \times 3 \times 0.77} = 1579 \text{ pc/hr/ln}$$

$$N = 4: v_p = \frac{V}{\text{PHF} \times N \times f_{HV}} = \frac{3465}{0.95 \times 4 \times 0.77} = 1184 \text{ pc/hr/ln}$$

Number of lanes:

N	FFS	v_p	LOS
2	63	2368	F
3	64.5	1579	D
4	66	1184	C

(Use Table 7-1 or speed–flow curve in Figure 7-7 to find LOS.)

4. MULTILANE HIGHWAY CAPACITY

4.1 Definitions and Characteristics

Multilane highways exist in a number of settings, from typical suburban communities leading to central cities or along high-volume rural corridors that connect two cities or significant activities generating a substantial number of daily trips. They generally have posted speed limits of between 40 and 55 mph. They usually have four or six lanes, often with physical medians or two-way left-turn lane (TWLTL), although they also may be undivided. Between two fixed interruption points, such as two signalized inter-

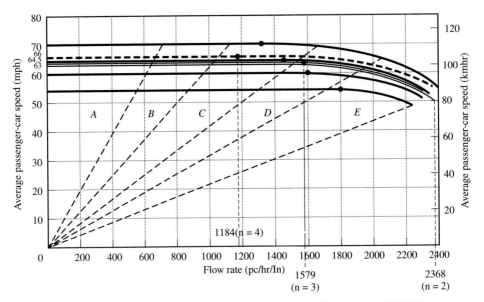

Figure 7-7 Example of Graphic Solution Using Speed-Flow Curves (NCHRP, 1995).

sections at a distance of more than 2 miles, multilane highways operate quite similarly to freeways, with uninterrupted flow conditions, except that the flow is not quite as efficient as one would observe on a freeway. Side and median friction, with vehicles entering and leaving the road and the existence of opposing vehicles on undivided multilane highways, are prevalent. Traffic signals spaced at 2 miles or less typically create urban arterial conditions.

4.2 Procedures

The prediction of level of service for a multilane highway involves three steps:

1. Determination of free-flow speed
2. Adjustment of volume
3. Determination of level of service

Free-flow speed is the theoretical speed of traffic as density approaches zero. It is the speed at which drivers feel comfortable traveling under the physical, environmental, and traffic conditions existing on an uncongested section of multilane highway. In practice, free-flow speed is determined by performing travel-time studies during periods of iow volume. The upper limit for low-volume conditions is considered 1400 passenger cars per hour per lane (pc/hr/ln) for the analyses.

Speed–flow and density–flow characteristics are shown in Figure 7-8. These relationships hold for a typical uninterrupted-flow segment on a multilane highway under either ideal or nonideal conditions in which free-flow speed is known. Figure 7.8(a)

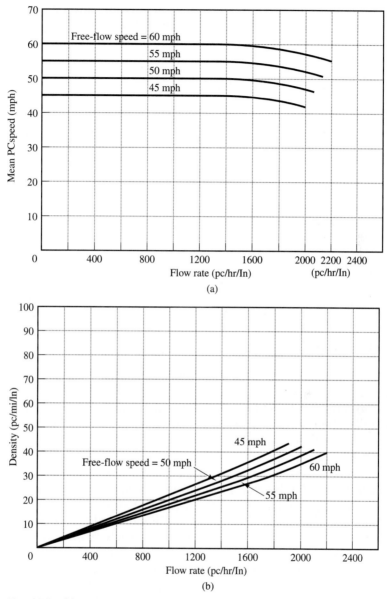

Figure 7-8   (a) Speed–Flow Relationships on Multilane Highways (TRB, 1994). (b) Density–Flow Relationships on Multilane Highways (TRB, 1994).

indicates that the speed of traffic is insensitive to traffic volume up to a flow rate of 1400 pc/hr/ln. It also shows that the capacity of a multilane highway under ideal conditions is 2200 pc/hr/ln for highways with a 60 mph free-flow speed. At flow rates between 1400 and 2200 pc/hr/ln, the speed on a multilane highway drops; for example, by 5 mph for a highway with a free-flow speed of 60 mph. Figure 7.8(b) shows that density varies continuously throughout the full range of flow rates. The capacity value of

2200 pc/hr/ln is representative of the maximum 15-min flow rate that can be accommodated under ideal conditions for 60-mph, free-flow highways.

Based on flow characteristics, ideal conditions for multilane highways are defined as follows:

1. Level terrain, with grades no greater than 1 to 2%
2. 12-ft lane widths
3. A minimum of 12 ft of total lateral clearance in the direction of travel. Clearances are measured from the edge of the traveled lanes (shoulders included) and of 6 ft or greater are considered to be equal to 6 ft
4. No direct access points along the highway
5. A divided highway
6. Only passenger cars in the traffic stream
7. A free-flow speed of 60 mph or more

The maximum sustained hourly flow rate, which forms the basis for defining capacity, is considered a function of density and therefore speed. Because free-flow speed is a direct input to the density computation, adjustments for geometric factors are applied to free-flow speed. Several traffic control, physical, and traffic conditions affect the the free-flow speed along a given highway. If speed enforcement is anticipated to affect operation often, the user may make local measurements to calibrate the relationship between the 85th-percentile speed and the free-flow speed. Design speed and posted speed limits may also impact free-flow speed.

Level-of-service (LOS) criteria for multilane highways are defined in terms of density. Various levels of service are applied to speed–flow curves presented in Figure 7-8(a) to give density boundary values. These LOS boundaries are represented in Figure 7-9 by slope lines, each corresponding to a constant value of density. Complete LOS criteria are given in Table 7-10. For average free-flow speeds of 60, 55, 50 and 45 mph, Table 7-10 gives the average travel speed, the maximum value of v/c, and the corresponding maximum service flow rate (MSF) for each level of service. Under ideal conditions, the speeds, v/c ratios, and MSF tabulated are expected to exist in traffic streams operating at the densities defined for each level of service.

The average of all passenger-car speeds measured in the field under low-volume conditions can be used directly as the free-flow speed if such measurements were taken at flow rates at or below 1400 pc/hr/ln. No adjustments are necessary as this speed reflects the net effect of all conditions at the site that influence speed, including lane width, lateral clearance, type of median, access points, posted speed limits, and horizontal and vertical alignment. Free-flow speed also can be estimated from 85th-percentile speed or posted speed limits, if it is not possible to measure directly in the field. Recent data for multilane highways indicate that mean free-flow speed under ideal conditions ranges from approximately 1 mph lower than the 85th-percentile speed when the latter is 40 mph to 3 mph lower when the 85th-percentile speed is 60 mph. Other research suggests that free-flow speed under ideal conditions is 7 mph higher

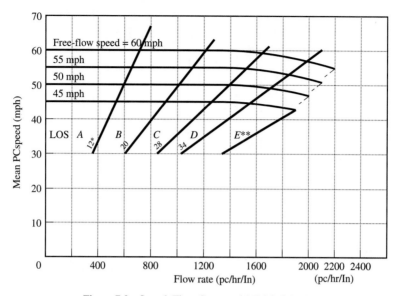

Figure 7-9   Speed–Flow Curves with LOS Criteria.

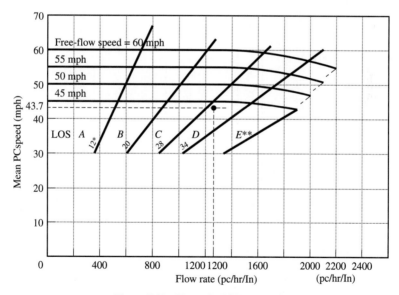

Figure 7-10   Example 5 (TRB, 1994).

*Maximum density for respective level of service. **Maximum densities for LOS E occur at volume-to-capacity ratio of 1.0. They are 40, 41, 43, and 45 pc/mi/ln at free-flow speeds of 60, 55, 50, and 45 mph respectively, TRB, 1994.

TABLE 7-10 LEVEL-OF-SERVICE CRITERIA FOR MULTILANE HIGHWAYS

Level of service	Free-flow speed															
	60 mph				55 mph				50 mph				45 mph			
	Max. density (pc/mi/ln)	Average speed (mph)	Max. v/c	Max. service-flow rate (pc/hr/ln)	Max. density (pc/mi/ln)	Average speed (mph)	Max. v/c	Max. service-flow rate (pc/hr/ln)	Max. density (pc/mi/ln)	Average speed (mph)	Max. v/c	Max. service-flow rate (pc/hr/ln)	Max. density (pc/mi/ln)	Average speed (mph)	Max. v/c	Max. service-flow rate (pc/hr/ln)
A	12	60	0.33	720	12	55	0.31	660	12	50	0.30	600	12	45	0.28	540
B	20	60	0.55	1200	20	55	0.52	1100	20	50	0.50	1000	20	45	0.47	900
C	28	59	0.75	1650	28	54	0.72	1510	28	50	0.70	1400	28	45	0.66	1260
D	34	57	0.89	1940	34	53	0.86	1800	34	49	0.84	1670	34	44	0.79	1500
E	40	55	1.00	2200	41	51	1.00	2100	43	47	1.00	2000	45	42	1.00	1900

Note: The exact mathematical relationship between density and v/c has not always been maintained at LOS boundaries because of the rounded values. Density is the primary determinant of LOS. LOS F is characterized by highly unstable and variable traffic flow. Prediction of accurate flow rate, density, and speed at LOS F is difficult.

than the speed limit for 40- to 50-mph speed limits and 5 mph higher for 50- to 55-mph speed limits.

When field data are not available, the free-flow speed can be estimated indirectly as follows:

$$\text{FFS} = \text{FFS}_I - F_M - F_{LW} - F_{LC} - F_A$$

where

$\text{FFS} = $ estimated free-flow speed (mph)
$\text{FFS}_I = $ estimated free-flow speed for ideal conditions (mph)
$F_M = $ adjustment for median type (from Table 7-11)
$F_{LW} = $ adjustment for lane width (from Table 7-12)
$F_{LC} = $ adjustment for lateral clearance (from Table 7-13)
$F_A = $ adjustment for access points (from Table 7-14)

The analyst should consider dividing the highway into homogeneous sections that reflect changes in median characteristics before applying adjustments as in Table 7-11. According to Table 7-12, the adjustment in mph increases as the lane width decreases from an ideal lane width of 12 ft. No data exist for lane widths less than 10 ft. Fixed obstructions whose lateral clearance effects should be considered using Table 7-13 include light standards, signs, trees, abutments, bridge rails, traffic barriers, and retaining walls. Total lateral clearance for Table 7-13 is defined as

$$\text{TLC} = \text{LC}_R + \text{LC}_L$$

TABLE 7-11   ADJUSTMENT FOR MEDIAN TYPE

Median type	Reduction in free-low speed (mph)
Undivided highways	1.6
Divided highways (including TWLTLs)	0.0

TABLE 7-12   ADJUSTMENT FOR LANE WIDTH

Lane width (ft)	Reduction in free-flow speed (mph)
10	6.6
11	1.9
12	0.0

TABLE 7-13   ADJUSTMENT FOR LATERAL CLEARANCE

Four-lane highways		Six-lane highways	
Total lateral clearance[a] (ft)	Reduction in free-flow speed (mph)	Total lateral clearance[a] (ft)	Reduction in free-flow speed (mph)
12	0.0	12	0.0
10	0.4	10	0.4
8	0.9	8	0.9
6	1.3	6	1.3
4	1.8	4	1.7
2	3.6	2	2.8
0	5.4	0	3.9

[a] Total lateral clearance is the sum of the lateral clearances of the median (if greater than 6 ft, use 6 ft) and shoulder (if greater than 6 ft, use 6 ft). Therefore, for analysis purposes, total lateral clearance cannot exceed 12 ft.

TABLE 7-14   ACCESS-POINT DENSITY ADJUSTMENT

Access points per mile	Reduction in free-flow speed (mph)
0	0.0
10	2.5
20	5.0
30	7.5
40 or more	10.0

where

TLC = total lateral clearance (ft)

LC_R = lateral clearance (ft) from the right edge of the traveled lanes to roadside obstructions (if greater than 6 ft, use 6 ft)

LC_L = lateral clearance (ft) from the left edge of the travel lanes to obstructions in the roadway median.

For undivided highways, there is no adjustment for the left-side lateral clearance as this is already accounted for in the median type. Therefore, in order to use Table 7-13 for undivided highways, the lateral clearance on the left edge is always 6 ft, as it is for roadways with TWLTLs.

The access-point density, for use in Table 7-14, for a divided roadway is found by dividing the total number of access points (intersections and driveways) on the right side of the roadway in the direction of travel being studied by the length of the section in miles. Only significant access points should be included. For one-way multilane highways, driveways on both sides of the roadway should be included. When data on the number of access points on a highway section are unavailable, the guidelines presented in Table 7-15 may be used. Research indicates that there are no significant dif-

TABLE 7-15   NUMBER OF ACCESS POINTS FOR
GENERAL DEVELOPMENT ENVIRONMENTS

Type of development	Access points per mile (one side of roadway)
Rural	0–10
Low-density suburban	11–20
High-density suburban	21 or more

ferences in free-flow speed relationships for commuter or noncommuter traffic conditions. Therefore, no adjustment for driver populations is generally recommended.

Adjustments for peak-hour factor and the heavy-vehicle adjustment factor must be made to hourly volume counts or estimates to arrive at the equivalent passenger-car flow rates used in LOS analysis. The adjustments are applied as follows:

$$v_p = \frac{V}{N \times \text{PHF} \times f_{HV}}$$

where

v_p = service flow rate (pc/hr/ln)

V = volume (number of vehicles passing a point in 1 hr)

N = number of lanes

PHF = peak-hour factor

f_{HV} = heavy-vehicle adjustment factor

The PHFs for multilane highways have been observed to range from 0.76 to 0.99. Lower values are typical of rural or off-peak conditions, whereas higher factors are typical of urban and suburban peak-hour conditions. Where local data are not available, 0.85 is a reasonable estimate of the PHF for rural multilane highways and 0.92 for suburban multilane highways.

Adjustment for the presence of heavy vehicles in the traffic stream applies for three types of vehicles: trucks, buses, and RVs. There is no evidence to indicate any distinct differences in performance characteristics between buses and trucks on multilane highways, and thus the total population is combined. Determination of heavy-vehicle adjustment factor requires two steps:

1. Find the equivalent factor for trucks and buses (E_T) and the RVs (E_R) for the prevailing operating conditions. Values of passenger-car equivalents are selected from Tables 7-16 through 7-19 for a variety of basic conditions.
2. Using values from step 1, compute the heavy-vehicle adjustment factor from Eq. 2.

Passenger-car equivalents are selected based on either extended general highway segments or specific grades. A long multilane highway segment may be classified as a general segment if no grades exceeding 3% are longer than 1/2 mile or grades of 3% or less do not exceed 1 mile. Such segments are further subclassified as level, rolling, and

TABLE 7-16 PASSENGER-CAR EQUIVALENTS ON EXTENDED
GENERAL MULTILANE HIGHWAY SEGMENTS

	Type of terrain		
Factor	Level	Rolling	Mountainous
E_T (trucks and buses)	1.5	3.0	6.0
E_R (RVs)	1.2	2.0	4.0

mountainous terrain. Values of E_T and E_R for extended general segments are selected from Table 7-16. Tables 7-17 and 7-18 give passenger-car equivalents for trucks and buses (E_T) and for RVs (E_R), respectively, on uniform upgrades. Table 7-17 is based on an average weight-to-horsepower ratio of 167 lb/hp, which is typical of the truck population currently found on multilane highways.

When several consecutive grades of different steepness form a composite grade, an average-grade technique is applied to compute a uniform grade, which is used to enter the tables. The average-grade technique computes the uniform grade based on the total rise from the beginning of the grade divided by the total horizontal distance over which the rise was accomplished. The average-grade technique is reasonably accurate for grades of 4000 ft or less, or no greater than 4%.

The level of service on a multilane highway can be determined directly from Figure 7-9 on the basis of the free-flow speed (FFS) and the service flow rate (v_p) in pc/hr/ln. The procedure is as follows:

1. Define and segment the highway as appropriate. The following conditions help define the segmenting of the highway:

- change in median treatment
- change in grade of 2% or more or a constant upgrade over 4000 ft long
- change in the number of travel lanes
- the presence of a traffic signal
- a significant change in the density of access points
- different speed limits
- the presence of a bottleneck condition

In general, the minimum length of study section should be 2500 ft, and the limits should be no closer than 1/4 mi from a signalized intersection.

2. On the basis of the actual free-flow speed on a highway segment, an appropriate speed–flow curve of the same shape as the typical curves in Figure 7-9 is drawn.

3. Locate the point on the horizontal axis corresponding to the appropriate flow rate (v_p) in pc/hr/pl and draw a vertical line.

4. Read up to the FFS curve identified in step 2 and determine the average travel speed at the point of intersection.

5. Determine the level of service on the basis of density region in which this point is located. The density can also be computed as

TABLE 7-17 PASSENGER-CAR EQUIVALENTS FOR TRUCKS AND BUSES ON UNIFORM UPGRADES

Grade (%)	Length (mi)	E_T [a]								
Percent trucks and buses		2	4	5	6	8	10	15	20	25
<2	All	1.5	1.5	1.5	1.5	1.5	1.5	1.5	1.5	1.5
2	$0-\frac{1}{4}$	1.5	1.5	1.5	1.5	1.5	1.5	1.5	1.5	1.5
	$\frac{1}{4}-\frac{1}{2}$	1.5	1.5	1.5	1.5	1.5	1.5	1.5	1.5	1.5
	$\frac{1}{2}-\frac{3}{4}$	1.5	1.5	1.5	1.5	1.5	1.5	1.5	1.5	1.5
	$\frac{3}{4}-1$	2.5	2.0	2.0	2.0	1.5	1.5	1.5	1.5	1.5
	$1-1\frac{1}{2}$	4.0	3.0	3.0	3.0	2.5	2.5	2.0	2.0	2.0
	$>1\frac{1}{2}$	4.5	3.5	3.0	3.0	2.5	2.5	2.0	2.0	2.0
3	$0-\frac{1}{4}$	1.5	1.5	1.5	1.5	1.5	1.5	1.5	1.5	1.5
	$\frac{1}{4}-\frac{1}{2}$	3.0	2.5	2.5	2.0	2.0	2.0	2.0	1.5	1.5
	$\frac{1}{2}-\frac{3}{4}$	6.0	4.0	4.0	3.5	3.5	3.0	2.5	2.5	2.0
	$\frac{3}{4}-1$	7.5	5.5	5.0	4.5	4.0	4.0	3.5	3.0	3.0
	$1-1\frac{1}{2}$	8.0	6.0	5.5	5.0	4.5	4.0	4.0	3.5	3.0
	$>1\frac{1}{2}$	8.5	6.0	5.5	5.0	4.5	4.5	4.0	3.5	3.0
4	$0-\frac{1}{4}$	1.5	1.5	1.5	1.5	1.5	1.5	1.5	1.5	1.5
	$\frac{1}{4}-\frac{1}{2}$	5.5	4.0	4.0	3.5	3.0	3.0	3.0	2.5	2.5
	$\frac{1}{2}-\frac{3}{4}$	9.5	7.0	6.5	6.0	5.5	5.0	4.5	4.0	3.5
	$\frac{3}{4}-1$	10.5	8.0	7.0	6.5	6.0	5.5	5.0	4.5	4.0
	>1	11.0	8.0	7.5	7.0	6.0	6.0	5.0	5.0	4.5
5	$0-\frac{1}{4}$	2.0	2.0	1.5	1.5	1.5	1.5	1.5	1.5	1.5
	$\frac{1}{4}-\frac{1}{3}$	6.0	4.5	4.0	4.0	3.5	3.0	3.0	2.5	2.0
	$\frac{1}{3}-\frac{1}{2}$	9.0	7.0	6.0	6.0	5.5	5.0	4.5	4.0	3.5
	$\frac{1}{2}-\frac{3}{4}$	12.5	9.0	8.5	8.0	7.0	7.0	6.0	6.0	5.0
	$\frac{3}{4}-1$	13.0	9.5	9.0	8.0	7.5	7.0	6.5	6.0	5.5
	>1	13.0	9.5	9.0	8.0	7.5	7.0	6.5	6.0	5.5
6	$0-\frac{1}{4}$	4.5	3.5	3.0	3.0	3.0	2.5	2.5	2.0	2.0
	$\frac{1}{4}-\frac{1}{3}$	9.0	6.5	6.0	6.0	5.0	5.0	4.0	3.5	3.0
	$\frac{1}{3}-\frac{1}{2}$	12.5	9.5	8.5	8.0	7.0	6.5	6.0	6.0	5.5
	$\frac{1}{2}-\frac{3}{4}$	15.0	11.0	10.0	9.5	9.0	8.0	8.0	7.5	6.5
	$\frac{3}{4}-1$	15.0	11.0	10.0	9.5	9.0	8.5	8.0	7.5	6.5
	>1	15.0	11.0	10.0	9.5	9.0	8.5	8.0	7.5	6.5

Note: If the length of grade falls on a boundary condition, the equivalent from the longer-grade category is used.
[a] Four- or six-lane highways.

TABLE 7-18 PASSENGER-CAR EQUIVALENTS FOR RECREATIONAL VEHICLES ON UNIFORM UPGRADES

Grade (%)	Length (mi)	E_R [a]								
Percent RVs		2	4	5	6	8	10	15	20	25
≤ 2	All	1.2	1.2	1.2	1.2	1.2	1.2	1.2	1.2	1.2
3	$0-\frac{1}{2}$	1.2	1.2	1.2	1.2	1.2	1.2	1.2	1.2	1.2
	$>\frac{1}{2}$	2.0	1.5	1.5	1.5	1.5	1.5	1.2	1.2	1.2
4	$0-\frac{1}{4}$	1.2	1.2	1.2	1.2	1.2	1.2	1.2	1.2	1.2
	$\frac{1}{4}-\frac{1}{2}$	2.5	2.5	2.0	2.0	2.0	2.0	1.5	1.5	1.5
	$>\frac{1}{2}$	3.0	2.5	2.5	2.0	2.0	2.0	2.0	1.5	1.5
5	$0-\frac{1}{4}$	2.5	2.0	2.0	2.0	1.5	1.5	1.5	1.5	1.5
	$\frac{1}{4}-\frac{1}{2}$	4.0	3.0	3.0	3.0	2.5	2.5	2.0	2.0	2.0
	$>\frac{1}{2}$	4.5	3.5	3.0	3.0	3.0	2.5	2.5	2.0	2.0
6	$0-\frac{1}{4}$	4.0	3.0	2.5	2.5	2.5	2.0	2.0	2.0	1.5
	$\frac{1}{4}-\frac{1}{2}$	6.0	4.0	4.0	3.5	3.0	3.0	2.5	2.5	2.0
	$>\frac{1}{2}$	6.0	4.5	4.0	4.0	3.5	3.0	3.0	2.5	2.0

Note: If a length of grade falls on a boundary condition, the equivalent from the longer-grade category is used.
[a] Four- or six-lane highways.

TABLE 7-19 PASSENGER-CAR EQUIVALENTS FOR TRUCKS ON DOWNGRADES

Downgrade (%)	Length (mi)	E_T [a]			
Percent trucks		5	10	15	20
< 4	All	1.5	1.5	1.5	1.5
4	≤ 4	1.5	1.5	1.5	1.5
4	> 4	2.0	2.0	2.0	1.5
5	≤ 4	1.5	1.5	1.5	1.5
5	> 4	5.5	4.0	4.0	3.0
6	≤ 2	1.5	1.5	1.5	1.5
6	> 2	7.5	6.0	5.5	4.5

[a] Four- or six-lane highways.

$$D = v_p/S$$

where:

D = density (pc/mi/ln)
v_p = service flow rate (pc/hr/ln)
S = average passenger-car travel speed (mph)

The level of service can then be determined from the density ranges shown in Table 7-10.

To use the procedures for design, a forecast of future traffic volumes has to be made and the general geometric and traffic control conditions, such as speed limits, must be estimated. With these data and a threshold level of service, an estimate of the number of lanes required for each direction of travel can be made.

4-3 Examples

Example 5

An undivided, 2-mile-long multilane segment of the Pacific Coast Highway is located in mountainous terrain and has a DDHV of 1500 veh/hr with a PHF = 0.95 during the tourist season. The traffic includes 5% trucks, 3% buses and 7% RVs. The segment has four 11.5-ft lanes, 3 ft of lateral clearance on both sides, and an average of 8 accesses/mile on each side. If the 85th-percentile speed is 50 mph, what is the LOS along this segment?

Solution

$$FFS = FFS_I - F_M - F_{LW} - F_{LC} - F_A$$

FFS_I = 85th-percentile speed − 2.0 mph (interpolated) = 50 − 2 = 48 mph
F_M = 1.6 (undivided; Table 7-11)
F_{LW} = 0.0 (12-ft lanes; Table 7-12)
F_{LC} = 0.7 (TLC = 6 + 3 = 9 ft; interpolate Table 7-13)
F_A = 2 (access/mile; Table 7-14)

$$FFS = 48 - 1.6 - 0.0 - 0.7 - 2 = 43.7 \text{ mph}$$

Service flow rate:

$E_T = 6, E_R = 4$ (Table 7-16)

$$f_{HV} = \frac{1}{1 + P_T(E_T - 1) + P_R(E_R - 1)} = \frac{1}{1 + 0.08(6 - 1) + 0.07(4 - 1)} = 0.621$$

$$v_p = \frac{V}{N \times PHF \times f_{HV}} = \frac{1500}{2 \times 0.95 \times 0.621} = 1271 \text{ pc/hr/ln}$$

Using v_p and FFS, read the level of service D as shown in Figure 7-10, or calculate the density as follows:

$$\text{Density} = \frac{v_p}{\text{average passenger-car speed}} = \frac{1271}{43.7} = 29.0 \text{ pc/mi/ln}$$

Using v_p, FFS, and the density, read the level of service D from Table 7-10.

Example 6

A segment of U.S. Hwy 101 near the Oregon coast is in rolling terrain and has an access density of 10 accesses/mile in the southerly direction and 4 accesses/mile in the northerly direction. The segment is a multilane divided highway with the following features: four 11-ft wide lanes, obstructions 4 ft away from the traveled lane on the right sides, and an 8-ft-wide median. The posted speed limit is 45 mph. If this segment carries a peak-hour demand of 2300 veh/hr in one direction with 10% trucks and a PHF of 0.90, what LOS can be expected in this segment?

Solution

$$FFS = FFS_I - F_M - F_{LW} - F_{LC} - F_A$$

FFS_I = posted speed limit + 7 mph = 45 mph + 7 mph = 52 mph

F_M = 0.0 (divided highway; Table 7-11)

F_{LW} = 1.9 mph (Table 7-12)

F_{LC} = 0.4 mph (TLC = 4 + 6 = 10 ft; Table 7-13)

F_A (south) = 2.5 mph (10 access/mile; Table 7-14)

F_A (north) = 1.0 mph (4 access/mile; Table 7-14)

$$FFS\ (\text{south}) = 52 - 0.0 - 1.9 - 0.4 - 2.5 = 47.2 \text{ mph}$$

$$FFS\ (\text{north}) = 52 - 0.0 - 1.9 - 0.4 - 1.0 = 48.7 \text{ mph}$$

Given:

Volume = 2300 vph   rolling terrain, 10% trucks

PHF = 0.90

N = 2 lanes each direction

Heavy-vehicle adjustment:

E_T = 3.0, for rolling terrain

$$f_{HV} = \frac{1}{1 + P_T(E_T - 1)} = \frac{1}{1 + 0.10(3 - 1)} = 0.833$$

Service flow rate:

$$v_p = \frac{V}{N \times \text{PHF} \times f_{HV}} = \frac{2300}{2 \times 0.90 \times 0.833} = 1534 \text{ pc/hr/ln}$$

$$\text{Density (south)} = \frac{v_p}{\text{average passenger-car speed}} = \frac{1534}{47.2} = 32.5 \text{ pc/mi/ln}$$

$$\text{Density (north)} = \frac{v_p}{\text{average passenger-car speed}} = \frac{1534}{48.4} = 31.7 \text{ pc/mi/ln}$$

Both north and south densities fall into LOS D, as found in either Table 7-10 or from speed–flow curves. See Figure 7.11.

5. TWO-LANE HIGHWAY CAPACITY

5.1 Characteristics and Parameters

Two-lane highways are important because they compose the predominant mileage of most highway systems. In the United States, only about 30% of all travel occurs on rural two-lane roads, even though this network comprises 80% of all paved rural highways. Generally, two-lane highways carry light traffic and experience few operational problems. For practical reasons, three parameters are used to describe service quality for two-lane highways:

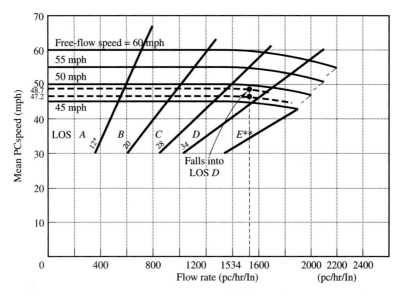

Figure 7-11   Example 6 (TRB, 1994).

1. Average travel speed represents the mobility function and equals the segment length of highways divided by the average travel time of vehicles traversing the segment in both directions.
2. Percent travel delay represents both mobility and accessibility functions and equals the average percent of time that vehicles are delayed while traveling in platoons due to their inability to pass. This delay occurs because faster vehicles wanting to pass slower vehicles require the use of the opposing lane, provided that sight distance and gaps in the opposing traffic stream permit such a maneuver. A surrogate measure of delay is headway in seconds.
3. Utilization of capacity is essentially the V/c ratio.

LOS criteria make use of all the parameters just mentioned, with percent time delay being the primary measure of service quality, and speed and capacity are secondary measures.

5.2 Ideal Conditions

The basic relationship between average travel speed, percent time delay, and flow are shown in Figure 7-12 under ideal traffic and roadway conditions. Ideal conditions include the following:

1. Design speed greater than or equal to 60 mph
2. Lane widths greater than or equal to 12 ft

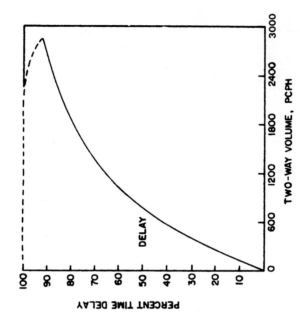

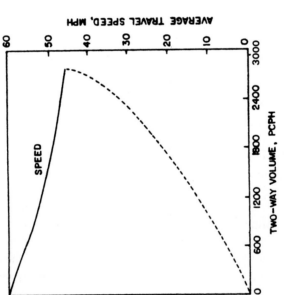

Figure 7-12 Speed–Flow and Percent Time Delay–Flow Relationships for Two-Lane Rural Highways (Ideal Conditions). (a) Relationship Between Average Speed and Flow on Two-Lane Highways. (b) Relationship Between Percent Time Delay and Flow on Two-Lane Highways (TRB, 1994).

3. Clear shoulders wider than or equal to 6 ft

4. No "no passing zones" on the highway

5. All passenger cars in the traffic stream

6. A 50/50 directional split of traffic

7. No impediments to through traffic due to traffic control or turning vehicles

8. Level terrain

Table 7-20 gives the level-of-service criteria for general terrain segments. The percentage time delay is indicated for each level of service. When a two-lane highway passes through a village or town, the speed is usually restricted, in which case the percentage of time delay and capacity utilization are the only meaningful indicators of level of service. For specific grade segments, the average upgrade speeds for different levels of service are given in Table 7-21. Downgrade operations are not addressed by these procedures. When motorists are able to drive at their own desired speeds, it represents the highest quality of traffic service. The highest volume attainable under level of service E defines the capacity of the highway; under ideal conditions, capacity is 2800 pc/hr, total in both directions. Capacity of two-lane highways is affected by directional split of traffic (Table 7-22).

5.3 Operational Analysis

The objective of operational analysis is generally the determination of level of service for an existing or projected facility operating under existing or projected traffic demand. Two-lane highway analysis is based on flow rates for a peak 15-minute period within the hour of interest, which is usually the peak hour.

$$\text{PHF} = \frac{V}{q_{15}} \tag{6}$$

where

q_{15} = flow rate for the peak 15-minute total for both directions (veh/hr)

V = full-hour volume for both directions (veh/hr)

PHF = peak-hour factor

Local field data should be used in all cases. Where these data are not available, factors tabulated in Table 7-23 can be used. Traffic data needed to apply the general terrain methodology include the two-way hourly volume, a peak-hour factor, and the directional distribution of traffic flow. The general relationship is as follows:

$$\text{SF}_i = 2800 \times \left(\frac{V}{c}\right)_i \times f_d \times f_w \times f_{HV} \tag{7}$$

TABLE 7-20 LEVEL-OF-SERVICE CRITERIA FOR GENERAL TWO-LANE HIGHWAY SEGMENTS

V/c ratio[a]

LOS	Percent time delay	Level terrain Avg[b] speed	0	20	40	60	80	100	Rolling terrain Avg[b] speed	0	20	40	60	80	100	Mountainous terrain Avg[b] speed	0	20	40	60	80	100
			Percent no-passing zones							Percent no-passing zones							Percent no-passing zones					
A	≤ 30	≥ 58	0.15	0.12	0.09	0.07	0.05	0.04	≥ 57	0.15	0.10	0.07	0.05	0.04	0.03	≥ 56	0.14	0.09	0.07	0.04	0.02	0.01
B	≤ 45	≥ 55	0.27	0.24	0.21	0.19	0.17	0.16	≥ 54	0.26	0.23	0.19	0.17	0.15	0.13	≥ 54	0.25	0.20	0.16	0.13	0.12	0.10
C	≤ 60	≥ 52	0.43	0.39	0.36	0.34	0.33	0.32	≥ 51	0.42	0.39	0.35	0.32	0.30	0.28	≥ 49	0.39	0.33	0.28	0.23	0.20	0.16
D	≤ 75	≥ 50	0.64	0.62	0.60	0.59	0.58	0.57	≥ 49	0.62	0.57	0.52	0.48	0.46	0.43	≥ 45	0.58	0.50	0.45	0.40	0.37	0.33
E	> 75	≥ 45	1.00	1.00	1.00	1.00	1.00	1.00	≥ 40	0.97	0.94	0.92	0.91	0.90	0.90	≥ 35	0.91	0.87	0.84	0.82	0.80	0.78
F	100	< 45	—	—	—	—	—	—	< 40	—	—	—	—	—	—	< 35	—	—	—	—	—	—

[a] Ratio of flow rate to an ideal capacity of 2800 pc/hr in both directions.
[b] Average travel speed of all vehicles (in mph) for highways with design speed ≥ 60 mph; for highways with lower design speeds, reduce speed by 4 mph for each 10-mph reduction in design speed below 60 mph; assumes that speed is not restricted to lower values by regulation.
Source: TRB, 1994.

TABLE 7-21   LEVEL-OF-SERVICE CRITERIA FOR SPECIFIC GRADES

Level of service	Average upgrade speed (mph)
A	≥ 55
B	≥ 50
C	≥ 45
D	≥ 40
E	$\geq 25\text{–}40$[a]
F	$< 25\text{–}40$[a]

[a] The exact speed at which capacity occurs varies with the percentage and length of grade, traffic composition, and volume; computational procedures are provided to find this value.
Source: TRB, 1994.

TABLE 7-22   CAPACITY OF TWO-LANE HIGHWAYS

Directional split	Total capacity (pc/hr)	Ratio of capacity to ideal capacity
50/50	2800	1.00
60/40	2650	0.94
70/30	2500	0.89
80/20	2300	0.83
90/10	2100	0.75
100/0	2000	0.71

TABLE 7-23   PEAK-HOUR FACTOR FOR TWO-LANE HIGHWAYS BASED ON RANDOM FLOW

(a) Level-of-service determinations

Total 2-way hourly volume (veh/hr)	Peak-hour factor, PHF	Total 2-way hourly volume (veh/hr)	Peak-hour factor, PHF
100	0.83	1000	0.93
200	0.87	1100	0.94
300	0.90	1200	0.94
400	0.91	1300	0.94
500	0.91	1400	0.94
600	0.92	1500	0.95
700	0.92	1600	0.95
800	0.93	1700	0.95
900	0.93	1800	0.95
		≥ 1900	0.96

(b) Service flow-rate determinations

Level of service:	A	B	C	D	E
Peak-hour factor:	0.91	0.92	0.94	0.95	1.00

Source: TRB, 1994.

where

SF_i = Total service flow rate in both directions for prevailing roadway and traffic conditions, for LOS i (veh/hr)

$(V/c)_i$ = ratio of flow rate to ideal capacity for LOS i from Table 7-20

f_d = adjustment factor for directional distribution from Table 7-24

f_w = adjustment factor for narrow lanes and restricted shoulder widths in traffic stream from Table 7-25

f_{HV} = adjustment factor for the presence of heavy vehicles in traffic stream
= $1/[1 + P_T(E_T - 1) + P_R(E_R - 1) + P_B(E_B - 1)]$ from Table 7-26

Operational Analysis of Grades of Significant Length. The analysis of extended specific grades on two-lane highways is more complex than for general terrain segments. Any sustained grade of more than 3% and longer than $\frac{1}{2}$ mile, and 3% or less and longer than 1 mile, is analyzed as a specific grade of significant length. In such sections, one finds the formation of platoons on the upgrade and passing becomes

TABLE 7-24 ADJUSTMENT FACTOR FOR DIRECTIONAL DISTRIBUTION ON GENERAL TERRAIN SEGMENTS

Directional distribution	100/0	90/10	80/20	70/30	60/40	50/50
Adjustment factor, f_d	0.71	0.75	0.83	0.89	0.94	1.00

Source: TRB, 1994.

TABLE 7-25 ADJUSTMENT FACTORS FOR THE COMBINED EFFECT OF NARROW LANES AND RESTRICTED SHOULDER WIDTH, f_w

Usable[a] shoulder width (ft)	12-ft lanes LOS A–D	12-ft lanes LOS[b] E	11-ft lanes LOS A–D	11-ft lanes LOS[b] E	10-ft lanes LOS A–D	10-ft lanes LOS[b] E	9-ft lanes LOS A–D	9-ft lanes LOS[b] E
≥ 6	1.00	1.00	0.93	0.94	0.84	0.87	0.70	0.76
4	0.92	0.97	0.85	0.92	0.77	0.85	0.65	0.74
2	0.81	0.93	0.75	0.88	0.68	0.81	0.57	0.70
0	0.70	0.88	0.65	0.82	0.58	0.75	0.49	0.66

[a] Where shoulder width is different on each side of the roadway, use the average shoulder width.
[b] Factor applies for all speeds less than 45 mph.
Source: TRB, 1994.

TABLE 7-26 AVERAGE PASSENGER-CAR EQUIVALENTS FOR TRUCKS, RVs, AND BUSES ON TWO-LANE HIGHWAYS OVER GENERAL TERRAIN SEGMENTS

Type of vehicle	Level of service	Type of terrain		
		Level	Rolling	Mountainous
Trucks, E_T	A	2.0	4.0	7.0
	B and C	2.2	5.0	10.0
	D and E	2.0	5.0	12.0
RVs, E_R	A	2.2	3.2	5.0
	B and C	2.5	3.9	5.2
	D and E	1.6	3.3	5.2
Buses, E_B	A	1.8	3.0	5.7
	B and C	2.0	3.4	6.0
	D and E	1.6	2.9	6.5

Source: TRB, 1994.

more difficult. LOS criteria are keyed to the average travel speed, which is most sensitive to grade.

LOS A: ≥ 55 mph

LOS B: ≥ 50 mph

LOS C: ≥ 45 mph

LOS D: ≥ 40 mph

LOS E: ≥ 25–40 mph

Part of the analysis of two-lane highways is the determination of capacity and the operating conditions under which it occurs.

The general relationship for sustained grades on two-lane highways is as follows:

$$SF_i = 2800 \times \left(\frac{V}{c}\right)_i \times f_d \times f_w \times f_g \times f_{HV} \qquad (8)$$

where the symbols are the same as before except for f_g is the adjustment factor for grades. This relationship for specific grades is generally not applied to grades of less than 3% or shorter than $\frac{1}{2}$ mile.

Table 7-27 provides the basic relationship governing values of V/c. Note that the V/c values are not specifically related to LOS, but to average upgrade speed. Also, a V/c of 1.00 does not signify capacity.

TABLE 7-27 VALUES OF V/c RATIO[a] VERSUS SPEED, PERCENT GRADE, AND PERCENT NO-PASSING ZONES FOR SPECIFIC GRADES[b]

Percent grade	Average upgrade speed (mph)	Percent no-passing zones					
		0	20	40	60	80	100
3	55	0.27	0.23	0.19	0.17	0.14	0.12
	52.5	0.42	0.38	0.33	0.31	0.29	0.27
	50	0.64	0.59	0.55	0.52	0.49	0.47
	45	1.00	0.95	0.91	0.88	0.86	0.84
	42.5	1.00	0.98	0.97	0.96	0.95	0.94
	40	1.00	1.00	1.00	1.00	1.00	1.00
4	55	0.25	0.21	0.18	0.16	0.13	0.11
	52.5	0.40	0.36	0.31	0.29	0.27	0.25
	50	0.61	0.56	0.52	0.49	0.47	0.45
	45	0.97	0.92	0.88	0.85	0.83	0.81
	42.5	0.99	0.96	0.95	0.94	0.93	0.92
	40	1.00	1.00	1.00	1.00	1.00	1.00
5	55	0.21	0.17	0.14	0.12	0.10	0.08
	52.5	0.36	0.31	0.27	0.24	0.22	0.20
	50	0.57	0.49	0.45	0.41	0.39	0.37
	45	0.93	0.84	0.79	0.75	0.72	0.70
	42.5	0.97	0.90	0.87	0.85	0.83	0.82
	40	0.98	0.96	0.95	0.94	0.93	0.92
	35	1.00	1.00	1.00	1.00	1.00	1.00
6	55	0.12	0.10	0.08	0.06	0.05	0.04
	52.5	0.27	0.22	0.18	0.16	0.14	0.13
	50	0.48	0.40	0.35	0.31	0.28	0.26
	45	0.49	0.76	0.68	0.63	0.59	0.55
	42.5	0.93	0.84	0.78	0.74	0.70	0.67
	40	0.97	0.91	0.87	0.83	0.81	0.78
	35	1.00	0.96	0.95	0.93	0.91	0.90
	30	1.00	0.99	0.99	0.98	0.98	0.98
7	55	0.00	0.00	0.00	0.00	0.00	0.00
	52.5	0.13	0.10	0.08	0.07	0.05	0.04
	50	0.34	0.27	0.22	0.18	0.15	0.12
	45	0.77	0.65	0.55	0.46	0.40	0.35
	42.5	0.86	0.75	0.67	0.60	0.54	0.48
	40	0.93	0.82	0.75	0.69	0.64	0.59
	35	1.00	0.91	0.87	0.82	0.79	0.76
	30	1.00	0.95	0.92	0.90	0.88	0.86

[a] Ratio of flow rate to ideal capacity of 2800 pc/hr, assuming that passenger-car operation is unaffected by grade.
[b] Interpolate for intermediate values of "percent no-passing zone"; round "percent grade" to the next-higher integer value.
Source: TRB, 1994.

Values of the directional distribution factor, f_d, are given in Table 7-28. Note that these values differ from those given for general terrain sections. The lane and shoulder width adjustment factor f_w is the same as for the general terrain segments (see Table 7-25). The grade factor, f_g, and the heavy vehicle factor, f_{HV}, are the result of a complex

TABLE 7-28 ADJUSTMENT FACTOR
FOR DIRECTIONAL DISTRIBUTION
ON SPECIFIC GRADES, f_d

Percent of traffic on upgrade	Adjustment factor
100	0.58
90	0.64
80	0.70
70	0.78
60	0.87
50	1.00
40	1.20
30	1.50

Source: TRB, 1994.

formulation. Passenger-car equivalent (PCE) values are tabulated, which reflect a standard mix of heavy vehicles, including 14% trucks, 4% recreational vehicles, and 0% buses. From the tabulation, two values can be found:

1. E = PCE value for the length and severity of grade and average upgrade speed under consideration
2. E_0 = PCE value for the average upgrade speed under consideration, but for a 0% grade

From these the adjustment factors f_g and f_{HV} are found as follows:

$$I_p = 0.02\,(E - E_0) \tag{9}$$

$$f_g = \frac{1}{1 + P_p I_p} \tag{10}$$

$$E_{HV} = 1 + (0.25 + P_{T/HV})(E - 1) \tag{11}$$

$$f_{HV} = \frac{1}{1 + P_{HV}(E_{HV} - 1)} \tag{12}$$

where

I_p = impedance factor for passenger cars on grades

P_p = proportion of passenger cars in the traffic stream

$P_{T/HV}$ = proportion of trucks in heavy vehicles (i.e., proportion of trucks divided by the total proportion of heavy vehicles in the traffic stream)

E_{HV} = passenger-car equivalent for the mix of heavy vehicles present on the section under study

Passenger-car equivalents for specific grades are given in Table 7-29.

A second relationship connecting capacity to critical speed is

TABLE 7-29 PASSENGER-CAR EQUIVALENTS FOR SPECIFIC GRADES ON TWO-LANE RURAL HIGHWAYS, E and E_0

Grade[a] (%)	Length of grade (mi)	Average upgrade speed (mph)					
		55.0	52.5	50.0	45.0	40.0	30.0
0	All	2.1	1.8	1.6	1.4	1.3	1.3
3	$\frac{1}{4}$	2.9	2.3	2.0	1.7	1.6	1.5
	$\frac{1}{2}$	3.7	2.9	2.4	2.0	1.8	1.7
	$\frac{3}{4}$	4.8	3.6	2.9	2.3	2.0	1.9
	1	6.5	4.6	3.5	2.6	2.3	2.1
	$1\frac{1}{2}$	11.2	6.6	5.1	3.4	2.9	2.5
	2	19.8	9.3	6.7	4.6	3.7	2.9
	3	71.0	21.0	10.8	7.3	5.6	3.8
	4	[b]	48.0	20.5	11.3	7.7	4.9
4	$\frac{1}{4}$	3.2	2.5	2.2	1.8	1.7	1.6
	$\frac{1}{2}$	4.4	3.4	2.8	2.2	2.0	1.9
	$\frac{3}{4}$	6.3	4.4	3.5	2.7	2.3	2.1
	1	9.6	6.3	4.5	3.2	2.7	2.4
	$1\frac{1}{2}$	19.5	10.3	7.4	4.7	3.8	3.1
	2	43.0	16.1	10.8	6.9	5.3	3.8
	3	[b]	48.0	20.0	12.5	9.0	5.5
	4	[b]	[b]	51.0	22.8	13.8	7.4
5	$\frac{1}{4}$	3.6	2.8	2.3	2.0	1.8	1.7
	$\frac{1}{2}$	5.4	3.9	3.2	2.5	2.2	2.0
	$\frac{3}{4}$	8.3	5.7	4.3	3.1	2.7	2.4
	1	14.1	8.4	5.9	4.0	3.3	2.8
	$1\frac{1}{2}$	34.0	16.0	10.8	6.3	4.9	3.8
	2	91.0	28.3	17.4	10.2	7.5	4.8
	3	[b]	[b]	37.0	22.0	14.6	7.8
	4	[b]	[b]	[b]	55.0	25.0	11.5
6	$\frac{1}{4}$	4.0	3.1	2.5	2.1	1.9	1.8
	$\frac{1}{2}$	6.5	4.8	3.7	2.8	2.4	2.2
	$\frac{3}{4}$	11.0	7.2	5.2	3.7	3.1	2.7
	1	20.4	11.7	7.8	4.9	4.0	3.3
	$1\frac{1}{2}$	60.0	25.2	16.0	8.5	6.4	4.7
	2	[b]	50.0	28.2	15.3	10.7	6.3
	3	[b]	[b]	70.0	38.0	23.9	11.3
	4	[b]	[b]	[b]	90.0	45.0	18.1
7	$\frac{1}{4}$	4.5	3.4	2.7	2.2	2.0	1.9
	$\frac{1}{2}$	7.9	5.7	4.2	3.2	2.7	2.4
	$\frac{3}{4}$	14.5	9.1	6.3	4.3	3.6	3.0
	1	31.4	16.0	10.0	6.1	4.8	3.8
	$1\frac{1}{2}$	[b]	39.5	23.5	11.5	8.4	5.8
	2	[b]	88.0	46.0	22.8	15.4	8.2
	3	[b]	[b]	[b]	66.0	38.5	16.1
	4	[b]	[b]	[b]	[b]	[b]	28.0

[a] Round "percent grade" to next-higher integer value.
[b] Speed not attainable on grade specified.
Source: TRB, 1994.

$$v_c = 25 + 3.75 \left(\frac{q_c}{1000} \right)^2 \tag{13}$$

where v_c is the critical speed at which capacity occurs (mph), and q_c is the flow rate at capacity (veh/hr).

Two-Lane Highway System Planning. When general planning and policy studies are needed to be done for a rural two-lane highway system, the procedure given in this section can be used. Table 7-30 contains a tabulation of appropriate AADTs that can be served by two-lane highways without violating stated levels of service during peak periods. These are given for a set of standard conditions and are valid only for general terrain sections. For each LOS, the related percent time-delay criteria were applied across all three types of terrain. The planning criteria also assumes a typical traffic mix of 14% trucks, 4% RVs, and no buses. A 60/40 directional split is used,

TABLE 7-30 MAXIMUM AADTs VERSUS LEVEL OF SERVICE AND TYPE OF TERRAIN FOR TWO-LANE RURAL HIGHWAYS[a]

	Level of service				
K factor	A	B	C	D	E
	Level terrain				
0.10	2,400	4,800	7,900	13,500	22,900
0.11	2,200	4,400	7,200	12,200	20,800
0.12	2,000	4,000	6,600	11,200	19,000
0.13	1,900	3,700	6,100	10,400	17,600
0.14	1,700	3,400	5,700	9,600	16,300
0.15	1,600	3,200	5,300	9,000	15,200
	Rolling terrain				
0.10	1,100	2,800	5,200	8,000	14,800
0.11	1,000	2,500	4,700	7,200	13,500
0.12	900	2,300	4,400	6,600	12,300
0.13	900	2,100	4,000	6,100	11,400
0.14	800	2,000	3,700	5,700	10,600
0.15	700	1,800	3,500	5,300	9,900
	Mountainous terrain				
0.10	500	1,300	2,400	3,700	8,100
0.11	400	1,200	2,200	3,100	7,300
0.12	400	1,100	2,000	3,100	6,700
0.13	400	1,000	1,800	2,900	6,200
0.14	300	900	1,700	2,700	5,800
0.15	300	900	1,600	2,500	5,400

[a] All values rounded to the nearest 100 veh/day. Assumed conditions include 60/40 directional split, 14% trucks, 4% RVs, no buses, and PHF values from Table 7-23. For level terrain, 20% no-passing zones were assumed; for rolling terrain, 40% no-passing zones; for mountainous terrain, 60% no-passing zones.
Source: TRB, 1994.

along with percent no-passing-zone values of 20%, 40%, and 60% for level, rolling, and mountainous terrain. Ideal geometrics of 12-ft lanes, 6-ft shoulders, and 60 mph design speed are used. The AADT volumes are obtained from the following equation:

$$\text{AADT}_i = \text{SF}_i \times \frac{\text{PHF}}{K} \tag{14}$$

where

AADT_i = maximum AADT for LOS i (veh/day)
 SF_i = maximum service flow eate for LOS i
  K = 30th-highest hourly volume factor (i.e., the proportion of AADT expected to occur in the design hour)
 PHF = 30th-highest hourly volume/AADT

5.4 Design and Operational Treatments

Two-lane highways can experience a number of operational and safety problems due to a variety of traffic, geometric, and environmental causes. One or more of the following alleviation techniques may be applied to reduce problems on two-lane highways.

1. Realignment to improve passing sight distance
2. Use of paved shoulders
3. Three-lane roadways with two lanes designated for travel in one direction
4. Three-lane road sections with continuous two-way median left-turn lanes
5. Three-lane roadway with reversible center lane
6. Special intersection treatments
7. Truck or heavy-vehicle climbing lanes
8. Turnouts
9. Short four-lane segments

5.5 Sample Calculations

Example 7

A segment of a rural two-lane highway has the following characteristics: 70-mph design speed; 11-ft lanes; 4-ft paved shoulders; rolling terrain; 20% no-passing zones; length 6 miles; 70/30 directional split; 10% trucks, 5% RVs, 1% buses, and 84% cars. What is the capacity of the section? What is the maximum flow rate that can be accommodated at LOS C?

Solution

$$\text{SF}_i = 2800 \times \left(\frac{V}{c}\right)_i \times f_d \times f_w \times f_{HV}$$

where

$$f_{HV} = \frac{1}{1 + P_T(E_T - 1) + P_R(E_R - 1) + P_B(E_B - 1)}$$

The following values are taken from tables:

$$\left(\frac{V}{c}\right)_C = 0.39 \text{ (Table 7-20; rolling terrain, 20\% no-passing zone)}$$

$$\left(\frac{V}{c}\right)_E = 0.94 \text{ (Table 7-20; LOS E)}$$

$$f_d = 0.89 \text{ (Table 7-24; 70/30 split)}$$

$$f_{wC} = 0.85 \text{ (Table 7-25; 11-ft lanes, 4-ft shoulders)}$$

$$f_{wE} = 0.92$$

$$E_T = 5.0 \text{ for LOS C; 5.0 for LOS E (Table 7-26; rolling terrain)}$$

$$E_R = 3.9 \text{ for LOS C; 3.3 for LOS E (Table 7-26)}$$

$$E_B = 3.4 \text{ for LOS C; 2.9 for LOS E (Table 7-26)}$$

$$P_T = 0.10; \qquad P_R = 0.05; \qquad P_B = 0.01 \text{ (given)}$$

$$f_{HV}(\text{LOS C}) = \frac{1}{1 + 0.10(5 - 1) + 0.05(3.9 - 1) + 0.01(3.4 - 1)}$$

$$= 0.64$$

$$f_{HV}(\text{LOS E}) = \frac{1}{1 + 0.10(5 - 1) + 0.05(3.3 - 1) + 0.01(2.9 - 1)}$$

$$= 0.651$$

$$\text{SF}_C = 2800 \times 0.39 \times 0.89 \times 0.85 \times 0.64 = 528 \text{ veh/hr}$$

$$\text{SF}_E = 2800 \times 0.94 \times 0.89 \times 0.92 \times 0.651 = 1903 \text{ veh/hr}$$

Example 8

A two-lane rural highway carries a peak-hour volume of 170 veh/hr and has the following characteristics: 60-mph design speed; 10-ft lanes; 4-ft shoulders; mountainous terrain; 80% no-passing zones; length 12 miles; 60/40 directional split; 15% trucks, 10% RVs, no buses, 75% passenger cars. At what LOS will the highway operate during peak periods?

Solution

Compare the actual flow rate to service-flow rates computed for each LOS. The actual flow rate is found as

$$q_{15} = \frac{V}{\text{PHF}}$$

where V is 170 veh/hr (given) and PHF is 0.87 (default value, Table 7-23; 200 veh/hr), giving

$$q_{15} = \frac{170}{0.87} = 195 \text{ veh/hr}$$

$$\text{SF}_i = 2800 \times \left(\frac{V}{c}\right)_i \times f_d \times f_w \times f_{HV}$$

$$f_{HV} = \frac{1}{1 + P_T(E_T - 1) + P_R(E_R - 1) + P_B(E_B - 1)}$$

where $V/c = 0.02$ for LOS A; 0.12 for LOS B; 0.20 for LOS C; 0.37 for LOS D; 0.80 for LOS E (Table 7-20, mountainous terrain, 80% no-passing zone).

$f_d = 0.94$ (Table 7-24; 60/40 split)

$f_w = 0.77$ for LOS A–D; 0.85 for LOS E (Table 7-25, 10-ft lanes, 4-ft shoulder)

$E_T = 7$ for LOS A; 10 for LOS B and C; 12 for LOS D and E (Table 7-26; mountainous terrain)

$E_R = 5$ for LOS A; 5.2 for LOS B–E

$P_T = 0.15$; $P_R = 0.10$ (given)

$$f_{HV} \text{ (LOS A)} = \frac{1}{1 + 0.05(7 - 1) + 0.10(5 - 1)} = 0.588$$

$$f_{HV} \text{ (LOS B and C)} = \frac{1}{1 + 0.05(10 - 1) + 0.10(5.2 - 1)} = 0.535$$

$$f_{HV} \text{ (LOS D and E)} = \frac{1}{1 + 0.05(12 - 1) + 0.10(5.2 - 1)} = 0.508$$

$SF_A = 2800 \times 0.02 \times 0.94 \times 0.77 \times 0.588 = 24$ veh/hr

$SF_B = 2800 \times 0.12 \times 0.94 \times 0.77 \times 0.535 = 130$ veh/hr

$SF_C = 2800 \times 0.20 \times 0.94 \times 0.77 \times 0.535 = 216$ veh/hr

$SF_D = 2800 \times 0.37 \times 0.94 \times 0.77 \times 0.508 = 401$ veh/hr

$SF_E = 2800 \times 0.80 \times 0.94 \times 0.85 \times 0.508 = 909$ veh/hr

If the actual flow rate of 195 veh/hr (which represents the flow rate during the peak 15-min flow) is compared to these values, it will be seen that it is between service flow rates for B and C. Therefore, the LOS for the highway is C for the conditions described.

Example 9

A rural two-lane highway in hilly terrain has a 5% grade for 1.5 miles. It consists of 12-ft lanes, 6-ft shoulders, 80% no-passing zones. The directional split is 60/40, 15% trucks, 6% recreational vehicles, 2% buses, and 77% passenger vehicles. What is the maximum volume that can be accommodated on the grade at a speed of 45 mph (LOS C, Table 7-21).

Solution

$$SF_i = 2800 \times \left(\frac{V}{c}\right)_i \times f_d \times f_w \times f_g \times f_{HV}$$

where

$f_g = 1/(1 + p_p I_p)$
$I_p = 0.02(E - E_0)$

$$f_{HV} = 1/[1 + P_{HV}(E_{HV} - 1)]$$
$$E_{HV} = 1 + (0.25 + P_{T/HV})(E - 1)$$

The following values are used in the calculations:

$$\left(\frac{V}{c}\right)_C = 0.72 \text{ (Table 7-27, 5\% grade, 80\% no-passing zone)}$$

$$f_d = 0.87 \text{ (Table 7-28, 60/40 split, 60\% upgrade)}$$

$$f_w = 1.00 \text{ (Table 7-25, 12-ft lanes, shoulder} = 6 \text{ ft)}$$

$$E = 6.30 \text{ (Table 7-29, 45 mph, 5\% for 1.5 miles)}$$

$$E_0 = 1.40 \text{ (Table 7-29, 45 mph, 0\% grade)}$$

$$P_{HV} = P_T + P_R + P_B = 0.15 + 0.06 + 0.02 = 0.23$$

$$P_{T/HV} = \frac{P_T}{P_{HV}} = \frac{0.15}{0.23} = 0.65$$

Compute f_g and f_{HV}:

$$I_p = 0.02(6.3 - 1.4) = 0.098$$

$$f_g = \frac{1}{1 + (0.77 \times 0.098)} = 0.93$$

$$E_{HV} = 1 + (0.25 + 0.65)(6.30 - 1) = 5.77$$

$$f_{HV} = \frac{1}{1 + 0.23(5.77 - 1)} = 0.48$$

$$SF_c \text{(for the peak 15 min)} = 2800 \times 0.72 \times 0.87 \times 1.0 \times 0.93 \times 0.48$$

$$= 783 \text{ veh/hr}$$

SUMMARY

The concepts of capacity and level of service of highway facilities have been utilized by engineers and planners since 1950. In this chapter, we dealt with the definitions, procedures, operation analysis, design, and planning of three major types of highway facilities: freeways, multilane highways, and two-lane highways. Only the basic details were provided. However, for more extensive details, consult the HCM (TRB, 1994).

REFERENCES

NATIONAL COOPERATIVE HIGHWAY RESEARCH PROGRAM (NCHRP) (1995). *Transportation Research Board (TRB), Revised Chapter 3: Basic Freeway Sections of the Highway Capacity Manual*, Draft Final, JHK & Associates, and Texas Transportation Institute, Texas A & M University (unpublished).

NATIONAL COOPERATIVE HIGHWAY RESEARCH PROGRAM (NCHRP 3-45) (1995). *Trans-portation Research Board (TRB), Speed-Flow Relationships for Basic Freeway Segments*, Final Report, JHK & Associates, and Texas Transportation Institute, Texas A & M University (unpublished).

TRANSPORTATION RESEARCH BOARD (TRB) (1994). *Highway Capacity Manual*, Special Report 209, National Research Council, Washington, DC.

EXERCISES

1. An eight-lane freeway is carrying a flow of 3000 veh/hr. Out of this flow, truck flow is 180 veh/hr (6%), recreational vehicle flow is 450 veh/hr (15%), and passenger bus flow is 360 veh/hr (12%). The rest of the traffic consists of passenger cars.
   (a) Determine the heavy-vehicle adjustment factor on a 2-mile section of the freeway with a grade of +3% and with obstructions 5 ft from both sides of the roadway.
   (b) If the width of each lane is 12 ft and the average interchange density adjacent to the section is 1 interchange/mile, determine the existing LOS and the maximum service flow at capacity (LOS E).

2. Consider the following data for a freeway: urban eight-lane freeway (PHF = 0.95); grade +2% for 2 miles; 12-ft lanes and full shoulders outside; extended section in rolling terrain with a local interchange density of 1.5 interchanges/mile; 6% trucks. Determine the SF for LOS B and D.

3. An urban radial freeway is to be designed on the basis of the following information: design year average annual daily traffic, AADT = 82,000 veh/day, with the peak-hour fraction of AADT (K) estimated at 10% and a peak directional proportion (D) of 0.55; 10% buses, 5% recreational vehicles, 10% trucks; peak-hour factor of the traffic is 0.90; level terrain—extended section, no lateral obstructions; design standard—LOS D. Determine the number of lanes needed if the estimated FFS = 65 mph.

4. A 1-mile segment of rural freeway has a +3% grade that follows a 5-mile segment of level terrain. The DDHV is 2500 veh/hr (weekday), including 20% trucks, with a PHF of 0.90. The lanes are 12 ft wide with no lateral obstructions on both sides of the freeway and the adjacent freeway segments have an average of 0.5 interchanges/mile. Determine the number of lanes required on this grade segment, including the climbing lane for trucks, for LOS B.

5. The equation $c = 2300\,NWB_c\,T_c$ can be used to roughly estimate highway capacity, where c is the capacity in each direction (veh/hr), N is the number of lanes in each direction, W is a design adjustment factor, B_c is the bus traffic adjustment factor, and T_c is the truck traffic adjustment factor. A highway has, initially, four lanes and is designed with the following adjustment factors: $W = 1$, $T_c = 1$, and $B_c = 1$. Calculate the change in capacity of the highway if one lane is used for buses only (i.e., $B_c = \frac{1}{4}$ for the bus-only lane).

6. Rework Example 7 with the design speed changed to 50 mph and the lane width increased to 12 ft.

7. A rural two-lane highway in level terrain presently connects two communities experiencing rapid growth. Future DDHV is expected to be 2300 veh/hr with 8% trucks and a PHF = 0.90. Design a multilane highway for LOS C to replace the two-lane highway, incorporating the following expected features: 10-ft-wide lanes, 3-ft clearance on both sides, a raised median, and 12 accesses/mile on each side. Assume an ideal free-flow speed of 60 mph.

8. A multilane highway needs to be designed to link a developing area of the Silicon Forest to a bedroom community several miles away. The highway is expected to carry a DDHV of 1800 veh/hr with 10% trucks and a PHF of 0.90. Make the following assumptions: 11.0-ft lanes, adequate shoulders, undivided highway, 20 access points per mile on each side, 50-mph posted speed limit, and rolling terrain. Design the highway for a LOS C.

9. An existing two-lane rural highway is expected to experience an estimated 4% traffic growth rate. The current ADT is 3000 veh/hr. The highway is located in rolling terrain and the K factor has been estimated to be 0.15. Determine (a) what the ADT will be in 12 years; (b) what the level of service will be 12 years hence.

10. A two-lane highway is being planned in mountainous terrain. with an estimated 30th-highest hour factor of 0.15. The state highway authorities recommend a LOS of C or better. What service flow can this highway handle?

11. A realigned section of a two-lane highway is proposed to be built in a mountainous terrain for a length of about 4 miles to carry a DDHV of 600 vehicles. Determine the level of service at which this section of highway will operate given the following information:

> Lane width = 11
> Shoulder width = 6 feet
> No passing zones = 20%
> Directional split = 60/40
> Trucks = 5%
> RV = 5%
> Passenger cars = 90%
> PHF = 0.90
> Use the operational level of analysis.

12. If the two-lane highway described in Exercise 11 is designed with a grade of 4%, determine the maximum hourly volume that can operate at an average speed of at least 45 mph.

13. An existing four-lane urban freeway has the following characteristics:

> 60-mph design speed
> Rolling terrain
> 11-ft lane width
> Median obstructions are at the edge of the travel lanes
> Roadside obstructions are 4 ft from the edge of the travel lanes
> 10% trucks, 3% RVs, 2% buses
> Peak-hour factor of 0.95
> Drivers are familiar users of the facility

Determine the service-flow rates and service volumes for each level of service. What level of service would be provided if demand on the freeway were 1800 veh/hr during the peak hour? 2200 veh/hr? 2800 veh/hr? For what time period do these levels of service apply?

14. A series of grades begins on a freeway segment entering a mountainous area. The section of freeway entering the grade is essentially level. The grade is composed of the following segments:

> 4% upgrade, 1 mile
>
> 3% upgrade, $\frac{1}{2}$ mile
>
> 5% upgrade, 1 mile
>
> 2% upgrade, $\frac{1}{2}$ mile
>
> 4% upgrade, 1 mile

Compute the average grade for this section. Is the average-grade technique applicable in this case? What is the equivalent composite grade for this sequence? To what point along the grade is the equivalent composite grade computed? Based on the equivalent composite grade and 15% trucks and 5% RVs in the traffic stream, compute the heavy-vehicle adjustment factor that would have to be applied to a capacity analysis of this section.

15. A multilane highway has the following characteristics:

> Four-lane rural highway, undivided design
>
> 12-ft lanes, 4-ft usable shoulder width
>
> Individual 3% grade 1.25 miles long
>
> 12% trucks, 2% buses, 10% RVs
>
> Assume an ideal free-flow speed of 55 mph
>
> Peak-hour factor of 0.85

Determine the maximum service flow, V (veh/hr), for levels of service C and E. If the AADT is 30,000, how many lanes must be provided for level-of-service C operation.

16. A four-lane highway has the following characteristics:

> Measured free-flow speed of 52 mph
>
> Located in a suburban area
>
> Carries 2000 vehicles per hour in the peak direction
>
> 11-ft lanes with no lateral clearances
>
> Located in rolling terrain
>
> 10% trucks, 5% buses
>
> Peak-hour factor of 0.90

At what level of service does the highway currently operate? If development in the area causes traffic to grow at a rate of 10% per year, when will the facility reach capacity? At the time the facility reaches capacity, what improvements would you recommend to improve operations to level of service D?

17. Determine the level of service provided on a 1-mile-long upgrade of 3% for a two-lane two-way roadway that is carrying a demand volume of 200 veh/hr. The roadway has 11-ft lanes and obstructions 4 ft from the edge of the pavement on both sides. This road has a passing sight distance greater than 1500 ft for 50% of its length. The composition of the traffic stream is 60% passenger cars, 25% RVs, and 15% trucks. Assume 60% upgrade and 40% downgrade. The roadway design speed is 60 mph.

Chapter 8

Intersection Control and Design

1. INTRODUCTION

Chapter 7 dealt with the basic concepts relating to the capacity and level of service of transportation facilities where there was uninterrupted traffic flow (e.g., freeways and rural two-lane highways). One of the most logical ways of continuing with that chapter would have been to discuss the capacity of intersections. However, it was felt that before such an attempt was made, it would be best to understand the fundamentals of how intersections are controlled and designed, and this is precisely what we describe in this chapter. The capacity and level of service of intersections are taken up in Chapter 9.

2. TYPES OF INTERSECTIONS

Intersections are an inevitable part of any street system. Driving around any city, one notices that a large majority of urban streets share an intersection, where drivers can decide whether to go straight or turn on to another street. A road or street intersection can be defined as the general area where two or more roads join or cross, including the roadway and roadside facilities for traffic movement within it (AASHTO, 1994).

Because an intersection has to be shared by everybody wanting to use it, it needs to be designed with great care, taking into consideration efficiency, safety, speed, cost of operation, and capacity. The actual traffic movement and its sequence can be handled by various means, depending on the type of intersection needed (AASHTO, 1994).

In general, there are three types of intersections: (1) intersection at grade, (2) grade separations without ramps, and (3) interchanges. The common *intersection at grade* is one where two or more highways join, with each highway radiating from an intersection and forming part of it. These approaches are referred to as *intersection legs*. Such intersections have their own limitations and use. Examples of at-grade intersections are shown in Figure 8-1. When it is necessary to accommodate high volumes of

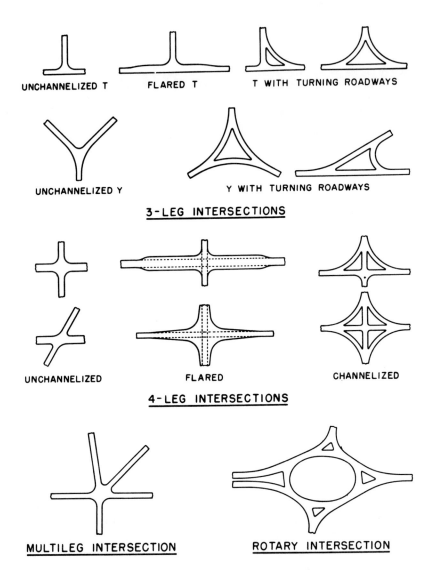

UNCHANNELIZED T   FLARED T   T WITH TURNING ROADWAYS

UNCHANNELIZED Y   Y WITH TURNING ROADWAYS

3-LEG INTERSECTIONS

UNCHANNELIZED   FLARED   CHANNELIZED

4-LEG INTERSECTIONS

MULTILEG INTERSECTION   **ROTARY INTERSECTION**

Figure 8-1   Examples of At-Grade Intersections.

traffic safely and efficiently through intersections, one resorts to through traffic lanes separated in grade, and this is generally referred to as an *interchange*. The basic types of interchanges are shown in Figure 8-2. When two highways or streets cross each other at a different grade, with no connections, the arrangement is referred to as a *grade separation*.

Details of the geometric design of at-grade intersections and interchanges are given in *A Policy on Geometric Design of Highways and Streets* published by the American Association of State Highway and Transportation Officials (1990, 1994), and this book should be consulted for a deeper insight into this important facet of design.

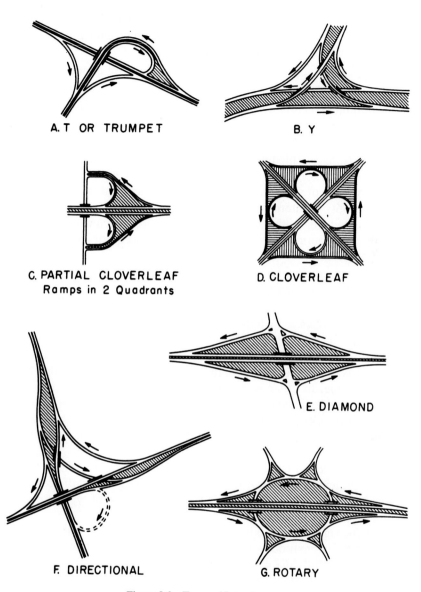

Figure 8-2   Types of Interchanges.

3. DESIGN CONSIDERATIONS AND OBJECTIVES

The objective of intersection design is to reduce the severity of potential conflicts between vehicles (including pedestrians) while providing maximum convenience and ease of movement to vehicles. Four basic elements are generally considered in the design of at-grade intersections:

1. Human factors, such as driving habits and decision and reaction times
2. Traffic considerations, such as capacities and turning movements, vehicle speeds, and size and distribution of vehicles
3. Physical elements, such as characteristics and use of abutting property, sight distance, and geometric features
4. Economics factors, such as costs and benefits and energy consumption

Insofar as interchanges are concerned, their type and design are influenced by many factors, such as highway classification, character and composition of traffic, design speed, and degree of access control. Interchanges are high-cost facilities, and because of the wide variety of site conditions, traffic volumes, and interchange layouts, the warrants that justify an interchange may differ at each location. AASHTO (1990, Chap. IX) provides details regarding grade separations and interchanges. The bottom line when considering adoption of an interchange is: Can the cost of an interchange be justified?

4. TRAFFIC CONTROL DEVICES

Traffic control devices include signs, movable barriers, and signals. All these can be used alone or in combination if necessary. They are the primary means of regulating, warning, or guiding traffic, on all streets and highways. Traffic control devices strive to provide safe and efficient functioning of intersections by separating conflicting vehicle streams in time. In other words, the right-of-way through an intersection, during a given period, is assigned to one or several streams of traffic. For example, yield or stop signs assign priority to particular traffic streams relative to other streams at the same intersection. A four-way stop sign establishes a rough first-come, first-served traffic control for intersections with light traffic.

The *Manual on Uniform Traffic Control Devices* (MUTCD) (FHWA, 1988) sets forth the principles that govern the design and usage of traffic control devices for all streets and highways open to public travel, regardless of type of class or the governmental agency having jurisdiction.

Traffic signs and markings are used for establishing a street or highway system that is clearly understood by its users: drivers and pedestrians. Specifically, traffic signs and markings fulfill the following purposes: the regulation of traffic (e.g., speed limits), turn prohibition, alerting and warning drivers and pedestrians regarding roadway conditions, and guiding traffic along appropriate routes to reach trip destinations through signs and markings. These purposes apply to all control devices, including signals, markings, and channelizations. Naturally, to be effective, control devices must meet the following basic requirements:

1. Fulfill a need
2. Command attention
3. Convey a clear, simple meaning

4. Command respect of road users

5. Give adequate time for proper response

The following criteria should be applied to ensure that these requirements are met. Signs should be of a proper design and should be placed properly and appropriately. They should be operated consistently and maintained routinely. Finally, there must be uniformity in application, so that recognition and understanding of these devices is easy and unambiguous.

Traffic signs fall into four broad areas of functional classification according to use:

1. *Regulatory signs* are used to impose legal restrictions applicable to particular locations. They inform drivers of certain laws and regulations, the violation of which constitutes a misdemeanor. There are four principal groups of regulatory signs, excluding those for pedestrians. They are right-of-way signs, the most common being stop and yield signs; speed signs; movement signs, such as turning or one-way signs; and parking signs.

2. *Warning signs* are used to call attention to hazardous conditions, actual or potential, that would otherwise not be readily apparent. Such signs require caution on the part of the driver and may call for a speed reduction or other maneuver. Typical conditions where warning signs are used include highway construction zones and approaches to intersections, merging areas, pedestrian crossings, and school zones.

3. *Guide or informational signs* provide directions to drivers and to various destinations. These are placed far enough ahead of intersections and interchanges to allow adequate time for drivers to make their routing decisions.

4. *Directional signs* on high-speed highways are used at interchanges associated with freeways.

5. CONFLICT AREAS AT INTERSECTIONS

Figure 8-3 shows vehicle streams and the merging, diverging, and crossing maneuvers for a simple four-leg intersection, and for a more complicated staggered intersection. Such diagrams are useful because the number and type of conflicts may indicate the accident potential of an intersection. In the case of a regular two-lane, two-way, four-leg intersection there are 16 potential crossing conflict points, eight merging and eight diverging conflict points. The staggered T-intersection shown in the figure serves about the same function as the four-leg intersection, and consists of only six potential crossing conflict points, three diverging and three merging conflict points. Is the staggered T-intersection, therefore, superior to the four-leg intersection? Not really. There are several other factors that play an important part in deciding the merits of adopting a particular type or design of intersection for a specific site.

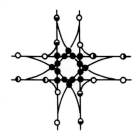

● 16 crossing conflicts
○　8 merging conflicts
◖　8 diverging conflicts

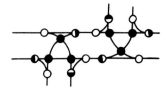

● 6 crossing conflicts
○ 6 merging conflicts
◖ 6 diverging conflicts

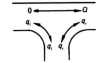

Vehicle streams at a three-way intersection

Figure 8-3   Vehicle Streams and the Merging, Diverging, and Crossing Maneuvers (Salter, 1974).

6. TYPES OF INTERSECTION CONTROLS

There are at least six principal ways of controlling traffic at intersections, depending on the type of intersection and the volume of traffic in each of the vehicle streams. The MUTCD (FHWA, 1988) provides guidelines for adopting any particular type of intersection control, in the form of warrants.

6.1 Stop Signs

Stop signs are warranted at intersections under the following conditions:

1. Intersection of a less important road with a main road, where application of the normal right-of-way rule is unduly hazardous
2. Intersection of a county road, city street, or township road with a state highway
3. Street entering a through highway or street

4. Unsignalized intersection in a signalized area
5. Unsignalized intersection where a combination of high speed, restricted view, and serious accident record indicates a need for control by the stop sign

Multiway (fourway or all-way) stops can be used as a safety measure at some locations where the volume on the intersecting roads is approximately equal and the following conditions exist:

1. An accident problem, as indicated by five or more reported accidents in a 12-month period, which may be corrected by a multiway stop installation.
2. (a) The total vehicular volume entering the intersection from all approaches averages at least 500 vehicles per hour for any 8 hours of an average day, and (b) the combined vehicular and pedestrian volume from the minor street or highway averages at least 200 units per hour for the same 8 hours, with an average delay to minor street vehicular traffic of at least 30 seconds per vehicle during the maximum hour, but (c) when the 85th-percentile approach speed of the major street traffic exceeds 40 mph, the minimum vehicular volume warrant is 70% of the foregoing requirements.
3. Where traffic signals are warranted, the multiway stop control can be used as an interim measure while arrangements are being made for installation of the signal.

6.2 Yield Signs

Yield signs are established as follows:

1. On a minor road at the entrance to an intersection when it is necessary to assign the right-of-way to the major road, but where a stop is not necessary at all times, and where the safe approach speed on the minor road exceeds 10 mph
2. On the entrance ramp to an expressway, where an adequate acceleration lane is not provided
3. Where there is a separate or channelized right-turn lane without an adequate acceleration lane
4. At any intersection where a problem can be possibly corrected by a yield sign installation
5. Within an intersection with a divided highway, where a stop sign is present at the entrance to the first roadway, and further control is necessary at the entrance to the second roadway. Median width between roadways must exceed 30 ft.

6.3 Intersection Channelization

Channelization is the separation or regulation of conflicting traffic movements into definite paths of travel by traffic islands or pavement markings to facilitate the safe and orderly movements of both vehicles and pedestrians. Proper channelization increases capacity, improves safety, provides maximum convenience, and instills driver confidence. Channelization is frequently used along with stop or yield signs or at signalized intersections.

Some basic principles to help design channelized intersections are as follows:

1. Motorists should be provided with channel lines that are easy to follow.
2. Sudden and sharp reverse curves should be avoided.
3. Areas of vehicle conflict should be reduced as much as possible.
4. Traffic streams that cross without merging and weaving should intersect at or near right angles.
5. Islands should be carefully selected and be as few as possible.
6. Overchannelization should be avoided, as it has proved to be counterproductive.

Figure 8-4 gives typical examples of channelized intersections. The *Intersection Channelization Design Guide* (TRB, 1985) and AASHTO (1994) provide further details on channelization.

6.4 Rotaries and Roundabouts

Rotaries and roundabouts are channelized intersections comprising a central circle surrounded by a one-way roadway. The basic difference between rotaries and roundabouts is that rotaries are generally signalized (as in Washington, D.C.), whereas roundabouts are not. Naturally, in the case of roundabouts, entering traffic yields to traffic already within.

Roundabouts generally have good safety records and traffic does not have to stop when traffic volumes are low. A well-designed roundabout should deflect the path of vehicles passing through an intersection by the use of a sufficiently large central island, properly designed approach islands, and staggering the alignment of entries and exits (Figure 8-1).

6.5 Uncontrolled Intersections

Where an intersection has no control device whatsoever, the operator of a vehicle approaching an intersection must be able to perceive a hazard in sufficient time to alter the vehicle's speed, as necessary, before reaching the intersection. The time needed to start decelerating is the driver's perception and reaction time and may be assumed to be 2.0 seconds. In addition, the driver needs to begin braking some distance from the intersection. This distance from the intersection, where a driver can first see a vehicle approaching on the intersecting road, is that which is traversed during 2.0 seconds for perception and reaction, plus an additional 1.0 second to actuate braking or to accelerate to regulate speed. By referring to Figure 8-5, the sight triangle is determined by the minimum distances along the road. For instance, if highway A has a speed limit of 50 mph and highway B has one of 30 mph, it would require an unobstructed sight triangle, with legs extending at least 220 ft and 130 ft, respectively, from the intersection,

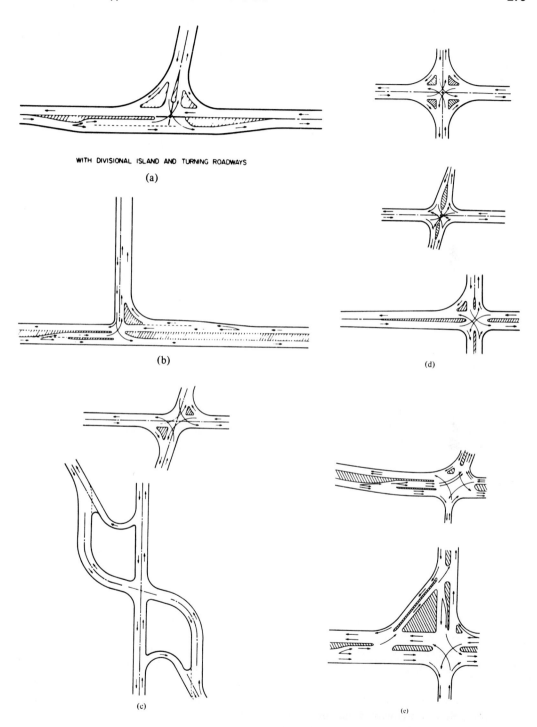

WITH DIVISIONAL ISLAND AND TURNING ROADWAYS

(a)

(b)

(c)

(d)

(e)

Figure 8-4   Typical Examples of Channelized Intersection (AASHTO, 1994).

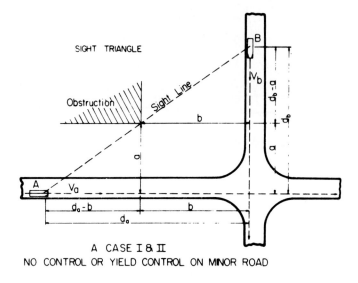

A CASE I & II
NO CONTROL OR YIELD CONTROL ON MINOR ROAD

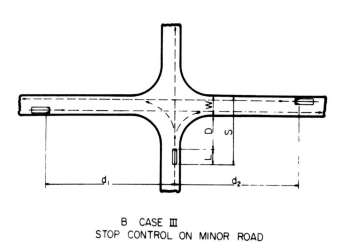

B CASE III
STOP CONTROL ON MINOR ROAD

Figure 8-5   Sight Triangle (AASHTO, 1994).

based merely on the average distance traveled in 3 seconds. These minimum distances will permit a vehicle on either road to change speeds before reaching the intersection, but this fact by itself does not imply that the intersection is safe.

There can be potential danger to vehicle operators on such intersections, especially when a succession of vehicles are approaching the intersection, when time is sufficient to avoid only a single vehicle. Because the distance covered in 3 seconds ranges from 70% of the safe stopping distance at 20 mph to only 36% at 70 mph, the use of sight triangles for design purposes must be approached with caution.

A safer design for such intersections should allow drivers on both highways to see the intersection and traffic in sufficient time to stop the vehicle before reaching the intersection. The safe stopping distances in this case are the same as those used for designing any other section of highway.

6.6 Traffic Signal Devices

One of the most important and effective methods of controlling traffic at an intersection is the use of traffic signals. The traffic signal is an electrically timed device that assigns the right-of-way to one or more traffic streams so that these traffic streams can pass through the intersection safely and efficiently. Traffic signals are appropriate for minimizing:

1. Excessive delays at stop signs and yield signs
2. Problems caused by turning movements
3. Angle and side collisions
4. Pedestrian accidents

Because traffic signaling devices are so important, the rest of this chapter is devoted to describing their design and application.

7. TRAFFIC SIGNALS

All power-operated devices (except signs) for regulating, directing, or warning motorists or pedestrians are classified as traffic signals. This section begins with a basic set of definitions connected with traffic signals and intersections and goes on to detail several methods of designing signal timing. The various sections provided on this subject are necessarily just the basics. Further details can be found in several references provided at the end of this chapter.

7.1 The Purposes of Traffic Signals

In general, a traffic signal is installed at an intersection:

- To improve overall safety
- To decrease average travel time through an intersection, and consequently increase capacity
- To equalize the quality of service for all or most traffic streams

Although traffic signals are installed on the basis of warrants, justification for their installation must be made in terms of safety, travel times, equity, pollution, and so on. More will be said about warrants in a later section.

Among the main advantages of traffic signals over sign control are the positive guidance that they provide to vehicle operators and pedestrians, which leaves less room for erroneous judgments on the part of drivers; flexibility, in the sense that allocation of right-of-way can be responsive to change in traffic flow; ability to assign priority treatment to some movements or vehicles; feasibility of coordinated control along streets or in area networks; and provision for continuous flow of a platoon of traffic through proper coordination at a specified speed along a given route. On the other hand, it has been observed that poorly designed traffic signals can cause increased accident frequency, excessive delay for vehicles on certain approaches, forcing motorists to adopt circuitous routes, and driver irritation.

7.2 Definitions Pertaining to Intersections and Traffic Signals

A number of terms used in this and other chapters need to be defined. They have been extracted from the *Traffic Engineering Handbook* (Pline, 1992), and the *Highway Capacity Manual* (TRB, 1994).

1. *Cycle (cycle length or cycle time):* any complete sequences of signal indications.
2. *Phase (signal phase):* the part of a cycle allocated to any combination of traffic movements receiving right-of-way simultaneously during one or more intervals.
3. *Interval:* the part or parts of the signal cycle during which signal indications do not change.
4. *Offset:* the time lapse, in seconds, between the beginning of a green phase at the intersection and the beginning of a green phase at the next intersection.
5. *Intergreen (clearance interval):* the time between the end of a green indication for one phase and the beginning of a green indication for another (Figure 8-6).

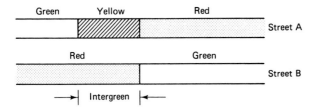

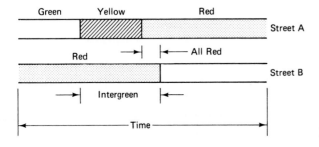

Figure 8-6   Intergreen.

6. *All-red interval:* the display time of a red indication for all approaches. In some cases, an all-red interval is used exclusively for pedestrians crossing very wide intersections (Figure 8-6).

7. *Peak-hour factor (PHF):* in the case of street intersections, the ratio of the number of vehicles entering the intersection during the peak hour to four times the number of vehicles entering during the peak 15-minute period. In the absence of field information a PHF value of 0.85 may be assumed, in which case, the traffic flow during the 15-minute period = $N/(4 \times 0.85)$ = $0.294N$, or approximately $0.3N$, where N is the peak-hour flow.

8. *Average departure headways:* observations made by Greenshields et al. (1947) show that for green intervals of 20 to 30 seconds, the average headway per vehicle is about 2.5 seconds.

9. *Passenger-car equivalents (PCEs):* to account for the adverse effects of commercial vehicles and turning movements on startup time (or average head way), it is customary to convert actual flows (given in mixed vehicles per hour) to an equivalent volume in straight-through passenger cars. Buses and trucks are assumed to be 1.5 PCE, and left-turn vehicles are approximately equal to 1.6 PCE.

In addition, the following terms will be used from time to time in this and subsequent chapters (TRB, 1994):

Approach: the portion of an intersection leg that is used by traffic approaching the intersection

Capacity: the maximum number of vehicles that has a reasonable expectation of passing over a given roadway or section of roadway in one direction during a given time period under prevailing roadway and traffic conditions

Critical volume: a volume (or combination of volumes) for a given street that produces the greatest utilization of capacity (e.g., needs the greatest green time) for that street, given in terms of passenger cars or mixed vehicles per hour per lane

Delay: the stopped time delay per approach vehicle (in seconds per vehicle)

Green time: the length of green phase plus its change interval, in seconds

Green ratio: the ratio of effective green time to the cycle length

Hourly volume: the number of mixed vehicles that pass a given section of a lane or roadway during a time period of an hour

Level of service: a measure of the mobility characteristics of an intersection, as determined by vehicle delay, and a secondary factor, volume/capacity ratio

Local bus: a bus having a scheduled stop at an intersection

Passenger-car volume: volume expressed in terms of passenger cars, following the application of passenger-car equivalency factor to vehicular volumes

Period volume: a design volume, based on the flow rate within the peak 15 minutes of the hour, and converted to an equivalent hourly volume

Through bus: a bus not having a designated stop at the intersection under analysis

Truck: a vehicle having six or more wheels (tires) on the pavement

7.3 Components of a Signal System

A signal installation consists of illuminated displays and a controlling mechanism. It may also include various vehicle-detecting devices or some other means for activation by demand (such as a push button for pedestrians desiring to cross a street).

Displays, or indications, are grouped into signal faces, each of which controls one or more traffic streams arriving from the same direction. A signal head contains one or more signal faces; it can be mounted on a post or suspended from a wire.

Signal indications differ by color, shape, and continuity. The colors used are: green, to give the right-of-way to one or a combination of traffic streams; red, to prohibit movement or to require a stop; amber, to regulate the switching of the right-of-way from one set of traffic streams to another or to advise caution. When there are special signals for pedestrians, these are in the form of illuminated letter messages or logos. Signal indications can be steady or flashing. As noted before, a flashing red indication has the same meaning as a stop sign, whereas a flashing amber allows one to proceed with caution. The flashing "walk" cautions a pedestrian that a vehicular stream is concurrently permitted to cross his or her line of movement. The flashing "don't walk" is the equivalent of an amber indication.

Signal controllers are electromechanical or electronic devices that regulate the length and sequence of signal indications at an intersection. Pretimed controllers operate with a fixed amount of time allotted to specific traffic movements in a fixed sequence; the timing is based on historical flow patterns at an intersection. Traffic-adjusted controllers are equipped to receive information on traffic flow patterns from various measuring devices at preset time intervals. This information is used to select one of several timing schemes stored in the controller's memory.

Traffic-actuated controllers also use some sensing devices to alter the length and/or the sequence of signal indications. Unlike traffic-adjusted controllers, however, they react to arrivals of individual vehicles rather than to changes in aggregate patterns of traffic at an intersection. Traffic-actuated timing schemes are usually constrained by specified minimum lengths of green indications for various traffic streams that can be marginally extended by vehicle arrivals up to specified maxima. Intersections can be controlled individually; alternatively, a sequence of intersections along a road could be connected and controlled as a group.

Detectors can be activated by the passage or the presence of a vehicle. A variety of physical principles are used for detection: pressure, distortion in a magnetic field, interruption of a light beam, a change in radar wave frequency, a change in inductance of a conducting loop, video detection using image processing techniques, and so on. They can be positioned above the road or on, inside, or under the road surface, or simply on the video screen for image processing. Detectors differ in investment and maintenance cost as well as in reliability.

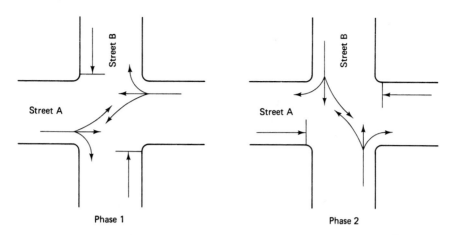

Figure 8-7   Signal Phases.

7.4 Elements of a Signal Timing System

At signalized intersections, some traffic streams are allowed to have a simultaneous right-of-way, while other streams are stopped. A signal phase is a period during which one or more movements concurrently are shown a green indication. Figure 8-7 shows a phasing diagram for an intersection of two streets.

Safety considerations dictate that a phase may be shared only by those traffic streams whose paths do not intersect. In practice, however, some conflicts are tolerated. In Figure 8-7, pedestrians and right- and left-turning vehicles enjoy concurrent green. Left turns are often allowed to clear an intersection through gaps in the traffic streams moving along the same street but in the opposite direction.

At some intersections, where both pedestrians and turning vehicles are numerous, it may be advantageous to provide a special phase where all pedestrian movements are permitted while all vehicles are stopped. Subsequent phases are meant for vehicles only.

The time between the end of a green indication for one phase and the beginning of a green for another is called *intergreen*, or a clearance interval. An amber indication is shown through the intergreen period followed by red. This is shown in Figure 8-6. When the computed clearance interval is long, a combination of amber and an all-red interval may be used instead. This is also illustrated in Figure 8-6.

Design of signal phases specifies a sequence of various phases following each other. A signal cycle is a part of that sequence. Note that safety (conflict avoidance) and the quality of service are the most important factors in designing signals.

A phasing diagram is developed by jointly considering intersection geometry (primarily, the number of lanes at each approach) and desired lines of movement through an intersection. Safety is the sole criterion in computing intergreen times, as shown in what follows. Factors affecting the length of intergreen include safe stopping distances, approach speeds of vehicles, walking speeds of pedestrians, and pavement widths.

When the sum of intergreens over all phases is subtracted from cycle time, what remains is the total green time per cycle. Selection of green times depends on whether to minimize the overall average travel time through an intersection, or to equalize demand and capacity over a given time period, or to minimize the maximum individual travel time through an intersection, and so on. Each objective may result in a different set of cycle times and green indications.

Calculation of the Intergreen Period.   An examination of Figure 8-6 indicates that the intergreen period, consisting of either an amber or an amber plus an all-red period, is necessary to alert motorists regarding the change from a green light to a red light. When the amber indication appears, drivers who are at a distance (from the stop line) greater than their stopping distance will be able to stop comfortably. Referring to Figure 8-8, those drivers who are nearer the stop line than their safe stop distance will accelerate and clear the intersection. Those who are at or near a safe stop distance away (the so-called "dilemma zone") should be able to (1) either stop or (2) accelerate (when near the stop line) and clear. The path of the vehicle covers a distance

$$S + W + l \qquad\qquad (1)$$

where

S = safe stopping distance
W = distance from the stop line until vehicle (rear) is clear
l = length of the vehicle

Therefore, the intergreen time is

$$l = \frac{S + W + l}{v} = t_r + \frac{v}{2f \times g} + \frac{W + l}{v} \qquad\qquad (2)$$

where

v = speed of the vehicle (ft/sec)
g = 32.2 ft/sec^2

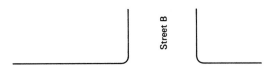

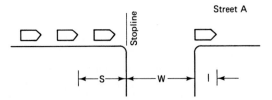

Figure 8-8   Clearance Distances.

t_r = perception-reaction time (1 to 2.5 sec)
f = 0.33 (wet pavement) to 0.62 (dry pavement) at 30 to 40 mph

W and l are measured in feet. If there is no all-red ($a = 0$),

$$I = A \tag{3}$$

that is, intergreen is equal to an amber indication. Now consider the amber indication time based on pedestrian movement, in the same direction as the car, crossing street B. If there are no pedestrian signals, assume that the last pedestrian will start crossing exactly as the amber indication comes on. The pedestrian clearance time is

$$R_i = \frac{W_j}{v_{\text{ped}}} \tag{4}$$

where W_j is the street width or up to the median, and v_{ped} is the walking speed of pedestrians (4 ft/sec). If R_i is longer than I_i, use R_i as I_i.

Example 1

Determine the optimal duration for the yellow phase to eliminate the *dilemma zone* at an intersection. It is assumed that the dilemma zone can be eliminated by adjusting only the yellow-light duration.

Solution

To stop a vehicle before the stop line, the driver has to start braking at a distance of

$$S = t_r v + \frac{v^2}{2a}$$

from the stop line, called the *stop zone*. The term t_r here is the driver's reaction time, v is the travel speed, and a is the deceleration rate.

On the other hand, an adequate yellow phase Y should be provided to allow a vehicle to clear the intersection if the vehicle is too close to the intersection when the yellow phase starts. The distance to the stop line from the vehicle, which is to clear the intersection during the yellow phase, is called the *go zone*, given by

$$G = vY - (W + l)$$

where W is the street width, and l is the vehicle length.

The dilemma zone is the distance between G and S, that is,

$$D = G - S = vY - (W + l) - (t_r v + v^2/2a)$$

To minimize D, we set $D = 0$, to get

$$Y = t_r + \frac{v}{2a} + \frac{W + l}{v} \tag{5}$$

which is the optimal duration for the yellow phase.

Discussion

The purpose of eliminating the dilemma zone is to help drivers set up a clear benchmark, or imaginary line, behind which the vehicle should brake, or beyond which the vehicle should go through the intersection, when the yellow phase starts. However, due to different geometric features of intersections, the dilemma zone of some intersections may not be totally removed by adjusting only the yellow phase. When there is no all-red ($a = 0$), the intergreen time is the same as the amber (yellow) phase.

The Shortest Cycle Based on Pedestrian Requirements.   In urbanized areas of a city where pedestrian circulation is important and necessary, it is the general practice to provide pedestrian control signals. When "walk/don't walk" indications are used, pedestrian phase length might govern the green indication for that approach. Pedestrians need a total time to cross an intersection of

$$Z + R_i \quad \text{seconds} \tag{6}$$

where Z is the initial period during which the "walk" signal is displayed, usually ≈ 7 seconds; and R_i is the clearance time for the last pedestrian who starts crossing the intersection when the pedestrian signal just begins to flash "don't walk."

As part or all of R_i can coincide with the intergreen time, the pedestrian phase requires only

$$P_i = Z + R_i - I_i \tag{7}$$

where P_i is the pedestrian phase (sec) during the green indication. (*Note:* It is assumed that R_i is longer than I_i; if I is the larger, $P_i = Z$.) It follows that

$$P_i = \min G_i \tag{8}$$

7.5 Signal Timing for Pretimed Isolated Signals

Over the years, traffic engineers have used several methods for designing pretimed isolated signals. In more recent times, a number of computer programs have been developed for rapid design. Three manual methods of design are presented here: (1) Homburger and Kell's method, (2) Pignataro's method, and (3) Webster's method.

Homburger and Kell's Method (Homburger et al., 1988).   This method utilizes traffic volumes as the basis for allocating time to approaches, keeping off-peak cycles as short as possible (40 to 60 seconds). Peak-hour cycles can be longer, favoring movement on the major street. The general procedure is as follows:

1. Select yellow change intervals between 3 to 5 seconds for speeds less than 35 mph to speeds greater than 50 mph.

2. Determine the need for additional clearance time using Eq. 5 and also ensure if an all-red phase is necessary.

$$Y = t_r + \frac{v}{2a} + \frac{W + l}{v}$$

assuming that a is a deceleration rate of 10 ft/sec^2.

3. Determine pedestrian clearance times, assuming pedestrian walking speed as 4 ft/sec.

4. Compute minimum green times. With pedestrian signals, the "walk" period should be at least 7 seconds.

5. Compute green times based on an approach volume in the critical lane on each street at peak hour.

6. Adjust the cycle length (sum of all greens and yellows) to the next-higher 5-second interval and redistribute extra green time.

7. Compute percentage values for all intervals.

Example 2

Time an isolated signal with pedestrian indications at the intersection of Pine and Oak: Pine is 56 ft wide, Oak is 40 ft wide. During the peak hour, the critical lane volumes are 350 and 250 veh/hr and approach speeds are 40 and 25 mph (58.7 and 36.7 ft/sec) for Pine and Oak, respectively (see Figure 8-E2).

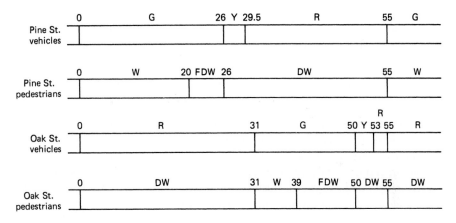

Figure 8-E2   Phases for Homburger and Kell's Method.

Solution

1. Select yellow change intervals.

$$\text{Pine: 3.5 sec} \qquad \text{Oak: 3 sec}$$

2. Calculate nondilemma clearance times.

$$\text{Pine: } 1 + \frac{58.7}{20} + \frac{40 + 20}{58.7} = 4.9 \text{ sec} \quad \text{or} \quad 5 \text{ sec}$$

$$\text{Oak: } 1 + \frac{36.7}{20} + \frac{56 + 20}{36.7} = 5.0 \text{ sec} \quad \text{or} \quad 5 \text{ sec}$$

Calculate all-red clearance intervals.

After Pine yellow: $5 - 3.5 = 1.5$ sec
After Oak yellow: $5 - 3 = 2$ sec

3. Determine pedestrian clearance times.

Pine (crossing Oak): 40/4 = 10 sec
Oak (crossing Pine): 56/4 = 14 sec
FDW (Pine) = 10 − 3.5 = 6.5 sec
FDW (Oak) = 14 − 3 = 11 sec

where FDW = flashing "don't walk."

4. Compute minimum green times (pedestrian clearance − yellow + walking minimum).

Pine: 10 − 3.5 + 7 = 13.5 sec; use 15-sec minimum
Oak: 14 − 3 + 7 = 18 sec

5. Compute green times (using Oak as critical minimum).

$$350(18)/250 = 25.2 \text{ sec} \approx 25 \text{ sec (Pine Street green)}$$

6. Adjust cycle length and redistribute extra green time.

Total cycle = 25 + 5 + 18 + 5 = 53 sec; use 55 sec
Extra green time = 55 − 53 = 2 sec; give 1 sec to Pine (= 26 sec) and 1 sec to Oak (= 19 sec)

7. Compute percentage values for all intervals for key settings.

Webster's Method. Webster utilized extensive field observations and computer simulation to establish an excellent procedure for designing traffic signals. A fundamental assumption of Webster's work is that of random vehicle arrivals. He developed the classic equation for calculating the average delay per vehicle on an intersection approach, and also derived an equation for obtaining the optimum cycle time that produces the minimum vehicle delay. Although the elements of Webster's method are presented in this section, readers should refer to Webster and Cobbe (1962) for further details.

Webster uses terminology that needs some explanation, so we begin this section by defining such terms as "saturation flow" and "lost time." This is followed by a description of average delay and the optimum cycle time. Several examples applying these concepts are included.

Saturation Flow, q_s, and Lost Time. A study of the discharge of vehicles across the stop bar of an intersection approach indicates that when the green period begins, vehicles take some time to start and accelerate to normal running speed, but after a few seconds, the vehicle queue discharges at a more-or-less constant rate called the *saturation flow* (see Figure 8-9). The saturation flow is one that would be obtained if there were a continuous queue of vehicles and they were given 100% green time. It is generally expressed in vehicles per hour of green time. It can be seen from the figure that the average rate of flow is lower during the first few seconds (while vehicles are accelerating to normal running speed) and also during the amber period (as some vehicles decide to stop and others continue to move on). It is convenient to replace

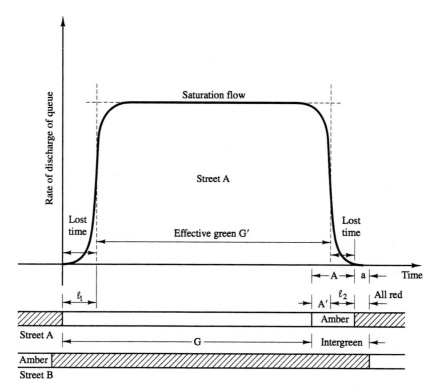

Figure 8-9 Saturation Flow.

the green and amber periods by an "effective green" period, during which the flow is assumed to take place at the saturation rate, combined with a "lost" time during which no flow takes place. This is a useful concept because capacity is then directly proportional to effective green time. In graphical terms, this means replacing the curve in the figure by a rectangle of equal area where the height of the rectangle is equal to the average saturation flow and its base is the effective green time. The difference between the effective green time and the combined green and amber periods is *lost time*.

Saturation flow and lost time can be measured on the road directly, and a method for doing this is as follows: Observe one lane, or the entire approach, as applicable, using a stopwatch to measure the number of vehicles crossing the stop bar, from the beginning of the green phase until the front bumper of the nth vehicle crosses the stop line during each successive 0.1-minute interval of green and amber. This procedure is best illustrated by an example.

Example 3

An intersection approach controlled by a fixed-time signal was observed for 15 saturated intervals, each of 0.1 minute, providing the following results.

Time (min) (intervals)	0	0.1	0.2	0.3	0.4	0.5
Number of saturated intervals		15	15	15	15	15	
Number of vehicles crossing stop bar		22	35	30	28	26	
Discharge per 0.1 min		1.46	2.33	2.00	1.86	1.73	

At the end of the amber period, there is an interval of less than 0.1 minute. The length of this period and the number of vehicles crossing the stop bar are noted. These intervals are referred to as the *last saturated intervals*. The observer noted the following:

Total duration of all the last saturated intervals = 50 sec

Total number of vehicles crossing stop line = 13

Discharge per 0.1 min during last saturated interval = $\dfrac{13}{50/60} \times \dfrac{1}{10} = 1.56$ vehicles

Estimate the lost time.

Solution

During the first and last saturated intervals, there is a loss of capacity, because vehicles are accelerating from a stationary position at the beginning of the green period and decelerating during the amber period. The flow during the remainder of the observed periods represents the maximum discharge possible.

$$\text{Saturation flow} = \frac{2.33 + 2.00 + 1.86 + 1.73}{4}$$

$$= 1.98 \text{ vehicles per 0.1 min}$$

$$= 1188 \text{ veh/hr}$$

The lost time at the beginning and end of the green period may be calculated with reference to Figure 8-E3. The number of vehicles represented by the rectangle *efij* is equal to the number of vehicles represented by the original bars of the histogram. Also, the number of vehicles represented by the area *dghk* is also equal to the number of vehicles represented by the four 0.1-min periods of saturated flows between *d* and *k*. Therefore,

Area *abdc* = area *efgd*

and   Area *hijk* = area *nmlk*

Therefore,

$ed \times 1.98 = 1.46 \times 0.1$

$\Rightarrow$ $ed = 0.074$ min

and $ce = 0.1 - 0.074 = 0.026$ min

Similarly,

$kj \times 1.98 = 1.56 \times 0.1$

$\Rightarrow$ $kj = 0.079$

and $jl = 0.1 - 0.079 = 0.021$ min

Therefore,

$$ce + jl = 0.026 + 0.021 = 0.047 \text{ min} = 2.82 \text{ sec}$$

that is, the lost time during the green phase = 2.82 seconds.

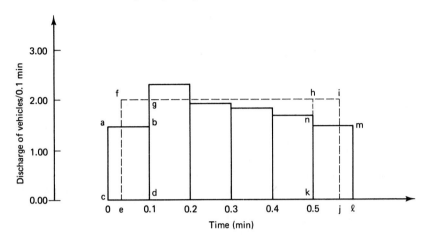

Figure 8-E3   Observed Discharge Across Stop Line.

Discussion

In this type of experimentation, it is important to note the following points:

1. If the flow on the approach is not saturated, observations should be discontinued until the flow once again reaches saturation level.
2. The distribution of trucks, buses, and other types of vehicles along with turning movements is also important and should be recorded.
3. The effect of large vehicles and motorcycles is accounted for (in British practice) by making the following conversions [in passenger-car units (PCUs)]:

   1 passenger car or light commercial vehicle = 1.00 PCU
   1 heavy or medium commercial vehicle   = 1.75 PCU
   1 bus   = 2.25 PCU
   1 motorcycle   = 0.33 PCU
   1 bicycle   = 0.20 PCU
      1 left-turning vehicle = 1.75 straight-ahead vehicles
4. Right-turn vehicles do not affect flow.

Average Delay and Optimum Cycle Time.   Webster's classic equation is

$$d = \frac{C(1 - \theta)^2}{2(1 - \theta x)} + \frac{x^2}{2q(1 - x)} - 0.65\left(\frac{C}{q^2}\right)^{1/3} x^{2+5\theta} \qquad (9)$$

where

d = average delay per vehicle on an approach (sec)
C = cycle time (sec)

θ = proportion of cycle length that is effectively green (g/C) for that phase

q = flow rate (veh/sec)

s = saturation flow (veh/sec) = 1800 veh/hr

x = degree of saturation

= ratio of actual flow to maximum flow through approach

= $q/\theta s$

To enable the delay to be estimated more easily, Eq. 9 can be written as

$$d = CA + \frac{B}{q} - D$$

A, B, and D have been calculated and tabulated by Webster for easy application.

Webster's model computes the approximate cycle length that minimizes total intersection delay as well as the effective green time for each approach, by differentiating the equation for the overall delay with respect to the cycle time. The result is

$$C_0 = \frac{1.5L + 5}{1.0 - (Y_1 + Y_2 + \cdots + Y_n)} = \frac{1.5L + 5}{1.0 - \sum\limits_{i=1}^{n} Y_i} \qquad (10)$$

where

C_0 = optimum cycle length (sec)

L = total lost time per cycle, generally taken as the sum of the total yellow and all-red clearance per cycle (sec)

Y = observed volume/saturation flow, for the critical approach in each phase

Note that this optimum cycle is the one that gives the highest ratio of flow to saturation flow.

The distribution of green time to each phase is proportional to the critical lane volumes on each phase. For a two-phase intersection the net green time is

$$G = C_0 - A_1 - A_2 - nL \qquad (11)$$

where

G = net green time (sec)

C_0 = optimum cycle length (sec)

A_1 = yellow change interval in phase 1 (sec)

A_2 = yellow change interval in phase 2 (sec)

n = number of phases ($n = 2$)

L = lost time per phase (sec)

Webster's equation can be utilized to select the cycle length for a given intersection. It yields cycle lengths (C_0) that are usually shorter than those that are obtained by other methods, which indicates that cycle lengths should be as short as possible. Webster also concluded that delay is not significantly increased by cycle-length variation in the range $0.75C_0$ to $1.5C_0$. However, his equation is sensitive to errors in estimates of vehicle flow rate and saturation flow.

Design of a Fixed-Time Signal for a Two-Phase Installation (No Exclusive Turning Lanes or Phases).   The following steps are indicated:

1. For each approach, calculate or measure the saturation flow, q_s.
2. For each approach, count the peak-hour volume in mixed traffic with known percent composition and percent turns; divide by the peak-hour factor; convert to design volume in straight-through passenger cars per hour, using the rough coefficients shown in Example 3.
3. For each approach, calculate the q/q_s ratio. For each street, choose the larger q/q_s for design. If q and q_s are per lane rather than per approach, choose the critical lane q/q_s ratio, following the same procedure.
4. Calculate the two intergreen periods.
5. Calculate the minimum green indications based on pedestrian requirements.
6. Calculate $C_{optimum}$.
7. Split the available effective green time between the two phases.
8. Check if the minimum green indications required by pedestrians are satisfied. If not, adjust upward.
9. Develop a table of all signal indication lengths, according to the following rules:
   (a) The minimum length of any green indication is 15 seconds.
   (b) The cycle length should be adjusted to the nearest higher length divisible by 5 (if $C < 90$ seconds) or 10 (if $C > 90$ seconds). Redistribute the extra green as described earlier.
   (c) All intervals should be integer percentage points of the cycle length.

Example 4

A simple four-leg intersection needs a fixed-time signal. The critical flows in the N–S and E–W directions are 600 and 400 veh/hr. Saturation flow is 1800 veh/hr and the lost time per phase is observed to be 5.2 seconds. Determine the cycle length and distribution of green (see Figure 8-E4).

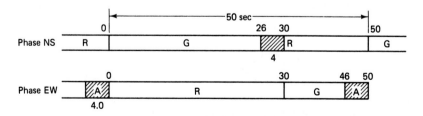

Figure 8-E4 Phase Diagrams for Example 4.

Solution

Assume an amber period of 4 seconds.

$$Y_i = \frac{\text{observed flow}}{\text{saturation flow}}$$

$$Y_1 = \frac{600}{1800} = 0.333 \qquad Y_2 = \frac{400}{1800} = 0.222$$

$$\Sigma\, Y_i = 0.333 + 0.222 = 0.555$$

$$C_0 = \frac{1.5L + 5.0}{1 - \Sigma\, Y_i} = \frac{1.5(2 \times 5.2) + 5}{1 - 0.555} = \frac{20.6}{0.445} = 46.3 \text{ sec}$$

Use $C_0 = 50$ sec.

$$G = C_0 - A_1 - A_2 - nl$$

$$= 50 - 4 - 4 - 2(5.2) = 31.6 \text{ sec}$$

$$\Sigma \text{ critical lane volumes} = 600 + 400 = 1000$$

$$G_{N-S} = \frac{600(31.6)}{1000} = 18.96$$

$$G_{E-W} = \frac{400(31.6)}{1000} = 12.64$$

$$\text{Phase}_{N-S} = \text{green} + \text{yellow} + \text{lost time}$$

$$= 18.96 + 4 + 5.2 = 28.16 \text{ sec} \approx 30 \text{ sec}$$

$$\text{Phase}_{E-W} = 12.64 + 4 + 5.2 = 21.84 \text{ sec} \approx 20 \text{ sec}$$

$$\text{Cycle length} = 30 + 20 = 50 \text{ sec}$$

Example 5

An intersection, as shown in Figure 8-E5, is to be provided with: five-phase operation, 7-second minimum green time (G), 4-second amber, and minimum flashing "don't walk" (FDW) time equal to the time required to walk from curb to the center of the farthest lane at a rate of 4 ft/sec.

Assumptions: 20-ft average vehicle length; 2.1-second average saturation flow headway; random arrival pattern. Use the Australian formula shown in what follows.

Solution

1. *Webster's optimum-cycle formula:*

$$C_0 = \frac{1.5L + 5}{1 - Y}$$

where

C_0 = optimum cycle to minimize delay (sec)
Y = volume/saturation flow for the critical approach in each phase

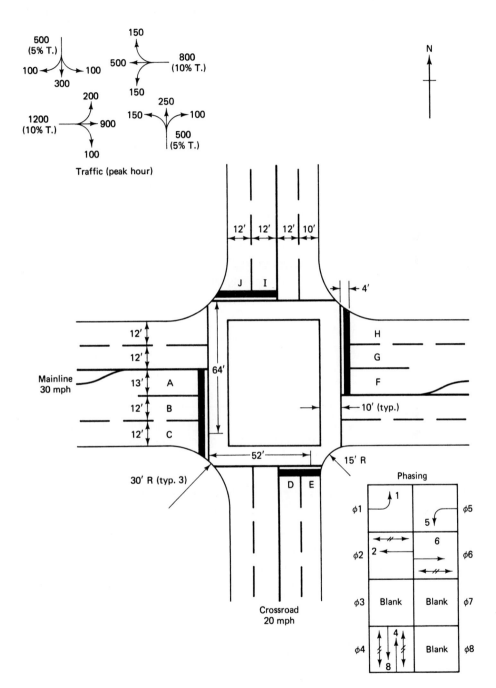

Figure 8-E5   Intersection for Example 5.

L = total lost time/cycle, generally taken as the sum of the total yellow and all-red clearance per cycle (sec)

2. *Australian left-turn factor formula:*

$$EL = \cfrac{1.5}{\cfrac{fnSG}{GnS - Q} - \cfrac{fQC}{GnS - Q} + \cfrac{4.5}{G}} \qquad (12)$$

where

EL = factor for converting unprotected LT vehicles to equivalent passenger car units

S = saturation flow (veh/hr/ln)

  = 3600/average headway

C = cycle length (sec)

n = number of opposing lanes

G = effective green time (sec), defined as the green plus amber time for a phase, minus startup and termination losses

Q = opposing straight-through and right-turning traffic (PCU/ph)

f = a function of the opposing flow (Q) that varies as follows:

Q	f
0	1.00
200	0.81
400	0.65
600	0.54
800	0.45

3. *Minimum-cycle calculation:* Assuming considerable pedestrian activity, the minimum-cycle length will be controlled by the minimum times set for phase 1 + 2 + 4 or by phase 5 + 6 + 4.

$$\text{Phase 1} = 7 \text{ sec}, G + 4 \text{ sec}, A \quad = 11 \text{ sec}$$
$$\text{Phase 2} = 7 \text{ sec}, W + 52/4 \text{ FDW} = 20 \text{ sec}$$
$$\text{Phase 4} = 7 \text{ sec}, W + 64/4 \text{ FDW} = 23 \text{ sec}$$
$$\text{Total} = 54 \text{ sec}$$

The total time for phases 5, 6, and 4 is the same.

4. *Optimum cycle:* Convert street volumes to equivalent passenger-car units (PCUs). Ideally, conversion factors should be developed from field measurements. However, for this example, the following factors have been assumed:

Straight-through passenger car: 1.0 PCU

LT passenger car during protected phase: 1.1 PCUs

RT passenger car with turning radius 25 ft: 1.1 PCUs

RT passenger car with turning radius < 25 ft: 1.3 PCUs

Truck or buses: 1.5 PCUs

LT passenger cars during a permissive phase are equal to the value obtained from the Australian formula. For this example, S = 1714, a 90-second cycle, and a 30-second effective green time for phase 4 have been assumed, resulting in a factor of 2.38 PCUs for the north approach and 2.61 PCUs for the south approach. Applying these factors to the traffic volumes results in the following equivalent passenger-car volumes:

Approach	Left	Straight	Right
West	231	945	116
East	173	525	173
North	244	308	113
South	401	256	133

Applying the approach volumes to the intersection lane configuration results in the following hourly volumes:

Approach	Lane	Phase	Movement	Volume
West	A	1	1	231
West	B	6	6	(945 + 116)/2 = 530
West	C	6	6	(945 + 116)/2 = 531
South	D	4	4	401
South	E	4	4	(256 + 133) = 389
East	F	5	5	173
East	G	2	2	(525 + 173)/2 = 349
East	H	2	2	(525 + 173)/2 = 349
North	I	4	8	244
North	J	4	8	421

The cycle length will be controlled by the critical lane volume for the phase combination of $(1 + 2 + 4)$ or $(5 + 6 + 4)$.

$$
\begin{aligned}
\text{Phase 1 lane A} &= 231 \\
\text{Phase 2 lane G or H} &= 349 \\
\text{Phase 4 lane D} &= 401 \\
\text{Total} &= 981 \\
\\
\text{Phase 5 lane F} &= 173 \\
\text{Phase 6 lane B or C} &= 530 \\
\text{Phase 4 lane D} &= 401 \\
\text{Total} &= 1104
\end{aligned}
$$

Phase $(5 + 6 + 4)$ will control the cycle length. The ratio of critical flows to saturation flow can now be calculated:

$$
Y = \frac{\text{total critical flows}}{\text{saturation flow}} = \frac{1104}{1714} = 0.644
$$

Lost time per cycle can now be calculated. Lost time consists of the time lost during queue startup and time lost during phase termination. Startup losses vary with the queue length as follows:

Queue (PCUs)	Stratup losses (sec)
1	1.7
2	2.7
3	3.3
4	3.6
5	3.7
6	3.7

Assuming a 90-second cycle, splits proportional to critical lane volumes, and a uniform arrival rate, the average queue length can be calculated. Average arrival rate (g):

Phase 5: 173 veh/ln/hr = 0.048 veh/ln/sec

Phase 6: 530 veh/ln/hr = 0.147 veh/ln/sec

Phase 4: 401 veh/ln/hr = 0.111 veh/ln/sec

Red time (R):

Phase 5: $[(1104 - 173)/1104] \times 90 = 76$ sec

Phase 6: $[(1104 - 530)/1104] \times 90 = 47$ sec

Phase 4: $[(1104 - 401)/1104] \times 90 = 57$ sec

Average queue length ($g \times R$):

Phase 5: $0.048 \times 76 = 3.6$ vehicles

Phase 6: $0.147 \times 47 = 6.9$ vehicles

Phase 4: $0.111 \times 57 = 6.3$ vehicles

Phase-termination losses include the time consumed while the last vehicle in the queue travels from the stop bar to a point safely beyond the traveled way of the next conflicting movement. It is assumed that these vehicles are traveling at the approach speed, except for the LT phase vehicles (phase 5), where 15 mph has been assumed.

Phase 5: $(50 + 20)/22 = 3.2$ sec

Phase 6: $(60 + 20)/44 = 1.8$ sec

Phase 4: $(75 + 20)/29 = 3.3$ sec

Combined startup and termination losses are as follows:

Phase 5: $3.6 + 3.2 = 6.8$ sec

Phase 6: $3.7 + 1.8 = 5.5$ sec

Phase 4: $3.7 + 3.3 = 7.0$ sec

Total = 19.3 sec = L

The optimum cycle to minimize delay

$$C_0 = \frac{1.5L + 5}{1 - Y} = \frac{(1.5 \times 19.3) + 5}{1 - 0.644} = 95 \text{ sec}$$

Splits are as follows:

Phase 5: $(173/1104) \times 95 = 15$ sec $(G + A)$
Phase 6: $(530/1104) \times 95 = 46$ sec $(G + A)$
Phase 4: $(401/1104) \times 95 = 34$ sec $(G + A)$

Based on a 95-second cycle and a $G = 27$ seconds $(34 - 7$ losses) for phase 4, the equivalent car factor for the south approach of 2.71 PCU results in a critical volume for lane D of 417 veh/hr. Adjusting the calculations for this volume results in an optimum cycle of 98 seconds. The final splits are

Phase 1: 25 sec   Phase 5: 15 sec
Phase 2: 37 sec   Phase 6: 47 sec
Phase 4: 36 sec   Phase 4: 36 sec

Based on these splits and cycle length, a capacity analysis will yield level-of-service values that can be used to evaluate alternative phase strategies. The Y value determined in the "Webster" analysis is related to volume/capacity ratio and level of service. A Y value of 0.70 or below indicates a level of service of C or better. A Y value greater than 0.70 indicates a level of service of D or worse. Anytime that Y exceeds 0.70, a serious capacity problem exists and measures should be taken to increase capacity.

Optimum cycle length also relates to level of service, although the correlation is not nearly as accurate as that provided by Y. As a general rule, if the optimum cycle calculation yields a cycle length in excess of 75 seconds for two critical phases, 100 seconds for three critical phases or 140 seconds for four critical phases, the intersection warrants more study and measures to increase capacity should be considered.

Assuming the example intersection as the base condition, other alternatives have been investigated in an effort to increase capacity. Each case is identified as follows:

Case 1: example intersection, five-phase operation

Case 2: example intersection, five-phase operation, add right-turn lane on west approach

Case 3: example intersection, six-phase split side street operation

Case 4: example intersection, six-phase split side street operation, add right-turn lane on west approach

Case 5: example intersection, six-phase split side street operation, add right-turn lane on west approach, widen side streets to provide left-turn lane and two through lanes

Case 6: example intersection, eight-phase operation, add right-turn lane on west approach, widen cross street to provide left-turn lane and two through lanes

Results of each investigation are as follows:

Case	Controller phases	Critical phases	Minimum cycle (sec)	Optimum cycle (sec)	Y
1	5	3	54	98	0.65
2	5	3	57	89	0.62
3	6	4	77	164	0.73
4	6	4	80	143	0.69
5	6	4	83	114	0.61
6	8	4	71	111	0.60

Cases 1, 2, 5, and 6 should be considered for improving intersection efficiency. Cases 3 and 4 should be rejected as feasible alternatives.

Pignataro's Method.   An alternative method of designing a four-approach two-phase signal cycle follows, as suggested by Pignataro (1973). By referring to definitions regarding PHF and headway, let

N_1 = major street critical lane flow, the number of vehicles in a single lane, N_1 being the largest lane flow of the two major street approaches

N_2 = minor street critical lane flow, similar to N_1

C = cycle length (sec)

S_1 = approximate average headway entering intersection among N_1

S_2 = corresponding average headway among N_2

Y_1 = vehicle clearance interval for N_1 (sec)

Y_2 = vehicle clearance interval for N_2 (sec)

K = number of signal cycles for a 15-minute period

Then the total time required to pass all vehicles through the intersection during the 15-minute period is

$$T = \frac{N_1 S_1 + N_2 S_2}{4(\text{PHF})} \tag{13}$$

and total time required for clearance interval is

$$K(Y_1 + Y_2) \tag{14}$$

Therefore,

$$T + K(Y_1 + Y_2) \le 15 \text{ min} \times 60 = 900 \text{ sec} \tag{15}$$

or, in the limiting condition,

$$K = \frac{900 - T}{Y_1 + Y_2} \tag{16}$$

But

$$K = \frac{900}{C} \tag{17}$$

By substituting Eqs. 13 and 17 in Eq. 16, we have

$$C_{\min} = \frac{Y_1 + Y_2}{1 - (N_1 S_1 + N_2 S_2)/3600(\text{PHF})} \tag{18}$$

or in the general case when there are three or more phases,

$$C_{\min} = \frac{\Sigma\, Y_i}{1 - \Sigma\, N_i\, S_i/3600(\text{PHF})} \tag{19}$$

Equation 18 can be further modified to reflect an exclusive pedestrian phase as follows:

$$C = \frac{Y_1 + Y_2 + P + Y_p}{1 - 0.000333(N_1 S_1 + N_2 S_2)} \qquad (20)$$

where P is the length of the pedestrian "walk" interval (sec), and Y_p is the length of the pedestrian clearance interval (sec).

Assuming that the average time spacing (headway) between vehicles on the major and minor approaches is about the same, the allocation of green time is made proportional to the critical lane volumes. Therefore,

$$G_1 + G_2 = C - (Y_1 + Y_2) \qquad (21)$$

and

$$G_1 = \frac{N_1}{N_1 - N_2}[C - (Y_1 + Y_2)] = \frac{C - (Y_1 + Y_2)}{1 + N_2/N_1} \qquad (22)$$

and

$$G_2 = C - (Y_1 + Y_2) - G_1 \qquad (23)$$

For intersections where S_1 and S_2 (the headways on major and minor streets) are not assumed to be the same, the green time should be allocated in the ratio of the product of critical lane flows and its corresponding headways:

$$\frac{G_1}{G_2} = \frac{N_1 S_1}{N_2 S_2} \qquad (24)$$

Example 6

An isolated intersection of Pine and Oak Streets needs a simple two-phase signal. The following data are available:

Pine Street: 56 ft wide, critical lane volume = 300 veh/hr, approach speed 40 mph
Oak Street: 40 ft wide, critical lane volume = 225 veh/hr, approach speed 25 mph
10% trucks and 15% left turns, PHF = 0.85
Pedestrian walking speed = 4 ft/sec

Solution

See Figure 8-E6.

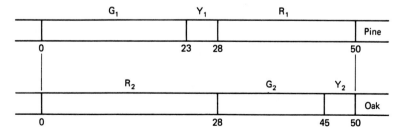

Figure 8-E6   Pignataro's Method.

1. Determine adjusted critical lane volumes: truck factor 1.5, LT 1.6.

   Pine Street: $0.15(1.6)(300) + 0.10(1.5)(300) + 0.75(1)(300) = 342$ veh/hr
   Oak Street: $0.15(1.6)(225) + 0.10(1.5)(225) + 0.75(1)(225) = 257$ veh/hr

2. Determine vehicle clearance intervals (nondilemma times).

$$Y = t + \frac{v}{2a} + \frac{W + l}{v}$$

   where

   $t = 1$ sec
   $a = 10$ ft/sec^2
   $l = 20$ ft

   Pine Street (crossing Oak): $Y = 1 + 58.7/20 + (40 + 20)/58.7 = 4.96$ sec (say, 5.0 sec)
   Oak Street (crossing Pine): $Y = 1 + 36.7/20 + (56 + 20)/36.7 = 4.91$ sec (say, 5.0 sec)

3. Determine pedestrian crossing times:

   Crossing Oak: $7 + 40/4 - 5 = 12$ sec $= G_1$
   Crossing Pine: $7 + 56/4 - 5 = 16$ sec $= G_2$

   where 7 sec is the assumed pedestrian startup time.

4. Determine approximate cycle length: Based on vehicle consideration (PHF $= 0.85$ and $s_1 = s_2 = 2.5$ sec):

$$C_{\min} = \frac{Y_1 + Y_2}{1 - (N_1 S_1 + N_2 S_2)/3600(\text{PHF})}$$

$$= \frac{4.96 + 4.91}{1 - (342 \times 2.5 + 257 \times 2.5)/(3600 \times 0.85)} = 19.3 \text{ sec}$$

   Based on pedestrian requirements:

$$C_{\min} = G_1 + G_2 + Y_1 + Y_2 = 12 + 16 + 4.96 + 4.91 = 38 \text{ sec}$$

   (This is the controlling value.)

5. Split the cycle:

$$G_1 = \frac{C - (Y_1 + Y_2)}{1 + N_2/N_1} = \frac{38 - (4.96 + 4.91)}{1 + 257/342} = 16 \text{ sec}$$

$$G_2 = \frac{C - (Y_1 + Y_2)}{1 + N_1/N_2} = \frac{38 - (4.96 + 4.91)}{1 + 342/257} = 12 \text{ sec}$$

6. Determine the cycle length:

   $C = 16 + 16 + 5 + 5 = 42$ sec (say, 45 sec, giving an extra 2 sec to Pine and 1 sec to Oak)

   $C = 18 + 17 + 5 + 5 = 45$ sec

7. Check the peak 15-minute requirement for each phase.

REQUIRED GREEN TIME

$$\frac{342}{4(0.85)} \times 2.5 = 251 \text{ sec}$$

$$\frac{257}{4(0.85)} \times 2.5 = 189 \text{ sec}$$

PROVIDED GREEN TIME

$$\frac{900}{45} \times 18 = 360 \text{ sec}$$

$$\frac{900}{45} \times 17 = 340 \text{ sec}$$

7.6 Traffic-Actuated Signals (Mitric, 1975)

The success of fixed-cycle signals depends on the degree to which actual traffic volumes agree with design flow rates built into the design process discussed before. A single cycle time and a single set of green indications would be difficult to justify for an entire 24-hour period, in view of the fluctuations in hourly traffic volumes. This problem can be partially solved by designing several fixed-cycle schemes for an intersection, each based on a traffic flow prevalent during a given part of the day.

Traffic-actuated signals represent an extension of the idea described before: Make the cycle time and green splits responsive to changes in traffic flow, down to the level of microchanges. That is, the length of green indication could be varied by the number of vehicle arrivals.

The simplest type of traffic-actuated installation has a detector located at a distance A ahead of the stop line at an intersection approach, and a controller sensitive to signals sent by the detector. At the beginning of a green phase, the maximum number of vehicles caught between the stop line and the detector (including a vehicle stopped exactly over the detector) is given by

$$\frac{A}{l + S_0} + 1 \tag{25}$$

where l is the average vehicle length, and S_0 is the distance between nearest points of consecutive stopped vehicles. The maximum indicated green for this approach is

$$G_{\min} = t_{sd} + \left(\frac{A}{l + S_0} + 1\right)\frac{1}{q_s} \tag{26}$$

where

q_s = saturation flow (pcu/hr)

t_{sd} = startup delay

The controller's memory starts at "blank" in each green phase. If within $G_{\min}$ seconds it does not receive any "calls" from the detector, signifying additional vehicle arrivals, the green may be switched to another phase. For each call, however, the controller will extend the green time by a fixed-time interval, called the *unit extension* of green, counted from the moment of the call. The unit extension must be long enough to allow a vehicle to cover the distance from the detector to the stop line moving at a given approach speed, v:

$$h = A/v \tag{27}$$

where h is the unit extension of green (also called a vehicle interval). Note that for every call that the controller receives, the green time is extended by h seconds, measured from the moment when that vehicle is detected. In other words, there is no accumulation of unused green time. Therefore, successive vehicles may be no more than h seconds apart to keep the green time at their approach. Because interval h is used to determine whether green will be extended or terminated for a given approach, it is sometimes called the *critical gap* (or *critical headway*). It is important to note that the critical gap and the unit extension of green are one and the same.

   If vehicles keep following each other at headways shorter or equal to the unit extension, the green time for this approach will be terminated only if it reaches a preset maximum based on maximum individual delay to vehicles on other approaches or on some other delay-based criterion. If all approaches are saturated, traffic-actuated installation operates in a fixed-cycle manner. The unit extension is clearly the most important parameter in the design of traffic-actuated signals, and the question is: How long should it be?

   Safety considerations dictate that a detector be located no closer than the safe stopping distance ahead of the stop line. This implies a minimum length of unit extension:

$$h_{\min} = \frac{S_{\text{safe}}}{v} \tag{28}$$

   Above this minimum, the length of the unit extension depends on traffic arrival rates. As with fixed-time signals, it can be selected by referring to average delay or number of stops. Intuitively, shorter unit extensions should be used when arrival rates are high, and vice versa. Experimental work shows that the optimal (average-delay-minimizing) unit extension ranges from 2 to 8 seconds, as a function of average traffic flow per phase. This is illustrated in Figure 8-10.

   It follows that simple traffic-actuated signals suffer from some of the same weaknesses as those of fixed-time signals. They will work well if the actual traffic flow matches the flow assumed when the unit extension of green was selected. An obvious direction of improvement is to design traffic-actuated signals with variable unit extension, varying with the length of green and/or with the changes in flow rate. Such improvements are, in fact, operational.

Timing of Fully-Actuated Traffic Signals (Mitric, 1975).   For isolated intersections (0.5 mile or more distant from adjacent signal-controlled intersection) that experience sharp fluctuations in flow during the day, the following procedure could be adopted.

1. For a given intersection approach, choose the headway, h, between vehicles that is just enough to hold the green indication. This is usually in the interval between 2 and 5 seconds. This headway is equal to the unit extension of green per vehicle.
2. Calculate the distance, A, upstream from the stop line at which the detector will be located, by the formula

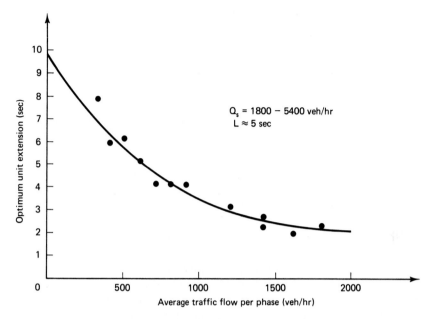

Figure 8-10   Average Traffic Flow per Phase versus Optimum Unit Extension.

$$A = v \times h \ (\text{ft}) \tag{29}$$

where v is the approach speed of vehicles (ft/sec), and h is the headway from step 1 (sec). This means that h seconds after the nth actuation the light will turn amber just as the nth vehicle is at the stop line, unless the $(n + 1)$th actuation occurred.

3. Calculate the length of the minimum green that should be long enough to pass all the vehicles that might be accumulated between the stop line and the detector, plus one vehicle.

$$G_{\min} = t_{sd} + \left(\frac{A}{l + S_0} + 1 \right) \left(\frac{3600}{q_s} \right) \tag{30}$$

where

$t_{sd} = $ startup delay (sec)
$A = $ distance between the stop line and detector (ft)
$l = $ vehicle length (ft)
$S_0 = $ distance between rear end and front of the consecutive stopped vehicles (ft)
$q_s = $ saturation flow (PCU/hr)

4. Calculate the elements of the lost time: intergreen (amber, all red) and amber effectively used as green, in the manner previously described.

5. Repeat the calculation for each approach; all components of the minimum cycle are now ready.

6. Determine the maximum cycle length and the corresponding green splits by using the Webster formula for the optimum cycle, where q values should be maximum design flows that can be expected during the day.

Assumption: A detector is the simplest device without a longitudinal dimension along the roadway. It records only that demand exists, not whether there are one, two, or more vehicles waiting.

Pedestrian crossing requirements can influence the setting of minimum greens if the situation so demands.

Authority for Signal Installation.   For traffic control signals to serve any useful purpose, their indications must be clearly understood and strictly observed. To achieve these objectives, traffic signals should be uniform, the authority for their installation unimpeachable, and their compliance legally enforceable. Thus, national standards have been developed in various countries for the installation and operation of traffic control signals, and the actions required of motorists and pedestrians are specified by statute or by local ordinance or resolution consistent with national standards. For example, suitable legislation establishing the authority for the installation, the meaning of the signal indications, and the required obedience to these indications by the road user are outlined in such documents as the *Uniform Vehicle Code*, the *Model Traffic Ordinance*, and specific vehicle codes, an example of which is the Highway Traffic Act of Ontario.

Benefits and Drawbacks.   When justified and properly designed, a traffic signal installation may achieve one or more of the following:

- Reduce the frequency of certain types of accidents, especially the right-angle type
- Effect orderly traffic movement
- Provide for the continuous flow of a platoon of traffic through proper coordination at a definite speed along a given route
- Allow other vehicles and pedestrians to cross a heavy traffic stream
- Control traffic more economically than by manual methods

Unjustified, ill-designed, improperly operated, or poorly maintained traffic signals may cause

- Increased accident frequency
- Excessive delay
- Disregard of signal indications
- Circuitous travel by alternative routes

Contrary to common belief, traffic signals do not always increase safety and reduce delay. Experience has indicated that although the installation of signals may result in a decrease in the number of right-angle collisions, it will, in many instances, result in an increase in rear-end collisions. Further, the installation of signals may not only increase overall delay, but also reduce intersection capacity. Consequently, it is of the utmost importance that the consideration of a signal installation and the selection of equipment be preceded by a thorough study of traffic and roadway conditions by an engineer experienced and trained in this field. This engineer should recognize that a signal should be installed only if the net effect, balancing benefits versus drawbacks, is to the public's advantage.

7.7 Warrant Criteria for Signal Control

Because of the complexity in deciding when and where a signal should be installed, it is apparent that some system of establishing the need for a signal installation at a particular location is necessary. Such a system has been established using a common denominator known as *signal warrants*. The eight warrants described briefly here are discussed in detail in the *Manual on Uniform Traffic Control Devices for Streets and Highways* (MUTCD; FHWA, 1988). These should be considered as a guide to the determination of the need tor traffic control signals rather than absolute criteria, and their use tempered with professional judgment based on experience and consideration of all related factors. For example, such factors as physical roadway features, age of pedestrians, or effect of adjacent signalized intersections may modify a decision based solely on the warrants.

Warrant 1: Minimum Vehicular Volume.   The minimum vehicular warrant is satisfied when, for each of any 8 hours of an average day, the traffic volumes given in Table 8-1 exist on the major street and on the higher-volume minor-street approach to the intersection. These major-street and minor-street volumes are for the same 8 hours. During each hour, the higher approach volume on the minor street is considered, regardless of its direction.

Warrant 2: Interruption of Continuous Traffic.   The interruption of continuous traffic warrant is satisfied when for each of any 8 hours of an average day, the traffic volumes given in Table 8-2 exist on the major street and on the higher-volume

TABLE 8-1   MINIMUM VEHICULAR VOLUME WARRANT 1

Number of lanes for moving traffic on each approach		Vehicles per hour on major street (total of both approaches)		Vehicle per hour on higher-volume minor-street approach (one direction only)	
Major street	Minor street	Urban	Rural	Urban	Rural
1	1	500	350	150	105
2 or more	1	600	420	150	105
2 or more	2 or more	600	420	200	140
1	2 or more	500	350	200	140

Source: FHWA, 1988.

TABLE 8-2   MINIMUM VEHICULAR VOLUMES FOR WARRANT 2

Number of lanes for moving traffic on each approach		Vehicles per hour on major street (total of both approaches)		Vehicle per hour on higher-volume minor-street approach (one direction only)	
Major street	Minor Street	Urban	Rural	Urban	Rural
1	1	750	525	75	53
2 or more	1	900	630	75	53
2 or more	2 or more	900	630	100	70
1	2 or more	750	525	100	70

Source: FHWA, 1988.

minor-street approach to the intersection and where the signal installation will not seriously disrupt progressive traffic flow. These major-street and minor-street volumes are for the same 8 hours. During each hour, the higher volume on the minor street is considered, regardless of its direction. A reduced volume, similar to that described under warrant 1, can be used in place of those shown in Table 8-2 on higher-speed roads or in smaller communities.

Warrant 3: Minimum Pedestrian Volume.   This warrant combines pedestrian volumes and available gaps in vehicular traffic at either intersection or midblock locations. The warrant is satisfied when, on an average day, there are 100 or more pedestrians for each of any four hours, or 190 or more pedestrians in one hour crossing the major street. These values may be reduced by as much as 50% where the predominant pedestrian crossing speed is below 3.5 ft per second. In addition, during the period when the pedestrian count is satisfied, there shall be less than 60 gaps (of adequate length for pedestrians to cross) per hour. This requirement applies to each direction separately on a divided street with a median wide enough for pedestrians to wait. A signal may not be warranted if the proposed site is included in coordinated traffic signal system that allows less than 60 gaps per hour as such a signal may provide fewer but longer gaps for pedestrians to cross safely.

This warrant applies only to locations where the nearest signalized intersection is more than 300 ft away and the proposed new installation will not unduly affect progressive flow. At midblock locations, curb parking should be prohibited at least 100 ft in advance of the crosswalk and for at least 20 ft beyond. Street lighting should be considered if the midblock location is used at night. Normal pedestrian signal indications shall be installed under this warrant and should utilize a traffic actuated signal controller and pedestrian detectors (push buttons). It shall operate within a coordinated scheme if the signal is in coordinated system.

Warrant 4: School Crossings.   The fourth warrant recognizes the unique problems related to children crossing a major street on the way to and from school, particularly near the school, and may be considered as a special case of the pedestrian warrant. A traffic control signal may be warranted at an established school crossing when a traffic engineering study of the frequency and adequacy of gaps in the vehicular traffic stream, as related to the number and size of groups of schoolchildren at the school crossing, shows that the number of adequate gaps in the traffic stream during the period when the children are using the crossing is less than the number of minutes in the same period.

Warrant 5: Progressive Movement.   The progressive movement warrant relates to the desirability of holding traffic in compact platoons. This warrant applies when the adjacent signals are so far apart that they do not provide the necessary degree of platooning and speed control on a one-way or two-way street. In addition, for a two-way street, the adjacent signals should constitute a progressive signal system. According to this warrant, the installation of a signal should not be considered where the resultant signal spacing would be less than 1000 ft (305 m).

Warrant 6: Accident Experience.   The accident experience warrant is satisfied when

1. An adequate trial of less restrictive remedies with satisfactory observance and enforcement has failed to reduce the accident frequency.
2. Five or more reported accidents of types susceptible to correction by traffic signal control have occurred within a 12-month period, each accident involving personal injury or property damage apparently exceeding the applicable requirements for a reportable accident.
3. There exists a volume of vehicular traffic not less than 80% of the requirements specified in warrants 1 or 2.
4. The signal installation will not seriously disrupt progressive traffic flow.

Warrant 7: Systems.   This warrant recognizes that traffic signal coordination can be two-dimensional. That is, progression along an important cross street can be as important as that along what would normally be called the major street. Both streets must be given equal consideration as major routes. The systems warrant is applicable when two or more major routes meet at a common intersection and the total existing or immediately projected entering volume is at least 1000 vehicles during the peak hour of a typical weekday or each of any 5 hours of a Saturday or Sunday.

Warrant 8: Combination of Warrants.   In exceptional cases, signals may occasionally be justified when no single warrant is satisfied, but where warrants 1 and 2 are satisfied to the extent of 80% or more of the stated values. The 80% requirement under this warrant is applied against 70% requirement for rural areas, making it 56% of that applicable to urban areas. Adequate trial of other remedial measures which cause less delay and inconvenience to traffic should precede installation of signals under this warrant.

Warrant 9: Four-Hour Volume.   This warrant uses 4-hour volumes instead of 8-hour volumes used in warrants 1, 2 and 8. The warrant defines curves representing vehicles per hour on the major street (both directions) and on the higher-volume minor-street approach (one direction only). This warrant is satisfied (for urban locations) when, for each of any 4 high hours of an average day, the plotted points fall above the specified curve for the existing combination of approach lanes.

The requirements are lower when the 85th percentile speed of major-street traffic exceeds 40 mph, or when the intersection lies within a built-up area of an isolated community with a population under 10,000. A separate series of curves is specified for 4-hour volume requirements of rural locations. A reference should be made to the Manual of Uniform Control Devices (MUTCD) (FHWA, 1988) for a complete discussion.

Warrant 10: Peak-Hour Delay.   For this warrant to be satisfied, all of the following conditions should be met:

1. The total delay experienced by the traffic, during the peak hour, on a side street controlled by a STOP sign equals or exceeds five vehicle-hours for a two-lane approach and four vehicle-hours for a one-lane approach.
2. The volume on the side-street approaches equals or exceeds 150 vph for a two-lane approach or 100 vph for a one-lane approach.
3. The total entering volume serviced during the peak hour equals or exceeds 800 vph for intersections with four or more approaches or 650 vph for intersections with three approaches.

Warrant 11: Peak-Hour Volume.   This is intended for intersections where minor-street traffic experiences undue delay or hazard in entering or crossing the main street. The warrant is based on a critical combination of main-street and cross-street volumes during any four consecutive 15-min periods during the peak hour of an average day. A series of curves is specified for traffic volume on major street (total for both directions) and the higher-volume minor street (vehicles per hour for one direction only) separately for urban and rural locations. A signal is warranted when measured volumes fall above the specified curve for a given combination of approach lanes. A reference should be made to MUTCD for a complete discussion.

7.8 Coordination of Traffic Signals

Some form of signal coordination is necessary on major streets having a series of intersecting streets in order that vehicles flow without having to stop at every intersection. Signals can be coordinated in several ways, but the three most common techniques are the simultaneous system, the alternative system, and the flexible progressive system.

1. *Simultaneous system.* In this technique, all signals along the coordinated length of the street display the same aspect to the same traffic stream at the same time. This system reduces capacity and encourages the tendency to travel at excessive speeds so as to pass as many signals as possible. It is best suited where the city blocks are short. Where turning traffic is light, it may have advantages for pedestrian movement. Some local control can be introduced using vehicle actuation, but a master controller keeps all the local controllers in step and imposes a common cycle time.

2. *Alternative system.* Here, alternative signals or groups of signals give opposite indications at the same time, which means that if a vehicle travels the distance between intersections in half the cycle time, it need not stop. The cycle time must be the same for all signals, so that the speed of progression is constant.

3. *Progressive system.* Two types of progressive systems are used. In the simple progressive system, the various signal faces controlling a given street give green indications in accordance with a time schedule to allow continuous operation of a platoon of vehicles to flow at a planned speed. In the flexible progressive system, the intervals at any signal may be adjusted independently to the traffic requirements and in which the green indications at separate signals may be started independently at the instant that will give the maximum efficiency. A master controller keeps the local controllers, which may be fixed-time or vehicle-actuated, in step.

Balanced Two-Way Signal Progression.   A balanced directional progression is often desired on streets during off-peak hours or during the entire day. In such cases, a simple method of obtaining the widest equal two-way "through" bands is desired. Purdy's (1967) method is one such technique that can be applied to a new or existing series of intersections. This quick-and-easy method is best described through an example.

Example 7

Design a signal progression that will give the maximum equal two-way "through" bandwidth given that speed 30 mph = 44.1 ft/sec; cycle = 60 sec; distances in feet: AB = 800, AC = 1800, AD = 2300, AE = 2900, AF = 4100; arterial green: A = 60%, B = 50%, C = 70%, D = 50%, E = 60%, F = 70%.

Solution

The most convenient way to assemble the data and calculations is in the form of a table.

| Line | Intersection | | | | | |
	A	B	C	D	E	F
1	0	800	1800	2300	2900	4100
2	0%	30%	68%	87%	110%	155%
3	0	−20	+18	−13	+10	+5
4	60%	50%	70%	50%	60%	70%
5	30/ −30	25/ −25	35/ −35	25/ −25	30/ −30	35/ −35
6	30/ −30	+5/ −45	53/ −17	12/ −38	40/ −20	40/ −30
7	30/ −30	+5/ −45	+3/ −67	12/ −38	40/ −20	40/ −30

Line 1: cumulative distance in feet to each intersection from base intersection A.

Line 2: speed = 30 mph = 44.1 ft/sec. Convert distance in feet to travel time in seconds and in terms of cycles (as a percentage). 1 cycle = 60 sec = 2646 ft/cycle (e.g., at B, 800/2646 = 30%).

Line 3: the algebraic numerical difference between the nearest multiple of 50% and the percentage appearing on line 2 (e.g., $x - 50\%, x - 100\%, x - 150\%$).

Line 4: percentage green as given.

Line 5: each green is placed in its initial reference position with respect to the common datum line.

Line 6: add lines 5 and 3 algebraically (numerator and denominator separately). The smallest "plus" = 5 and the smallest "minus" = 17. The through band in both directions = 5 + 17 = 22% of cycle.

Line 7: it may be necessary to adjust values in the numerator and denominator by adding +50/+50 or −50/−50 to any values in line 6 to obtain a wider bandwidth. For example, *C* can be $(+53/-17) + (-50/-50) = +3/-67$. In this case, the bandwidth works out to 3 + 20 = 23% of the cycle.

From Figure 8-11 and the preceding procedure, steps for drawing the time–speed diagram follow:

1. Use half of green time at *A* as a benchmark. The band at *A* will have a portion below the benchmark equal to the smallest "plus" and a portion above the benchmark equal to the smallest "minus" of the cycle time.
2. Draw a band in both directions according to the speed limit of the vehicles.
3. Find the offset time for each intersection by subtracting the line 7 values from line 2 values. If line 2 is more than 100, subtract 100 and use the remainder as the line 2 value.
4. Draw the signal intervals based on these offsets for each intersection.
5. Note that offsets are given in percentages, and then converted to seconds (e.g., $-30 \times 60 = 18$ secs).

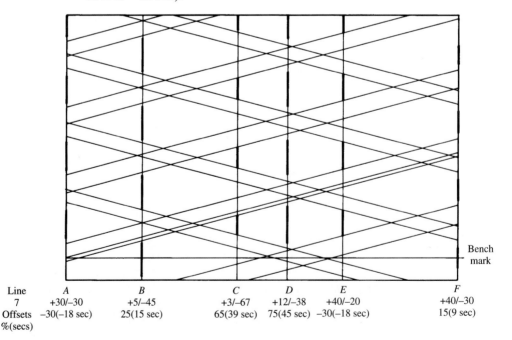

Line	*A*	*B*	*C*	*D*	*E*	*F*
7	+30/−30	+5/−45	+3/−67	+12/−38	+40/−20	+40/−30
Offsets %(secs)	−30(−18 sec)	25(15 sec)	65(39 sec)	75(45 sec)	−30(−18 sec)	15(9 sec)

Figure 8-11   Purdy's Method.

SUMMARY

Traffic control devices are an important part of an engineer's kit of tools for dealing with the efficient movement of vehicles and pedestrians along streets and through intersections. In this chapter, much emphasis has been given to the general topic of traffic signals, because they are much more difficult to design and operate than are other methods of traffic control. First, the basis for understanding the components of traffic signals was provided, followed by a theoretical examination of signal systems. A number of practical field methods of designing isolated intersection signals were given, followed by a brief description of designing traffic-actuated signals.

Eleven warrants specified by the *Manual on Uniform Traffic Control Devices* (FHWA, 1988) were explained briefly, followed by a short description of a method to coordinate signals on an arterial in progression.

For those inclined to obtain a more comprehensive knowledge of intersection design and control would do well to refer to Drew (1968), FHWA (1985), and JHK & Associates (1980).

REFERENCES

AMERICAN ASSOCIATION OF STATE HIGHWAY AND TRANSPORTATLON OFFICLALS (AASHTO) (1990). *A Policy on Geometric Design of Highways and Streets*, AASHTO, Washington, DC.

AMERICAN ASSOCIATION OF STATE HIGHWAY AND TRANSPORTATION OFFICIALS (AASHTO) (1994). *A Policy on Geometric Design of Highways and Streets* (Metric), AASHTO, Washington, DC.

DREW, DONALD R. (1968). *Traffic Flow Theory and Control*, McGraw-Hill, New York.

FEDERAL HIGHWAY ADMINISTRATION (FHWA) (1985). *Traffic Control Systems Handbook*, U.S. Department of Transportation, Washington, DC.

FEDERAL HIGHWAY ADMINISTRATION (FHWA) (1988). *Manual on Uniform Traffic Control Devices for Streets and Highways*, U.S. Department of Transportation, Washington, DC.

GREENSHIELDS, B. D., D. SHAPIRO, and E. L. ERICKSON (1947). *Traffic Performance at Urban Street Intersections*, Technical Report 1, Bureau of Highway Traffic, Yale University, New Haven, CT.

HOMBURGER, W. S. ET AL. (1996). *Fundamentals of Traffic Engineering*, 14th Ed., University of California, Berkeley.

JHK & ASSOCIATES (1980). *Design of Urban Streets*, Technology Sharing Report 80-204, U.S. Department of Transportation, Washington, DC.

MITRIC, S. (1975). *Transportation Engineering* notes, Ohio State University, Columbus.

PIGNATARO, L. J. (1973). *Traffic Engineering Theory and Practice*, Prentice-Hall, Englewood Cliffs, NJ.

PLINE, J. S. (Ed.) (1992). *Traffic Engineering Handbook*, 4th Ed., Prentice Hall, Englewood Cliffs, NJ.

PURDY, R. J. (1967). *Balanced Two-way Signal Progression, Traffic Engineering*, Institute of Traffic Engineers, Washington, DC.

SALTER, R. J. (1974). *Highway Traffic Analysis and Design*, Addison-Wesley, Reading, MA.

TRANSPORTATION RESEARCH BOARD (TRB) (1985). *Intersection Channelization Design Guide*, NCHRP Report 279, National Research Council, Washington, DC.

TRANSPORTATION RESEARCH BOARD (TRB) (1994). *Highway Capacity Manual*, Special Report 209, National Research Council, Washington, DC.

WEBSTER, F. V., and B. M. COBBE (1962). *Traffic Signals*, Road Research Technical Paper 56, Her Majesty's Stationery Office, London.

EXERCISES

1. List measures of performance that you would use to compare stop-sign control with signal control.

2. Examine an intersection in your city that has a high accident record. Redesign this intersection, introducing basic concepts of channelization. Write a short report describing the logic and criteria you adopted.

3. What are the criteria used to select the length of amber phases at an intersection?

4. Compare short and long signal cycles. What are the criteria for selecting the shortest and the longest cycles?

5. Assuming a deceleration of 9 ft/sec^2, an intersection width of 50 ft, a perception-reaction time of 2.0 sec, and an average vehicle length of 20 ft, plot the minimum duration of the yellow light versus the approach speed. What is the minimum duration of the yellow light?

6. Refer to Example 8-2 in the text. If the critical lane volumes during the peak hour are 500 and 300 veh/hr, and the approach speeds are 35 and 25 mph for Pine and Oak Streets, respectively, compute the revised cycle length and green times. Sketch the phase diagram for this case. Use Homburger and Kell's method.

7. Refer to Example 8-4 in the text. It is estimated that the critical flows in the N–S and E–W directions are 900 and 450 veh/hr respectively, and the lost time per phase is 6.5 seconds. Determine the cycle length and the distribution of green. Sketch a phase diagram showing the splits. Use Webster's method.

8. Develop elements of the shortest cycle based entirely on pedestrian movements, assuming no special "walk/don't walk" signals.

9. What is the basis for splitting total available green time per cycle among various phases?

10. An isolated intersection at Ohio and Michigan streets needs a two-phase signal. Ohio is 48 ft wide, with a critical lane volume of 450 veh/hr; Michigan is 36 ft wide, with a critical lane volume of 380 veh/hr. Both streets carry 15% trucks, have 15% left turns, have a PHF of 0.90, and have approach speeds of 35 mph. Design the signal system by all three methods described in the text. Assume lost times of 3 sec/phase and average headways of 2.5 sec.

11. An approach to a signalized intersection has a maximum posted speed of 35 mph, the yellow-light duration = 4 sec, the all-red duration = 1.5 sec, average vehicle length = 20 ft, deceleration = 9 ft/sec^2, intersection width = 45 ft, distance from stop bar to far edge of intersection curb = 48 ft, and perception-reaction time = 2.5 sec. Examine these data and write a detailed report regarding the intersection's safety and suitability. What recommendations would you offer to the city traffic engineer?

12. (a) A major street in Cincinnati needs progressive signaling favoring movement from its intersection with Ludlow to McMillan. The signal cycle lengths are all 60 sec, with a uniform cycle split of 60/40%. The recommended speed is 30 mph. Draw a time–space diagram. Adopt a horizontal scale of 1 in. = 600 ft for distance and a vertical scale of 1 in. = 40 sec for time. The center of the green interval at Ludlow should be adopted as the benchmark. Mark the signal offsets. What is the green bandwidth? Also show movements in the opposite direction.

   (b) If this major street is now to be designed for progressive signaling with balanced flow in both directions, maintaining the same speed and cycle split as before, draw the new time–space diagram. The center of green interval at Ludlow should be adopted as the benchmark. What is the maximum bandwidth in each direction? Mark the offsets in each direction.

(c) Write a short report describing how you solved the problem.

Cross streets	Distance between streets (ft)
Ludlow St.	0
Dixsmyth St.	870
Capitol St.	800
Riddle St.	1310
University St.	620
O.C. St.	1450
Calhoun St.	400
McMillan St.	240

13. Observed and saturation flows at a two-phase signal-controlled intersection are given. Both intergreen periods are 8 sec. and lost times are 3 sec per phase. Approach widths are 48 ft and 60 ft for the E–W and N–S directions, respectively. Make suitable assumptions and design the cycle time and green distribution by all three methods. Compare the results and write a report.

	N	S	E	W
Flow, q (veh/hr)	900	750	700	800
Saturation flow, s (veh/hr)	1800	1800	1600	1600

14. Rework Example 8-6 in the text using Pignataro's method if the critical volumes are 500 and 300 veh/hr, and the approach speeds are 35 and 25 mph for Pine and Oak Streets, respectively. Other details remain the same.

15. A new intersection planned for a suburban area needs to be designed with the following details:

	N–S approaches	E–W approaches
Curb-to-curb width	60 feet	56 feet
Peak-hour approach volumes		
Through	500 North	400 East
	600 South	350 West
Right turn	100 North	120 East
	130 South	100 West
Left turn	100 North	75 East
	95 South	125 West
PHF	0.90	0.90
Passenger cars	95%	95%
Trucks	5%	5%
Pedestrians	200 North	250 East
	250 South	300 West

Make whatever assumptions you feel appropriate, and then design a suitable signal system, indicating the cycle length and phase lengths by applying (a) Homburger and Kell's method, (b) Webster's method, and (c) Pignataro's method.

Chapter 9

At-Grade Intersection Capacity and Level of Service

1. INTRODUCTION

The basic concepts of capacity and level of service were covered in Chapter 7, which also included details of designing transportation facilities where uninterrupted flow occurred. Chapter 8 dealt with intersection design and control, and the information contained in that chapter is a good foundation for dealing with the capacity and level of service of signalized and unsignalized intersections, a topic dealt with in this chapter. Most of the material contained in this chapter has been drawn from the *Highway Capacity Manual* (TRB, 1994).

An at-grade road intersection is a complicated and complex part of a highway system. It is here that the majority of vehicle and pedestrian conflicts occur, resulting invariably in delays, accidents, and congestion. At-grade intersections are generally controlled by signals, in which case they are known as signalized intersections. Unsignalized intersections, which make up the vast majority of at-grade junctions in any street system, are controlled by stop and yield signs, to assign the right-of-way to one street or approach.

Generally, the capacity of a highway is dependent on the geometric characteristics of the facility, together with the composition of the traffic stream using the facility. The capacity of a highway is thus relatively stable. On the other hand, when we consider at-grade intersections on the highway, we are considering additional elements. For instance, in the case of a signalized intersection, we are introducing the concept of time; a traffic signal allocates time among traffic movements using the same space. With unsignalized intersections, controlled by stop and yield signs, the distribution of gaps in the major street traffic stream, coupled with the judgment of drivers in selecting gaps in the major stream, fixes the capacity of the controlled legs of the intersection.

2. CAPACITY AND LEVEL OF SERVICE

The concepts of capacity and level of service are central to the analysis of intersections as they are for all types of facilities. In intersection analysis, however, these two concepts are not as strongly correlated as they are for other facility types. In Chapter 7, it was shown that the analysis of uninterrupted flow facilities yielded a determination of both the capacity and the level of service of the facility. In the case of signalized intersections, the two are analyzed separately, because they are not simply related to each other. It is critical to note, however, that both capacity and level of service must be fully considered to evaluate the overall operation of a signalized intersection (TRB, 1994).

Capacity analysis of intersections results in the computation of V/c ratios for individual movements and a composite V/c ratio for the sum of critical movements or lane groups within the intersection. The V/c ratio is the actual or projected rate of flow on an approach or designated group of lanes during a peak 15-minute interval divided by the capacity of the approach or designated group of lanes. Level of service is based on the average stopped delay per vehicle for various movements within the intersection. Although V/c affects delay, there are other parameters that more strongly affect it, such as the quality of progression, length of green phase, and cycle lengths. Thus, for any given V/c ratio, a range of delay values may result, and vice versa. For this reason, both the capacity and level of service of the intersection must be examined carefully. These two concepts are discussed in detail in the following sections (TRB, 1994).

2.1 Signalized Intersections

Capacity at intersections is defined for each lane group. A lane group is made up of one or more lanes that accommodate traffic and have a common stop line and capacity shared by all vehicles. The lane group capacity is the maximum rate of flow for the subject lane group that may pass through the intersection under prevailing traffic, roadway, and signalization conditions. The rate of flow is generally measured or projected for a 15-min period, and capacity is stated in vehicles per hour (vph).

Traffic conditions include volumes on each approach, the distribution of vehicles by movement (left, through, right), the vehicle-type distribution within each movement, the location of and use of bus stops within the intersection area, pedestrian crossing flows, and parking movements within the intersection area.

Roadway conditions include the basic geometries of the intersection, including the number and width of lanes, grades, and lane-use allocations (including parking lanes).

Signalization conditions include a full definition of the signal phasing, timing, type of control, and an evaluation of signal progression on each approach.

It is not only the allocation of green time that has a significant impact on capacity and operations at a signalized intersection, but the manner in which turning movements are accommodated within the phase sequence as well. Signal phasing can provide for either protected or permitted turning movements.

A permitted turning movement is made through a conflicting pedestrian or opposing vehicle flow. Thus, a left-turn movement that is made at the same time as the opposing through movement is considered to be "permitted," as is a right-turn movement made at the same time as pedestrian crossing in a conflicting crosswalk.

Protected turns are those made without these conflicts, such as turns made during an exclusive left-turn phase or a righ-turn phase during which conflicting pedestrian movements are prohibited.

3. CAPACITY OF SIGNALIZED INTERSECTIONS

Capacity of signalized intersections is based on the concept of saturation flow and saturation flow rates. Saturation flow rate is defined as the maximum rate of flow that can pass through a given lane group under prevailing traffic and roadway conditions, assuming that the lane group had 100% of real time available as effective green time and is expressed in units of vehicles per hour of effective green time (vphg). The flow ratio is defined as the ratio of actual or projected flow rate for the lane group, V, to the saturation flow rate, s. The flow ratio is $(V/s)_i$ for lane group i.

The capacity of the lane group is

$$c_i = s_i(g_i/C) \tag{1}$$

where

c_i = capacity of lane group i (vph)
s_i = saturation flow for lane group i (vphg)
g_i/C = green ratio for lane group i
C = cycle length

The ratio of flow rate to capacity, V/c, is X; therefore,

$$X_i = \left(\frac{V}{c}\right)_i = \frac{V_i}{s_i(g_i/C)}$$

$$X_i = \frac{V_iC}{s_ig_i} = \frac{(V/s)_i}{g_i/C} \tag{2}$$

where

X_i = V/c ratio for lane group i
V_i = actual or projected demand flow rate for lane group i (vph)
s_i = saturation flow rate for lane group i (vphg)
g_i = effective green time for lane group i (sec)
C = cycle time of signal (sec)

Values of X_i range from 1.00, when the flow rate equals capacity, to 0.00, when the flow rate is zero. Another capacity concept of utility in the analysis of signalized intersections is the critical V/c ratio, X_c, which is a V/c ratio for the intersection as a whole, considering only the lane groups that have the highest flow ratio, V/s, for a given signal phase. For example, in a two-phase signal, opposing approaches move during the same green time, although one of these two approaches will require more green time than the other (i.e., it will have a higher flow ratio). This would be the critical approach for the signal phase. Thus, each signal phase will have a critical lane group that determines the green time requirements for the phase. Where signal phases overlap, the identification of these critical approaches becomes complex.

The critical (V/c) ratio for the intersection is defined in terms of critical lane groups or approaches.

$$X_c = \sum_i \left(\frac{V}{s}\right)_{ci} \frac{C}{C - L} \tag{3}$$

where

$$X_c = \text{critical } (V/c) \text{ ratio for the intersection}$$
$$\Sigma_i \, (V/s)_{ci} = \text{the summation of flow ratios for all critical lane groups}$$
$$C = \text{cycle length (sec)}$$
$$L = \text{total lost time per cycle, computed as the sum of startup and change}$$
interval lost time minus the portion of the change interval used by vehicles on the critical lane group for each signal phase

This equation is useful in evaluating the overall intersection with respect to the geometries and total cycle length provided, and is also useful in estimating signal timings where they are not known or specified by local policies or procedures.

4. LEVEL OF SERVICE FOR SIGNALIZED INTERSECTIONS

LOS for signalized intersections is defined in terms of delay. Delay is a measure of driver discomfort, frustration, fuel consumption, and lost travel time. Specifically, LOS criteria are stated in terms of the average stopped delay per vehicle for a 15-minute analysis period, and are given in Table 9-1.

TABLE 9-1 LEVEL-OF-SERVICE CRITERIA FOR SIGNALIZED INTERSECTIONS

Level of service	Stopped delay per vehicle (sec)
A	≤ 5.0
B	5.1–15.0
C	15.1–25.0
D	25.1–40.0
E	40.1–60.0
F	> 60.0

Source: TRB, 1994.

4.1 Capacity/LOS Relationship

This relationship is not a simple one. Although in previous chapters, LOS E has always been defined to be capacity (i.e., $V/c = 1$), this is not the case for signalized intersections. It is possible to have delays in the range of LOS F (unacceptable) while the V/c ratio is less than 1.00. Delays can occur when, for example, the cycle length is long or the signal progression is poor.

4.2 Level of Analysis

HCM (TRB, 1994) presents two levels of analysis. The primary methodology is operational analysis, which provides for a full analysis of capacity and LOS. It is used to evaluate alternative traffic demands, geometric designs, and/or signal plans. The other method is used in planning analysis. At this level, only capacity is addressed, because detailed information needed to estimate delay is not available. The planning method generates a projection of the status of the intersection with respect to its capacity and an approximation of a signal timing plan. By combining this approximation with appropriate values for other parameters, it is possible to extend the planning analysis into the level of the operational analysis.

Operational analysis can be used to

1. Solve for LOS, given the details of intersection flows, signalization, and geometrics.
2. Solve for allowable service flow rates for selected LOS, given the details of signalization and geometrics.
3. Solve for signal timing, given the desired LOS and details of flows and geometrics.
4. Solve for basic geometrics (number of lanes), given the desired LOS and details of flow and signalization.

In most cases, this type of analysis would consider various alternatives on a trial-and-error basis.

4.3 Operational Analysis

This analysis is complex and therefore the analysis procedures are done in a modular fashion, as shown in Figure 9-1. Details of the modules follow:

Input module. This is in the form of worksheet that allows all the variables needed for analysis to be easily summarized.

Volume-adjustment module. All hourly volumes are converted to peak flow rates by dividing by the appropriate peak-hour factors. The intersection is divided into lane groups for analysis. All exclusive turning lanes are considered as separate lane groups.

Saturation flow-rate module. This module computes the saturation flow rate for each lane group under prevailing roadway, traffic, and control conditions.

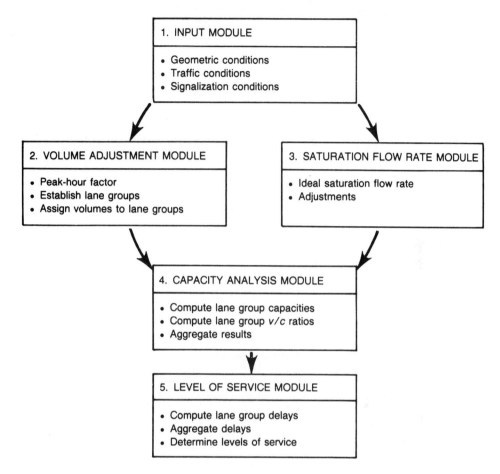

Figure 9-1   Operational Analysis Procedure (TRB, 1994).

Capacity analysis module. Here, capacities and V/c ratios for each lane group are computed, as is a critical V/c ratio for the intersection as a whole, given that the signal timing is known.

Level-of-service module. This module estimates the average stopped delay per vehicle in each lane group, and aggregates these estimates for approaches and for the intersection as a whole.

Input Module.   Figure 9-2 provides a summary of the input information required for operational analysis. Note some of the special information needed, such as

- Pedestrian flows in all crosswalks
- Number of parking movements per hour into and out of parking spaces within 250 ft of the intersection

TYPE OF CONDITION	PARAMETER	SYMBOL
Geometric Conditions	Area Type	CBD or other;
	Number of Lanes	N
	Lane Widths, ft	W
	Grades, %	+ (Upgrade)
	Existence of Exclusive LT or RT Lanes	– (Downgrade)
	Length of Storage Bay, LT or RT Lanes	L_s
	Parking Conditions	Y or N
Traffic Conditions	Volumes by Movement, vph	V_i
	Peak-Hour Factor	PHF
	Percent Heavy Vehicles	$\%HV$
	Conflicting Pedestrian Flow Rate, peds/hr	$PEDS$
	Number of Local Buses Stopping in Intersection	N_B
	Parking Activity, pkg maneuvers/hr	N_m
	Arrival Type	AT
Signalization Conditions	Cycle Length, sec	C
	Green Times, sec	G_i
	Yellow change interval	Y
	All-red clearance interval	AR
	Actuated vs Pretimed Operation	A or P
	Pedestrian Push-Button?	Y or N
	Minimum Pedestrian Green	G_p
	Phase Plan	

Figure 9-2   Input Data Needed for Each Analysis Lane Group (TRB, 1994).

- Arrival type for each approach (details follow)
- Percent grade on all approaches

Five arrival types are defined for the dominant arrival flow as follows:

Type 1: a dense platoon arriving at the intersection at the beginning of the red phase. This is the worst platoon condition.

Type 2: a moderately dense platoon arriving during the middle of the red phase or a dispersed platoon arriving throughout the red phase, a condition somewhat better than type 1.

Type 3: random arrivals in which the main platoon contains less than 40% of the lane group volume. This is an average condition.

Type 4: a moderately dense platoon arriving during the middle of the green phase or a dispersed platoon arriving throughout the green phase, and is a moderately favorable condition.

Type 5: a dense platoon arriving at the beginning of the green phase. This is the most favorable condition.

A dense platoon would contain over 80% of the lane group volume, and a dispersed platoon from 40 to 80%. The arrival type can best be observed on site and should be determined as accurately as possible, because it will have a significant impact on delay estimates and LOS determination. The following ratio is useful:

$$R_p = P(C/g) \qquad (4)$$

where

R_p = platoon ratio
P = percent of all vehicles in the movement arriving during the green phase
C = cycle length
g = effective green time for the movement

P can be observed in the field, whereas g and C are computed from the signal timing. Table 9-2 gives approximate ranges of R_p related to arrival type.

Complete information regarding signalization is needed, which includes a phase diagram giving the phase plan, cycle length, green times, and change intervals. Actuated phases must be identified, including the existence of push-button pedestrian-actuated phases. If pedestrian timing requirements exist, the minimum green times for the phase should be indicated and must be provided for in signal timing. The minimum green time for a phase may be estimated as

$$g_p = 7.0 + \frac{W}{4.0} - Y \qquad (5)$$

where

g_p = minimum green time (sec)
W = distance from the curb to the center of the farthest travel lane on the street being crossed, or to the nearest pedestrian refuge island (ft)
Y = change interval (yellow + all-red time) (sec). It is assumed that the 15th-percentile walking speed of pedestrian crossing a street is 4.0 ft/sec

TABLE 9-2   RELATIONSHIP BETWEEN ARRIVAL TYPE AND PLATOON RATIO (R_p)

Arrival type	Range of platoon ratio (R_p)	Default value (R_p)	Progression quality
1	≤ 0.50	0.333	Very poor
2	> 0.50 and ≤ 0.85	0.667	Unfavorable
3	> 0.85 and ≤ 1.15	1.000	Random arrivals
4	> 1.15 and ≤ 1.50	1.333	Favorable
5	> 1.50 and ≤ 2.00	1.667	Highly favorable
6	> 2.00	2.000	Exceptional

Source: TRB, 1994.

Where signal phases are actuated, the cycle length and green times will vary from cycle to cycle in response to demand. For analysis purposes, field observations can be recorded for the same time periods as volume and averages taken. In the absence of field data, use may be made of default values, with some caution. Table 9-3 presents default values.

Volume-Adjustment Module.   To account for unbalanced lane utilization, three major analytical steps are performed in the volume-adjustment module, as described in what follows.

Adjust Volumes to Reflect Peak Flow Rates.   Convert demands stated as hourly volumes to flow rates for the peak 15-minute period within the hour, using an appropriate peak-hour factor (PHF), which may be defined for the intersection as a whole, for each approach, or for each movement.

TABLE 9-3   DEFAULT VALUES FOR USE IN OPERATIONAL AND PLANNING ANALYSIS

Characteristic	Default value
Traffic	
Ideal saturation flow rate	1,900 pcphgpl
Conflicting pedestrian volume	None: 0 peds/hr
(assume none unless field data	Low: 50 peds/hr
indicate otherwise)	Moderate: 200 peds/hr
	High: 400 peds/hr
Percent heavy vehicles	2
Grade	0%
Number of stopping buses	0/hr
Parking conditions	No parking
Parking maneuvers	20/hr where parking exists
Arrival type	
Lane groups with through movements	3 if isolated
	4 if coordinated[a]
Lane groups without through movements	3
Peak-hour factor	0.90
Lane-utilization factor	See Table 9-4
Facility and traffic signal	
Signal type	Pretimed
Cycle length range	60-120 sec
Lost time	3.0 sec/phase
Yellow plus all-red	4.0 sec/phase
Area type	Non-CBD
Lane width	12 ft

[a] Better arrival types are often possible with favorable progression design.
Source: TRB, 1994.

$$V_p = \frac{V}{\text{PHF}} \qquad (6)$$

where

V_p = flow rate during 15-minute period (vph)

V = hourly volume (vph)

PHF = peak-hour factor

It is recommended that 15-minute flows be observed in the field, because all intersection movements may not peak at the same time.

Determine Lane Groups for Analysis.   A lane group is defined as one or more lanes of an intersection approach serving one or more traffic movements. Grouping may be done as follows:

1. An exclusive left-turn lane(s) should be designated as a separate group unless there is also a shared left-through lane present, in which case distribution of traffic volumes between the movements is considered in defining the lane groups. The same is true of an exclusive right-turn lane as well. The same applies to right-turn lanes.
2. On approaches with exclusive right-turn and/or left-turn lanes, all other lanes on the approach would be included in a single lane group.
3. Where an approach with more than one lane includes a lane that may be used by both left-turn and through vehicles, judgment should be used as to whether to consider it an exclusive left-turn lane, depending on the distribution of left turn to through vehicles.

This determination cannot be made effectively until the proportion of left turns in the shared lane has been computed. If the computed proportion of the left turns in the shared lane equals or exceeds 1.0 (i.e., 100%), the shared lane must be considered a de facto left-turn lane.

Initially, the lane groups are established on the basis stated before in the Volume-Adjustment Module Worksheet. The proportion of left turns in the shared lane will be determined later as a part of the Saturation Flow Rate Module. For multiple-lane groups, P_L is computed as the proportion of left turns in the left-hand lane of the lane group. If this value is determined to be 1.0 or higher, the lane groups for the approach should be reassigned showing this left-hand lane as an exclusive left-turn lane (a de facto left-turn lane), because it is occupied entirely by left-turning vehicles. This requires redoing all of the computations for this approach beginning with the Volume-Adjustment Module Worksheet where the lane groups were initially established.

Where two or more lanes are included in a lane group for purposes of analysis, all subsequent calculations treat these lanes as a single entity. Figure 9-3 shows some common lane groups for purposes of analysis. The operation of a shared left-turn and through lane with a left turn permitted is complex.

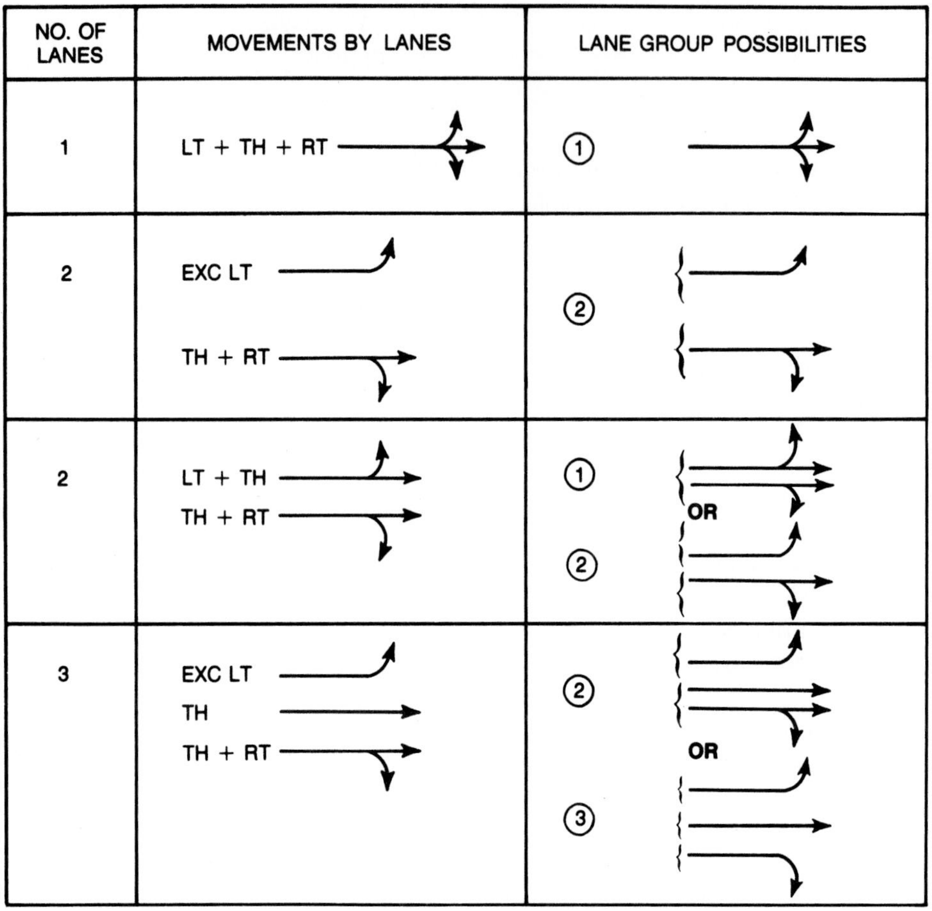

NO. OF LANES	MOVEMENTS BY LANES	LANE GROUP POSSIBILITIES
1	LT + TH + RT	①
2	EXC LT TH + RT	②
2	LT + TH TH + RT	① OR ②
3	EXC LT TH TH + RT	② OR ③

Figure 9-3 Typical Lane Groups for Analysis (TRB, 1994).

Adjust for Lane Distribution. Flow rates in each lane group are now adjusted to reflect unequal lane utilization. Flow will not divide equally when more than one lane is used.

$$V = V_g \times U \tag{7}$$

where

V = adjusted demand flow rate for the lane group (vph)

V_g = unadjusted demand flow rate for the lane group (vph)

U = lane-utilization factor (Table 9-4)

The lane-utilization factor is calculated as follows:

$$U = V_{g1}N/V_g \tag{8}$$

TABLE 9-4   DEFAULT LANE UTILIZATION FACTORS

Lane group movements	No. of lanes in lane group	Percent of traffic in most heavily traveled lane	Lane-utilization factor (U)
Through or shared	1	100.0	1.00
	2	52.5	1.05
	3 [a]	36.7	1.10
Exclusive left turn	1	100.0	1.00
	2 [a]	51.5	1.03
Exclusive right turn	1	100.0	1.00
	2 [a]	56.5	1.13

[a] If lane group has more lanes than the number shown in this table, it is recommended that surveys be made or the largest U factor shown for that type of lane group be used.
Source: TRB, 1994.

where

V_{g1} = unadjusted demand flow rate on the single lane with the highest volume in the lane group

N = number of lanes in the lane group

Use of the lane-utilization factor is resorted to when it is desirable to analyze the worst of two or more lanes; otherwise, the factor is set to 1.00. Actual lane distribution should be used, if known, in the computation of the lane-utilization factor. Field observations are important when the number of available lanes changes near the intersection. When average conditions exist or traffic distribution on a lane group is not known, the default values in Table 9-4 may be used.

When a right turn on red (RTOR) is permitted, the right-turn volume may be reduced by the volume of right-turning vehicles moving on the red phase. This is generally done on the basis of hourly volumes before converting to flow rates. At an existing intersection, the number of right turns on red should be determined by field observations. For both the shared lane and the exclusive right-turn lane conditions, the number of right turns on red should be subtracted from the right-turn volume before the analysis of lane group capacity or level of service. In the absence of field data, it is preferable for most purposes to utilize the right-turn volumes directly without a reduction for the number of right turns on red except when a protected left turn exists on the cross street. In this case, right turns on red may be taken as equal to the shadowing left-turn volume per lane.

Saturation Flow Module.   The saturation flow rate for each lane group is computed. The saturation flow rate is the flow in vehicles per hour that could be accommodated by the lane group assuming that the green phase was always available to the lane group (i.e., that the green ratio, g/C, was 1.00). Computations begin with the selection of an "ideal" saturation flow rate, usually 1900 passenger cars per hour of green time per lane (pcphgpl), and adjustment of this value for a variety of prevailing conditions that are not ideal.

$$s = s_0 N f_w f_{HV} f_g f_p f_{bb} f_a f_{RT} f_{LT} \tag{9}$$

where

$s =$ saturation flow rate for the subject lane group, expressed as a total for all lanes in the lane group under prevailing conditions (vphg)

$s_0 =$ ideal saturation flow rate per lane, usually 1900 (pc/hr green/ln)

$N =$ number of lanes in the lane group

$f_w =$ adjustment factor for lane width; 12-ft lanes are standard; given in Table 9-5

$f_{HV} =$ adjustment factor for heavy vehicles in the traffic stream; given in Table 9-6

TABLE 9-5 ADJUSTMENT FACTOR FOR AVERAGE LANE WIDTH (f_w)

Average lane width, W (ft)	Lane-width factor, f_w
8	0.867
9	0.900
10	0.933
11	0.967
12	1.000
13	1.033
14	1.067
15	1.100
16	1.133

Note: $f_w = 1 + (W - 12)/30$; $W \geq 8$ (if $W > 16$, a two-lane analysis may be considered).
Source: TRB, 1994.

TABLE 9-6 ADJUSTMENT FACTOR FOR HEAVY VEHICLES (f_{HV})

Percent heavy vehicles, % HV	Heavy–vehicle factor, f_{HV}
0	1.000
2	0.980
4	0.962
6	0.943
8	0.926
10	0.909
15	0.870
20	0.833
25	0.800
30	0.769
35	0.741
40	0.714
45	0.690
50	0.667
75	0.571
100	0.500

Note: $f_{HV} = 100/[100 + \% \text{ HV}(E_T - 1)]$; $0 \leq \% \text{ HV} \leq 100$, where $E_T = 2.0$ passenger cars per heavy vehicle.
Source: TRB, 1994.

f_g = adjustment factor for approach grade; given in Table 9-7

f_p = adjustment factor for the existence of a parking lane adjacent to the lane group and the parking activity in that lane; given in Table 9-8

f_{bb} = adjustment factor for the blocking effect of local buses stopping within the intersection area; given in Table 9-9

TABLE 9-7 ADJUSTMENT FACTOR FOR GRADE (f_g)

Type	Grade, %G Percent	Grade factor, f_g
Downhill	−6 or less	1.030
	−4	1.020
	−2	1.010
Level	0	1.000
Uphill	+2	0.990
	+4	0.980
	+6	0.970
	+8	0.960
	+10 or more	0.950

Note: $f_g = 1 - \%G/200; -6 \leq \%G \leq +10$.
Source: TRB, 1994.

TABLE 9-8 ADJUSTMENT FACTOR FOR PARKING (f_p)

No. of lanes in lane group, N	No parking	No. of parking maneuvers per hour, N_m				
		0	10	20	30	40[a]
1	1.000	0.900	0.850	0.800	0.750	0.700
2	1.000	0.950	0.925	0.900	0.875	0.850
3[a]	1.000	0.967	0.950	0.933	0.917	0.900

Note: $f_p = (N - 0.1 - 18N_m/3600)/N; 0 \leq N_m \leq 180; f_p \geq 0.05$.
[a] Use formula for more than three lanes or more than 40 maneuvers per hour.

TABLE 9-9 ADJUSTMENT FACTOR FOR BUS BLOCKAGE (f_{bb})

No. of lanes in lane group, N	No. of buses stopping per hour, N_B				
	0	10	20	30	40[a]
1	1.000	0.960	0.920	0.880	0.840
2	1.000	0.980	0.960	0.940	0.920
3[a]	1.000	0.987	0.973	0.960	0.947

Note: $f_{bb} = (N - 14.4N_B/3600)/N; 0 \leq N_B \leq 250; f_{bb} \geq 0.05$.
[a] Use formula for more than three lanes or more than 40 buses stopping per hour.

TABLE 9-10 ADJUSTMENT FACTOR
FOR AREA TYPE (f_a)

Type of area	Area-type factor, f_a
CBD	0.90
All other area	1.00

Source: TRB, 1994.

f_a = adjustment factor for area type; given in Table 9-10

f_{RT} = adjustment factor for right turns in the lane group; given in Table 9-11

f_{LT} = adjustment factor for left turns in the lane group; given in Table 9-12, or computed as described in following sections

More accurate results are obtained from a use of the measured values of the prevailing saturation flow rate rather than an estimate.

Adjustment Factors.   Each factor accounts for the impact of one or several prevailing conditions that are different from the ideal conditions for which the ideal saturation flow rate of 1900 pcphgpl applies. The lane width factor, f_w, accounts for the deleterious impact of narrow lanes on saturation flow rate, and allows for an increased flow on wide lanes. Twelve-foot lanes are the standard.

The lane-width factor may be calculated with caution for lane widths greater than 16 ft, or an analysis using two lanes may be conducted. Use of two lanes will result in higher saturation flow rate, but the analysis should reflect the way in which the width is actually used or expected to be used.

The effects of heavy vehicles and grades are treated by separate factors, f_{HV} and f_g, respectively. Their separate treatment recognizes that passenger cars are affected by approach grades, as are heavy vehicles. The heavy-vehicle factor accounts for the additional space occupied by these vehicles and for the differential in the operating capabilities of heavy vehicles with respect to passenger cars.

The passenger-car equivalent (E_T) used for each heavy vehicle is 2.0 passenger-car units (PCUs) and is reflected in the formula.

The grade factor accounts for the effect of grades on the operation of all vehicles.

The parking factor, f_p, accounts for the frictional effect of a parking lane on flow in adjacent lane groups, as well as for the occasional blocking of an adjacent lane by vehicles moving into and out of parking spaces.

Each maneuver, either in or out, is assumed to block traffic in the lane next to parking maneuvers for an average of 18 seconds. The number of maneuvers per hour in parking areas directly adjacent to the lane group and within 250 ft upstream from the stop line should be used. A practical limit of 180 is used if more parking maneuvers occur per hour. If the parking is adjacent to an exclusive-turn-lane group, the factor only applies to that lane group. On a one-way street, parking on the left side will affect the leftmost lane group. If parking is on both sides of a single-lane group, as in a one-way street with no exclusive-turn lanes, the number of maneuvers used is the total for both sides of the lane group. Note that parking conditions with zero maneuvers are not the same as no parking.

TABLE 9-11A ADJUSTMENT FACTOR FOR RIGHT TURNS (f_{RT}): FORMULAS

Cases 1–6: Exclusive/shared lanes and protected/permitted phasing

$f_{RT} = 1.0 - P_{RT}[0.15 + (\text{PEDS}/2100)(1 - P_{RTA})]$
$0.0 \le P_{RT} \le 1.0$

Proportion of RT in lane group = 1.00 for excl. RT lane (Cases 1–3); < 1.00 for shared lane (Cases 4–6).

$0.0 \le P_{RTA} \le 1.0$

Proportion of RT using protected phase = 1.00 for complete protection—no peds; < 1.00 for permitted with conflicting peds.

$0 \le \text{PEDS} \le 1700$

Volume (peds/hr) of peds conflicting with RT (if PEDS > 1700, use 1700).

$f_{RT} \ge 0.05$

Case 7: Single-lane approach (all traffic on approach in a single lane, as defined in Figure 9-3)

$f_{RT} = 0.90 - P_{RT}[0.135 + (\text{PEDS}/2100)]$
$0 \le P_{RT} \le 1.0$

Proportion of RT in lane group.

$0 \le \text{PEDS} \le 1700$

Volume (peds/hr) of peds conflicting with RT (use 0 if RT is completely protected).

$f_{RT} = 1.00$ if $P_{RT} = 0.0$
$f_{RT} \ge 0.05$

Case	Range of variable values			Simplified formula
	P_{RT}	P_{RTA}	PEDS	
1 Excl. RT lane; prot. RT phase	1.0	1.0	0	0.85
2 Excl. RT lane; perm. RT phase	1.0	0.0	0–1700	$0.85 - (\text{PEDS}/2100)$
3 Excl. RT lane; prot. + perm. RT phase	1.0	0–1.0	0–1700	$0.85 - (\text{PEDS}/2100)(1 - P_{RTA})$
4 Shared RT lane; prot. RT phase	0–1.0	1.0	0	$1.0 - P_{RT}[0.15]$
5 Shared RT lane; perm. RT phase	0–1.0	0.0	0–1700	$1.0 - P_{RT}[0.15 + (\text{PEDS}/2100)]$
6 Shared RT lane; prot. + perm. RT phase	0–1.0	0–1.0	0–1700	$1.0 - P_{RT}[0.15 + (\text{PEDS}/2100)(1 - P_{RTA})]$
7 Single-lane approach	0–1.0	—	0–1700	$0.9 - P_{RT}[0.135 + (\text{PEDS}/2100)]$

Source: TRB, 1994.

TABLE 9-11B ADJUSTMENT FACTOR FOR RIGHT TURNS: FACTORS

			Proportion of RTs in Lane Group, P_{RT}					
			Cases 4, 5, 6					Cases 1, 2, 3
Case	P_{RTA}	PEDS	0	0.2	0.4	0.6	0.8	1.0
2 and 5	0	0	1.00	0.970	0.940	0.910	0.880	0.850
		50 (low)	1.00	0.965	0.930	0.896	0.861	0.826
		100	1.00	0.960	0.921	0.881	0.842	0.802
		200 (mod.)	1.00	0.951	0.902	0.853	0.804	0.755
		400 (high)	1.00	0.932	0.864	0.796	0.728	0.660
		800	1.00	0.894	0.788	0.681	0.575	0.469
		1200	1.00	0.856	0.711	0.567	0.423	0.279
		≥ 1700	1.00	0.808	0.616	0.424	0.232	0.050
	0.20	0	1.00	0.970	0.940	0.910	0.880	0.850
		50 (low)	1.00	0.966	0.932	0.899	0.865	0.831
		100	1.00	0.962	0.925	0.887	0.850	0.812
		200 (mod.)	1.00	0.955	0.910	0.864	0.819	0.774
		400 (high)	1.00	0.940	0.879	0.819	0.758	0.698
		800	1.00	0.909	0.818	0.727	0.636	0.545
		1200	1.00	0.879	0.757	0.636	0.514	0.393
		≥ 1700	1.00	0.840	0.681	0.521	0.362	0.202
3 and 6	0.40	0	1.00	0.970	0.940	0.910	0.880	0.850
		50 (low)	1.00	0.967	0.934	0.901	0.869	0.836
		100	1.00	0.964	0.929	0.893	0.857	0.821
		200 (mod)	1.00	0.959	0.917	0.876	0.834	0.793
		400 (high)	1.00	0.947	0.894	0.841	0.789	0.736
		800	1.00	0.924	0.849	0.773	0.697	0.621
		1200	1.00	0.901	0.803	0.704	0.606	0.507
		≥ 1700	1.00	0.873	0.746	0.619	0.491	0.364
	0.60	0	1.00	0.970	0.940	0.910	0.880	0.850
		50 (low)	1.00	0.968	0.936	0.904	0.872	0.840
		100	1.00	0.966	0.932	0.899	0.865	0.831
		200 (mod)	1.00	0.962	0.925	0.887	0.850	0.812
		400 (high)	1.00	0.955	0.910	0.864	0.819	0.774
		800	1.00	0.940	0.879	0.819	0.758	0.698
		1200	1.00	0.924	0.849	0.773	0.697	0.621
		≥ 1700	1.00	0.905	0.810	0.716	0.621	0.526
	0.80	0	1.00	0.970	0.940	0.910	0.880	0.850
		50 (low)	1.00	0.969	0.938	0.907	0.876	0.845
		100	1.00	0.968	0.936	0.904	0.872	0.840
		200 (mod)	1.00	0.966	0.932	0.899	0.865	0.831
		400 (high)	1.00	0.962	0.925	0.887	0.850	0.812
		800	1.00	0.955	0.910	0.864	0.819	0.774
		1200	1.00	0.947	0.894	0.841	0.789	0.736
		≥ 1700	1.00	0.938	0.875	0.813	0.750	0.688
1 and 4	1.00	0	1.00	0.970	0.940	0.910	0.880	0.850
7	—	0	1.00	0.873	0.846	0.819	0.792	0.765
		50 (low)	1.00	0.868	0.836	0.805	0.773	0.741
		100	1.00	0.863	0.827	0.790	0.754	0.717
		200 (mod)	1.00	0.854	0.808	0.762	0.716	0.670
		400 (high)	1.00	0.835	0.770	0.705	0.640	0.575
		800	1.00	0.797	0.694	0.590	0.487	0.384
		1200	1.00	0.759	0.617	0.476	0.335	0.194
		≥ 1700	1.00	0.711	0.522	0.333	0.144	0.050

Source: TRB, 1994.

TABLE 9-12 ADJUSTMENT FACTOR FOR LEFT TURNS (f_{LT})

Case	Type of lane group	Left-turn factor, f_{LT}
1	Exclusive LT lane; Protected phasing	0.95
2	Exclusive LT lane; Permitted phasing	Special procedure; see worksheet in Figure 9-4 or 9-5
3	Exclusive LT lane; Protected-plus-permitted phasing	Apply Case 1 to protected phase; Apply Case 2 to permitted phase
4	Shared LT lane; Protected phasing	$f_{LT} = 1.0/(1.0 + 0.05 P_{LT})$

	Proportion of left turns, P_{LT}					
	0.00	0.20	0.40	0.60	0.80	1.00
Factor	1.00	0.99	0.98	0.97	0.96	0.95

Case	Type of lane group	Left-turn factor, f_{LT}
5	Shared LT lane; Permitted phasing	Special procedure; see worksheet in Figure 9-4 or 9-5
6	Shared LT lane; Protected-plus-permitted phasing	$f_{LT} = (1400 - V_0)/[(1400 - V_0) + (235 + 0.435 V_0) P_{LT}]; \ V_0 \leq 1220$ vph; $f_{LT} = 1/(1 + 4.525 P_{LT}); \ V_0 \geq 1220$ vph

Opposing volume, V_0	Proportion of Left Turns, P_{LT}					
	0.00	0.20	0.40	0.60	0.80	1.00
0	1.00	0.97	0.94	0.91	0.88	0.86
200	1.00	0.95	0.90	0.86	0.82	0.78
400	1.00	0.92	0.85	0.80	0.75	0.70
600	1.00	0.88	0.79	0.72	0.66	0.61
800	1.00	0.83	0.71	0.62	0.55	0.49
1000	1.00	0.74	0.58	0.48	0.41	0.36
1200	1.00	0.55	0.38	0.29	0.24	0.20
≥ 1220	1.00	0.52	0.36	0.27	0.22	0.18

Source: TRB, 1994.

SUPPLEMENTAL WORKSHEET FOR PERMITTED LEFT TURNS				
*** For Use Where the Subject Approach is Opposed by a Multilane Approach ***				
APPROACH	EB	WB	NB	SB
Enter Cycle Length, C				
Enter Actual Green Time For Lane Group, G				
Enter Effective Green Time For Lane Group, g				
Enter Opposing Effective Green Time, g_o				
Enter Number of Lanes in Lane Group, N				
Enter Number of Opposing Lanes, N_o				
Enter Adjusted Left-Turn Flow Rate, v_{LT}				
Enter Proportion of Left Turns in Lane Group, P_{LT}				
Enter Adjusted Opposing Flow Rate, v_o				
Enter Lost Time per Phase, t_L				
Compute Left Turns per Cycle: LTC = v_{LT} C/3600				
Compute Opposing Flow per Lane, Per Cycle: $v_{olc} = v_o$ C/(3600 N_o)				
Determine Opposing Platoon Ratio, R_{po} (Table 9-2 or Eq [9-4])				
Compute g_f ** = G exp $(-0.882$ LTC$^{0.717})$ − t_L, $g_f \le g$				
Compute Opposing Queue Ratio: $qr_o = 1 - R_{po}$ (g_o/C)				
Compute g_q, using equation 9-13, $g_q \le g$				
Compute g_u: $g_u = g - g_q$ if $g_q \ge g_f$ $g_u = g - g_f$ if $g_q < g_f$				
Compute $f_s = (875 - 0.625 v_o)/1000$, $f_s \ge 0$				
Compute P_L† = P_{LT} [1 + {($N-1$)g/($f_s g_u$ + 4.5)}]				
Determine E_{L1} (Figure 9-6)				
Compute $f_{min} = 2(1 + P_L)/g$				
Compute f_m: $f_m = [g_f/g] + [g_u/g][1/\{1 + P_L(E_{L1} - 1)\}]$ min = f_{min}; max = 1.00				
Compute $f_{LT} = [f_m + 0.91 (N - 1)]N$ ‡				

** For special case of single-lane approach opposed by multilane approach, see text.

† If $P_L \ge 1$ for shared left-turn lanes with $N > 1$, then assume de facto left-turn lane and redo calculations.

‡ For permitted left turns with multiple exclusive left-turn lanes $f_{LT} = f_m$.

Figure 9-4 Supplemental Worksheet for Permitted Left Turns: Multilane Approach (TRB, 1994).

The bus blockage factor, f_{bb}, accounts for the impacts of local transit buses stopping to discharge or pick up passengers at a near-side or far-side stop within 250 ft of the stop line (upstream or downstream). This factor should be used only when stopping buses block traffic flow in the subject lane group. A practical limit of 250 ft is used if more buses per hour exist. Where local transit buses are believed to be a major factor

SUPPLEMENTAL WORKSHEET FOR PERMITTED LEFT TURNS				
*** For Use Where the Subject Approadl is Opposed by a Single lane Approach ***				
APPROACH	EB	WB	NB	SB
Enter Cycle Length, C				
Enter Actual Green Time For Lane Group, G				
Enter Effective Green Time For Lane Group, g				
Enter Opposing Effective Green Time, g_o				
Enter Number of Lanes in Lane Group, N				
Enter Adjusted Left-Turn Flow Rate, v_{LT}				
Enter Proportion of Left Turns in Lane Group, P_{LT}				
Enter Proportion of Left Turns in Opposing Flow, P_{LTo}				
Enter Adjusted Opposing Flow Rate, v_o				
Enter Lost Time per Phase, t_L				
Compute Left Turns per Cycle: LTC $= v_{LT}\, C/3600$				
Compute Opposing Flow per Lane, per Cycle: $v_{olc} = v_o\, C/3600$				
Determine Opposing Platoon Ratio, R_{po} (Table 9-2 or Eq [9-4])				
Compute g_f** $= G \exp\,(-0.882\ \mathrm{LTC}^{0.629}) - t_L,\ g_f \leq g$				
Compute Opposing Queue Ratio: $qr_o = 1 - R_{po}\,(g_o/C)$				
Compute $g_q, = 4.943\ v_{olc}^{0.762}\ qr_o^{1.061} - t_L\ g_q \leq g$				
Compute g_u: $g_u = g - g_q$ if $g_q \geq g_f$ $g_u = g - g_f$ if $g_q < g_f$				
Compute $n = (g_q - g_f)/2,\ n \geq 0$				
Compute $P_{THo} = 1 - P_{LTo}$				
Determine E_{L1} (Figure 9-7)				
Compute $E_{L2}\,(1 - P_{THo}{}^n)/P_{LTo}$				
Determine $f_{\min} = 2(1 + /P_{LT})/g$				
Compute f_{LT}** $= f_m = [g_f/g] + [(g_q - g_f)/g]\,[1\,/\{1 + P_{LT}(E_{L2} - 1)\}]$ $+ [g_u/g]\,[1\,/\{1 + P_{LT}(E_{L1} - 1)\}]$ $\min = f_{\min};\ \max = 1.00$				

** For special case of multilane approach opposed by single-lane approach or when $g_f > g_q$, see text.

Figure 9-5 Supplemental Worksheet for Permitted Left Turns: Single-Lane Approach (TRB, 1994).

in intersection performance, Chapter 10 may be consulted for a more precise method of quantifying this effect. The factor used here assumes an average blockage time of 14.4 seconds during a green indication.

The area type factor, f_a, accounts for the relative inefficiency of business area intersections in comparison to those in other locations. This is due primarily to the complexity and general congestion of the environment in business areas.

The right-turn factor, f_{RT}, depends on a number of variables, including the following:

1. Whether right turns are made from an exclusive or shared lane
2. Type of signal phasing (protected, permitted, or protected plus permitted); a protected right-turn phase has no conflicting pedestrian movements
3. Volume of pedestrians using the conflicting crosswalk
4. Proportion of right turns using a shared lane
5. Proportion of right turns using the protected portion of a protected plus permitted phase

Item 5 should be determined by field observation, but can be grossly estimated from the signal timing. This is done by assuming that the proportion of right-turning vehicles using the protected phase is approximately equal to the proportion of the turning phase that is protected. Where right turn on red (RTOR) is permitted, the right-turn volume may be reduced as described in the discussion of the Volume Adjustment Module.

The left-turn factor, f_{LT}, is based on similar variables, including the following:

1. Whether left turns are made from exclusive or shared lanes
2. Type of phasing (protected, permitted, or protected plus permitted)
3. Proportion of left-turning vehicles using a shared lane group
4. Opposing flow rate when permitted left turns are made

The left-turn adjustment factor is 1.0 if the lane group does not include any left turns. When a left turn is not opposed at any time by through vehicles but encounters conflicting pedestrian movements, use the adjustment procedure for right turns. If no conflicting pedestrian movements are present, a normal protected left-turn adjustment should be performed.

The turn factors basically account for the fact that these movements cannot be made at the same saturation flow rates as through movements. They consume more of the available green time, and consequently, more of the lane group's available capacity.

All of the adjustment factors described are given in Tables 9-5 through 9-12.

Special Procedure: Left-turn Adjustment Factors for Permitted Phasing.
The adjustment factors in Tables 9-11 and 9-12 reflect seven different cases under which turns may be made. These are

Case 1: Exclusive lane with protected phasing

Case 2: Exclusive lane with permitted phasing

Case 3: Exclusive lane with protected-plus-permitted phasing

Case 4: Standard lane with protected phasing

Case 5: Standard lane with permitted phasing

Case 6: Shared lane with protected-plus-permitted phasing

Case 7: Single-lane approaches (right turn only)

When permitted left turns exist, either from shared lanes or from exclusive lanes, their impact on the intersection operations is quite complicated. The procedure outlined in this section is applied to preceding cases 2, 3, and 5. The left-turn adjustment factor for the lane from which permitted left turns are made can be stated as follows:

$$f_m = \frac{g_f}{g} + \frac{g_u}{g}\left[\frac{1}{1 + P_L(E_{L1} - 1)}\right] \tag{10}$$

where

g_f = portion of effective green until the first left-turning vehicle arrives in a shared lane

g_u = portion of effective green during which left-turning vehicles select gaps through the unsaturated opposing flow

E_{L1} = through-car equivalent for each left-turning vehicle

P_L = proportion of left-turning vehicles in the shared lane

In addition, to account for sneakers, a practical minimum value of one sneaker per cycle may be assumed. The estimated number of sneakers per cycle may be computed as $(1 + P_L)$. Assuming an approximate average headway of 2 seconds per vehicle in an exclusive lane on a protected phase, the practical minimum value of f_m may be estimated as $2(1 + P_L)/g$.

For multilane groups, the impact of left turns on a shared lane must be extended to include their impact on the entire lane group. This can be achieved by the following relationship:

$$f_{LT} = [f_m + 0.91(N - 1)]/N \tag{11}$$

where

f_{LT} = left-turn adjustment factor applied to a total lane group from which left turns are made

f_m = left-turn adjustment factor applied only to the lane from which left turns are made

When a single (or double) exclusive-permitted left-turn lane is involved, $f_{LT} = f_m$.

To implement this model, it is necessary to estimate the subportions of the effective green phase, g_f, g_q, and g_u. This is based on regression relationships as follows:

1. Compute g_f:

$$g_f = G\exp(-0.882\,LTC^{0.717}) - t_L \qquad \text{(shared-permitted left-turn lanes)} \tag{12}$$

$$g_f = 0.0 \qquad \text{(exclusive-permitted left-turn lanes)}; 0 \le g_f \le g$$

where

G = actual green time for the permitted phase (sec)

LTC = left turns per cycle (vpc), computed as $v_{LT}C/3600$

v_{LT} = adjusted left-turn flow rate (vph)

C = cycle length (sec)

t_L = lost time per phase (sec)

2. Compute g_q:

$$g_q = \frac{v_{olc}qr_o}{0.5 - [v_{olc}(1 - qr_o)/g_o]} - t_L \tag{13}$$

$$v_{olc}(1 - qr_o)/g_o \leq 0.49$$

$$0.0 \leq g_q \leq g$$

where

v_{olc} = adjusted opposing flow rate per lane per cycle computed as $v_o C/3600N_o$ (vplpc)

v_o = adjusted opposing flow rate (vph)

N_o = number of opposing lanes

qr_o = opposing queue ratio, that is, the proportion of opposing flow rate originating in opposing queues, computed as $1 - R_{po}(g_o/C)$

R_{po} = platoon ratio for the opposing flow, obtained from Table 9-2 based on opposing arrival type

g_o = effective green for the opposing flow (sec)

3. Compute g_u:

$$g_u = g - g_q \qquad \text{when } g_q \geq g_f$$

$$g_u = g - g_f \qquad \text{when } g_q < g_f$$

where g is the effective green time for subject-permitted left turn (sec). When the first left-turning vehicle does not arrive until after the opposing queue clears ($g_q < g_f$), an effective adjustment factor of 1.0 is applied throughout g_f and a factor based on E_{L1} thereafter.

4. Compute P_L (proportion of left turns in shared lane):

$$P_L = P_{LT}\left[1 + \frac{(N - 1)g}{f_s g_u + 4.5}\right] \tag{14}$$

$$f_s = (875 - 0.625v_o)/1000 \qquad f_s \geq 0$$

where

P_{LT} = proportion of left turns in the lane group

N = number of lanes in the lane group

f_s = left-turn saturation factor

When an exclusive-permitted left turn is involved, $P_L = P_{LT} = 1.0$. Also, it should be checked at this stage if P_L is 1.0 or higher for multilane lane groups. If so, the lane groups for this approach should be reassigned showing this left-hand lane as an exclusive left-turn lane.

5. Select the appropriate value of E_{L1} from Figure 9-6 on the basis of the opposing flow rate, v_o, and the number of opposing lanes, N_o. Exclusive-turn lanes or their flow rates are not included in N_o or v_o.

6. Compute f_m using Eq. 10.

7. Compute f_{LT} using Eq. 11.

Total no. of signal phases	Type of left turn lane	No. of opposing lanes	Opposing Flow, V_o						
			0	200	400	600	800	1000	≥ 1200
2	Shared	1	1.05	2.0	3.3	6.5	16.0*	16.0*	16.0*
		2	1.05	1.9	2.6	3.6	6.0	16.0*	16.0*
		≥ 3	1.05	1.8	2.5	3.4	4.5	6.0	16.0*
	Exclusive	1	1.05	1.7	2.6	4.7	10.4*	10.4*	10.4*
		2	1.05	1.6	2.2	2.9	4.1	6.2	10.4*
		≥ 3	1.05	1.6	2.1	2.8	3.6	4.8	10.4*
More than 2	Shared	1	1.05	2.2	4.5	11.0*	11.0*	11.0*	11.0*
		2	1.05	2.0	3.1	4.7	11.0*	11.0*	11.0*
		≥ 3	1.05	2.0	2.9	4.2	6.0	11.0*	11.0*
	Exclusive	1	1.05	1.8	3.3	8.2*	8.2*	8.2*	8.2*
		2	1.05	1.7	2.4	3.6	5.9	8.2*	8.2*
		≥ 3	1.05	1.7	2.4	3.3	4.6	6.8	8.2*

Figure 9-6   Through-car equivalents, E_{L1}, for permitted left turns (TRB, 1994).

The basic model for single-lane approaches opposed by single-lane approaches is as follows:

$$f_{LT} = \frac{g_f}{g} + \frac{g_q - g_f}{g}\left[\frac{1}{1 + P_{LT}(E_{L2} - 1)}\right] + \frac{g_u}{g}\left[\frac{1}{1 + P_{LT}(E_{L1} - 1)}\right] \qquad (15)$$

As in the multilane case, the opposing single-lane model has no term to account for sneakers, but has a practical minimum value of $f_{LT} = 2(1 + P_{LT})/g$.

To implement this model, the subportions of the effective green phase are estimated as follows:

1. Compute g_f:

$$g_f = G\exp(-0.860LTC^{0.629}) - t_L \qquad (16)$$

2. Compute g_q:

$$g_q = 4.943v_{olc}^{0.762}\, qr_o^{1.061} - t_L; \qquad 0.0 \leq g_q \leq g \qquad (17)$$

3. Compute g_u the same way as for multilane approaches.
4. Select the appropriate value of E_{L1} from Figure 9-6 on the basis of the opposing flow rate, v_o, and the number of opposing lanes, N_o.
5. Compute E_{L2}:

$$E_{L2} = (1 - P_{THo}^n)/P_{LTo} \qquad (18)$$

where

P_{LTo} = proportion of left turns in opposing single-lane approach

P_{THo} = proportion of through and right-turning vehicles in opposing single-lane approach, computed as $1 - P_{LTo}$

n = maximum number of opposing vehicles that could arrive during $g_q - g_f$, computed as $(g_q - g_f)/2$. Note that n is subject to a minimum value of zero.

6. Compute f_{LT} using Eq. 15.

Special Cases for Permitted Left Turns. A single-lane approach opposed by a multilane approach and vice versa are two special cases. When the subject lane is a single-lane approach and is opposed by a multilane flow, even if it is single through lane and exclusive left-turn lane, opposing left turns will not open gaps in the opposing flow. Thus, the special structure of the single-lane model does not apply. The multiple model is used instead, except that $f_{LT} = f_m$. However, g_f is computed using the single-lane equation.

When the subject approach is multilane and the opposing flow is single-lane, the opposing left turns could conceivably open gaps for subject left turners. The single model may be applied with some revisions as follows:

- g_f should be computed using the multilane equation:

$$g_f = G \exp(-0.882 LTC^{0.717}) - t_L$$

- P_L must be estimated and substituted for P_{LT} in the single-lane model. P_L may be estimated from P_{LT} using the multilane equations:

$$f_s = (875 - 0.625 v_o)/1000$$

and

$$P_L = P_{LT}\left\{1 + \left[\frac{(N-1)g}{f_s g_u + 4.5}\right]\right\}$$

- $f_{LT} \neq f_m$. Conversion must be made using the multilane equations, except when the subject approach is a dual left-turn lane.

$$f_{LT} = [f_m + 0.91(N-1)]/N$$

Worksheets that may be used to assist in implementing special models for permitted left-turn movements are presented in Figures 9-4 and 9-5, and in the illustrative examples. Care should be used when using these worksheets as they do not account for the modifications that must be made to analyze single-lane approaches opposed by multilane approaches, and vice versa.

Capacity Analysis Module. Computational results of previous modules are used to compute key capacity variables, including the following:

1. Flow ratio for each lane group
2. Capacity of each lane group
3. V/c ratio of each lane group
4. Critical V/c ratio for the overall intersection

Flow ratios are computed by dividing the adjusted demand flow, v, computed in the volume-adjustment module by the adjusted saturation flow rate, s, computed in the saturation flow-rate module.

The capacity of each lane group is computed from Eq. 1:

$$c_i = s_i \left(\frac{g_i}{C} \right)$$

If the signal timing is not known, a timing plan will have to be estimated or assumed to make these computations. The V/c ratio for each lane group is computed directly, by dividing the adjusted flows by the capacities computed before, as in Eq. 2:

$$X_i = \frac{V_i}{c_i}$$

The final capacity parameter of interest is the critical V/c ratio, X_c, for the intersection. It is computed from Eq. 3 as follows:

$$X_c = \sum_i \left(\frac{V}{s} \right)_{ci} \frac{C}{C - L}$$

This ratio indicates the proportion of available capacity that is being used by vehicles in critical lane groups. If this ratio exceeds 1.00, one or more of the critical lane groups will be oversaturated. It is an indication that the intersection design, cycle length, phase plan, and/or signal timing is inadequate for the existing or projected demand. A ratio of less than 1.00 indicates that the design, cycle length, and phase plan are adequate to handle all critical flows without demand exceeding capacity, assuming that green times are proportionally assigned. Where phase splits are not proportional, some movement demands may exceed movement capacities even where the critical V/c ratio is less than 1.00.

The computation of this ratio requires that critical lane groups be identified. During each signal phase, one or more lane groups are given the green. One lane group will have the most intense demand, will determine the amount of green time needed, and would be the critical lane group. The v/s ratio for the lane group is the normalized measure of demand intensity. Where there are no overlapping signal phases in the signal design, the determination of critical lane groups is straightforward. Overlapping phases (concurrent phase timing) complicate matters, as various lane groups may move in several phases of the signal. The following guidelines may be used in determining critical lane groups:

1. Where phases do not overlap:
   (a) There will be one critical lane group for each signal phase.

(b) The lane group with the highest flow ratio, v/s, of those lane groups moving in a given signal phase is critical.

(c) The critical lane group v/s ratios are summed for use in computing X_c.

2. When phases overlap, the critical path must conform to the following rules:
   (a) One critical lane group must be moving at all times during the signal cycle, excluding lost times.
   (b) No more than one critical lane group may be moving at any time in the signal cycle.
   (c) The critical path has the highest sum of v/s ratios.

In summary a few key points are stated here:

1. A critical V/c ratio of greater than 1.00 indicates that the signal and geometric design cannot accommodate the combination of critical flows at the intersection. Conditions possibly can be improved by a combination of increased cycle length, changes in the phasing plan, and/or basic changes in geometrics.
2. When there is a large variation in the V/c ratio for the lane groups, although V/c ratios are acceptable, it is evident that the green times are not proportionally distributed and reallocation is suggested.
3. If permitted left turns result in extreme reductions in saturation flow rates for applicable lane groups, protected phasing may be considered.
4. If the critical ratio exceeds 1.0, it is unlikely that the existing geometric and signal design can accommodate the demand. Changes in either or both should be considered.
5. Where V/c ratios are unacceptable and signal phasing includes protective phasing for turn movements, it is probable that geometric changes may improve conditions.

Level-of-Service Module. Results from the volume-adjustment, saturation flow-rate, and capacity analysis modules are used in the level-of-service module to find the average stopped delay per vehicle for each lane group, and then averaged for approaches and the intersection as a whole. Level of service is directly related to the delay value, and is found from Table 9-1.

Delay Assuming Random Arrivals. The delay for each lane group is found using the following equations:

$$d = d_1 DF + d_2 \tag{19}$$

$$d_1 = 0.38C \frac{(1 - g/C)^2}{1 - (g/C)[\min(X, 1.0)]} \tag{20}$$

$$d_2 = 173X^2 \{(X - 1) + [(X - 1)^2 + mX/c]^{0.5}\} \tag{21}$$

where

d = average stopped delay per vehicle for the lane group (sec/veh)
d_1 = uniform delay (sec/veh)

d_2 = incremental delay (sec/veh)

DF = delay adjustment factor for quality of progressive and control type

C = cycle length (sec)

X = V/c ratio for the lane group

c = capacity of the lane group (vph)

g = effective green time for lane group (sec)

m = an incremental delay calibration term representing the effect of arrival type and degree of platooning

Equation 19 predicts the average stopped delay per vehicle for an assumed random arrival pattern of approaching vehicles. Equation 20 accounts for uniform delay, the delay that occurs if arrival demand in the subject lane group is uniformly distributed over time. Equation 21 estimates the incremental delay of random arrivals over uniform arrivals, and for the additional delay due to cycle failures. The equation yields reasonable results for values of X between 0.0 and 1.0. Where oversaturation occurs for long periods (> 15 min), it is difficult to accurately estimate delay, because spillbacks may extend to adjacent intersections. The equation may be used with caution for values of X up to the lesser of 1.2, or 1/PHF, but delay estimates for higher values are not recommended. Oversaturation, that is $X > 1.0$, is an undesirable condition that should be ameliorated if possible.

As a practical consideration, the traffic volume must not exceed the capacity of an uninterrupted lane group (i.e., 100% green time), or the basic computational methodology will not apply. This condition will occur if X is greater than C/g.

Delay-Adjustment Factor, DF.   The impact of control type (control factor, CF) and signal progression (progression factor, PF) on delay is accounted for through the use of the delay-adjustment factor. Only one or the other is applied, as the two effects are mutually exclusive. Table 9-13 indicates the appropriate value of DF for all possible control modes. As the advantage of progression and traffic-actuated control diminishes at very high V/c ratios, DF is only applied to d_1. Because traffic-actuated controllers are able to adjust timing from cycle to cycle, they are generally able to accommodate the same traffic volumes as pretimed controllers with slightly less delay. CF in Table 9-13 provides for that. Table 9-13 may also be used to determine PF as a function of the arrival type based on g/C, and the default values of R_p and f_p associated with each arrival type. The incremental-delay calibration term, m, represents the effect of arrival type and degree of platooning on the incremental-delay term.

Aggregating Delay Estimates.   The procedure for delay estimation yields the average stopped delay per vehicle for each lane group. It is also desirable to aggregate these values to provide average delay for an intersection approach and for the intersection as a whole. In general, this is done by computing weighted averages, where the lane group delays are weighted by the adjusted flows in the lane groups.

Thus, the delay for an approach is computed as

$$d_A = \frac{\Sigma d_i V_i}{\Sigma V_i} \tag{22}$$

TABLE 9-13   UNIFORM DELAY (d_1) ADJUSTMENT FACTOR (*DF*)

Controller-type adjustment factor (*CF*)

Controller type	Noncoordinated intersections	Coordinated intersections
Pretimed (no traffic-actuated lane groups)	1.0	PF as computed below
Semiactuated:		
Traffic-actuated lane groups	0.85	1.0
Nonactuated lane groups	0.85	PF as computed below
Fully actuated (all lane groups traffic-actuated)	0.85	Treat as semiactuated

Progression adjustment factor (*PF*)
$PF = (1 - P)f_p/(1 - g/C)$ (see Note)

Green ratio (g/C)	Arrival type (AT)					
	AT-1	AT-2	AT-3	AT-4	AT-5	AT-6
0.20	1.167	1.007	1.000	1.000^3	0.833	0.750
0.30	1.286	1.063	1.000	0.986	0.714	0.571
0.40	1.445	1.136	1.000	0.895	0.555	0.333
0.50	1.667	1.240	1.000	0.767	0.333	0.000
0.60	2.001	1.395	1.000	0.576	0.000	0.000
0.70	2.556	1.653	1.000	0.256	0.000	0.000
Default, f_p	1.00	0.93	1.00	1.15	1.00	1.00
Default, R_p	0.333	0.667	1.000	1.333	1.667	2.000
Incremental delay calibration term, m	8	12	16	12	8	4

Note: 1. Tabulation is based on default values of f_p and R_p.
    2. $P = R_p g/C$ (may not exceed 1.0).
    3. *PF* may not exceed 1.0 for AT-3 through AT-6.
Source: TRB, 1994.

where

d_A = delay for approach A (sec/veh)

d_i = delay for lane group i (on approach A)(sec/veh)

V_i = adjusted flow for lane group i (veh/hr)

Approach delays then can be further averaged to provide the average delay for the intersection:

$$d_I = \frac{\Sigma d_A V_A}{\Sigma V_A} \tag{23}$$

where d_I is the average delay per vehicle for the intersection (sec/veh), and V_A is the adjusted flow for approach A (veh/hr).

Level-of-Service Determination.   Intersection level of service is directly related to the average stopped delay per vehicle. Once delays have been estimated for each lane group and aggregated for each approach and the intersection as a whole, Table 9-1 is consulted, and the appropriate levels of service are determined.

Interpretation of Results.   The results of an operational analysis will yield two key results that must be considered:

1. The V/c ratios for each lane group and for the intersection as a whole
2. Average stopped-time delays for each group and approach, and for the intersection as a whole and the levels of service that correspond

Any V/c ratio greater than 1.00 is an indication of actual or potential breakdowns, and is a condition requiring amelioration. Where the critical V/c ratio is less than 1.00 but some lane groups have V/c ratios greater than 1.00, the green time is generally not appropriately apportioned, and a retiming using the existing phasing should be attempted.

A critical V/c ratio greater than 1.0 indicates that the overall signal and geometric design provides inadequate capacity for the existing or projected flows. In this situation, any or all of the following improvements may be considered:

1. Basic changes in intersection geometry (number and use of lanes)
2. Increases in the signal cycle length if it is determined to be too short
3. Changes in the signal phase plan

As delays can be high even when V/c ratios are low, a careful examination of signal progression, cycle length, or both is generally appropriate.

Suggestions for Specifying Design Elements.   The geometric design of an intersection involves several critical decisions involving the number and use of lanes to be provided on each approach. The following guidelines are useful:

1. Left-turn lanes may be provided to accommodate heavy left-turn movements without disruption to through and right-turning vehicles.
2. Where fully protected left-turn phasing is provided, an exclusive left-turn lane should be provided.
3. Where space permits, use of a left-turn lane should be considered where left-turn volumes exceed 100 veh/hr.
4. Where left-turn volumes exceed 300 veh/hr, provision of a double left-turn lane should be considered.
5. The length of the storage bay should be sufficient to handle the turning traffic without reducing the safety or capacity of the approach.
6. An exclusive right-turn lane should be considered when the right-turn volume also exceeds 300 veh/hr.
7. The number of lanes required on an approach depends on a variety of factors, including the signal design. In general, enough main roadway lanes should be

provided such that the total of through plus right-turn volume (plus left-turn volume, if present) does not exceed 450 veh/hr/ln.

8. Where lane widths are unknown, the 12-ft standard lane width should be assumed, unless known restrictions prevent this.

9. Parking conditions should be assumed to be consistent with local practice in the area.

10. For analysis purposes, where no information exists, no curb parking and no local buses should be assumed.

4.4 Planning Analysis

For planning analysis, a detailed treatment of the operation of traffic signal is not necessary; nor does the accuracy of the available data justify a precise analysis. An approximate analysis of the level of service is recommended with assumed values for most of the required data. Table 9-3 contains recommended default values for several data items. The only site-specific data required are the traffic volumes, number of lanes for each movement, and a minimal description of the signal design, and other operating parameters. The concept of the planning method is to apply the required approximations to the input data and not the computational procedure. The computational method involves the summation of conflicting critical lane volumes for the intersection. The computations themselves depend on the traffic signal phasing, which in turn depends on the type of protection assigned to each left turn. The critical volume summation divided by the computed intersection capacity represents the critical V/c ratio, X_{cm}. It is possible to evaluate the operational status of the intersection for planning purposes based on X_{cm} without assigning a level of service. Table 9-14 expresses the operational status as "over," "at," "near," or "under" capacity.

A reasonable and effective signal timing plan for the intersection, which emerges from the critical volume summation, along with the assumed operating parameters, provides all the data required to apply the full operational analysis. However, planning analysis should not be taken beyond the evaluation of intersection status if the traffic volumes are rough approximations of future conditions.

Field Data Requirements.   Information has to be developed through judgment or observation about parking, signal coordination, and treatment of left turns for the

TABLE 9-14   INTERSECTION STATUS
CRITERIA FOR SIGNALIZED INTERSECTION
PLANNING ANALYSIS

Critical v/c ratio (X_{cm})	Relationship to probable capacity
$X_{cm} \leq 0.85$	Under capacity
$0.85 < X_{cm} \leq 0.95$	Near capacity
$0.95 < X_{cm} \leq 1.00$	At capacity
$X_{cm} > 1.00$	Over capacity

Source: TRB, 1994.

intersection under consideration for improvement or for a proposed new one. The following assumptions are recommended:

1. Both parking and no-parking conditions may be analyzed if a decision is not already made and used to develop a parking plan.

2. On the major street, signal coordination may be assumed if the upstream intersection is less than 2000 ft away. On the minor street, the corresponding distance reduces to 1200 ft.

3. The actual left-turn treatment should be used. If this is unknown, local policies or practice should provide a guide.

4. Split-phase operation should be considered only under certain geometric and left-turn treatment conditions. Split-phase operation provides complete separation between movements in opposing directions by allowing all movements in only one direction to proceed at the same time.

5. Unless the intersection is known to be within the central business district (CBD), the "other" category should be used.

6. The use of a peak-hour factor (PHF) is desirable and a value of 0.9 may be used.

7. The design cycle length should be used if it is known. Default values of 60 sec minimum and 120 sec maximum are recommended.

8. Lane-utilization factors of 1.0 are suggested.

9. No pedestrian conflicts are considered.

For a detailed discussion of this methodology, reference should be made to the *Highway Capacity Manual* (TRB, 1994, Chapter 9). It is, however, anticipated that a computerized version of the technique would be employed in most practical applications. Computation through worksheets is extremely time-consuming because of the detailed nature of the process.

5. OTHER ANALYSES

As noted previously, the computational procedures in this chapter emphasize the estimation of level of service (delay) based on known or projected traffic demands, signalization, and geometric design. Other computational applications include

1. Determination of V/c ratios and service flow rates associated with selected levels of service, given a known signalization and geometric design

2. Determination of signal timing parameters when known inputs are a selected level of service, demand flow rates, and geometric design

3. Determination of geometric parameters (number of lanes, lane-use allocations, etc.), given a selected level of service, demand flow rates, and signalization

Example 1

(a) One approach to a signalized intersection has the following conditions: two approach lanes 10.5 ft wide; 8% heavy-truck traffic; 4% downgrade; CBD location; 10 buses stopping per hour; 20 (street-side) parking maneuvers per hour; type 2 arrivals; through traffic only; 55-sec cycle length; and green ratio = 0.55. Determine the capacity of this approach.

(b) What is the LOS for a flow rate of 1100 veh/hr assuming equal lane utilization for the two lanes?

Solution

(a) $s = s_0 N f_w f_{HV} f_g f_p f_{bb} f_a f_{RT} f_{LT}$

$\qquad = 1900 \times 2 \times 0.95 \times 0.926 \times 1.02 \times 0.90 \times 0.98 \times 0.90 \times 1.0 \times 1.0$

$\qquad = 2707$ veh/hr green

$\qquad c_i = s_i \dfrac{g}{C} = 2707 \times 0.55 = 1489$ veh/hr

(b) $d = d_1 DF + d_2$

$\qquad d_1 = 0.38C \dfrac{(1 - g/C)^2}{(1 - g/C)[\min(X, 1.0)]} = 0.38\,(0.55)\dfrac{(1 - 0.55)^2}{(1 - 0.55)\,(0.739)} = 12.72$

$\qquad$ where $X = \dfrac{V_i}{c_i} = \dfrac{1100 \text{ veh/hr}}{1489 \text{ veh/hr}} = 0.739$

$\qquad d_2 = 173 X^2 \{(X - 1) + [(X - 1)^2 + mX/c_i]^{0.5}\}$

$\qquad\quad = 173\,(0.739)^2 \{(0.739 - 1) + [(0.739 - 1)^2 + 12\,(0.739)/1489]^{0.5}\} = 1.056$

$\qquad d = 12.72 \times 1.318 + 1.056 = 17.8$ sec/veh

From Table 9-1, the level of service is C.

Example 2

The intersection of Main Street and Third Avenue is illustrated in Figure 9-E2(a), which is the input module worksheet for this intersection. It is a four-leg intersection with a two-phase, pretimed signal on a 70-sec cycle. The objective is to analyze the capacity and level of service of the existing intersection, according to the operational method of the *Highway Capacity Manual* (TRB, 1994).

Solution

See Figures 9-E2(b) to 9-E2(g).

Comments

With the level of service F on the eastbound (EB) approach, the intersection operation is marginal. The EB v/c ratio of 1.108 is almost at the maximum allowable limit of

---

INPUT MODULE WORKSHEET

Intersection: **MAIN ST & THIRD AVE** _____   Date: _____

Analyst: **TY REYNOLDS** _____   Time Period Analyzed: **4-5PM** Area Type: ☒ CBD ☐ Other

Project No: _____   City/State: _____

VOLUME AND GEOMETRICS

THIRD
N/S STREET

SB TOTAL: 600

+ (10 RTOR)
20
700
30
WB TOTAL: 750

15'

50 510 40
+ (25 RTOR)

NORTH

LOST TIME PER PHASE (sec): **3**

11'
11'

IDENTIFY IN DIAGRAM:

1. Volumes
2. Lanes, lane widths
3. Movements by lane
4. Parking (PKG) location
5. Bay storage lengths
6. Islands (physical or painted)
7. Bus stops

11'
11'

MAIN E/W STREET
370
30
20
+ (10 RTOR)
15'
420
NB TOTAL

65
620
35

EB TOTAL: 720
+ (17 RTOR)

TRAFFIC AND ROADWAY CONDITIONS

Approach	Grade %	% HV	Adj. Pkg. Lane		Buses (N_b)	PHF	Conf. Peds. (peds./hr)	Pedestrian Button		Arr. Type
			Y or N	N_m				Y or N	Min. Timing	
EB	0	5	N	—	0	0.9	100	N	10.0	4
WB	0	5	N	—	0	0.9	100	N	10.0	2
NB	0	5	N	—	0	0.9	100	N	13.6	3
SB	0	5	N	—	0	0.9	100	N	13.6	3

Grade: + up, − down
HV: veh. with more than 4 wheels
N_m: pkg. maneuvers/hr

N_b: buses stopping/hr
PHF: peak-hour factor
Conf. Peds.: Conflicting peds./hr

Min. Timing: min. green for
pedestrian crossing
Arrival Type: Type 1-6, or P

PHASING

D I A G R A M

Timing	G = **27** Y + AR = **3**	G = **37** Y + AR = **3**	G = Y + AR =	G = Y + AR =	G = Y + AR =	G = Y + AR =	G = Y + AR =	G = Y + AR =
Pretimed or Actuated	**P**	**P**						

Protected turns   Permitted turns   ------------ Pedestrian   Cycle Length **70** Sec

(a)

Figure 9-E2(a)   Input Module Worksheet.

VOLUME ADJUSTMENT MODULE WORKSHEET

1 Appr.	2 Mvt.	3 Mvt. Volume (vph)	4 Peak Hour Factor PHF	5 Flow Rate v_p (vph) [3]/[4]	6 Lane Group	7 Flow Rate in Lane Group v_g (vph)	8 Number Of Lanes N	9 Lane Utilization Factor U Table 9-4	10 Adj. Flow v (vph) [7]*[9]	11 Prop. of LT or RT P_{LT} or P_{RT}
	LT	65	0.90	72						
EB	TH	620	0.90	689	LTR	800	2	1.05	840	$P_{LT}=0.090$ $P_{RT}=0.049$
	RT	35	0.90	39						
	LT	30	0.90	33						
WB	TH	700	0.90	778	LTR	833	2	1.05	875	$P_{LT}=0.040$ $P_{RT}=0.026$
	RT	20	0.90	22						
	LT	30	0.90	33						
NB	TH	370	0.90	411	LTR	466	1	1.00	466	$P_{LT}=0.071$ $P_{RT}=0.047$
	RT	20	0.90	22						
	LT	40	0.90	44						
SB	TH	510	0.90	567	LTR	667	1	1.00	667	$P_{LT}=0.066$ $P_{RT}=0.084$
	RT	50	0.90	56						

(b)

Figure 9-E2(b)   Volume Adjustment Worksheet.

1/PHF, or 1.11. Beyond this point, the queue that forms during the peak 15 min could not be expected to clear during the peak hour, because the hourly volume will exceed capacity and the delay estimation model would not be valid. Some further consideration of the operation at this intersection may include reallocation of green time to produce equitable service to vehicles on all approaches or repeat the analysis without the lane-utilization factors, which reflected the resulting operation in the worst lane. The latter reconsideration assumes that the drivers on the EB approach would change lanes to maintain equilibrium in lane distribution.

Example 3

Figure 9-E3(a) indicates the expected design volume and layout of a proposed improvement being planned for traffic in the year 1999 at the signalized intersection of Oak and Pine streets. Assume that there will be no conflicting pedestrians, no heavy vehicles, no bus stops, and no street-side parking. All lanes will be 12 ft wide. Progression is of arrival type 3. Lost time is 3.0 sec/phase. The change interval is 4.0 sec/phase. The left turns will be

SATURATION FLOW RATE MODULE WORKSHEET

1 Appr.	2 Lane Group Mvt.	3 Ideal Sat. Flow (pcphgpl)	4 No. of Lanes N	ADJUSTMENT FACTORS								13 Adj. Set Flow Rate s (vphg)
				5 Lane Width f_w Table 9-5	6 Heavy Veh f_{HV} Table 9-6	7 Grade f_g Table 9-7	8 Pkg. f_p Table 9-8	9 Bus Blockage f_{bb} Table 9-9	10 Area Type f_a Table 9-10	11 Right Turn f_{RT} Table 9-11	12 Left Turn f_{RL} Table 9-12	
EB		1900	2	0.96	0.95	1.00	1.00	1.00	0.90	0.99	0.63	1963
WB		1900	2	0.96	0.95	1.00	1.00	1.00	0.90	0.99	0.80	2510
NB		1900	1	1.10	0.95	1.00	1.00	1.00	0.90	0.89	0.78	1252
SB		1900	1	1.10	0.95	1.00	1.00	1.00	0.90	0.88	0.88	1395

(c)

Figure 9-E2(c)   Saturation Flow-Rate Module Worksheet.

protected. Determine if the capacity of the proposed design will suffice, using the planning level analysis.

Solution

For planning analysis, some of the assumptions that must be made about the signal phasing are included in the statement of the problem to assist the reader. All left turns are heavy and face heavy opposing traffic and therefore will need to be protected as proposed. As the proposed improvement is in the near term, local policies should be looked at to see if a trend is emerging toward protected-only phasing. Also, crash rates at this site or nearby intersections could favor left-turn protection.

SUPPLEMENTAL WORKSHEET FOR PERMITTED LEFT TURNS				
*** For Use Where the Subject Approach is Opposed by a Multilane Approach ***				
APPROACH	EB	WB	NB	SB
Enter Cycle Length, C		70	70	
Enter Actual Green Time For Lane Group, G		27	27	
Enter Effective Green Time For Lane Group, g		27	27	
Enter Opposing Effective Green Time, g_o		27	27	
Enter Number of Lanes in Lane Group, N		2	2	
Enter Number of Opposing Lanes, N_o		2	2	
Enter Adjusted Left-Turn Flow Rate, v_{LT}		72	33	
Enter Proportion of Left Turns in Lane Group, P_{LT}		0.090	0.049	
Enter Adjusted Opposing Flow Rate, v_o		875	840	
Enter Lost Time per Phase, t_L		3	3	
Compute Left Turns per Cycle: LTC = v_{LT} C/3600		1.4	0.642	
Compute Opposing Flow per Lane, Per Cycle: $v_{olc} = v_o$ C/(3600 N_o)		8.507	8.167	
Determine Opposing Platoon Ratio, R_{po} (Table 9-2 or Eq [9-4])		0.667	1.333	
Compute g_f ** = G exp (–0.882 LTC$^{0.717}$) – t_L, $g_f \le g$		5.786	11.210	
Compute Opposing Queue Ratio: $qr_o = 1 - R_{po}$ (g_o/C)		0.743	0.436	
Compute g_q, using equation 9-13, $g_q \le g$		12.084	8.521	
Compute g_u: $g_u = g - g_q$ if $g_q \ge g_f$ $g_u = g - g_f$ if $g_q < g_f$		14.916	15.790	
Compute $f_s = (875 - 0.625 v_o)$/1000, $f_s \ge 0$		0.328	0.350	
Compute P_L† = P_{LT} [1 + {$(N-1)g$/($f_s g_u + 4.5$)}]		0.349	0.148	
Determine E_{L1} (Figure 9-6)		9.75	8.00	
Compute f_{min} = 2(1 + P_L)/g		0.099	0.085	
Compute f_m: $f_m = [g_f/g] + [g_u/g][1/\{1 + P_L(E_{L1} - 1)\}]$ min = f_{min}; max = 1.00		0.351	0.704	
Compute $f_{LT} = [f_m + 0.91 (N - 1)]N$ ‡		0.631	0.807	

** For special case of single-lane approach opposed by multilane approach, see text.

† If $P_L \ge 1$ for shared left-turn lanes with $N > 1$, then assume de facto left-turn lane and redo calculations.

‡ For permitted left turns with multiple exclusive left-turn lanes $f_{LT} = f_m$.

(d)

Figure 9-E2(d)   Supplemental Worksheet for Permitted Left Turns: Multilane Approach.

Comments

Generated solution is based on Highway Capacity Software (HCS, 1996). See Figures 9-E3(b) and 9-E3(c). The intersection status is indicated as near capacity, with a critical V/c ratio of 0.92. The estimated cycle length is 120 sec. The near-capacity rating could

SUPPLEMENTAL WORKSHEET FOR PERMITTED LEFT TURNS				
*** For Use Where the Subject Approach is Opposed by a Single lane Approach ***				
APPROACH	EB	WB	NB	SB
Enter Cycle Length, C			70	70
Enter Actual Green Time For Lane Group, G			37	37
Enter Effective Green Time For Lane Group, g			37	37
Enter Opposing Effective Green Time, g_o			37	37
Enter Number of Lanes in Lane Group, N			1	1
Enter Adjusted Left-Turn Flow Rate, v_{LT}			33	44
Enter Proportion of Left Turns in Lane Group, P_{LT}			0.071	0.066
Enter Proportion of Left Turns in Lane Group, P_{LTo}			0.066	0.071
Enter Adjusted Opposing Flow Rate, v_o			667	466
Enter Lost Time per Phase, t_L			3	3
Compute Left Turns per Cycle: LTC $= v_{LT}$ $C/3600$			0.642	0.856
Compute Opposing Flow per Lane, Per Cycle: $v_{olc} = v_o$ $C/3600$			12.969	9.061
Determine Opposing Platoon Ratio, R_{po} (Table 9-2 or Eq [9-4])			1.00	1.00
Compute g_f ** $= G \exp (-0.860 \text{ LTC}^{0.629}) - t_L, g_f \leq g$			16.301	13.963
Compute Opposing Queue Ratio: $qr_o = 1 - R_{po} (g_o/C)$			0.471	0.471
Compute $g_q = 4.943 \ v_{olc}^{0.762} \ qr_o^{1.061} - t_L, g_q \leq g$			12.671	8.924
Compute g_u: $g_u = g - g_q$ if $g_q \geq g_f$ $g_u = g - g_f$ if $g_q < g_f$			20.699	23.037
Compute $n = (g_q - g_f)/2, n \geq 0$			0	0
Compute $P_{THo} = 1 - P_{LTo}$			0.934	0.929
Determine E_{L1} (Figure 9-6)			9.683	4.356
Compute $E_{L2} = (1 - P_{THo}^n)/P_{LTo}$			0	0
Determine $f_{min} = 2(1 + P_{LT})/g$			0.058	0.058
Compute f_{LT} ** $= f_m = [g_f/g] + [g_q - g_f)/g]$ $[1/\{1 + P_{LT}(E_{L2} - 1)\}] + [g_u/g]$ $[1 + P_{LT}(E_{L1} - 1)]$ min $= f_{min}$; max $= 1.00$			0.787	0.887

** For special case of multilane approach opposed by single-lane approach or when $g_f > g_q$, see text.

Figure 9-E2(e)   Supplemental Worksheet for Permitted Left Turns: Single-Lane Approach.

be interpreted to mean that it is fairly certain that demand will almost equal capacity as the forecast volumes are for not-too-distant future.

6. SIGNALIZATION DETAILS

Although the subject of traffic signals was dealt with in great detail in Chapter 8, there are a few more items that need to be repeated, reinforced, and added now that the capacity of intersections has been discussed. Three topics are dealt with in this section: (1) the type of signal controller to be used, (2) the phase plan to be adopted, and (3) the allocation of green time among the various phases.

CAPACITY ANALYSIS MODULE WORKSHEET

1 Lane Group Mov'ts.	2 Phase Type (P,S,T)	3 Adj. Flow Rate (v)	4 Adj. Sat. Flow Rate (s)	5 Flow Ratio (v/s) [3]/[4]	6 Green Ratio g/C	7 Lane Group Capacity (c) [4]×[6]	8 Lane Group v/c Ratio (X) [3]/[7]	9 Critical Lane Grp. [*]
EB								
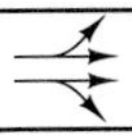	T	840	1963	0.428	0.386	758	1.108	✳
WB								
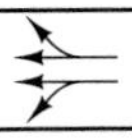	T	875	2510	0.349	0.386	969	0.903	
NB								
	T	468	1252	0.372	0.529	662	0.704	
SB								
	T	667	1395	0.478	0.529	738	0.904	✳

1 Permitted left turns subject to minimum capacty of $(1 + P_L)$ $(3600/C)$ in column 7.

Cycle length, C_____ sec. $X_c = Y \times C/(C\text{-}L) =$ _____

Lost Time per Cycle, L_____ sec. $Y = \text{Sum } (v/s)_d =$ _____

Figure 9-E2(f)   Capacity Analysis Module Worksheet.

6.1 Type of Signal

There are, in general, three types of signals available: (1) pretimed, (2) semiactuated, and (3) fully actuated. Their characteristics are as follows:

LOS MODULE WORKSHEET

Cycle	sec	First Term Delay			Second Term Delay			Lane Group		Approach	
1 Lane Group Movements	2 v/c Ratio X	3 Green Ratio g/C	4 Uniform Delay d_1 sec/veh	5 Delay Adj. Factor DF	6 Lane Grp. Capacity C vph	7 d_2 Cal. Term m	8 Incremental Delay d_2 sec/veh	9 Delay [4]×[5] +[8] sec/veh	10 LOS	11 Delay sec/veh	12 LOS
EB	1.108	0.386	16.33	0.908	758	12	59.2	74.0	F	74.0	F
WB	0.903	0.386	15.39	1.126	969	12	6.6	23.9	C	23.9	C
NB	0.704	0.529	9.40	1.000	662	16	2.4	11.8	B	11.8	B
SB	0.904	0.529	11.31	1.000	738	16	10.4	21.7	C	21.7	C

Intersection Delay **36.2** sec\veh   Intersection LOS **D**

Figure 9-E2(g)   LOS Module Worksheet.

PRETIMED

- Preset times and phases
- Cycle lengths are constant
- Simple and inexpensive
- Less efficient in fluctuating demand
- Effective in progressive signal system
- Can be operated on different timing plans

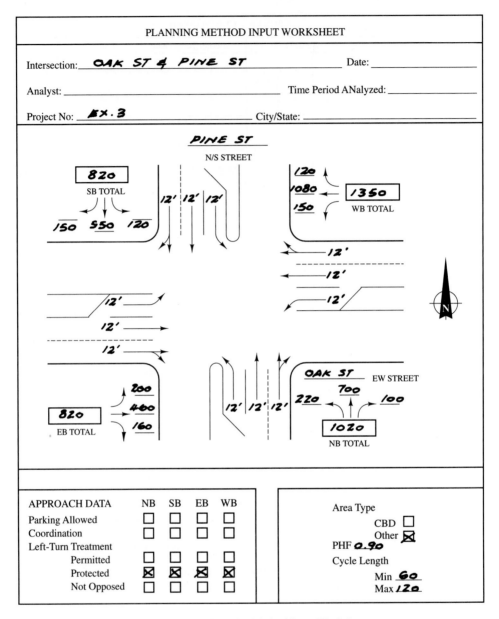

Figure 9-E3(a) Planning Method Input Worksheet.

SEMIACTUATED

- Vehicle detector on minor street
- Green is always on major street unless minor-street actuation occurs
- Two-phase plan is common
- Cycle length may vary from cycle to cycle

HIGHWAY CAPACITY MANUAL SIGNALIZED INTERSECTION PLANNING METHOD LANE VOLUME WORKSHEET

File name: Date: 7 - 19 - 1996   Time Period:

(E/W): Oak St.   (N/S): Pine St.   Analyst:

Peak hour factor: .9   Comment: Example 3

	EAST BOUND	WEST BOUND	NORTH BOUND	SOUTH BOUND
LEFT TURN MOVEMENT				
1. LT volume	200	150	220	120
2. Opposing mainline volume	1200	620	700	800
3. Number of exclusive LT lanes	1	1	1	1
Cross Product [2] * [1]	240000	93000	154000	96000
Left Lane Configuration (E=Excl, S=Shrd):	E	E	E	E
Left Turn Treatment Type:	Prot	Prot	Prot	Prot
4. LT adjustment factor	.95	.95	.95	.95
5. LT lane vol	211	158	232	126
RIGHT TURN MOVEMENT				
Right Lane Configuration (E=Excl, S=Shrd)	S	S	S	S
6. RT volume	160	120	100	150
7. Exclusive lanes	N/A	N/A	N/A	N/A
8. RT adjustment factor	.85	.85	.85	.85
9. Exclusive RT lane volume	0	0	0	0
10. Shared lane vol	188	141	118	176
THROUGH MOVEMENT				
11. Thru volume	460	1080	700	550
12. Parking adjustment factor	1	1	1	1
13. No. of thru lanes including shared	2	2	2	2
14. Total approach volume	648	1221	818	726
15. Prop. of left turns in lane group	0	0	0	0
16. Left turn equivalence	N/A	N/A	N/A	N/A
17. LT adj. factor:	N/A	N/A	N/A	N/A
18. Through lane volume	324	610	409	363
19. Critical lane volume	324	610	409	363
Left Turn Check (if [16] > 8)				
20. Permitted left turn sneaker capacity: 7200/Cmax	N/A	N/A	N/A	N/A

Figure 9-E3(b)   Signalized Intersection Planning Method Lane-Volume Worksheet.

HIGHWAY CAPACITY MANUAL SIGNALIZED INTERSECTION PLANNING METHOD SIGNAL OPERATIONS WORKSHEET

File name: Date: 7 - 19 - 1996   Time Period:

(E/W): Oak St.   (N/S): Pine St.   Analyst:

	EAST BOUND	WEST BOUND	NORTH BOUND	SOUTH BOUND
Phase Plan Selection from Lane Volume Worksheet				
Critical through-RT vol: [19]	324	610	409	363
LT lane vol: [5]	211	158	232	126
Left turn protection: (P/U/N)	P	P	P	P
Dominant left turn: (Indicate by '*')	*		*	

		EAST BOUND	WEST BOUND	NORTH BOUND	SOUTH BOUND
Selection Criteria based on the	Plan 1:	U	U	U	U
specified left turn protection	Plan 2a:	U	P	U	P
	Plan 2b:	P	U	P	U
* Indicates the dominant	Plan 3a:	*P	P	*P	P
left turn for each opposing pair	Plan 3b:	P	*P	P	*P
	Plan 4:	N	N	N	N

Phase plan selected (1 to 4)   3a   3a

MIn. cycle (Cmin)   60   Max. cycle (Cmax)   120

Timing Plan		EAST-WEST			NORTH-SOUTH		
	Value	Ph 1	Ph 2	Ph 3	Ph 1	Ph 2	Ph 3
Movement codes		EWL	ETL	EWT	NSL	NTL	NST
Critical phase vol [CV]		158	53	610	126	106	363
Critical sum [CS]	1416						
CBD adjustment [CBD]	1						
Reference sum [RS]	1539						
Lost time/phase [PL]		3	0	3	3	0	3
Lost time/cycle [TL]	12						
Cycle length [CYC]	120						
Green time		15.1	4	49.5	12.6	8.1	30.7
Critical v/c ratio [Xcm]	0.92						
Status	Near capacity.						

Figure 9-E3(c) Signalized Intersection Planning Method Signal Operations Worksheet.

- Good for low to moderate side-street demand
- Usually installed when insufficient gaps occur in major stream
- Can be used in overall progressive signal system

FULLY ACTUATED

- All approaches have vehicle detectors
- Each phase is subject to minimum and maximum green time
- Some phases can be "skipped" if no demand is detected
- A phase is terminated when there are no further actuations within a specific time interval or when the maximum green time has been reached
- Cycle length varies from cycle to cycle
- Most flexible form of signal control
- Most efficient use of available green time
- Best used at isolated intersections that are not coordinated with other signals
- May be used in progressive signal system

Local policies, guidelines, standards, demands, and budgets are the primary factors in selecting a controller.

6.2 Phase Plans

An appropriate selection of a phase plan is most critical for efficient movement of vehicles, and this includes the determination of the number of phases and their sequence. It is best to adopt a simple two-phase control unless conditions need something more elaborate. Remember that change intervals between phases contribute to lost time per cycle.

Figure 9-7 shows a number of common phase plans that may be used with pretimed or actuated controllers. The two-phase control is the most straightforward and simple. Multiphase control is adopted where one or more left or right turns are determined to require protected phasing. Usually, the left-turn movements require a protected phase. Local agencies have guidelines that require protected left-turn control when such left-turn volumes exceed 100 to 200 veh/hr. Left-turn phasing may also be considered where the speed of opposing traffic is greater than 40 mph. Figure 9-7(b) shows a three-phase plan in which an exclusive left-turn phase is provided for both left-turn movements on the major street. It is followed by a through and right-turn phase for the major street, during which left turns may be permitted on an optional basis. Figure 9-7(c) is commonly referred to as "leading and lagging green" phasing. The initial phase is a through and left-turn phase for one direction of the major street, followed by a through phase for both directions during which left turns in both directions may be permitted on an optional basis. The direction of flow started in the first phase is then stopped, providing the opposing direction with a through plus left-turn phase. The final phase provides all movements on the minor street. Such phasing is extremely flexible. Figure 9-7(d) is a four-phase sequence if both streets need left-turn phases.

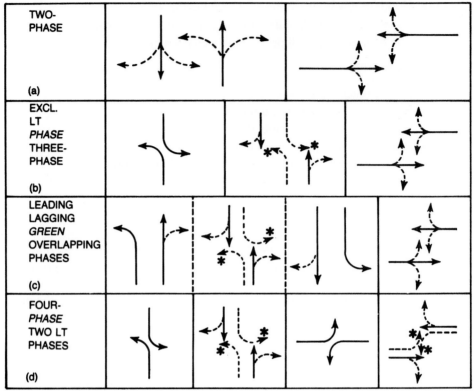

* Optional movement

Figure 9-7   Common Phase Plans That May Be Used with Pretimed or Actuated Controllers (TRB, 1994).

6.3 Allocation of Green Time

Once a phase plan has been decided on, the allocation of green time can be estimated using

$$X_i = \frac{V_i C}{s_i g_i} = \frac{V_i/s_i}{g_i/C} \tag{2}$$

$$X_c = \sum_i \left(\frac{V}{s}\right)_{ci} \frac{C}{C - L} \tag{3}$$

Therefore,

$$C_{\min} = \frac{LX_c}{X_c - \Sigma (V/s)_{ci}}$$

and

$$g_i = \frac{V_i C}{s_i X_i} = (V_i/s_i)(C/X_i)$$

where

$$C = \text{cycle length (sec)}$$
$$L = \text{lost time per cycle (sec)}$$
$$X_c = \text{critical } V/c \text{ ratio for the intersection}$$
$$X_i = V/c \text{ ratio for lane group } i$$
$$(V/s)_i = \text{flow ratio for lane group } i$$
$$g_i = \text{effective green time for lane group } i \text{ (sec)}$$

Cycle lengths and green times may be estimated using these relationships, flow ratios computed as part of the capacity analysis module, and desired v/c ratios.

Example 4

A two-phase signal has flow ratios in the E–W direction $(V/s) = 0.40$ and the N–S direction $(V/s) = 0.30$, and the assumed lost time is equal to the change intervals of 4.5 sec/phase, or 9 sec/cycle. Estimate the minimum and actual cycle length and green times.

Solution

The minimum cycle length may be computed using $X_c = 1.00$.

$$C_{min} = \frac{LX_c}{X_c - \Sigma(V/s)_{ci}}$$

$$= \frac{9(1)}{1.0 - (0.4 + 0.3)} = 30 \text{ sec}$$

If a V/c of no more than 0.8 were desired, then

$$C = \frac{9(0.8)}{0.8 - (0.4 + 0.3)}$$

$$= 72 \approx 70 \text{ sec} \quad \text{(rounding to the nearest 5 sec)}$$

The actual critical V/c ratio provided by a 70-second cycle is

$$X_c = \sum_i \left(\frac{V}{s}\right)_i \frac{C}{C - L}$$

$$= (0.4 + 0.3)\left(\frac{70}{70 - 9}\right) = 0.8$$

If we allocate the green so that the V/c ratios for critical movements in each phase are equal, then

$$g_i = \left(\frac{V}{s}\right)\left(\frac{C}{X_i}\right)$$

$$g_1 = 0.4\left(\frac{70}{0.8}\right) = 35 \text{ sec}$$

$$g_2 = 0.3\left(\frac{70}{0.8}\right) = 26 \text{ sec}$$

$$\text{Total} \quad = 61 \text{ sec}$$

plus lost time of 9 sec = 70 sec.

Another way to allocate green time is to assign the minimum green to minor approach and the balance to the major approach.

$$g_2 = 0.3\left(\frac{70}{1.0}\right) \quad = 21 \text{ sec}$$

$$g_1 = 70 - 9 - 21 = 40 \text{ sec}$$

Plus a lost time of 9 sec = 70 sec.

Discussion

There are a number of policies operating in local agencies and cities. It is best to follow some rational method for allocating green time on a uniform basis.

Example 5

An intersection controlled by a semiactuated signal has $(V/s)_1 = 0.60$ and $(V/s)_2 = 0.12$ for the main and side streets. The signal is supposed to operate at a critical V/c ratio of 0.85. Estimate a suitable cycle time and allocate the greens appropriately. Assume that lost time = 9 sec.

Solution

Semiactuated signals are assumed to make efficient use of available green time. Any excess green time can be allotted to the major street.

$$C = \frac{LX_c}{X_c - \Sigma(V/s)_i}$$

$$= \frac{9(0.85)}{0.85 - (0.6 + 0.12)} = 58.85 \text{ sec}$$

Because the signal is semiactuated, the cycle length is not rounded off. The green time for the side street is estimated using a value $X_2 = 1.0$.

$$g_2 = 0.12\left(\frac{58.85}{1.0}\right) = 7.06 \text{ sec}$$

$$g_1 = 58.85 - 9 - 7.06 = 42.79 \text{ sec}$$

Plus a lost time of 9 sec = 58.85 sec.

Discussion

The cycle length and green times for semiactuated signals vary from cycle to cycle and the preceding computations reflect the average timing.

Example 6

A fully actuated signal has lane groups, flow ratios, and phasing, as shown in Figure 9-E6. A V/c ratio of 0.95 is assumed, and the lost time can be taken as 3 sec/phase. Note that although the change intervals for the two left-turn movements occur at different times in the cycle, they contribute only once to the total cycle lost time.

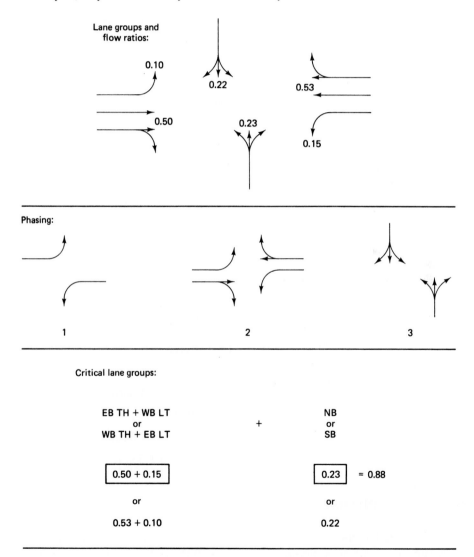

Figure 9-E6   Timing an Actuated Signal with Phase Overlap (for Example 6).

Solution

$$C = \frac{LX_c}{X_c - \Sigma(V/s)_i}$$

$$= \frac{9(0.95)}{0.95 - 0.88} = 122 \text{ sec}$$

This figure will not be rounded because the signal is actuated. Green times will be computed based on the constant $V/c = 0.95$, because it is assumed that the signal assigns green times proportionally.

The eastbound left turn is permitted only in phases 1 and 2, and the green time for these phases is controlled by the flow ratio for this movement.

$$g_1(\text{EBLT}) = 0.10\left(\frac{122}{0.95}\right) = 12.84 \text{ sec}$$

Similarly, the westbound (WB) left-turn green time is

$$g_1(\text{WBLT}) = 0.15\left(\frac{122}{0.95}\right) = 19.26 \text{ sec}$$

Also, the eastbound through movement is allowed in phases 2 and 3.

$$g_2(\text{WBTH}) = 0.53\left(\frac{122}{0.95}\right) = 68.06 \text{ sec}$$

$$g_2(\text{EBTH}) = 0.50\left(\frac{122}{0.95}\right) = 64.21 \text{ sec}$$

The minor-street movement is controlled by the northbound flow ratio:

$$g_3 = 0.23\left(\frac{122}{0.95}\right) = 29.53 \text{ sec}$$

$$g_1(\text{EBLT}) = 12.84 \text{ sec}$$
$$g_1(\text{WBLT}) = 19.26 \text{ sec} \qquad \text{Controls}$$
$$g_2(\text{WBTH}) = 68.06 \text{ sec} \qquad \text{Controls}$$
$$g_2(\text{EBTH}) = 64.21 \text{ sec}$$
$$g_3 = 29.53 \text{ sec} \qquad \text{Controls}$$
$$19.26 + 3 + 68.06 + 3 + 29.53 + 3 = 125.85 \text{ sec}$$

7. UNSIGNALIZED INTERSECTIONS

Unsignalized intersections make up a great majority of at-grade junctions in any street system. Stop and yield signs are used to assign the right-of-way, but drivers have to use their judgment to select gaps in the major street flow to execute crossings and turn movements at two-way and yield-controlled intersections. Two methods are discussed in this section: Blunden (1961) and the *Highway Capacity Manual* (TRB, 1994).

In general, stop- and yield-sign-controlled intersections are seldom a menacing capacity problem. Before such a problem becomes serious, action has already been taken to signalize the intersection.

7.1 Blunden's Method

The analysis of stop- and yield-sign-controlled approaches has been investigated by a number of researchers, especially Blunden. The capacity of such intersections depends on the traffic flow in the major stream and the confidence of individual drivers to cross this major stream. In formulating an acceptable expression to calculate flow rate, the pattern of gaps in the major stream is assumed to follow a Poisson distribution. Also, it is assumed that all drivers, on average, will accept a minimum gap. An expression that indicates the number of vehicles per hour that can be "absorbed" by the major traffic stream is given by Blunden (1961) as

$$q_{max} = \frac{Ve^{-VT/3600}}{1 - e^{-Vt/3600}} \tag{24}$$

where

$\quad q_{max}$ = maximum flow rate from the controlled approach

$\qquad V$ = total traffic volumes on the uncontrolled street in both directions

$\qquad T$ = minimum gap acceptable to the first driver on the side street

$\qquad t$ = additional time required for a second driver to follow the first driver into the intersection when a large gap occurs

Several implications are involved while considering use of this expression:

1. Vehicles from the minor street must wait until a suitable gap occurs in the major stream.
2. If a sufficiently large gap opens up in the major stream, the second and subsequent vehicles accept a smaller acceptance gap t, as compared to T.
3. The mix of vehicle types, such as trucks, buses, and so on, must be taken into account.
4. When good sight distance prevails, the stop sign can be replaced by a yield sign, where vehicles from the minor street can coast into the major stream without coming to a full stop, if a gap is available.
5. The width of the major street should be taken into account for fixing the values of T and t.
6. Values of T vary between 5 and 10 sec; values of t may be between 3 and 5 seconds.

Table 9-15 gives the maximum traffic volumes from the controlled approach using the previous expression given for values of T ranging between 5 and 8 sec and values of t varying between 3 and 5 sec.

TABLE 9-15   MAXIMUM FLOW RATES FROM A CONTROLLED APPROACH INTO
AN INTERSECTION WITH UNINTERRUPTED MAJOR-STREET TRAFFIC

Type of side-street control	Assumed value of T (sec)	t (sec)	Total two-way uninterrupted flow (veh/hr) 800	1000	1200	1400	1600
			Maximum flow rate from controlled approach (pc/hr)				
Stop	8	5	200	140	100	75	45
	7	5	250	190	140	110	80
	6	5	315	250	200	160	125
Yield	7	3	350	250	185	135	95
	6	3	[a]	335	255	200	150
	5	3	[a]	440	360	290	235

Vehicle type	Estimated passenger-car equivalent				
Bus[b]	2	3	3	4	4
Medium truck[c]	4	5	6	8	10
Truck-trailer[d]	6	7	9	12	16

[a] Maximum flow rate exceeds half of major-street two-way volume; hence, this situation is unlikely to occur.
[b] Vehicle length, 40 ft; acceleration rate, 4 ft/sec^2.
[c] Vehicle length, 30 ft; acceleration rate, 2 ft/sec^2.
[d] Vehicle length, 60 ft; acceleration rate, 2 ft/sec^2.
Source: Homburger et al., 1996.

Example 7

A two-lane street is expecting to generate 500 vehicles per hour during the peak hour, comprising 60% right turners and 40% left turners (Figure 9-E7). The corresponding flow on the main street is 850 and 950 veh/h. If $T = 6$ sec and $t = 5$ sec, analyze whether this intersection is satisfactory.

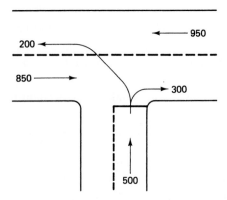

Figure 9-E7   Sketch for Example 7.

Solution

$$q_{max} = \frac{Ve^{-VT/3600}}{1 - e^{-Vt/3600}}$$

$$q_{RT} = \frac{850 \exp(-850 \times 6/3600)}{1 - \exp(-850 \times 5/3600)} = 297 \text{ veh/hr}$$

This situation is just marginal (297 < 300):

$$q_{LT} = \frac{1800 \exp(-1800 \times 6/3600)}{1 - \exp(-1800 \times 5/3600)} = 98 \text{ veh/hr}$$

This situation is most unsatisfactory (98 < 200). We would expect long delays. It would relieve the condition if a separate RT lane were provided so that RT drivers are not delayed by LT drivers.

7.2 *Highway Capacity Manual* Method (TRB, 1994)

Capacity analysis at two-way stop-controlled (TWSC) utilizes a clear description and understanding of the interaction of drivers on the minor or stop-controlled approach with the drivers on the major street. Figure 9-8 illustrates the relative priority of streams at both T- and four-leg intersections. The gap acceptance method employed in the procedure used in determining the capacity of these intersections computes the potential capacity of each minor traffic stream in accordance with the following equation:

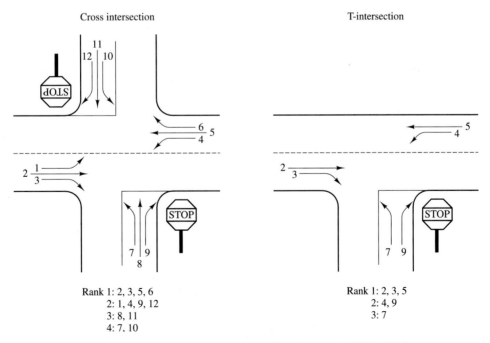

Figure 9-8   Traffic Streams at an Unsignalized Intersection (TRB, 1994).

$$c_{p,x} = \frac{3600}{t_f} e^{-\frac{\left[\sum_y V_{c,y}\right] t_0}{3600}}$$

$$(25)$$

where

$c_{p,x}$ = potential capacity of minor movement x (pcph)

$V_{c,y}$ = volume of traffic in conflicting stream y (pcph)

$t_0 = t_g - (t_f/2)$

t_g = critical gap, that is, the minimum time interval that allows intersection entry to one minor-stream vehicle (sec)

t_f = follow-up time (time interval between consecutive departures of two vehicles from a minor street under a continuous queue condition).

Values of t_g and t_f for passenger cars are given in Table 9-16. The nature of the conflicting movements at TWSC intersections is depicted in Figure 9-9. Vehicles per hour are converted to passenger cars per hour using Table 9-17. The potential capacity, $c_{p,x}$,

TABLE 9-16   CRITICAL GAPS t_g AND FOLLOW-UP TIMES t_f FOR TWSC INTERSECTIONS

	Critical gap t_g		
Vehicle maneuver	Two-lane major road	Four-lane major road	Follow-up time t_f (sec)
Left turn, major street	5.0	5.5	2.1
Right turn, minor street	5.5	5.5	2.6
Through traffic, minor street	6.0	6.5	3.3
Left turn, minor street	6.5	7.0	3.4

Note: The critical gap and follow-up time values presented in this table reflect data obtained on roadways where the average approach speed of the major-street through vehicles approximated 30 mph. In cases where no better data are available, these same values may be used to approximate t_g and t_f for roadways with approach speeds other than 30 mph.
Source: TRB, 1994.

TABLE 9-17   PASSENGER-CAR EQUIVALENTS FOR TWSC INTERSECTIONS

	Grade (%)				
Type of vehicle	−4	−2	0	+2	+4
Motorcycles	0.3	0.4	0.5	0.6	0.7
Passenger cars	0.8	0.9	1.0	1.2	1.4
SU/RVs[a]	1.0	1.2	1.5	2.0	3.0
Combination vehicles[b]	1.2	1.5	2.0	3.0	6.0
All vehicles[c]	0.9	1.0	1.1	1.4	1.7

[a] Single-unit trucks and recreational vehicles.
[b] Includes tractor-trailer combinations and buses.
[c] If vehicle composition is unknown, these values may be used as an approximation.
Source: TRB, 1994.

Subject movement	Conflicting traffic, $V_{c,x}$	Illustration
1. RIGHT TURN from minor street $(V_{c,9})$	$1/2(V_3)^{①}+ V_2^{②}$	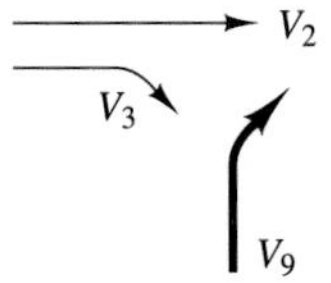
2. LEFT TURN from major street $(V_{c,4})$	$V_2 + V_3^{③}$	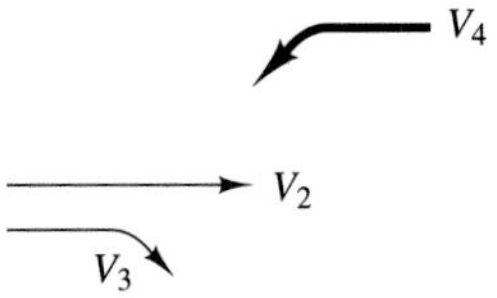
3. THROUGH MOVEMENT from minor street $(V_{c,8})$	$1/2(V_3)^{①}+ V_2 + V_1$ $+ V_6^{③}+ V_5 + V_4$	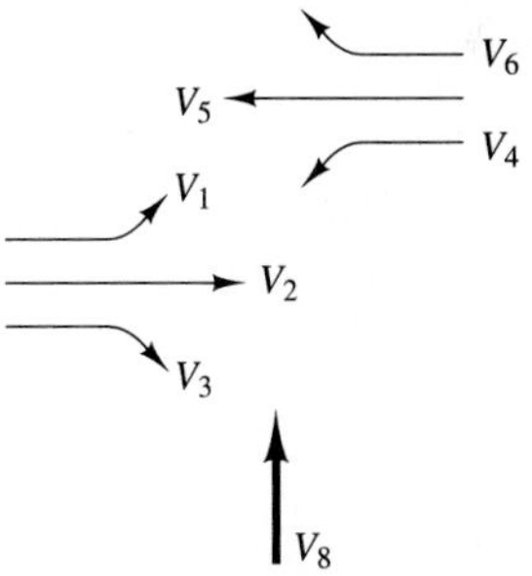
4. LEFT TURN from minor street $(V_{c,7})$	$1/2(V_3)^{①}+ V_2 + V_1$ $+1/2(V_6)^{⑤}+ V_5 + V_4$ $+1/2(V_{11} + V_{12}^{④})$	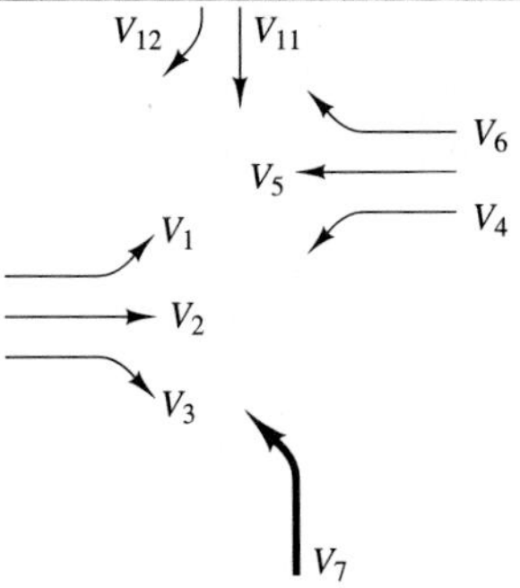

① Where a right-turn lane is provided on major street, and/or where V_3 is STOP-/YIELD-controlled, eliminate V_3.

② V_2 Includes only the volume in the right hand lane.

③ Where the right-turn is STOP- or YIELD-controlled, eliminate V_3, V_6.

④ V_{12} should be eliminated on mulit-lane major streets.

⑤ Where a right-turn lane is provided on major street, and/or where V_6 is STOP- or YIELD-controlled, and/or on mulit-lane major streets, eliminate V_6.

(a)

Figure 9-9(a)   Definition and Computation of Conflicting Traffic Volumes for Two Minor Approaches (TRB, 1994).

Subject movement	Conflicting traffic, $V_{c,x}$	Illustration
5. RIGHT TURN from minor street $(V_{c,12})$	$1/2(V_6)^{⑥} + V_5^{⑦}$	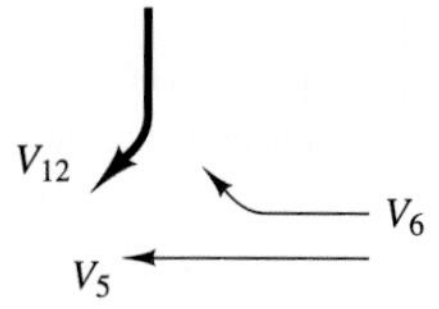
6. LEFT TURN from major street $(V_{c,1})$	$V_5 + V_6^{⑤}$	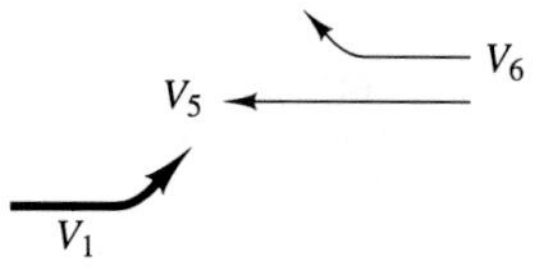
7. THROUGH MOVEMENT from minor street $(V_{c,11})$	$1/2(V_6)^{⑥} + V_5 + V_4 + V_3^{⑧} + V_2 + V_1$	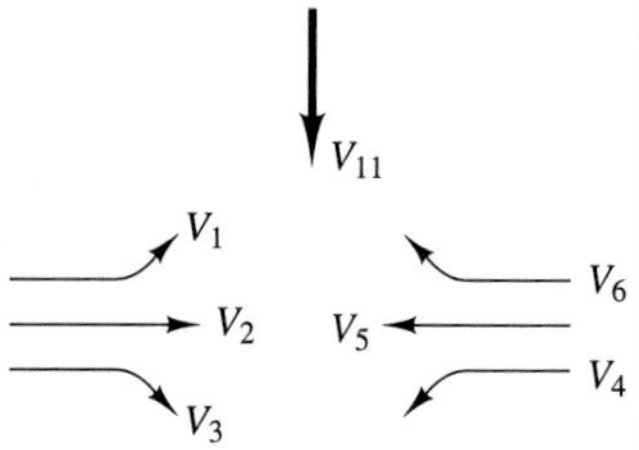
8. LEFT TURN from minor street $(V_{c,10})$	$1/2(V_6)^{⑧} + V_5 + V_4 +1/2(V_3)^{⑨} + V_2 + V_1 +1/2(V_8 + V_9^{⑨})$	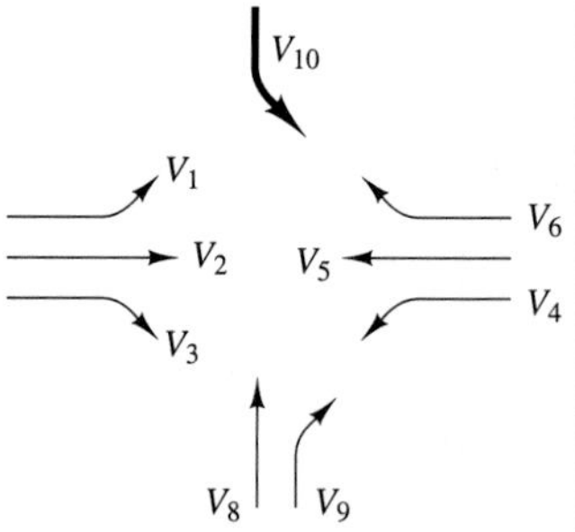

Ⓞ Where a right-turn lane is provided on major street, and/or where V_5 is STOP-/YIELD-controlled, eliminate V_5.

⑦ V_5 Includes only the volume in the right hand lane.

⑧ Where the right-turn is STOP- or YIELD-controlled, eliminate V_6, V_3.

⑨ V_9 should be eliminated on mulit-lane major streets.

⑩ Where a right-turn lane is provided on major street, and/or where V_3 is STOP- or YIELD-controlled, and/or on mulit-lane major streets, eliminate V_3.

(b)

Figure 9-9(b)   Definition and Computation of Conflicting Traffic Volumes for Two Minor Approaches (TRB, 1994).

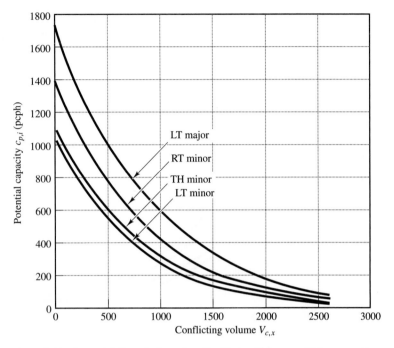

Figure 9-10   Potential Capacity Based on Conflicting Volume and Movement Type (Two-Lane Roadways) (TRB, 1994).

of the individual minor traffic streams is given in Figure 9-10 for a two-lane major road and in Figure 9-11 for a four-lane major road. The potential capacity of a movement is defined under ideal conditions for a specific subject movement, assuming the following conditions:

1. Traffic from nearby intersections does not back up.
2. A separate lane is provided for the exclusive use of each minor-street movement. A separate lane is also provided for the exclusive use of each major-street left-turn movement.
3. No other movements of ranks 2, 3, or 4 impede the subject movement.

   Impedance Effects.   Major traffic streams of rank 1 are assumed to be unimpeded by any of the minor traffic stream movements. Minor-street movements of rank 2 must yield only to major road through and right-turning traffic streams of rank 1, so the movement capacity of all rank 2 traffic streams is equal to the potential capacity:

$$c_{m,j} = c_{p,j} \qquad (26)$$

where j are movements of rank 2 priority.

   As minor traffic streams of rank 3 must yield to ranks 1 and 2, the probability of major street left-turning vehicles (rank 2) waiting for an acceptable gap at the same

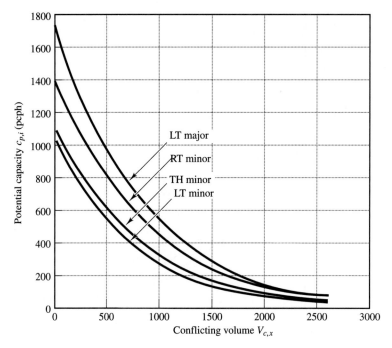

Figure 9-11 Potential Capacity Based on Conflicting Volume and Movement Type (Four-Lane Roadways) (TRB, 1994).

time as vehicles of rank 3 should be examined. The probability that the major street will operate in a queue-free state is given by

$$P_{o,j} = 1 - \frac{v_j}{c_{m,j}} \tag{27}$$

where $j = 1, 4$ (major-street left-turn movements of rank 2).

The capacity adjustment factor for all rank 3 is

$$f_k = \prod_j (p_{o,j}) \tag{28}$$

where

$p_{o,j}$ = probability that conflicting rank 2 movement j will operate in a queue-free state
k = rank 3 movements only

The movement capacity for the rank 3 movements can be computed as

$$c_{m,k} = (c_{p,k}) f_k \tag{29}$$

Rank 4 movements have the potential to be impeded by the queues of three higher-ranked traffic streams. However, it is recognized that not all these probabilities are independent of each other. Applying a simple product of the probabilities may

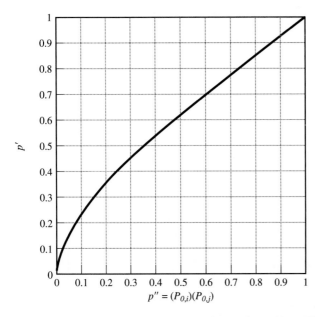

Figure 9-12   Adjustment to the Major Left, Minor Through Impedance Factor (TRB, 1994).

result in an overestimate of the impeding effect. Figure 9-12 can be used to adjust for the overestimate caused by the statistical dependence between queues in streams of ranks 2 and 3.

The movement capacity for the minor-street left-turn movements of rank 4 can be determined from

$$c_{m,1} = f_1 c_{p,1} \tag{30}$$

The methodology just described assumes that each minor-street movement has the exclusive use of a lane. Refinements are made to this analysis for the shared-lane capacity using procedures described in the *Highway Capacity Manual* (TRB, 1994, Chapter 10). The level-of-service criteria are given in Table 9-18.

For a 15-min analysis period, an estimate of average total delay is given by Eq. 31, which is graphically depicted in Figure 9-13 for a discrete range of capacities.

TABLE 9-18   LEVEL-OF-SERVICE CRITERIA FOR TWSC INTERSECTIONS

Level of service	Average total delay (sec/veh)
A	≤ 5
B	> 5 and ≤ 10
C	> 10 and ≤ 20
D	> 20 and ≤ 30
E	> 30 and ≤ 45
F	> 45

Source: TRB, 1994.

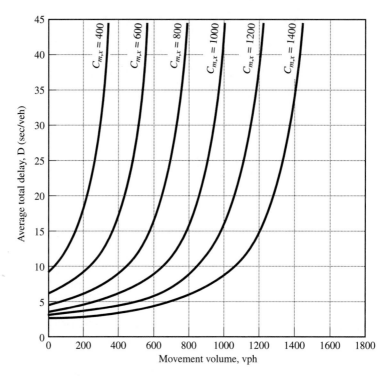

Figure 9-13   Average Total Delay Based on Conflicting Volume and Movement Capacity (15-Min Analysis Period)

$$D = \frac{3600}{C_{m,x}} + 900\,T\left\{\frac{V_x}{C_{m,x}} - 1 + \left[\left(\frac{V_x}{C_{m,x}} - 1\right)^2 + \frac{(3600/C_{m,x})\,(V_x/C_{m,x})}{450\,T}\right]^{1/2}\right\} \quad (31)$$

where

D = average total delay (sec/veh)

V_x = volume for movement x, expressed as an hourly flow rate

$C_{m,x}$ = capacity of movement x, expressed as an hourly flow rate

T = analysis period (hr) (for a 15-min period, use $T = 0.25$)

The analysis of two-way stop-controlled (TWSC) intersections is generally applied to existing locations either to evaluate operational conditions under present traffic demand or to estimate the impact of anticipated new demand. The methodology yields a level of service and an estimate of average total delay. Design applications are treated as trial-and-error computations. Field data requirements for input are as follows:

1. Volumes by movement for the hour of interest
2. Vehicle classification
3. Peak-hour factor
4. Number of lanes on the major street
5. Number and use of lanes on the minor-street approaches

6. Grade of all approaches

7. Other geometric features such as channelization, angle of intersection, acceleration lanes, and so on

8. Type of control on the minor-street approaches

As the methodology is based on prioritized use of gaps by vehicles at a TWSC intersection, it is important that computations be made in a precise order. The sequence of computation is the same as the priority of gap use, and movements are considered in the following order:

1. Right turns from the minor street

2. Left turns from the major street

3. Through movements from the minor street

4. Left turns from the minor street

Worksheets are provided for four-leg and T-type intersections and their use is illustrated through worked examples.

Example 8

The intersection of Eighth Avenue and Pine Street is a four-leg two-way stop-controlled (TWSC) intersection. The input worksheet is shown in Figure 9-E8(a). Determine the delay and level of service for each approach and for the overall intersection.

Solution

The solution is given through worksheets. See Figures 9-E8(b) and 9-E8(c). A computer printout based on Highway Capacity Software (HCS, 1996) is also included. See Figures 9-E8(d) and 9-E8(e). A close match between the two solutions is obvious. An increasing trend towards use of computer software is indicated as the analysis becomes labor intensive with worksheets.

Example 9

The intersection of First Avenue and Lincoln Street is a stop-controlled T-intersection. The analysis worksheet for this intersection is shown in Figure 9-E9. Determine the delay and level of service for each approach and for the overall intersection.

Solution

The procedure is similar to TWSC intersections except that it is made simpler because many movements and conflicts are removed. The solution is illustrated through the use of a worksheet.

7.3 All-Way Stop-Controlled Intersections

All-way stop-controlled (AWSC) intersections represent another type of unsignalized intersection. The methodology for the capacity and level-of-service analysis of AWSC intersections has been introduced in the United States recently (TRB, 1994). The methodology analyses each approach independently. AWSC intersections require that every vehicle stop at the intersection before proceeding. Because each driver must stop, the judgment as to enter the intersection is a function of the traffic conditions on the

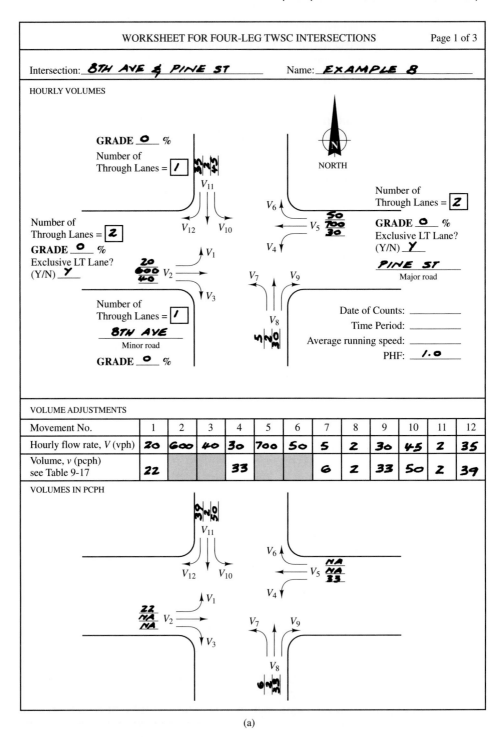

(a)

Figure 9-E8(a)   Worksheet for Four-Leg TWSC Intersections (1 of 3).

WORKSHEET FOR FOUR-LEG TWSC INTERSECTIONS		
STEP 1: RT from Minor Street	$\overset{\curvearrowright}{}$ V_9	$\overset{\curvearrowleft}{}$ V_{12}
Conflicting flows: V_c (Fig. 9-6) Potential Capacity: $c_{p,i}$ (Fig. 9-7, 8) Movement Capacity: c_m Prob. of Queue-free State: $p_{0,i}$	$V_{c,9} = 1/2V_3^① + V_2^②$ $\underline{20} + \boldsymbol{300} = \boldsymbol{320}$ vph $c_{p,9} = \underline{\boldsymbol{950}}$ pcph $c_{m,9} = c_{p,9} = \underline{\boldsymbol{950}}$ pcph $p_{0,9} = 1 - v_9/c_{m,9} = \underline{\boldsymbol{0.97}} = 1 - \frac{\boldsymbol{33}}{\boldsymbol{950}}$	$V_{c,12} = 1/2V_6^⑥ + V_5^⑦$ $\underline{25} + \boldsymbol{350} = \boldsymbol{375}$ vph $c_{p,12} = \underline{\boldsymbol{890}}$ pcph $c_{m,12} = c_{p,12} = \boldsymbol{890}$ pcph $p_{0,12} = 1 - v_{12}/c_{m,12} = \underline{\boldsymbol{0.96}} = 1 - \frac{\boldsymbol{39}}{\boldsymbol{890}}$
STEP 2: LT from Major Street	$\overset{\curvearrowleft}{}$ V_4	$\overset{\nearrow}{}$ V_1
Conflicting flows: V_c (Fig. 9-6)	$V_{c,4} = V_2 + V_3^③$ $\boldsymbol{600} + \underline{\boldsymbol{40}} = \boldsymbol{640}$ vph	$V_{c,1} = V_5 + V_6^⑧$ $\boldsymbol{700} + \underline{\boldsymbol{50}} = \boldsymbol{750}$ vph
Potential Capacity: $c_{p,i}$ (Fig. 9-7, 8) Movement Capacity: $c_{m,i}$ Prob. of Queue-free State: $p_{0,i}$ Major Left Shared Lane Prob. of Queue-free State: $p^*_{0,i}$	$c_{p,4} = \underline{\boldsymbol{780}}$ pcph $c_{m,4} = c_{p,4} = \underline{\boldsymbol{780}}$ pcph $p_{0,4} = 1 - v_4/c_{m,4} = \underline{\boldsymbol{0.96}} = 1 - \frac{\boldsymbol{33}}{\boldsymbol{780}}$ $p^*_{0,4} = 1 - \dfrac{1 - p_{0,4}}{1 - \left(\frac{v_5}{S_5} + \frac{v_6}{S_6}\right)} = \underline{\boldsymbol{NA}}$	$c_{p,1} = \underline{\boldsymbol{680}}$ pcph $c_{m,1} = c_{p,1} = \underline{\boldsymbol{680}}$ pcph $p_{0,1} = 1 - v_1/c_{m,1} = \underline{\boldsymbol{0.97}} = 1 - \frac{\boldsymbol{22}}{\boldsymbol{680}}$ $p^*_{0,1} = 1 - \dfrac{1 - p_{0,1}}{1 - \left(\frac{v_2}{S_2} + \frac{v_3}{S_3}\right)} = \underline{\boldsymbol{NA}}$
STEP 3: TH from Minor Street	$\uparrow$ V_8	$\downarrow$ V_{11}
Conflicting flows: V_c (Fig. 9-6) Potential Capacity: $c_{p,i}$ (Fig. 9-7, 8) Capacity Adjustment Factor due to Impeding Movements: f_i Movement Capacity: $c_{m,i}$ Prob. of Queue-free State: $p_{0,i}$	$V_{c,8} = 1/2V_3^① + V_2 + V_1 + V_6^③ + V_5$ $+ V_4$ $\boldsymbol{20} + \boldsymbol{600} + \boldsymbol{20} + \underline{\boldsymbol{50}} +$ $\boldsymbol{700} + \underline{\boldsymbol{30}} = \boldsymbol{1420}$ vph $c_{p,8} = \underline{\boldsymbol{160}}$ pcph $f_8 = p_{0,4} \times p_{0,1} = \underline{\boldsymbol{0.93}}$ (shared lane use p^*) $c_{m,8} = c_{p,8} \times f_8 = \underline{\boldsymbol{149}}$ pcph $p_{0,8} = 1 - v_8/c_{m,8} = \underline{\boldsymbol{0.99}} = 1 - \frac{\boldsymbol{2}}{\boldsymbol{149}}$	$V_{c,11} = 1/2V_6^⑥ + V_5 + V_4 + V_3^⑧ + V_2$ $+ V_1$ $\boldsymbol{25} + \boldsymbol{700} + \boldsymbol{30} + \underline{\boldsymbol{40}} +$ $\boldsymbol{600} + \underline{\boldsymbol{20}} = \boldsymbol{1415}$ vph $c_{p,11} = \underline{\boldsymbol{160}}$ pcph $f_{11} = p_{0,4} \times p_{0,1} = \underline{\boldsymbol{0.93}}$ (shared lane use p^*) $c_{m,11} = c_{p,11} \times f_{11} = \underline{\boldsymbol{149}}$ pcph $p_{0,11} = 1 - v_{11}/c_{m,11} = \underline{\boldsymbol{0.99}} = 1 - \frac{\boldsymbol{2}}{\boldsymbol{149}}$
STEP 4: LT from Minor Street	$\overset{\curvearrowleft}{}$ V_7	$\overset{\curvearrowright}{}$ V_{10}
Conflicting flows: V_c (Fig. 9-6) Potential Capacity: $c_{p,i}$ (Fig. 9-7, 8) Major Left, Minor through Impedance Factor: p''_i Major Left, Minor through Adjusted Impedance Factor: p'_i Capacity Adjustment Factor due to Impeding Movements: f_i Movement Capacity: $c_{m,i}$	$V_{c,7} = 1/2(V_3)^① + V_2 + V_1 + 1/2(V_3)^⑤$ $+ V_5 + V_4 + 1/2(V_{11} + V_{12}^④)^0$ $\boldsymbol{20} + \boldsymbol{600} + \boldsymbol{20} + \underline{\boldsymbol{0}}$ $\boldsymbol{700} + \underline{\boldsymbol{30}} + \underline{\boldsymbol{1}} = \boldsymbol{1371}$ vph $c_{p,7} = \underline{\boldsymbol{140}}$ pcph $p''_7 = p_{0,11} \times f_{11} = \underline{\boldsymbol{0.92}}$ $p'_7 = \underline{\boldsymbol{0.93}}$ (Fig. 9-9) $f_7 = p'_7 \times p_{0,12} = \underline{\boldsymbol{0.89}}$ $c_{m,7} = f_7 \times c_{p,7} = \underline{\boldsymbol{125}}$ pcph	$V_{c,10} = 1/2(V_6)^⑥ + V_5 + V_4 + 1/2(V_3)^⑩$ $+ V_2 + V_1 + 1/2(V_8 + V_9^⑨)^0$ $\boldsymbol{25} + \boldsymbol{700} + \boldsymbol{30} + \underline{\boldsymbol{0}}$ $\boldsymbol{600} + \underline{\boldsymbol{20}} + \underline{\boldsymbol{1}} = \boldsymbol{1376}$ vph $c_{p,10} = \underline{\boldsymbol{140}}$ pcph $p''_{10} = p_{0,8} \times f_8 = \underline{\boldsymbol{0.92}}$ $p'_{10} = \underline{\boldsymbol{0.93}}$ (Fig. 9-9) $f_{10} = p'_{10} \times p_{0,9} = \underline{\boldsymbol{0.90}}$ $c_{m,10} = f_{10} \times c_{p,10} = \underline{\boldsymbol{126}}$ pcph

Figure 9-E8(b)   Worksheet for Four-Leg TWSC Intersections (2 of 3).

WORKSHEET FOR FOUR-LEG TWSC INTERSECTIONS	Page 3 of 3

SHARED LANE CAPACITY

$$c_{SH} = \frac{v_i + v_j}{(v_i + c_{mi}) + (v_j + c_{mj})} \qquad \text{Where 2 movements share a lane}$$

$$c_{SH} = \frac{v_i + v_j + v_k}{(v_i + c_{mi}) + (v_j + c_{mj}) + (v_k + c_{mk})} \qquad \text{Where 3 movements share a lane}$$

MINOR STEET APPROACH MOVEMENTS 7, 8, 9

Movement	v(pcph)	c_m(pcph)	c_{SH}(pcph)	Avg. Total Delay (Fig. 9-10)	LOS	Avg. Total Delay for the approach, D_A
7	6	125	426.4	30.2	9.3	B	9.3
8	2	149		24.5		
9	33	950		3.9		

MINOR STEET APPROACH MOVEMENTS 10, 11, 12

Movement	v(pcph)	c_m(pcph)	c_{SH}(pcph)	Avg. Total Delay (Fig. 9-10)	LOS	Avg. Total Delay for the approach, D_A
10	50	126	200.4	46.2	32.1	E	32.1
11	2	149		24.5		
12	39	890		4.2		

MAJOR STEET APPROACH MOVEMENTS 1, 4

Movement	v(pcph)	c_m(pcph)	c_{SH}(pcph)	Avg. Total Delay (Fig. 9-10)	LOS	Avg. Total Delay for the approach, D_A
1	22	680		5.5	B	0.2
4	33	780		4.8	A	0.2

Avg. Total Delay for the intersection

$$\frac{9.3\,(37) + 32.1\,(82) + 0.2\,(660) + 0.2\,(780)}{20 + 600 + 40 + 30 + 700 + 50 + 5 + 2 + 30 + 45 + 2 + 35} = 2.1$$

$$D_1 = \frac{D_{A,1}\, V_{A,1} + D_{A,2}\, V_{A,2} + D_{A,3}\, V_{A,3} + D_{A,4}\, V_{A,4}}{V_1 + V_2 + V_3 + \ldots + V_{11} + V_{12}}$$

(c)

① Where a right-turn lane is provided on major street, and/or where V_3 is STOP-/YIELD-controlled, eliminate V_3

② V_2 includes only the volume in the right hand lane.

③ Where the right-turn is STOP- or YIELD-controlled, eliminate V_3, V_6

④ V_6, V_{12} should be eliminated on multilane major streets.

⑤ Where a right-turn lane is provided on major street, and/or where V_6 is STOP-/YIELD-controlled, and/or on multi-lane major streets, eliminate V_6

⑥ Where a right-turn lane is provided on major street, and/or where V_6 is STOP-/YIELD-controlled, eliminate V_6

⑦ V_5 includes only the volume in the right hand lane.

⑧ Where the right-turn is STOP- or YIELD-controlled, eliminate V_6, V_3

⑨ V_3, V_9 should be eliminated on multilane major streets.

⑩ Where a right-turn lane is provided on major street, and/or where V_3 is STOP-/YIELD-controlled, and/or on multi-lane major streets, eliminate V_3

Figure 9-E8(c)   Worksheet for Four-Leg TWSC Intersections (3 of 3).

```
HCS: Unsignalized Intersections   Release 2.1c    EX_08.HC0      Page 1
=======================================================================
Portland State University, Dept. Of Civil Engineering
P/O BOX 751
PORTLAND, OR   97201-0000
Ph: (503) 725-4244
=======================================================================
Streets: (N-S) 8th Ave.                     (E-W) Pine St.
Major Street Direction.... EW
Length of Time Analyzed... 15 (min)
Analyst..................
Date of Analysis......... 7/24/96
Other Information........
Two-way Stop-controlled Intersection
=======================================================================
            | Eastbound   | Westbound   | Northbound  | Southbound
            | L    T    R  | L    T    R  | L    T    R  | L    T    R
            |---- ---- ----|---- ---- ----|---- ---- ----|---- ---- ----
No. Lanes   | 1    2  < 0 | 1    2  < 0 | 0  > 1  < 0 | 0  > 1  < 0
Stop/Yield  |            N|            N|             |
Volumes     | 20  600   40| 30  700   50|  5    2   30| 45    2   35
PHF         |  1    1    1|  1    1    1|  1    1    1|  1    1    1
Grade       |      0      |      0      |      0      |      0
MC's (%)    |             |             |             |
SU/RV's (%) |             |             |             |
CV's (%)    |             |             |             |
PCE's       |1.10         |1.10         |1.10 1.10 1.10|1.10 1.10 1.10

--------------------------------------------------------------------------



                        Adjustment Factors

Vehicle                          Critical           Follow-up
Maneuver                         Gap (tg)           Time (tf)
-----------------------------------------------------------------
Left Turn Major Road               5.50               2.10
Right Turn Minor Road              5.50               2.60
Through Traffic Minor Road         6.50               3.30
Left Turn Minor Road               7.00               3.40
```

Figure 9-E8(d)   Worksheet for TWSC Intersection.

other approaches. Although the rules of the road suggest that the driver on the right has the right-of-way, the actual operation is somewhat more complex. Average total delay per vehicle is the recommended measure of effectiveness for evaluating the performance of AWSC intersections. The procedure for analysis is applied to existing or forecast geometric and traffic conditions. Field data requirements include the following:

- Traffic volumes in vehicles per hour for the peak hour, including turning movements, for each intersection approach

```
HCS: Unsignalized Intersections    Release 2.1c    EX_08.HC0      Page 2
========================================================================
                Worksheet for TWSC Intersection
-------------------------------------------------------------
Step 1: RT from Minor Street                    NB            SB
-------------------------------------------------------------
Conflicting Flows: (vph)                        320           375
Potential Capacity: (pcph)                      953           894
Movement Capacity: (pcph)                       953           894
Prob. of Queue-Free State:                      0.97          0.96
-------------------------------------------------------------
Step 2: LT from Major Street                    WB            EB
-------------------------------------------------------------
Conflicting Flows: (vph)                        640           750
Potential Capacity: (pcph)                      777           678
Movement Capacity: (pcph)                       777           678
Prob. of Queue-Free State:                      0.96          0.97
-------------------------------------------------------------
Step 3: TH from Minor Street                    NB            SB
-------------------------------------------------------------
Conflicting Flows: (vph)                        1420          1415
Potential Capacity: (pcph)                      161           162
Capacity Adjustment Factor
 due to Impeding Movements                      0.93          0.93
Movement Capacity: (pcph)                       149           150
Prob. of Queue-Free State:                      0.99          0.99
-------------------------------------------------------------
Step 4: LT from Minor Street                    NB            SB
-------------------------------------------------------------
Conflicting Flows: (vph)                        1372          1376
Potential Capacity: (pcph)                      140           140
Major LT, Minor TH
 Impedance Factor:                              0.91          0.91
Adjusted Impedance Factor:                      0.93          0.93
Capacity Adjustment Factor
 due to Impeding Movements                      0.89          0.90
Movement Capacity: (pcph)                       125           126
-------------------------------------------------------------

                Intersection Performance Summary

                                      Avg.    95%
            Flow    Move   Shared    Total   Queue            Approach
            Rate    Cap     Cap      Delay   Length   LOS      Delay
  Movement  (pcph) (pcph) (pcph) (sec/veh)  (veh)            (sec/veh)
  --------  ------ ------ ------ ------- ------- -----  ---------
  NB  L        6    125 >
  NB  T        2    149 >    427     9.3     0.3     B       9.3
  NB  R       33    953 >

  SB  L       50    126 >
  SB  T        2    150 >    201    31.9     2.0     E      31.9
  SB  R       39    894 >

  EB  L       22    678            5.5     0.0     B       0.2
  WB  L       33    777            4.8     0.0     A       0.2

              Intersection Delay  =      2.1 sec/veh
```

Figure 9-E8(e)   Worksheet for TWSC Intersection.

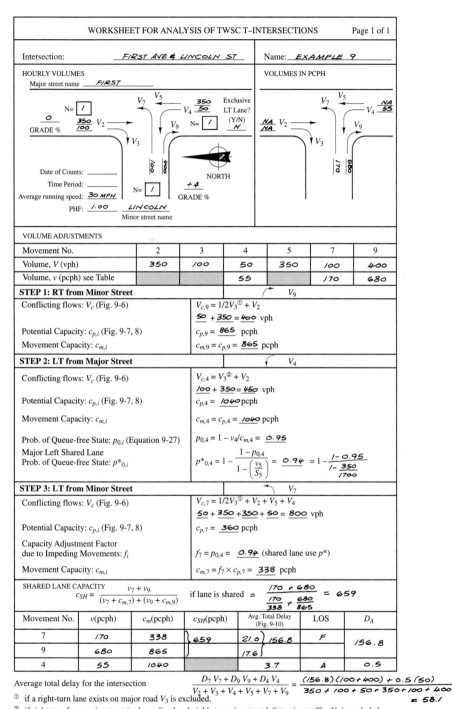

WORKSHEET FOR ANALYSIS OF TWSC T–INTERSECTIONS   Page 1 of 1

Intersection: _FIRST AVE & LINCOLN ST_   Name: _EXAMPLE 9_

HOURLY VOLUMES
Major street name _FIRST_

Exclusive LT Lane? (Y/N) _N_

VOLUMES IN PCPH

Date of Counts: _____
Time Period: _____
Average running speed: _30 MPH_
PHF: _1.00_   _LINCOLN_
Minor street name

GRADE % _0_

N= _1_

GRADE % _+4_

NORTH

VOLUME ADJUSTMENTS

Movement No.	2	3	4	5	7	9
Volume, V (vph)	350	100	50	350	100	400
Volume, v (pcph) see Table			55		170	680

STEP 1: RT from Minor Street V_9

Conflicting flows: V_c (Fig. 9-6)

$V_{c,9} = 1/2 V_3^{①} + V_2$
$50 + 350 = 400$ vph

Potential Capacity: $c_{p,i}$ (Fig. 9-7, 8) $c_{p,9} = 865$ pcph

Movement Capacity: $c_{m,i}$ $c_{m,9} = c_{p,9} = 865$ pcph

STEP 2: LT from Major Street V_4

Conflicting flows: V_c (Fig. 9-6)

$V_{c,4} = V_3^{②} + V_2$
$100 + 350 = 450$ vph

Potential Capacity: $c_{p,i}$ (Fig. 9-7, 8) $c_{p,4} = 1040$ pcph

Movement Capacity: $c_{m,i}$ $c_{m,4} = c_{p,4} = 1040$ pcph

Prob. of Queue-free State: $p_{0,i}$ (Equation 9-27) $p_{0,4} = 1 - v_4/c_{m,4} = 0.95$

Major Left Shared Lane
Prob. of Queue-free State: $p^*_{0,i}$

$p^*_{0,4} = 1 - \dfrac{1 - p_{0,4}}{1 - \left(\dfrac{v_5}{S_5}\right)} = 0.94 = 1 - \dfrac{1 - 0.95}{1 - \dfrac{350}{1700}}$

STEP 3: LT from Minor Street V_7

Conflicting flows: V_c (Fig. 9-6)

$V_{c,7} = 1/2 V_3^{①} + V_2 + V_5 + V_4$
$50 + 350 + 350 + 50 = 800$ vph

Potential Capacity: $c_{p,i}$ (Fig. 9-7, 8) $c_{p,7} = 360$ pcph

Capacity Adjustment Factor
due to Impeding Movements: f_i $f_7 = p_{0,4} = 0.94$ (shared lane use p^*)

Movement Capacity: $c_{m,i}$ $c_{m,7} = f_7 \times c_{p,7} = 338$ pcph

SHARED LANE CAPACITY

$c_{SH} = \dfrac{v_7 + v_9}{(v_7 + c_{m,7}) + (v_9 + c_{m,9})}$ if lane is shared $= \dfrac{170 + 680}{\dfrac{170}{338} + \dfrac{680}{865}} = 659$

Movement No.	v(pcph)	c_m(pcph)	c_{SH}(pcph)	Avg. Total Delay (Fig. 9-10)	LOS	D_A
7	170	338	659	21.0 ⎫ 156.8	F	156.8
9	680	865		17.6 ⎭		
4	55	1040		3.7	A	0.5

Average total delay for the intersection $\dfrac{D_7 V_7 + D_9 V_9 + D_4 V_4}{V_2 + V_3 + V_4 + V_5 + V_7 + V_9} = \dfrac{(156.8)(100 + 400) + 0.5(50)}{350 + 100 + 50 + 350 + 100 + 400}$
$= 58.1$

① if a right-turn lane exists on major road V_3 is excluded.
② if right turn from major street is channelized and yields to major street left turning traffic, V_3 is excluded.

Figure 9-E9   Worksheet for Analysis of TWSC T-Intersections.

- Peak-hour factor
- Number of lanes on each approach

The methodology is based on a set of four worksheets:

1. Input Worksheet
2. Volume Summary Worksheet
3. Capacity Analysis Worksheet
4. Level-of-Service Worksheet

A worked example is given to illustrate the methodology and the use of worksheets. Reference should be made to Chapter 10 of the *Highway Capacity Manual* (TRB, 1994) for a detailed treatment of the subject.

Example 10

The intersection of Molalla Avenue and Main Street is a four-leg all-way stop-controlled (AWSC) intersection. There is a single travel lane in each direction. The input worksheet is shown in Figure 9-E10(a). Determine the delay and level of service for each approach and for the overall intersection.

Solution

The operational analysis proceeds through the worksheets in systematic fashion. See Figures 9-E10(b) through 9-E10(d).

SUMMARY

The highway network consists of links and nodes, links representing segments of highways where more-or-less uninterrupted traffic flow is usually prevalent, and nodes representing at-grade intersections controlled by signals, stop signs, or yield signs. In some cases, such intersections may not be provided with any control device. In this chapter and the preceding one, we have dealt with intersection control, design, capacity, and level of service. The complexity of understanding all the details of at-grade intersections is somewhat overwhelming. The current *Highway Capacity Manual* (TRB, 1994), from which most of the material for this chapter has been drawn, marks yet another significant departure from the previous editions. Indeed, the most dramatic changes in the revision presented in the 1994 *HCM* are contained in the sections for signalized and unsignalized intersections.

The methodology adopted in finding the capacity and level of service of signalized intersections is based on critical movement analysis, which focuses on the identification and analysis of those movements, in particular lanes or groups of lanes, that control the requirements for green-time allocation. The concept of delay has been adopted as the measure for the level of service of intersections. Analysis of both V/c and delay levels is required to get a full picture of intersection operation. Also, it may be noted that capacity and level-of-service analyses of signalized intersections are somewhat independent of one another.

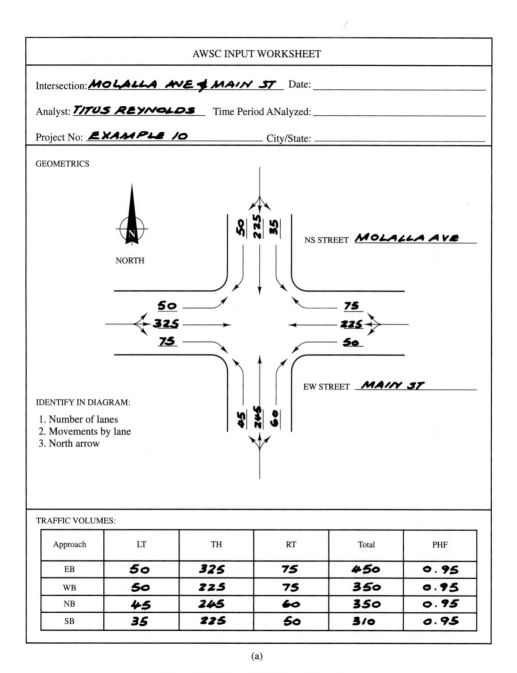

AWSC INPUT WORKSHEET

Intersection: *MOLALLA AVE & MAIN ST* Date: _____

Analyst: *TITUS REYNOLDS* Time Period ANalyzed: _____

Project No: *EXAMPLE 10* _____ City/State: _____

GEOMETRICS

NORTH

NS STREET *MOLALLA AVE*

EW STREET *MAIN ST*

IDENTIFY IN DIAGRAM:

1. Number of lanes
2. Movements by lane
3. North arrow

TRAFFIC VOLUMES:

Approach	LT	TH	RT	Total	PHF
EB	50	325	75	450	0.95
WB	50	225	75	350	0.95
NB	45	245	60	350	0.95
SB	35	225	50	310	0.95

(a)

Figure 9-E10(a) AWSC Input Worksheet.

AWSC VOLUME SUMMARY WORKSHEET				Page 2 of 4	
Step	Calculation	EB	WB	NB	SB
(1)	LT Volume	50	50	45	35
(2)	TH Volume	325	225	245	225
(3)	RT Volume	75	75	60	50
(4)	Peak Hour Factor	0.95	0.95	0.95	0.95
(5)	LT Flow Rate, (1)/(4)	53	53	47	37
(6)	TH Flow Rate, (2)/(4)	342	237	258	237
(7)	RT Flow Rate, (3)/(4)	79	79	63	53
(8)	Approach Flow Rate, (5) + (6) + (7)	474	369	368	327
(9)	Proportion LT, (5)/(8)	0.112	0.145	0.128	0.113
(10)	Proportion RT, (7)/(8)	0.167	0.214	0.171	0.162
(11)	Opposing Approach (Direction)	WB	EB	SB	NB
(12)	Conflicting Approaches (Directions)	NB,SB	NB,SB	EB,WB	EB,WB
(13)	Subject Approach Flow Rate	474	369	368	327
(14)	Opposing Approach Flow Rate	369	474	327	368
(15)	Conflicting Approach Flow Rate	695	695	843	843
(16)	Total Intersection Flow Rate, (13) + (14) + (15)	1538	1538	1538	1538
(17)	Proportion, Subject Approach Flow Rate, (13)/(16)	0.308	0.240	0.239	0.213
(18)	Proportion, Opposing Approach Flow Rate, (14)/(16)	0.240	0.308	0.213	0.239
(19)	Proportion, Conflicting Approach Flow Rate, (15)/(16)	0.452	0.452	0.548	0.548
(20)	LT, Opposing Approach	50	50	35	45
(21)	RT, Opposing Approaches	75	75	50	60
(22)	LT, Conflicting Approaches	80	80	100	100
(23)	RT, Conflicting Approaches	110	110	150	150
(24)	Proportion LT, Opposing Approach, (20)/(14)	0.136	0.105	0.107	0.122
(25)	Proportion RT, Opposing Approach, (21)/(14)	0.203	0.158	0.153	0.163
(26)	Proportion LT, Conflicting Approach, (22)/(15)	0.115	0.115	0.119	0.119
(27)	Proportion RT, Conflicting Approach, (23)/(15)	0.158	0.158	0.178	0.178

(b)

Figure 9-E10(b)   AWSC Volume Summary Worksheet.

AWSC CAPACITY ANALYSIS WORKSHEET				Page 3 of 4
Step		EB	WB	NB	SB

Step		EB	WB	NB	SB
(1)	Proportion, Subject Approach Flow Rate	0.308	0.240	0.239	0.213
(2)	Proportion, Opposing Approach Flow Rate	0.240	0.308	0.213	0.239
(3)	Lanes on Subject Approach	1	1	1	1
(4)	Lanes on Opposing Approach	1	1	1	1
(5)	+ 1000 x (1)	308	240	239	213
(6)	+ 700 x (2)	168	216	149	167
(7)	+ 200 x (3)	200	200	200	200
(8)	− 100 x (4)	−100	−100	−100	−100
(9)	(5) + (6) + (7) + (8)	576	556	488	480
(10)	Proportion LT, Opposing Approach	0.136	0.105	0.107	0.122
(11)	Proportion RT, Opposing Approach	0.203	0.158	0.153	0.163
(12)	Proportion LT Conflicting Approaches	0.115	0.115	0.119	0.119
(13)	Proportion RT Conflicting Approaches	0.158	0.158	0.178	0.178
(14)	− 300 x (10)	−40.8	−31.5	−32.1	−36.6
(15)	+ 200 x (11)	40.6	31.6	30.6	32.6
(16)	− 300 x (12)	−34.5	−34.5	−35.7	−35.7
(17)	+ 300 x (13)	47.4	47.4	53.4	53.4
(18)	(14) + (15) + (16) + (17)	12.7	13.0	16.2	13.7
(19)	Approach Capacity, (9) + (18)	589	569	504	494

(c)

Figure 9-E10(c)   AWSC Capacity Worksheet.

Analysis is done at two levels: operational and planning levels. Operational-level analysis provides for a full and detailed analysis and can be used to evaluate alternative geometric designs and/or signal plans. At the planning analysis level, only capacity is addressed, because detailed information is not available at this stage.

In this chapter, we also dealt with unsignalized intersections, namely, two-way, all-way stop-controlled and T-intersections, which are prominent through display of stop and yield signs. Methods described here cover the range of classical queueing and probability theories to the recently advanced procedures. Understanding of unsignalized intersections and their operation continues to make rapid advances through current research (NCHRP 3-4, 1996; NCHRP, 1996).

	AWSC LEVEL-OF-SERVICE WORKSHEET				Page 4 of 4
Step	Calculation	EB	WB	NB	SB
(1)	Approach Flow Rate	*474*	*369*	*368*	*327*
(2)	Approach Capacity	*589*	*569*	*504*	*494*
(3)	Volume/Capacity Ratio, (1)/(2)	*0.805*	*0.649*	*0.730*	*0.662*
(4)	Average Total Delay = exp[3.8 x (3)]	*21.3*	*11.8*	*16.0*	*12.4*
(5)	Level-of-Service (from table)	*D*	*C*	*C*	*C*

$$Intersection\ Total\ Delay = \frac{\Sigma\ (Vehicle\ Total\ Delay \times Volume)}{\Sigma\ Volume} = 15.9\ SEC/VEH$$

$$= \frac{21.3\,(450) + 11.8\,(350) + 16.0\,(350) + 12.4\,(310)}{450 + 350 + 350 + 310}$$

Level-of-Service (Intersection) = *C*

Level-of-Service	Average Total Delay, sec/veh
A	≤ 5
B	> 5 and ≤ 10
C	> 10 and ≤ 20
D	> 20 and ≤ 30
E	> 30 and ≤ 45
F	> 45

Figure 9-E10(d)   AWSC Level-of-Service Worksheet.

REFERENCES

BLUNDEN, W. ROSS (1961). *The Microscopic Models—Vehicle Speed, Inter-vehicle Headway, and Counting Distributions*, University of California Press, Berkeley.

HIGHWAY CAPACITY SOFTWARE (1996). HCS2.S, McTrans, University of Florida, Transportation Research Center, Gainesville.

HOMBURGER, W. S., ET AL. (1996). *Fundamentals of Traffic Engineering*, 14th Ed., Insti-

tute of Transportation Studies, University of California, Berkeley.

NATIONAL COOPERATIVE HIGHWAY RESEARCH PROGRAM, NCHRP (1996). *Unsignalized Intersections*, Draft Report, Chapter 10, Transportation Research Board, Washington, DC.

NATIONAL COOPERATIVE HIGHWAY RESEARCH PROGRAM, NCHRP 3-46 (1996). *Capacity and Level of Service at Unsignalized Intersec-*

tions, Final Report, Transportation Research Board, University of Idaho, Kittelson and Associates, Ruhr University, Queensland University of Technology.

TRANSPORTATION RESEARCH BOARD (1994). *Highway Capacity Manual*, Special Report 209, National Research Council, Washington, DC.

EXERCISES

1. Determine the saturation flow at a two-lane intersection approach in the CBD, where the curb lane is for through traffic only and the median lane is shared by both through and left-turn traffic. The left-turn traffic is about 30% of the traffic in the median lane. No cross-street pedestrians are allowed when the turns are permitted. The lanes are each 11 ft wide on a +2% grade. There are no parking lanes or bus stops in the intersection area. The traffic stream consists of about 6% heavy vehicles. Protected phasing is used.

2. Rework Example 2 with the following traffic volumes:

	EB	WB	NB	SB
Left	60	50	40	45
Through	700	550	500	500
Right	50	20	30	15
RTOR	15	10	20	10

3. Calculate the left-turn adjustment factor for permitted phasing using the "special procedure" described in the text for an EB approach having the following characteristics: cycle time = 70 sec; effective green time = 36 sec; lost time = 3.0 sec/phase; change interval = 4.0 sec/phase; PHF = 0.95; arrival type = 3; and located in CBD. The approach consists of two through lanes and an exclusive left-turn lane. The opposing approach has the same lane configuration. Both approaches have exclusive left-turn flow rates of 80 veh/hr, which amounts to 8% of the total EB approach flow rate and 9% of the total WB approach flow rate. The EB right turns amount to 5% of the total EB approach flow rate and the WB right turns amount to 6% of the total WB approach flow rate. The intersection has a two-phase signal cycle.

4. If everything in Exercise 3 remains unchanged except that the left turns are now made from a shared lane (i.e., there is no exclusive left-turn lane), determine the left-turn adjustment factor for the approach.

5. The intersection of Grand Avenue and 4th Street is shown in Figure 9-P5, which is the input module worksheet for this exercise. It is a four-leg intersection with a two-phase pretimed signal having a 65-sec cycle. Analyze the capacity and level of service of this intersection, and recommend changes to the signal and/or geometric design if the current situation is unacceptable. The intersection is located in the central city. An operational analysis is needed.

6. Figure 9-P6 shows the estimated traffic demand at the intersection of Oak Street and Pine Boulevard in the next 10 years. Oak Street is a one-way facility having two travel lanes and two side-street parking lanes. There is no parking on Pine Boulevard. Determine a suitable three-phase signal plan and check if this plan is acceptable without geometric improvement. The intersection is located in the suburbs. A critical V/c ratio of 0.85 (or lower) is required. Assume lost time = 3.0 sec/phase.

7. The intersection of Johnson Street and Main Street is shown in Figure 9-P7 and the anticipated traffic for the design year is indicated. Determine the adequacy of the geometric and signal design to accommodate the demand.

INPUT WORKSHEET

Intersection: **Grand Avenue + 4th Street** _____ Date: _____

Analyst: _____ Time Period Analyzed: _____ Area Type: ☐ CBD ☐ Other

Project No.: _____ City/State: _____

VOLUME AND GEOMETRICS

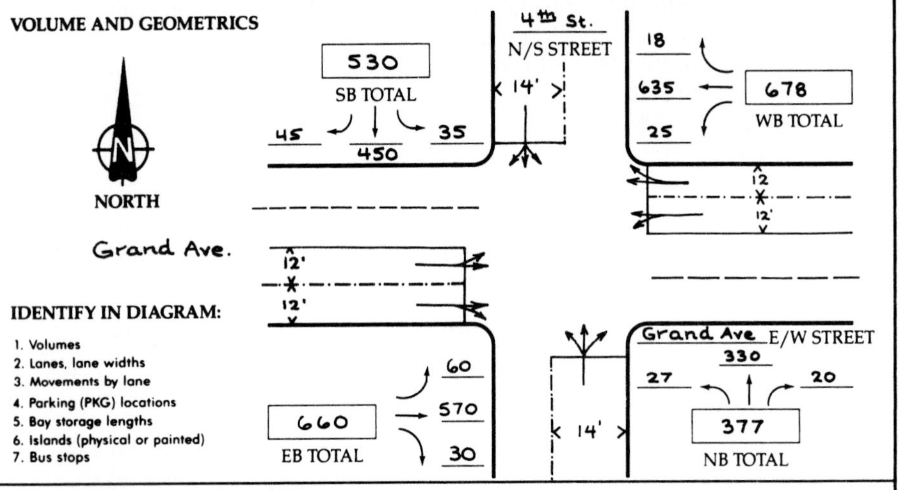

4th St.
N/S STREET

```
          530
        SB TOTAL
   45  ↙↓↘  35
         450
```

```
18
635 ←  678
          WB TOTAL
25
      12
      12'
```

NORTH

Grand Ave.

```
12'
12'
```

Grand Ave E/W STREET

IDENTIFY IN DIAGRAM:

1. Volumes
2. Lanes, lane widths
3. Movements by lane
4. Parking (PKG) locations
5. Bay storage lengths
6. Islands (physical or painted)
7. Bus stops

```
              60
    660    → 570
  EB TOTAL    30
```

```
  Grand Ave E/W STREET
       330
   27  ↓  20
      377
    NB TOTAL
```

TRAFFIC AND ROADWAY CONDITIONS

Approach	Grade (%)	% HV	Adj. Pkg. Lane Y or N	N_m	Buses (N_B)	PHF	Conf. Peds. (peds./hr)	Pedestrian Button Y or N	Min. Timing	Arr. Type
EB	O	5	N	–	O	0.85	80	N	9	4
WB	O	5	N	–	O	0.85	80	N	9	2
NB	O	5	N	–	O	0.85	80	N	14	3
SB	O	5	N	–	O	0.85	80	N	14	3

Grade: + up, − down N_B: buses stopping/hr   Min. Timing: min. green for
HV: veh. with more than 4 wheels   PHF: peak-hour factor   pedestrian crossing
N_m: pkg. maneuvers/hr   Conf. Peds: Conflicting peds./hr   Arr. Type: Type 1-5

PHASING

D I A G R A M								
Timing G = Y + R =	G = Y + R =	G = Y + R =	G = Y + R =	G = Y + R =	G = Y + R =	G = Y + R =	G = Y + R =
Pretimed or Actuated **P**	**P**							

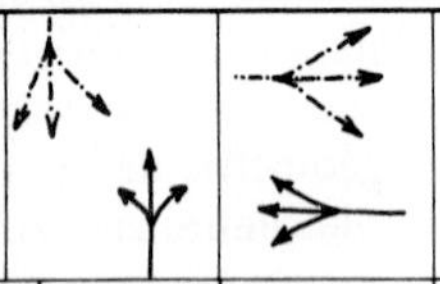

↙ Protected turns   ↗ Permitted turns   - - - - - Pedestrian   Cycle Length_____ Sec

Figure 9-P5   (TRB, 1994).

INPUT WORKSHEET

Intersection: Pine Blvd. + Oak Street Date: ____________

Analyst: ____________ Time Period Analyzed: ____________ Area Type: ☐ CBD ☐ Other

Project No.: ____________ City/State: ____________

VOLUME AND GEOMETRICS

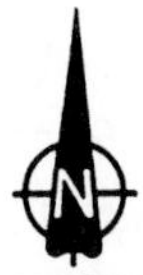
NORTH

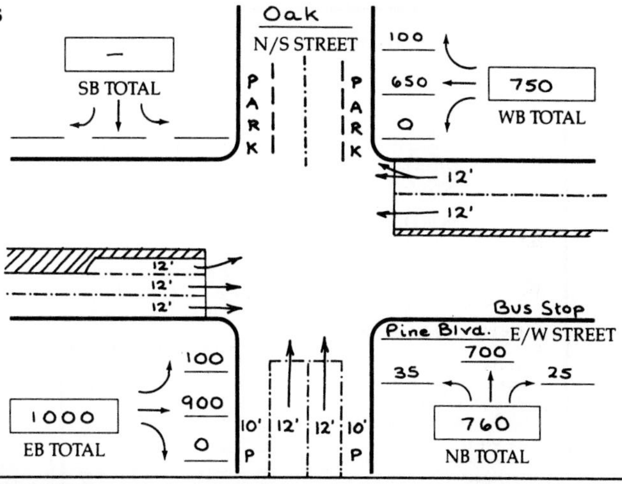
Oak
N/S STREET
–
SB TOTAL
P A R K
P A R K
100
650
750
0
WB TOTAL
12'
12'
12'
12'
12'
Bus Stop
Pine Blvd. E/W STREET
700
35
25
760
NB TOTAL
100
900
1000
0
EB TOTAL
10'
12'
12'
10'
P
P

IDENTIFY IN DIAGRAM:

1. Volumes
2. Lanes, lane widths
3. Movements by lane
4. Parking (PKG) locations
5. Bay storage lengths
6. Islands (physical or painted)
7. Bus stops

TRAFFIC AND ROADWAY CONDITIONS

Approach	Grade (%)	% HV	Adj. Pkg. Lane Y or N	Adj. Pkg. Lane N_m	Buses (N_B)	PHF	Conf. Peds. (peds./hr)	Pedestrian Button Y or N	Pedestrian Button Min. Timing	Arr. Type
EB	−1	10	N	–	10	0.95	30	N	13	3
WB	+1	10	N	–	10	0.95	30	N	13	3
NB	−0	5	Y	10	0	0.95	30	N	18	3
SB	–	–	–	–	–	–	–	–	–	–

Grade: + up, − down
HV: veh. with more than 4 wheels
N_m: pkg. maneuvers/hr
N_B: buses stopping/hr
PHF: peak-hour factor
Conf. Peds: Conflicting peds./hr
Min. Timing: min. green for pedestrian crossing
Arr. Type: Type 1-5

PHASING

DIAGRAM

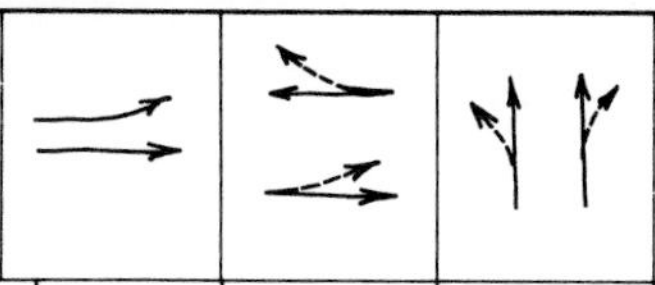

Timing	G = Y + R =	G = Y + R =	G = Y + R =	G = Y + R =	G = Y + R =	G = Y + R =	G = Y + R =	G = Y + R =
Pretimed or Actuated	P	P	P					

Protected turns — Permitted turns - - - - - - - Pedestrian — Cycle Length ______ Sec

Figure 9-P6 (TRB, 1994).

INPUT WORKSHEET

Intersection:_____ Date:_____

Analyst:_____ Time Period Analyzed:_____ Area Type: ☐ CBD ☐ Other

Project No.:_____ City/State:_____

VOLUME AND GEOMETRICS

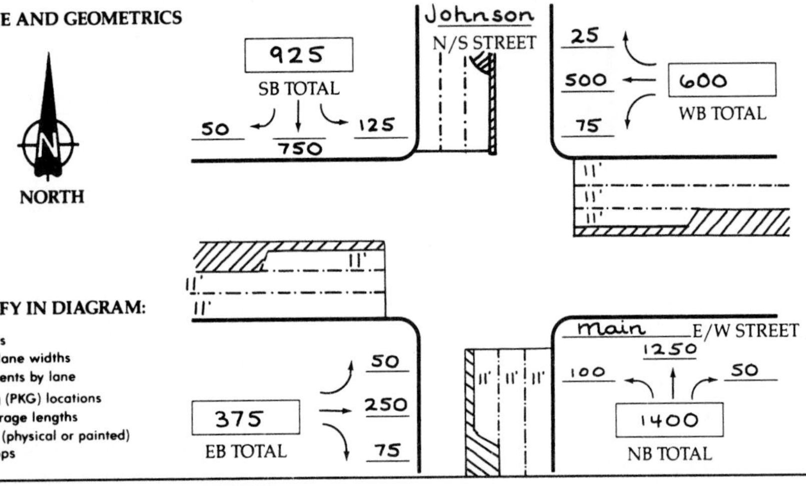

Johnson N/S STREET

925 SB TOTAL

50 ↙ ↓ ↘ 125
750

25

500 ← 600 WB TOTAL
75 ↓

11'
11'
11'

11'
11'
11'

IDENTIFY IN DIAGRAM:

1. Volumes
2. Lanes, lane widths
3. Movements by lane
4. Parking (PKG) locations
5. Bay storage lengths
6. Islands (physical or painted)
7. Bus stops

375 EB TOTAL

↱ 50
→ 250
↳ 75

main E/W STREET
1250
100 ← ↑ → 50
1400 NB TOTAL

TRAFFIC AND ROADWAY CONDITIONS

Approach	Grade (%)	% HV	Adj. Pkg. Lane Y or N	Adj. Pkg. Lane N_m	Buses (N_B)	PHF	Conf. Peds. (peds./hr)	Pedestrian Button Y or N	Pedestrian Button Min. Timing	Arr. Type
EB	O	5	N	–	O	0.85	100	Y	20	3
WB	O	5	N	–	O	0.85	100	Y	20	3
NB	O	5	N	–	O	0.85	100	Y	20	3
SB	O	5	N	–	O	0.85	100	Y	20	3

Grade: + up, − down N_B: buses stopping/hr   Min. Timing: min. green for pedestrian crossing
HV: veh. with more than 4 wheels   PHF: peak-hour factor
N_m: pkg. maneuvers/hr   Conf. Peds: Conflicting peds./hr   Arr. Type: Type 1-5

PHASING

DIAGRAM

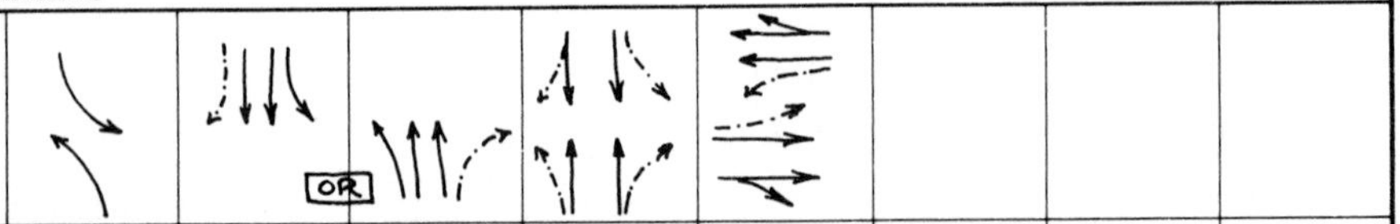

OR

Timing	G = Y + R =	G = Y + R =	G = Y + R =	G = Y + R =	G = Y + R =	G = Y + R =	G = Y + R =	G = Y + R =
Pretimed or Actuated	A	A	A	A	A			

↗ Protected turns   ↗ Permitted turns   ------- Pedestrian   Cycle Length_____Sec

Figure 9-P7 (TRB, 1994).

8. An isolated signalized intersection is planned for an urban area that will have the following traffic demand. Determine a suitable intersection layout and phasing sequence that will provide a condition of under capacity. Assume that the PHF will be 0.90 and a pretimed signal will be used. Use a planning level analysis.

	EB	WB	NB	SB
Left	100	125	150	200
Through	500	400	900	1000
Right	100	120	250	200

9. Refer to Exercise 4. The following additional data are now available: EB and NB approaches have a +3% grade, and the WB and SB approaches have a −3% grade; all lanes are 12 ft wide; traffic stream consists of 5% heavy vehicles; no side-street parking; no bus stops; no pedestrians. The NB and SB approaches consist of a single lane. The volumes for these approaches were found to be

	NB	SB
Left	35	15
Through	425	500
Right	25	30

A critical V/c ratio of 0.90 or lower is desired. Is this achieved? Determine the overall level of service for the intersection using the operation analysis procedure.

10. Using the procedure for allocation of green time described in this chapter (TRB, 1994), find the cycle length and green times for the following cases:
    (a) An intersection with a two-phase pretimed signal has flow ratios in the E–W direction of 0.50 and the N–S direction of 0.25, and the assumed lost time is equal to the change interval of 4 sec for each phase (8 sec/cycle).
    (b) A semiactuated signal controls an intersection where the flow ratios in the E–W and N–S directions are $V/s = 0.65$ and 0.10, respectively. The signal is designed to operate at a critical V/c of 0.85. Assume a lost time of 9 sec/cycle.

11. Refer to Example 10. The four-leg intersection of Molalla Avenue and Main Street, shown in Figure 9-E10(a), is an all-way stop-controlled intersection. If Main Street is widened and a second through lane is added (resulting in two-shared lanes), determine the delay and level of service for each approach and for the overall intersection.

12. The input module worksheet for the intersection of Moss Avenue and Ryder Street is shown in Figure 9-P12. It is a four-leg intersection with a two-phase pretimed signal of 100-sec cycle. Analyze the capacity and level of service of this intersection using operational analysis. Recommend changes to the signal and/or geometric design if the current situation is unacceptable.

13. The phasing in the intersection of Exercise 12 has been changed. The left turns are now given a 7-sec protected phase (cycle length = 100 sec). Analyze the capacity and level of service of this intersection using operational analysis.

14. The input module worksheet for the intersection of 5th Avenue and James Street is shown in Figure 9-P14. It is a four-leg intersection with street-side parking on James Street. Analyze the capacity and level of service of this intersection using operational analysis.

INPUT MODULE WORKSHEET

Intersection: _MOSS AVE & RYDER ST_____   Date: _____

Analyst: _____   Time Period Analyzed: _____   Area Type: ☐ CBD ☒ Other

Project No: _PROBLEM 12_____   City/State: _____

VOLUME AND GEOMETRICS

LOST TIME PER PHASE (sec): **3**

IDENTIFY IN DIAGRAM:

1. Volumes
2. Lanes, lane widths
3. Movements by lane
4. Parking (PKG) locations
5. Bay storage lengths
6. Islands (physical or painted)
7. Bus stops

TRAFFIC AND ROADWAY CONDITIONS

Approach	Grade (%)	% HV	Adj. Pkg. Lane Y or N	Adj. Pkg. Lane N_m	Buses (N_b)	PHF	Conf. Peds. (peds./hr)	Pedestrian Button Y or N	Pedestrian Button Min. Timing	Arr. Type
EB	O	3	N	—	O	0.9	O	N	—	3
WB	O	3	N	—	O	0.9	O	N	—	3
NB	O	5	N	—	O	0.9	O	N	—	5
SB	O	5	N	—	O	0.9	O	N	—	5

Grade: + up, – down
HV: veh. with more than 4 wheels
N_m: pkg. maneuvers/hr

N_b: buses stopping/hr
PHF: peak-hour factor
Conf. Peds.: Conflicting peds./hr

Min. Timing: min. green for pedestrian crossing
Arrival Type: Type 1-6, or P

PHASING

Timing	$G=$ 46	$G=$ 46	$G=$	$G=$	$G=$	$G=$	$G=$	$G=$
	$Y+AR=$ 4	$Y+AR=$ 4	$Y+AR=$	$Y+AR=$	$Y+AR=$	$Y+AR=$	$Y+AR=$	$Y+AR=$
Pretimed or Actuated	P	P						

Protected turns   Permitted turns   Pedestrian   Cycle Length _100_ Sec

Figure 9-P12   Input Module Worksheet (TRB, 1994).

INPUT MODULE WORKSHEET

Intersection: **5TH AVE & JAMES ST** _____ Date: _____

Analyst: _____ Time Period Analyzed: _____ Area Type: ☐ CBD ☒ Other

Project No: _____ City/State: _____

VOLUME AND GEOMETRICS

NORTH

LOST TIME PER PHASE (sec): **3**

IDENTIFY IN DIAGRAM:

1. Volumes
2. Lanes, lane widths
3. Movements by lane
4. Parking (PKG) location
5. Bay storage lengths
6. Islands (physical or painted)
7. Bus stops

TRAFFIC AND ROADWAY CONDITIONS

Approach	Crade %	% HV	Adj. Pkg. Lane		Buses (N_s)	PHF	Conf. Peds. (peds./hr)	Pedestrian Button		Arr. Type
			Y or N	N_m				Y or N	Min. Timing	
EB	2	2	N	—	4	0.95	50	Y	15	4
WB	-2	2	N	—	2	0.95	50	Y	15	4
NB	0	5	Y	40	—	0.95	50	Y	20	3
SB	0	5	Y	25	—	0.95	50	Y	20	3

Grade: + up, – down   Nb: buses stopping/hr   Min. Timing: min. green for
HV: veh. with more than 4 wheels   PHF: peak-hour factor   pedestrian crossing
Nm: pkg. maneuvers/hr   Conf. Peds.: Conflicting peds./hr   Arrival Type: Type 1-6, or P

PHASING

DIAGRAM								
Timing G= **10** Y+AR= **4**	G= **44** Y+AR= **4**	G= **24** Y+AR= **4**	G= Y+AR=	G= Y+AR=	G= Y+AR=	G= Y+AR=	G= Y+AR=	
Pretimed or Actuated **P**	**P**	**P**						

_____↑ Proteced turns	- - - -↑ Permitted turns	- - - - - - Pedestrian	Cycle Length **90** Sec

Figure 9-P14 Input Module Worksheet (TRB, 1994).

Chapter 10

Public Passenger Transportation

1. INTRODUCTION

Thus far, the major thrust of this book has been to present various aspects of highway engineering, although attempts have been made to connect these aspects to other modes of transportation as well. This is perhaps an appropriate place to introduce the topic of public transportation systems. These are modes of passenger transportation that are open for public use.

Public transportation in the United States has to deal with at least two substantially different problems with respect to mobility requirements. First, the problem of basic mobility comes to the forefront. The ability to provide mobility for those who are not able to provide their own private transportation is a perplexing problem, because it involves different needs and service markets, such as the economically disadvantaged and handicapped, and those who cannot drive. The second problem is equally troublesome. Public transportation or mass transit involves the movement of large numbers of people between a relatively small number of given locations. The situation in urbanized areas of this country is just the opposite. The travel demand is for a small number of trips between a diverse number of locations. The nature of this low-density travel pattern results in relatively high unit costs, whereas the ability of the transit system user to pay for the high cost is limited. Therefore, fare-box revenues are insufficient to offset transit expenses, and subsidies from the government are essential to keep the transit system alive and operating.

An understanding of the role of public transportation in the total picture of urban transportation is essential, and therefore this chapter begins with a brief history of public transit in the United States. This introductory history is followed by definitions and classifications used in public transit as well as its operational and service characteristics. The passenger-carrying capacity of an urban transit system is examined in a manner similar to the highway mode. Finally, rail and bus operation design is outlined and compared.

Public transportation systems is a fascinating area of study. What is included in this chapter is just the barebones, with the hope that inquisitive readers will expand their knowledge via the references provided at the end of this chapter. It is pretty certain that what happens to public transportation in the future will significantly influence urban development and the quality of life in this country (Meyer, Kain, and Wohl, 1965; Pushkarev and Zupan, 1977; U.S. DOT, 1980; Meyer and Gómez-Ibáñez, 1981).

2. HISTORICAL DEVELOPMENT OF URBAN TRANSPORTATION (U.S. DOT, 1977)

Throughout recorded history, human beings have strived for freedom, independence, and mobility. The first two—freedom and independence—are guaranteed because of the democratic setup of this country. Mobility is provided through modern technology and the current state of personal affluence, although one notices deterioration in this respect. Most people can travel when, where, how, and with whomever they wish. A small percentage does not enjoy this luxury.

To better understand the current situation, a brief review of the history of urban travel is described and a chronology of urban transit is shown in Figure 10-1. For centuries, urban residents traveled within the city by walking. Transportation was not a major problem because cities were small, and all points within the city were accessible on foot within a reasonable period of time. Only the rich enjoyed the luxury of riding.

With the industrial revolution of the nineteenth century, cities began to grow and expand in population, and with this growth came the need for traveling greater distances, which in turn increased the demand for improved means of transportation. Thus, animal-drawn vehicles became commonplace as a means of public and private transportation. This brought about new dimensions in transportation and a new form of traffic congestion.

In the days of animal-drawn vehicles, most city streets were not paved. To provide a smoother-riding surface and to permit one horse to do the work of several, many transit companies built street railways for horse-drawn transit vehicles. This horse-drawn "tram" was typical of those used in New York City in the mid-nineteenth century. However, the advent of rails did not increase the speed or the range of horse-drawn vehicles.

Andrew Hallidie sought a better power source for transit vehicles; he developed the cable car. Although cable cars were not necessarily faster than horse-drawn vehicles, they eliminated the need of maintaining a large herd of horses and the manure they produced. Elimination of horses thus solved a very significant pollution problem! These two advantages brought about the installation of cable cars in several large cities around the nation, San Francisco being the first in 1873.

Cable cars were limited in speed and distance and were soon replaced by electric streetcars. Electric streetcar systems appeared in nearly every large city in America and even in some cities of less than 5000 population. There is no doubt that the streetcar was a very significant development in urban transportation because it provided a

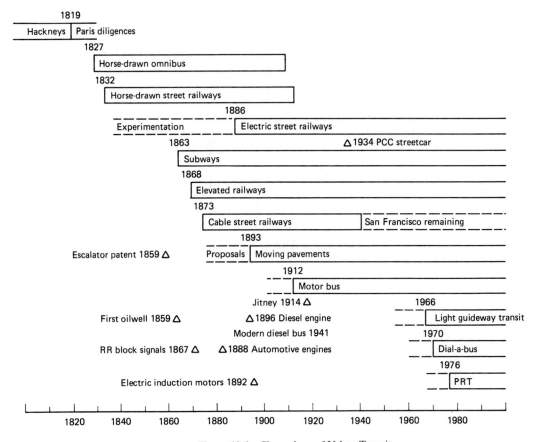

Figure 10-1   Chronology of Urban Transit.

longer range and higher operating speeds. Even so, streetcars did not eliminate congestion. With the advent of the streetcar, many cities soon discovered that some streets were too narrow to accommodate streetcar lines and still serve wagon and buggy traffic, so they built transit facilities above the street. For example, the 9th Avenue Elevated in New York City was opened in about 1875. The vehicles were initially steam powered, but were later converted to electric power. Some elevated facilities, such as the Chicago "El," are still in service today.

Because transit vehicles using elevated facilities were free of the normal traffic interferences at the ground level, they could travel faster; thus rail rapid transit was born before the turn of the century. Some cities objected to the poor aesthetics of the elevated facilities and chose to build similar facilities underground; and the subway was introduced. Boston began its subway system about 1885, and New York City opened the first segment of its system in 1904. Still, congestion was not eliminated.

The motorbus increased in popularity as advancements in technology improved operational capability. The steam-powered double-deck bus was used in London in the early twentieth century. Passengers could ride on top, exposed to the elements, for about half-fare. It is significant to note that motorbuses became popular in London a full 20 years before they did in the United States. About 1920, transit companies began replacing streetcar lines with bus lines because of the inherent flexibility of the bus as a transit vehicle. Buses were more compatible with other traffic, and routes could be altered as demand changed. Even so, the use of buses did not eliminate congestion. Today, about three-fourths of transit ridership is carried by buses. The buses of today are usually air-conditioned and operate at reasonably good travel speeds, depending on traffic conditions, number of stops, and preferential treatment.

The major impact on urban transportation began in 1893 when the Duryea Brothers built the first gasoline-powered automobile. Little did they realize that they were unleashing a power that would totally revolutionize the transportation and life-styles of the world. Originally a rich man's novelty, the automobile quickly became a device of utility because it satisfied man's inborn desire for mobility. Rapidly advancing technology, including assembly-line production methods, soon provided a dependable automobile and one that was priced within the financial capabilities of most Americans. North America soon became a nation on wheels, and mobility became a way of life.

Since the 1920s, automobile travel has experienced phenomenal growth, and progressive countries have tried valiantly to keep ahead of this growth with the development of new road systems. In the United States, the first major road development was the Primary System, created as a part of the Federal Aid Highway Act of 1921. The Primary System represented only 7% of the nation's rural mileage, so it was soon necessary to supplement it with other systems. As a result, the Federal Aid Secondary System was established in 1944. Also, in 1944, the Federal Aid Urban Extension system was created to provide continuity of the Federal Aid Primary and Secondary Systems through urban areas. These three systems are commonly referred to as the ABC systems. A major milestone in the 1944 Act was the establishment of a system to be known as the National System of Interstate and Defense Highways connecting principal metropolitan areas, thus forming the backbone of the country's highway system.

In 1956, the greatest of all peacetime public work efforts, the Interstate Highway System, was begun, resulting in the construction of about 42,000 miles of freeways. This system ushered in the era of intercity highway travel and the introduction of interstate commercial trucking as a feasible competitor to the railroads. Federal (90%) and state (10%) funds have been used for constructing the Interstate system, obtained primarily from taxes in gasoline sales. There have been several acts passed connected with highways and more recently with public transit. For instance, the Federal Aid Highway Act of 1973 allowed local officials in urban areas to decline the use of highway money from the Highway Trust Fund for highway improvement and use it instead for public transit. The idea is to have a truly "balanced" transportation system by not concentrating development in just one mode.

These changes are evidence that attempts have been made to satisfy the ever-increasing demand for travel by developing new systems and building more facilities. However, the demands have increased more rapidly than the country has been able to develop and to build new facilities.

Many feel that it is time to draw the line—to require some "sacrifices" to make the system more efficient. Alternative travel modes have been proposed as possible substitutes for the private automobile. We can recall several that were tried, only to be replaced by transportation modes more compatible with our preferred life-style. A return to rail rapid transit is suggested as a solution to urban transportation problems. But rail rapid transit often requires a capital outlay greater than that for any freeway system, and in all but a few instances, it is difficult to justify the economic investment in view of a decline in transit ridership. A review of the causes of this decline would be worthwhile.

It is the general impression that the decline in transit ridership has occurred only since World War II. That is because data generally are plotted only for that period. To get the true picture of the decline, one should plot ridership data going back to about 1900. Then it will be noted that the decline actually began in the 1920s. Parenthetically, this corresponds with expanded usage of the automobile. We may note that transit ridership stabilized to some extent during the Depression of the 1930s and then peaked again during the war years, when there was a serious shortage of automobiles and fuel.

Obviously, the major decline in mass transit has occurred since World War II. Some felt that fare increases caused the decline, and have proposed lower fares or no-fare transit service. It is true that the decline since World War II generally corresponds to fare increases, but this is more in the nature of a syndrome. It is important to remember that the decline in ridership began in the 1920s and was interrupted only by the unfortunate circumstances of a drastic economic depression followed by a major war.

The number of automobiles has increased proportionately with population since 1920. The ratio of persons per automobile has changed drastically—20 to 1 in 1920 as compared to 2 to 1 in 1980. Increasing personal income, or personal affluence, is probably the primary reason why Americans have been able to afford private transportation.

Another major factor in the cause of declining transit ridership is the type of housing made possible by the various transportation modes. Urban residents always have preferred the comparatively isolated domain afforded by the modern-day single-family residence. Such a life-style could not be achieved because of the limitations imposed by early transportation modes. Note that during the era of horse-drawn and cable-car transportation, major cities began to mushroom in population. To utilize the existing transportation system, these additional residents had to crowd into about the same land area that the cities had occupied prior to their rapid increase in growth—that is, the land area that could be served by horse-drawn and cable-car transportation. The net result was high-density residential areas.

The electric streetcar provided higher operating speeds and greater ranges, so it allowed the cities to expand outward. Row houses, such as those in San Francisco, were typical of residences built along new streetcar lines. The electric streetcar was readily accepted, not only because of its better transportation service, but also because it permitted a more desired form of housing.

When the motorbus and the automobile came along, they had the flexibility to expand into the open space between these corridors and to go even farther out. People readily accepted them because of the opportunity to achieve single-family dwelling units in and around large cities. The current life-style has in fact created a greater dependence on the automobile and lessened the need for mass transit.

Thus, it appears that the major factors contributing to the decline of transit usage are as follows:

1. Public affluence
2. Increasing availability of automobiles
3. A desire for low-density housing
4. Government policies connected with housing and highway development, indirectly encouraging the use of the private automobile

These factors have combined to create a life-style in which transit may never return to its former role as the primary means of urban travel. Yet there does appear to be an increasing need for some form of public transportation. Rail rapid transit is not new; it was actually developed before the turn of the century. But now there is a renewed interest in rail rapid transit. Its success in the modern-day setting depends on the success of new systems, such as those in Toronto, Montreal, San Francisco, Washington, D.C., and Atlanta.

Toronto was the first of these systems, opening its subway in the 1950s with modern stations and comfortable vehicles. The ridership was initially large and is still growing. Montreal opened a modern fixed-guideway rapid transit system just prior to Expo '67. Both systems are considered as successes.

Before a conclusion is reached that the experience of these two Canadian cities is evidence that rail rapid transit can be successful everywhere, one should note some of the differences in life-styles. Residents of both Toronto and Montreal live at much higher densities than do those in many cities of the United States. Toronto has numerous high-rise apartment buildings, whereas most Montreal residents live in row houses containing six or more families each. Both Canadian cities have a history of high transit usage. Even though the subway stations are spaced about $\frac{1}{2}$ mile apart to be in easy walking distance, surveys show that about three out of every four subway passengers ride to or from the subway station on streetcars or buses. Probably as many buses come to one subway station in Montreal each day as come to many of the U.S. downtowns. Even though their total transit ridership is increasing, they are encountering financial difficulties, and subway ridership seems to be declining.

One of the more recent fully operational systems to open in the United States is BART, the Bay Area Rapid Transit system, serving the San Francisco–Oakland area. It is, indeed, more modern in appearance than any previous system and was designed to be more automated. A trainman starts and stops the train; otherwise, it is computer-controlled. The vehicles are spacious, comfortable, and modern in appearance with upholstered seats and carpeted floors. In an effort to attract as many automobile drivers as possible, the BART system installed park-and-ride lots at each of its outlying stations. BART is indeed a beautiful system. It cost $1.7 billion, or about $21 million per mile! Other systems built in the United States include the Metro System in Washington, D.C., the cost of which is over $50 million per mile.

It should be realized that transit systems in existing urban settings do not pay their way through the fare box. Thus, it appears that the implementation of transit relates to the goals and objectives of the community more than purely economic considerations. In making decisions relating to urban transportation, the decision makers must realize that rail systems have high initial costs, require an extended time period for implementation, are fairly inflexible, and require high-density development to be cost effective. For these reasons, many cities are considering other ways of providing mass transportation into the center of the city.

Bus semirapid transit provides an expedient means of mass transportation. Such a system can be implemented in a much shorter time at a lower capital investment because it can utilize the existing highway network. The quality of bus rapid transit is largely a function of the line-haul trip. To make the systems comparable, steps must be taken to reduce the effects of prevailing congestion on bus travel. This is being done in several projects around the country. An exclusive busway is provided in the median of the Shirley Highway leading into Washington, D.C.

In many cities, freeway lanes are reserved for the exclusive use of buses and other high-occupancy vehicles. In still other instances, contraflow freeway lanes enable buses to take advantage of unused capacity in the off-peak direction. At ramps onto freeways, buses are given priority treatment to reduce delay. Buses bypass the waiting queue of passenger cars by way of a turnout or a separate ramp.

Cities are also implementing procedures to speed up the operation of buses in the downtown area, such as reserving lanes exclusively for buses during the peak hours. This allows the buses to keep moving rather than being stopped in traffic jams. Others are designating some streets as "bus streets," exclusively for the operation of buses.

Where do we go from here? Although this appears to be a question for the crystal ball gazer, there are a few facts that may shed some light on the subject. First, the desire for personal mobility will not be denied as long as we have the affluence to support it. Even in the face of fossil-fuel limitations and environmental concerns, individual mobility will prevail.

We have a tremendous investment in our system of streets and highways, and we will continue to develop and to use this system for personal transportation. No doubt, there will be changes in the way this system is used and in the vehicles using it. For example, the personal vehicle is shrinking in size and weight while the transit vehicle is being enlarged to carry more passengers.

There is obviously a maximum amount of urban space that can be reserved for transportation. As the demand for this space gets greater, we will see that efficient compromises take place. For example, most of our cities today experience congestion in the densely developed commercial areas. On the other end of the trip, there is virtually no congestion. Such a condition is conducive to mixed-mode operations. The individual may use his or her private vehicle for travel to an intermediate terminal point and then complete the remainder of the trip by a transit mode.

The solutions to urban transportation problems are not simple. They cannot be simply legislated, regulated, or purchased. Solutions are achieved by first identifying the problems, setting objective solutions acceptable to society, and applying all available resources to the achievement of these objectives. In the following sections, an attempt is made to cover a large number of topics geared to broaden the reader's perspective of public transport, including some elementary methods of designing rail and bus systems.

3. MASS TRANSIT DEFINITIONS AND CLASSIFICATIONS

Before we get into details of transit system operations, service, and characteristics, it is best to define a number of terms used in urban public transportation. In this section, we also describe transit service characteristics. Over the years, there has been a lot of controversy as to what is meant by a mode of transportation. The classification of modes can be done for transit based on three characteristics: right-of-way (R/W), technology, and type of service, as suggested by Vuchic (1981).

The right-of-way (R/W) is the strip of land on which the transit vehicle operates. There are three basic R/W categories, distinguished by the degree of separation from other traffic:

Category A: "grade-separated" or "exclusive." It is a fully controlled R/W without grade crossings or any legal access by other vehicles. In some ways, this category resembles a freeway system.

Category B: includes R/W types that are longitudinally physically separated from other traffic, but with grade crossing for vehicles and pedestrians, including regular street intersections. A light-rail system that crosses a few streets at the surface falls into this category.

Category C: surface streets with mixed traffic. Most bus systems and streetcar systems fall into this category.

Table 10-1 illustrates classification of transit modes on the basis of technology and R/W category.

The technology of transit modes is concerned with the mechanical features of the vehicles and the riding surface. There are at least four important characteristics of transit modes. First is the support between the vehicle and the riding surface: rubber tires on bituminous roadways and steel wheels on steel rails. The steering or guidance of vehicles is another characteristic. A third characteristic is the method of propulsion. And last, there is the means of regulating or controlling the vehicles longitudinally. For example, in the case of an automobile, a driver needs to control the speed and spacing of his or her vehicle in relation to other vehicles in the stream to avoid accidents.

Transit service can be classified into three groups by the types of routes and trips served:

1. Short haul is low-speed service within small areas with high travel density, such as central business districts.

2. City transit, which is the most common type, serves people needing transport in the city.

TABLE 10-1 CLASSIFICATION OF URBAN PUBLIC TRANSPORT MODES [a]

R/W category	Technology			
	Highway— driver-steered	Rubber-tired— guided, partially guided	Rail	Special
C	Paratransit Shuttle bus Regular bus Express bus (on streets)	Trolleybus	Streetcar Cable car	Ferryboat Hydrofoil Helicopter
B	Semirapid bus	Dual mode[*]	Light-rail transit	
A	Bus on busway only[*]	Rubber-tired RT Rubber-tired monorails Automated guided transit GRT PRT[*]	Light rapid transit Schwebebahn Rail rapid transit Regional rail	Incline Aerial tramway Continuous short- haul systems

[a] Modes used extensively are shown in italic type. Modes that are not operational are designated by asterisks.
Source: Vuchic, 1981. Reproduced by permission of the publisher.

3. Regional transit serves long trips, makes few stops, and generally has high speeds. Rapid rail and express bus systems fall into this category.

Another way of classifying transit service is by its stopping schedule, such as local service and express service. In a way it is closely connected to speeds and population densities. Still another classification refers to time of operation, such as peak-hour service or special-purpose service. An overview of mode classification and characteristics is given in Figure 10-2.

4. TRANSIT SYSTEM OPERATIONS, SERVICE, AND CHARACTERISTICS

Transit operations include such activities as scheduling, crew rostering, running and supervision of vehicles, fare collection, and system maintenance. They produce transportation that is offered to potential users. Vuchic (1981) defines several terms used in transit practice. Transit service is the transit system as experienced by its actual and potential users. Transit system characteristics are classified in four categories.

1. System performance refers to the entire set of performance elements, the most important of which are
   (a) Service frequency (f), number of transit unit departures per hour.
   (b) Operating speed (v_o), speed of travel on the line that passengers experience.
   (c) Reliability, expressed as percentage of vehicle arrivals with less than a fixed-time deviation from schedule (e.g., 4 minutes).
   (d) Safety, measured by the number of fatalities, injuries, and property damage per 100 million passenger-km (passenger-mi) or a similar unit.
   (e) Line capacity (c), the maximum number of persons that transit vehicles can carry past a point along the line.
   (f) Productive capacity (P_c), the product of operating speed and line capacity. As a composite indicator incorporating one basic element affecting passenger speed (speed), and one affecting operator (capacity), productive capacity is a very convenient performance indicator for mode comparisons.
   (g) Productivity, the quantity of output per unit of resource [e.g., vehicle-km(-mi), space-km(-mi) per unit of labor, operating cost, fuel, R/W width, etc.].
   (h) Utilization, also the ratio of output to input, but of the same unit, for example, person-km/space-km (person-mi/space-mi) offered.
2. Level of service (LOS) is the overall measure of all service characteristics that affect users. LOS is a basic element in attracting potential users to the system. Major factors comprising LOS can be divided into two groups:
   (a) Performance elements that affect users, such as operating speed, reliability, and safety.
   (b) Service quality (SQ), consisting of qualitative elements of service, such as convenience and simplicity of using the system, riding comfort, aesthetics, cleanliness, and behavior of passengers.

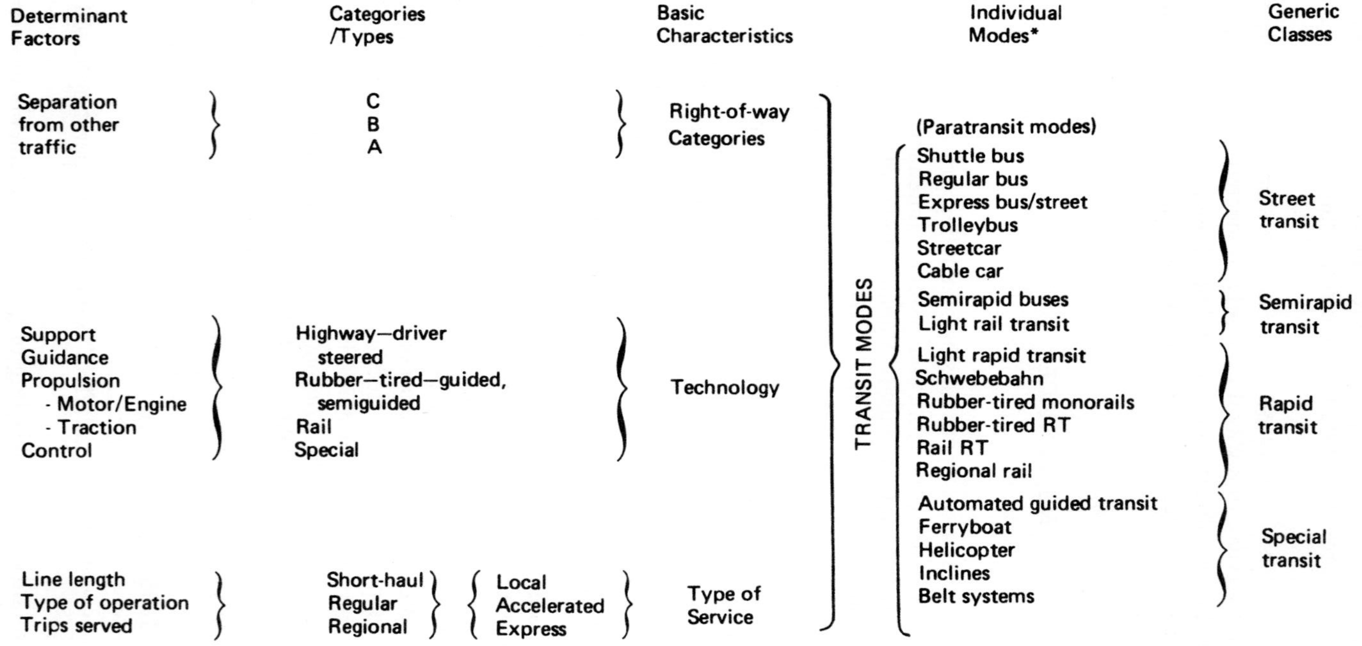

Determinant Factors	Categories/Types	Basic Characteristics	Individual Modes*	Generic Classes
Separation from other traffic	C B A	Right-of-way Categories	(Paratransit modes) Shuttle bus Regular bus Express bus/street Trolleybus Streetcar Cable car	Street transit
			Semirapid buses Light rail transit	Semirapid transit
Support Guidance Propulsion - Motor/Engine - Traction Control	Highway—driver steered Rubber—tired—guided, semiguided Rail Special	Technology	Light rapid transit Schwebebahn Rubber-tired monorails Rubber-tired RT Rail RT Regional rail	Rapid transit
Line length Type of operation Trips served	Short-haul Regular Regional Local Accelerated Express	Type of Service	Automated guided transit Ferryboat Helicopter Inclines Belt systems	Special transit

TRANSIT MODES

*The list is not exhaustive.

Figure 10-2 An Overview of Transit Mode Definition, Classification, and Characteristics (Vuchic, 1981). Reproduced by permission of the publisher.

3. Impacts are the effects that transit service has on its surroundings and the entire area it serves. They may be positive or negative. Short-run impacts include reduced street congestion, changes in air pollution, noise, and aesthetics along a new line. Long-run impacts consist of changes in land values, economic activities, physical form, and social environment of the city.

4. Costs are usually divided into two major categories: investment costs (or capital costs) are those required to construct or later make permanent changes in the physical plant of the transit system. Operating costs are costs incurred by regular operation of the system.

Evaluation and comparative analysis of transit systems must include all four categories: performance, LOS, impacts, and costs of each system. The preferred mode is usually not the one with the highest performance or lowest costs, but the one with the most advantageous "package" or combination of the four.

5. FAMILY OF REGULAR TRANSIT MODES

Table 10-2 gives broad ranges of the basic technical, operational, and systems characteristics of the most important modes, classified into the three generic classes: street transit, semirapid, and rapid transit. The private automobile is also included for comparison. A brief description of the various transit modes follows:

Taxis: automobiles operated by a driver and hired by users for individual trips, tailored entirely to the user's desire.

Dial-a-ride or dial-a-bus: minibuses or vans directed from a central dispatching office. Passengers call the office and give their origin, destination, and desired time of travel. The office plans the bus routings to maximize the number of passengers on a single trip.

Jitneys: privately owned automobiles or vans that operate generally on a fixed route, but may deviate in some cases, without fixed schedules.

Subscription buses: buses with paid drivers, operating between, say, a residential neighborhood and a particular employment area, involving some route deviation for minor collection and distribution patterns at either end of the trip.

Carpools: prearranged ride-sharing services where parties of two or more people travel together in a car on a regular basis. It is private transport and therefore cannot be organized, scheduled, or regulated by an agency, but it can be encouraged by employers.

Vanpools: privately or publicly provided vans (7- to 15-seaters) transporting groups of persons to and from work on a daily basis. They need a somewhat formal organization for vehicle purchase, maintenance, and driving.

Regular buses: buses operating along fixed routes on fixed schedules. Vehicles vary from minibuses (20 to 35 spaces) to articulated buses (up to 130 spaces).

TABLE 10-2 TECHNICAL, OPERATIONAL, AND SYSTEM CHARACTERISTICS OF URBAN TRANSPORT MODES[a]

Characteristic	Unit[b]	Private		Street transit		Semirapid transit		Rapid transit	
		Auto on street	Auto on freeway	RB	SCR	SRB	LRT	RRT	RGR
1. Vehicle capacity, C_v	sp/veh	4–6 total, 1.2–2.0	usable	40–120	100–180	40–120	110–250	140–280	140–210
2. Vehicles/transit unit	veh/TU	1	1	1	1–3	1	1–4	1–10	1–10
3. Transit unit capacity	sp/TU	4–6 total, 1.2–2.0	usable	40–120	100–300	40–120	110–600[c]	140–2000	140–1800
4. Maximum technical speed, v	km/hr	40–80	80–90	40–80	60–70	70–90	60–100	80–100	80–130
5. Maximum frequency f_{max}[d]	TU/hr	600–800	1500–2000	60–120	60–120	60–90	40–90	20–40	10–30
6. Line capacity, C	sp/hr	720–1050[c]	1800–2600[c]	2400–8000	4000–15,000	4000–8000	6000–20,000	10,000–40,000	8000–35,000
7. Normal operating speed, v_o	km/hr	20–50	60–90	15–25	12–20	20–40	20–45	25–60	40–70
8. Operating speed at capacity, v_o^C	km/hr	10–30	20–60	6–15	5–13	15–30	15–40	24–55	38–65
9. Productive capacity, P_c	(sp-km/hr²) ×10⁻³	10–25[e]	50–120	20–90	30–150	75–200	120–600	400–1800	500–2000
10. Lane width (one-way)[f]	m	3.00–3.65	3.65–3.75	3.00–3.65	3.00–3.50	3.65–3.75	3.40–3.75	3.70–4.30	4.00–4.75
11. Vehicle control[f]	—	Man./vis.	Man./vis.	Man./vis.	Man./vis.	Man./vis.	Man./vis.-sig.	Man.-aut./sig.	Man.-aut./sig.
12. Reliability	—	Low-med.	Med.-high	Low-med.	Low-med.	High	High	Very high	Very high
13. Safety	—	Low	Low-med.	Med.	Med.	High	High	Very high	Very high
14. Station spacing	m	200–500	—	200–500	250–500	350–800	350–800	500–2000	1200–4500
15. Investment cost per pair of lanes	($/km) ×10⁻⁶	0.2–2.0	2.0–15.0	0.1–0.4	1.0–2.0	3.0–9.0	3.5–12.0	8.0–25.0	10.0–25.0

[a] RB, regular bus; SCR, streetcar; SRB, semirapid bus; LRT, light-rail transit; RRT, rail rapid transit; RGR, regional rail.
[b] Metric conversion: 1 km = 0.62 mi. Abbreviations: sp, spaces; veh, vehicles; TU, transit unit.
[c] Values for C and P_c are not necessarily products of the extreme values of their components because these seldom coincide.
[d] For auto, lane capacity; for transit, line (station) capacity.
[e] For private auto, capacity is product of average occupancy (1.2–1.3) and f_{max} because all spaces cannot be utilized.
[f] Abbreviations are for: manual, visual, signal, and automatic.
Source: Vuchic, 1981. Reproduced by permission of the publisher.

Express buses: provide fast, comfortable travel on long routes with widely spaced stops.

Trolley buses: same as regular buses except that they are propelled by electric power, and therefore constrained to operate only where power lines exist.

Street cars or tramways: electrically powered rail transit vehicles operating on streets along with other vehicles, riding on steel rails flush with the street.

Semirapid buses: regular or high-performance buses operating on routes that include substantial sections of R/W categories A and B. Buses on busways consist of an exclusive busway on the freeway or in the median utilized by a great number of bus routes. They typically represent commuter transit.

Light-rail transit: mode utilizing predominantly reserved, but not necessarily grade-separated R/W. It is electrically propelled. A fuller description is given later in this chapter.

Rapid transit: includes the following: (1) Light-rail rapid transit consisting of light-rail vehicles operating on R/W category A only. (2) Rubber-tired rapid transit consisting of moderately large vehicles, supported or guided by rubber tires running on wooden, steel, or concrete surfaces. (3) Rail rapid transit, typically having four-axle rail vehicles operating in trains of up to 10 cars on fully controlled R/W category A with high speed, reliability, and capacity. Some are highly automated. (4) Regional rail operated on a long route with few stations at high speed on exclusive R/W category A (Vuchic, 1981).

5.1 Bus Transit Systems

These include motor buses and trolley buses operating on public streets. The local transit authority usually prescribes the routes, frequencies, fares, and stops. A uniform flat fare or one based on zones or distances is charged. The vehicles can carry anywhere from 12 to 240 passengers and a mixture of standing and seated passengers can be accommodated. Local services may entail frequent stops, whereas express service may require only a few stops. Service standards are perceived in terms of reliability, frequency, journey time, and quality of ride.

Bus transit enjoys the advantage of being quite flexible in meeting changes in demand, with virtually no cost. Extensions, expansions, and new routes can be introduced with little effort and low cost. Trolley bus systems, however, do not have this same flexibility because they are constrained by overhead electric transmission lines.

Transit systems using buses with a seating capacity of 40 to 120 are capable of carrying from 2400 to 15,000 passengers in mixed traffic. Journey speed (including stops) in mixed traffic is likely to be in the range 12 to 25 km/hr; where several lanes are available in the same street, one can expect volumes of between 25,000 and 30,000 bus passengers per hour in one direction.

Journey speed and capacity can be enhanced by utilizing reserved bus lanes. Maximum bus transit performance can be provided by exclusive busways in which buses are physically separated from other traffic by medians or barriers, with grade separation or priority at intersections. Volumes in excess of 30,000 passengers per hour per

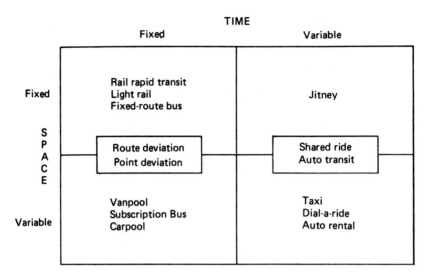

Figure 10-3 Temporal–Spatial Characteristics of Urban Transport Alternatives (Gray and Hoel, 1979). Reproduced by permission of the publisher.

lane with speeds up to 30 km/hr can be achieved with off-line stations and multiple boarding platforms. Naturally, there are many variations between the two operating extremes: mixed traffic and exclusive busways (Ismail and Khisty, 1974).

This is perhaps the best place to discuss the role of *paratransit*. The term applies to small passenger vehicles operating informally on a fare-paying basis, and serving in some places as an alternative to regular bus transit services. Of the many classifications possible, the one that helps to capture the essence of paratransit alternatives is shown in Figure 10-3 differentiated by time and space characteristics. Although conventional transit operates on a fixed-schedule (time) and fixed-route (space) basis (quadrant 1), paratransit operates in the environment of fixed space–variable time (jitney), or fixed time–variable space (vanpool/carpool), or variable time–variable space (taxi, dial-a-ride). Paratransit systems are capable of offering (1) personalized door-to-door service, (2) shared service with routes determined by individual passengers, or (3) regular service along fairly well-defined routes, similar in some respects to bus transit. The general service characteristics of paratransit modes are shown in Table 10-3. Notice that the private automobile and conventional transit are shown on the extreme left and right, respectively. The comparison is self-explanatory. Generally, operators of paratransit are free to choose vehicles, routes, frequency, and hours of operation, although fares may be regulated and certain congested routes barred to paratransit (Kirby et al., 1974; Armstrong-Wright, 1986).

5.2 Light-Rail Transit Systems

A *light-rail transit* (LRT) *system* is a generic term embracing a wide range of electrically powered vehicles running on steel rails. At one extreme are streetcars operating on tracks sharing the roadway with cars and buses; the other could include LRT metros

TABLE 10-3 GENERAL SERVICE CHARACTERISTICS BY MODE

	Private auto	Hire and drive services — Daily and short-term rental car	Hail or phone services — Taxi	Hail or phone services — Dial-a-ride	Paratransit modes — Jitney	Prearranged ride-sharing services — Car pool	Prearranged ride-sharing services — Subscription bus	Conventional transit
Direct route (DR) or route deviations (RD)?	DR →	DR →	DR →	RD	RD	RD	RD	RD
Door-to-door?	Yes →	Maybe →	Yes	Yes →	No →	Yes →	Maybe →	No
Travel time spent as passenger (P) or driver (D)?	D	D →	P	P	P →	P/D →	P →	P
Ride shared (S), or personal (P)?	P	P →	P/S	S	S	S	S	S
System routes fixed (F), semifixed (S), or variable (V)?	V	V	V →	V	S →	S	S →	F
Access determined by prior arrangement (A), fixed schedule (F), phone (P), street hailing (H), or at user's discretion (U)?	U	U →	H/P →	P →	H →	A →	A	F
Vehicle parking required (PR) or not (NP)?	PR	PR →	NP →	NP	NP →	PR →	PR/NP →	NP
Convenient for baggage?	Yes	Yes →	Yes →	Maybe →	Maybe →	Maybe →	Maybe →	No

Source: Kirby et al., 1974.

operating on exclusive rights-of-way. Passengers usually board from the road surface or from low platforms.

Streetcars (SCRs) or trams usually run on fixed rails flush with the roadway streets in mixed traffic. The vehicles carry about 100 to 300 sitting and standing passengers and are comparatively simple to operate and maintain. Routing is constrained by the alignment, and rerouting is expensive. SCRs operating in mixed traffic can carry from 4000 to 15,000 passengers per track per hour at journey speeds of about 12 to 20 km/hr.

LRT systems operate along streets, but may be provided with exclusive rights-of-way over all or part of their routes. Grade-separated and priority signalized routes at intersections are common. The system operates with trains of one to four cars. A typical LRT with two cars has a capacity of 500 passengers. LRT is intended to provide high-capacity service, fast and frequent. Technologically, LRTs are superior to streetcars where signaling and control is concerned.

Where LRT systems operate on exclusive rights-of-way, capacity can be as high as 20,000 passengers per hour per track at speeds of between 15 and 40 km/hr. Because of these facts, LRT has been adopted by a number of metropolitan areas across the world (Armstrong-Wright, 1986; Vuchic, 1981).

5.3 Rapid Rail Transit Systems

Rapid rail transit (RRT) *systems*, called *metros*, the *underground*, or the *tube*, operate on exclusive rights-of-way and at relatively high speeds and thus provide the highest line capacity available. Elevated and underground rights-of-way are common. Flat fares, zone fares, or distance-based fares are collected through automatic or other ticketing systems. Four to 10 cars per train is not uncommon. A typical train with, say, six cars may have a capacity of 1500 passengers, seated and standing. Operating at headways of 2 minutes and with a speed of 100 km/h, the line capacity can be 70,000 passengers per hour per line.

RRTs usually require sophisticated signaling and control devices to maintain high speeds and frequencies with very high safety standards. Rapid loading and unloading of passengers is achieved by providing high-level platforms. Ventilation systems and escalators are needed for underground operations. The costs of construction, maintenance, and operation are enormous and changes in routing are almost impossible.

To maximize patronage, it is generally necessary to supplement RRTs with feeder systems such as buses and LRTs and other flexible modes. The reliability of RRTs is very high, particularly when it is underground (Armstrong-Wright, 1986; Khisty and Kaftanski, 1981; Pushkarev and Zupan, 1977).

6. CLASSIFICATION OF URBAN TRANSPORT DEMANDS

Hutchinson (1974) has classified travel demand into five broad categories:

1. Trips along radial routes focused basically in the central business district (CBD)
2. Trips along circumferential routes

3. Trips between and within local areas

4. Trips within the CBD

5. Trips connecting major activity centers

Guenther (1971) has classified the spectrum of public transport system by routing and scheduling types (Figure 10-4). The diagram not only identifies conventional transit systems such as rail, bus, and taxi but puts demand-responsive systems in proper perspective. In recent years, demand-responsive systems have been promoted and adopted in many cities in the United States and Canada.

For purposes of definition, paratransit systems represent modes of semipublic and public transportation that are considered intermediate between private transportation and conventional transit. Table 10-4 shows a classification system of urban transport by type of usage. The taxi, for example, is a paratransit system. Demand-responsive systems are a subset of paratransit, because they change with the desires of individual users. Dial-a-ride, for example, is a demand-responsive system.

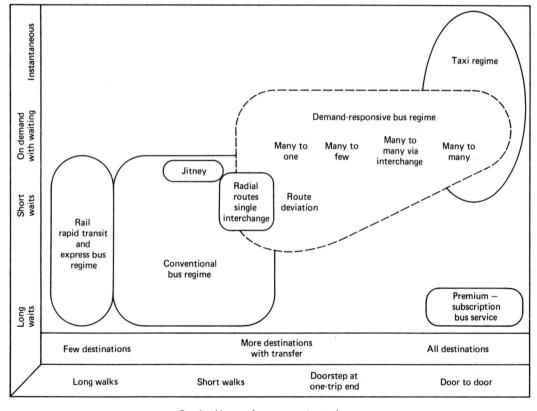

Figure 10-4   Public Transport System Classified by Routing and Scheduling Types (Guenther, 1971).

TABLE 10-4 CLASSIFICATION OF PASSENGER TRANSPORT BY TYPE OF USAGE

Characteristic \ Usage type	Private — Individual	Private — Group	For-hire — Individual	For-hire — Group	Public or common carrier
Common designation	Private transport		Paratransit		Transit
Service availability	Owner		Public		Public
Service supplier	User		Carrier		Carrier
Route determination	User (flexible)		User	User (carrier)	Carrier (fixed)
Time-schedule determination	User (flexible)		User	User (carrier)	Carrier (fixed)
Cost-price	User absorbs		Fixed rate		Fixed fare
Carrier type	Individual			Group	
Modes	Automobile; Motorcycle; Bicycle; Walking	Carpools; Vanpools	Taxi; Rented car	Dial-a-ride; (Jitney); Charter bus	Street transit (bus, trolleybus, streetcar); Semirapid transit (semirapid bus, light-rail transit); Rapid transit (rail, rubber-tired, regional rail); Special and proposed modes
Optimum (but not exclusive) domain of operation: Area density	Low-medium	Origin: low / Destination: high	Low		High-medium
Routing	Dispersed	Radial	Dispersed		Concentrated (radial)
Time	Off-peak	Peak only	All times		Peak
Trip purposes	Recreation, shopping, business	Work only	Business		Work, school, business

Source: Vuchic, 1981. Reproduced by permission of the publisher.

7. CAPACITY AND LEVEL OF SERVICE OF URBAN TRANSIT

7.1 Capacity of Urban Transit

The passenger-carrying capacity of an urban transit route is the maximum number of people that can be carried past a given location during a given period of time, under specified operating conditions, without unreasonable delay, hazard, or restriction, and with reasonable certainty. Passenger-carrying capacity is naturally a less precise measure than vehicle-carrying capacity. The latter is a measure of the maximum number of vehicles capable of passing a point during a given period of time multiplied by the maximum number of passengers that can be transported in each vehicle. For example, the person capacity of a freeway lane with bus and car traffic under prevailing conditions of flow can be estimated as

$$c_p = f' O_1 + [(1800 - 1.5f')O_2] \qquad (1)$$

where

f' = number of buses per hour
O_1 = bus occupancy
O_2 = car occupancy
c_p = person capacity (people/hr)

A freeway lane where $f' = 28$, $O_1 = 45$, and $O_2 = 1.25$ would have a capacity of

$$c_p = 28 \times 45 + [1800 - (28 \times 1.5)]1.25 = 3457 \text{ people/hr}$$

The same freeway approaching the downtown area, where the level of service has deteriorated, would naturally have a capacity quite different from the example because the maximum freeway capacity of 1800 veh/hr/ln (without buses) may not prevail (TRB, 1985).

Example 1

An urban freeway during the peak hour carries 5200 cars with an average vehicle occupancy of 1.25 persons and 35 buses with 45 passengers each. In addition, there are 90 vanpools carrying 10 passengers each (including the driver). Calculate the person flow. What percentage of passenger flow is represented by cars, vans, and buses?

Solution

	Veh/hr	Persons/veh	Persons/hr	Percent of vehicles	Percent of persons
Cars	5200	1.25	6500	97.65	72.42
Vanpools	90	10.00	900	1.69	10.02
Buses	35	45.00	1575	0.66	17.56
Total	5325		8975	100.0	100.00

The total person flow is 8975 in the peak hour. Cars, which represent 97.65% of the total vehicles, carry only 72.42% of the passengers. Vanpools and buses, which represent just 1.69% and 0.66% of the vehicles, carry 10.02% and 17.56% of the passengers, respectively.

The passenger capacity of a transit line depends on four major factors:

1. The maximum number of vehicles per transit unit (bus, car, train)
2. The passenger capacity of the individual transit vehicles
3. The minimum possible headway, or time spacing between individual vehicles or trains
4. The number of movement channels, loading positions, or station platform capacity

Factors 1, 2, and 4 are independent of one another. Factor 3 is influenced by the other three.

Just as street capacities are based on critical intersection along a route, transit capacities are generally governed by critical stops where passenger boarding or alighting takes place, or where vehicles terminate or turn around. Transit operators and planners recognize the complexity of determining transit capacity for a particular technology or system and are therefore content in calculating the realistic rate of flow that can be obtained.

7.2 Determinants of Capacity

Table 10-5 defines important terms that relate to transit capacity. Although it has been pointed out that passenger capacity is governed by four basic factors, there are many variables that influence these factors, and they are given in Table 10-6.

The capacity of a transit line can be estimated from the following equations:

$$c_v = \frac{3600R}{h} = \frac{3600R}{D + t_c} \tag{2}$$

$$c_p = nSc_v = \frac{3600nSR}{D + t_c} \tag{3}$$

where

c_v = vehicles/hr per channel (maximum)
c_p = people/hr per channel (maximum)
h = headway between successive vehicles (sec)
t_c = clearance between successive vehicles (sec)
D = dwell time at major stops (sec)
S = passengers per vehicle
n = vehicles per unit ($n = 1$ for buses; $n = 1$ to 11 for rail)

TABLE 10-5 IMPORTANT TERMS IN TRANSIT CAPACITY

Clearance time: all time losses at a stop other than passenger dwell times, in seconds. It can be viewed as the minimum time, in seconds, between one transit vehicle leaving a stop and the following vehicle entering (i.e., the clearance time between successive buses should not be less than 15 sec).

Crush capacity: the maximum number of passengers that can be physically accommodated on a transit vehicle. It is also defined as level of service F. It can be viewed as an "offered" capacity, because it cannot be achieved on all vehicles for any sustained period of time.

Dwell time: the time, in seconds, that a transit vehicle is stopped for the purpose of serving passengers. It includes the total passenger service time plus the time needed to open and close doors.

Interrupted flow: transit vehicles moving along a roadway or track and having to make service stops at regular intervals.

Maximum load point: the point, actually section, along a transit route at which the greatest number of passengers is being carried.

Passenger service time: the time, in seconds, that is required for a passenger to board or alight from a transit vehicle.

Person capacity: the maximum number of persons that can be carried past a given location during a given time period under specified operating conditions without unreasonable delay, hazard, or restriction. Usually measured in terms of persons per hour.

Person level of service: the quality of service offered the passenger within a transit vehicle, as determined by the available space per passenger.

Productive capacity: a measure of efficiency or performance. The product of passenger capacity along a transit line and speed.

Seat capacity: the number of passenger seats on a transit vehicle.

Standees: the number of standing passengers on a transit vehicle. The ratio of total passengers carried to the number of seats during a specified time period is called the *load factor*. The *percent standees* represents the number of standing passengers expressed as a percentage of the number of seats. A transit vehicle with 40 seats and 60 passengers has a load factor of 1.5 and 50% standees.

Uninterrupted flow: transit vehicles moving along a roadway or track without stopping. This term is most applicable to transit service on freeways or on its own right-of-way.

Source: TRB, 1985.

$$R = \text{reduction factor for dwell time and arrival variation}$$
$$= 0.833 \text{ for bus operation on city streets}$$

If g = green time (sec) and C = cycle time (sec) for transit running on signalized streets, then

$$c_p = \frac{(g/C)3600nSR}{(g/C)D + t_c} \tag{4}$$

Example 2

A three-car light-rail transit line operates on city streets through signalized intersections. The $g/C = 0.45$ and the passenger dwell time is 90 sec. How many persons can the train carry? (Assume 170 passengers/car; train clearance time is 50 sec.)

TABLE 10-6 FACTORS THAT INFLUENCE TRANSIT CAPACITY

Vehicle characteristics:
  Allowable number of vehicles per transit unit (i.e., single unit bus, or several units-cars per train)
  Vehicle dimensions
  Seating configuration and capacity
  Number, location, width of doors
  Number and height of steps
  Maximum speed
  Acceleration and deceleration rates
  Type of door actuation control

Right-of-way characteristics:
  Cross-section design (i.e., number of lanes or tracks)
  Degree of separation from other traffic
  Intersection design (at grade or grade separated, type of traffic controls)
  Horizontal and vertical alignment

Stop characteristics:
  Spacing (frequency) and duration
  Design (on-line or off-line)
  Platform height (high-level or low-level loading)
  Number and length of loading positions
  Method of fare collection (prepayments, pay when entering vehicle; pay when leaving vehicle)
  Type of fare (single-coin, penny, exact)
  Common or separate areas for passenger boarding and alighting
  Passenger accessibility to stops

Operating characteristics:
  Intercity versus suburban operations at terminals
  Layover and schedule adjustment practices
  Time losses to obtain clock headways or provide driver relief
  Regularity of arrivals at a given stop

Passenger traffic characteristics:
  Passenger concentrations and distribution at major stops
  Peaking of ridership (i.e., peak-hour factor)

Street traffic characteristics:
  Volume and nature of other traffic (on shared right-of-way)
  Cross traffic at intersections if at grade

Method of headway control:
  Automatic or by driver trainman
  Policy spacing between vehicles

Source: TRB, 1985.

Solution

$$c_p = \frac{(g/C)3600nSR}{(g/C)D + t_c}$$

$g/C = 0.45$, $n = 3$, $S = 170$, $R = 0.833$, $D = 90$, and $t_c = 50$

Therefore,

$$c_p = \frac{0.45 \times 3600 \times 3 \times 170 \times 0.833}{(0.45 \times 90) + 50} = 7604 \text{ passengers}$$

Discussion

The train clearance time includes the minimum separation between two trains plus the time for a train to clear the stop. Usually, 20 to 30 seconds is the minimum headway between trains. The length of the train is, say, $75 \times 3 = 225$ ft. Assuming that the train accelerates from rest to 20 mph (30 ft/sec), the average speed is 15 ft/sec and therefore about $225/15 = 15$ seconds is needed for clearance. Total clearance is $30 + 15 = 45$ seconds. The assumption of 50 seconds is acceptable. The passenger capacity of about 7600 passengers/hr is based on the assumption that there are two major stops, passengers pay fares by prepayment, and that the alighting and boarding time is not excessive.

The capacity of a rail line is determined by station capacity or way capacity, whichever is smaller; usually, station capacity governs. Capacity depends on (1) car size and train-station length, (2) allowable standees as determined by scheduling policy, and (3) minimum headway (spacing) between trains. The minimum headway is a function not only of dwell time at major stations, but also train length, acceleration and deceleration ratio, and train control systems. Theoretical approaches of calculating the minimum spacing are used, and these have been discussed briefly in Chapter 5. However, a more general practice is to obtain the minimum spacing between trains based on actual experience, station dwell times, and signal control systems.

Passenger capacity in the peak direction during peak hours can be estimated from the following equations:

$$\text{passengers/hr} = \frac{\text{trains}}{\text{hr}} \times \frac{\text{cars}}{\text{train}} \times \frac{\text{seats}}{\text{car}} \times \frac{\text{passengers}}{\text{seat}} \tag{5}$$

or

$$\text{passengers/hr} = \frac{\text{cars}}{\text{hr}} \times \frac{\text{seats}}{\text{car}} \times \frac{\text{passengers}}{\text{seat}} \tag{6}$$

or basing the equations on allowable levels of passenger space:

$$\text{passengers/hr} = \frac{\text{trains}}{\text{hr}} \times \frac{\text{cars}}{\text{train}} \times \frac{\text{ft}^2}{\text{car}} \div \frac{\text{ft}^2}{\text{passenger}} \tag{7}$$

Equation 7 derives passenger capacity that is independent of the seating configuration and is related directly to the area of the car.

Example 3

A rail rapid transit operates 10 four-car trains per track per hour. Schedule loads average 1.90 passengers per seat. How many people can the line carry? Cars are 80 ft long and can seat 80 people.

Solution

$$\text{Passengers/hr} = \frac{\text{trains}}{\text{hr}} \times \frac{\text{cars}}{\text{train}} \times \frac{\text{seats}}{\text{car}} \times \frac{\text{passengers}}{\text{seat}}$$

$$= 10 \times 4 \times 80 \times 1.9 = 6080 \text{ persons/hr}$$

The precise number of passengers/hr will vary depending on the type of equipment used, the arrangement of the seats, and the operating policy.

7.3 Level of Service

The level-of-service concept is much more complex to apply in the case of transit as compared to highways. Alter has suggested the use of six levels of service: basic accessibility, travel time, reliability, directness of service, frequency of service, and passenger density. All of these factors must be combined to evaluate a composite level of service (Yu, 1982, p. 213).

7.4 System Performance Evaluation

A variety of indicators describe efficiency and effectiveness in transit. Table 10-7 summarizes some of the important indicators. Two aspects of level of service are important from a capacity perspective: (1) the number of passengers per vehicle, and (2) the number of vehicles per hour. Capacity-related level-of-service criteria should reflect both. Figure 10-5 illustrates this two-dimensional nature of urban transit capacity.

8. OPERATIONAL DESIGN

8.1 Rail Operation Design

The purpose of this section is to give an elementary idea of how the concepts of capacity, volume, headway, and safety considerations are used in rail and bus operation design. We have seen that volume may be defined as the number of vehicles passing a fixed point on the guideway in a unit of time. Volume is related to headway given by

$$V = \frac{3600}{h} \qquad (8)$$

Similarly, capacity is related to headway as given by the equation

$$c_v = \frac{3600}{h_m} \qquad (9)$$

where c_v, is the theoretical vehicular capacity or maximum volume (veh/hr), and h_m is the minimum headway (sec). Theoretical passenger capacity is given by

$$c_p = pNc_v = \frac{3600pN}{h_m} \qquad (10)$$

where

  c_p = theoretical passenger line capacity (number of passengers)
  p = vehicles/train
  N = maximum passenger per vehicle

TABLE 10-7   SYSTEM PERFORMANCE INDICATORS

Indicator	Construction	Focus
	Efficiency	
Operating cost per revenue vehicle-mile	$\dfrac{\text{Total operating costs}}{\text{Total revenue vehicle-miles}}$	Cost efficiency
Operating cost per revenue vehicle-hours	$\dfrac{\text{Total operating costs}}{\text{Total revenue vehicle-hours}}$	
Revenue vehicle-hours per employee	$\dfrac{\text{Total revenue vehicle-hours}}{\text{Total system employees}}$	Labor productivity
Revenue vehicle-miles per employee	$\dfrac{\text{Total revenue vehicle-miles}}{\text{Total employees}}$	
Revenue miles per vehicle	$\dfrac{\text{Total vehicle miles}}{\text{Total vehicles}}$	
Revenue hours per vehicle	$\dfrac{\text{Total revenue vehicle hours}}{\text{Total revenue vehicles}}$	Vehicle utilization
Total passengers per vehicle	$\dfrac{\text{Total passengers}}{\text{Total vehicles}}$	
Energy consumption per revenue vehicle-mile	$\dfrac{\text{Total energy consumption}}{\text{Total revenue vehicle-miles}}$	Energy efficiency
Energy consumption per revenue vehicle-hour	$\dfrac{\text{Total energy consumption}}{\text{Total revenue vehicle-hours}}$	
	Effectiveness	
Revenue passengers per revenue vehicle-mile	$\dfrac{\text{Total revenue passengers}}{\text{Total revenue vehicle miles}}$	Service utilization measures
Revenue passengers per service area population	$\dfrac{\text{Total passenger trips}}{\text{Total population of service area}}$	
Percentage of population served	$\dfrac{\text{Total service area population}}{\text{Total coverage area population}}$	Accessibility
System reliability	$\dfrac{\text{Total trips on time}}{\text{Total trips}}$	Quality of service
Vehicle revenue miles per square mile of served area	$\dfrac{\text{Total vehicle revenue miles}}{\text{Total square miles of served area}}$	
	Efficiency effectiveness	
Operating ratio	$\dfrac{\text{Total operating revenues}}{\text{Total operating costs}}$	Financial performance
Cost per passenger trip	$\dfrac{\text{Total operating costs}}{\text{Total passenger trips}}$	
Cost per passenger-mile	$\dfrac{\text{Total operating costs}}{\text{Total passenger miles}}$	

Source: Yu, 1982.

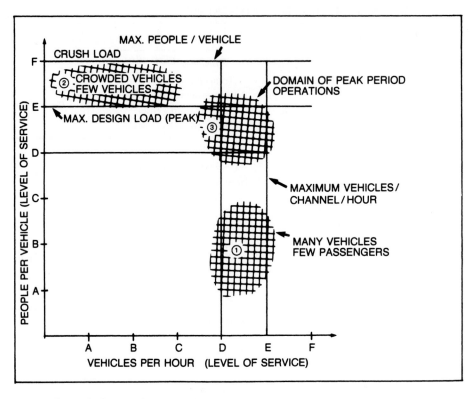

Figure 10-5   Two-Dimensional Nature of Transit Travel Level of Service as Related to Transit Capacity (TRB, 1985).

If practical vehicular capacities are to be considered, the ratio of practical to theoretical vehicular line capacities is introduced. This ratio, called the *guideway utilization factor*, is denoted by α. Therefore, actual vehicle capacity is given by

$$c_a = \frac{3600\alpha}{h_m} \tag{11}$$

A load factor is usually used to express the percentage of vehicle occupancy; hence,

$$\text{Actual passenger capacity} = \frac{3600 \times \sigma \times \alpha \times p \times N}{h_m} \tag{12}$$

where σ is the load factor. When the load factor $\sigma = 1$, it denotes that the vehicle is fully occupied. The maximum number of passengers who can theoretically be squeezed into a vehicle is called its *crush load*, and therefore the load factor can exceed 1.00 during, say, rush hours.

     In rail design particularly, safe stopping distance is a major concern. A safety factor is used for safe design on the *brick-wall-stop (BWS) concept*. Say, for example, that when the lead vehicle on a track stops instantaneously, the following vehicle must be able to stop safely, with a factor of safety of K. K can be assumed to be 1.5. The following vehicle in such cases is considered to stop with constant deceleration. Based on this principle,

$$\text{Minimum headway, } h_m = \frac{Kv_o}{2d} + \frac{pL}{v_o} \tag{13}$$

where

$v_o =$ cruise speed (ft/sec)
$L =$ vehicle length (ft)
$d =$ deceleration rate (ft/sec^2)

Therefore, we can now write the theoretical capacity equation as

$$c_p = \frac{3600pN}{Kv_o/2d + pL/v_o} \tag{14}$$

For maximum capacity, we differentiate c_p with respect to v_o and obtain

$$v_o = \sqrt{\frac{2pLd}{K}} \qquad \text{for maximum capacity} \tag{15}$$

and if we substitute v_o in Eq. 13, we get

$$h_m = \sqrt{\frac{2pLK}{d}} \tag{16}$$

and the maximum theoretical capacity $= 2546\,N\sqrt{pd/LK}$.

The basic line-capacity equation tells us that if we wish to increase the line capacity, we could achieve it in one of five ways (Yu, 1982):

1. Increase the number of passengers carried by each vehicle.
2. Increase the length of the trains.
3. Decrease the minimum allowable headway.
4. Improve the load factor.
5. Improve the guideway utilization.

Example 4: Rail Operation Design

A transit authority needs to design a rapid rail line to meet peak-hour demand of 10,000 passengers per hour, with a required speed of 35 to 40 ft/sec (24 to 27 mph). The following assumptions are made: deceleration 2 ft/sec^2, safety factor $K = 1.35$; minimum headway = 120 sec; maximum headway = 240 sec; load factor = 0.9; guideway utilization factor = 0.6, station platform limit = 10 vehicles (maximum); car length = 70 ft; car capacity = 130 passengers. How many cars should a train consist of to provide adequate passenger volume capacity? What will be the corresponding headway?

Solution

1. Determine headway.

$$c_x = 3600 \times \alpha \times \sigma \times p_x \times \frac{N}{h_s}$$

$$10{,}000 = 3600 \times 0.6 \times 0.9 \times p_x \times \frac{130}{h_x}$$

Therefore, $p_x = 0.03937 h_x$

p_x (veh/train)	h_x (headways; sec)	
1	25.27	
2	50.54	
3	75.82	
4	101.09	———————
5	126.36	Possible range
6	151.63	
7	176.91	(min. h_x = 120 sec)
8	202.18	(max. h_x = 240 sec)
9	227.45	———————
10	252.72	

2. Examine computed headways and train size. From the brick-wall-stop (BWS) concept:

$$v_0 = \sqrt{\frac{2pLd}{K}} \quad \text{and} \quad h_x = \sqrt{\frac{2pLK}{d}}$$

Number of cars per train, p	Speed, v_0	Time headways, h_0 (sec)	
		Computed (h_x)	BWS (h_x)
1	14.4	25.27	9.72
2	20.3	50.54	13.75
3	24.8	75.82	16.84
4	28.8	101.09	19.44
5	32.2	126.36[a]	21.73
6	35.2[a]	151.63[a]	23.81
7	38.1[a]	176.91[a]	25.71
8	40.7	202.18[a]	27.49
9	43.1	227.65[a]	29.20

[a] Acceptable.

3. Evaluate. Examination of the preceding table should be based on three criteria: (a) computed speed should be in the range 35 to 40 ft/sec; (b) minimum headway = 120 sec; (c) BWS h_0 should be less than 120 sec.

4. *Conclusion.* Six- or seven-car trains are all right.

Six-car train: speed 35.28 ft/sec, h_x = 151.63 sec
Seven-car train: speed 38.10 ft/sec, h_x = 176.91 sec

Discussion

These results are meant for peak-hour service. Naturally, for off-peak hours, the train lengths will be different, depending on what policy headways are needed.

8.2 Bus Operation Design

The design of a bus route is somewhat different from rail design operation and the differences will be evident from the description that follows. Ultimately, an operation plan would contain information regarding the adopted headway, cycle time, terminal time, fleet size, and the commercial speed.

Again, the capacity of a transit route is the product of the passenger capacity per vehicle and the maximum number of vehicles that can travel on that route. The last term is usually the capacity of the busiest stop on the route. An expression for the headway of a bus stop is

$$h_m = 2t_d/60 \tag{17}$$

where h_m is the minimum headway between buses (min), and t_d is the average dwell time (sec). Here dwell time is the total time spent by a bus at a stop. Dwell time is calculated from one of the following formulas:

$$t_d = \begin{cases} aA + bB + C & \text{(for two-way flow through busiest door)} \\ aA + C & \text{(for one-way flow, alighting)} \\ bB + C & \text{(for one-way flow, boarding)} \end{cases} \tag{18}$$

where

  a, b = average alighting and boarding time per passenger in seconds, respectively; a = 1.5 to 2.0 sec, b = 2.5 to 3.5 sec

  A, B = number of alighting and boarding passengers, respectively

  C = clearance time = lost time in opening and closing doors, or to traffic delays when bus is ready to leave; C is usually 15 sec

The frequency of service is given by

$$f = \frac{n}{N} \tag{19}$$

where

  f = frequency (buses/hr) required

  n = demand for service (passengers/hr)

  N = maximum number of passengers per bus

Usually, the bus company decides the minimum headway, and this figure is set in multiples of 7.5 or 10 minutes for the sake of coordinating the operation of several bus routes operating.

The capacity of a bus route is governed by four factors: the street capacity, the bus station platform capacity, the vehicle capacity, and the headway. Each of the first three factors is independent of one another, and the headway is influenced by all three. Vehicle capacity depends on two factors: seating capacity and standing capacity. A load factor is often used to measure seat availability, and a load factor of 1.0 indicates that every seat is occupied.

The passenger capacity of a bus is given by

$$c_t = c_a + \alpha c_b \tag{20}$$

where

c_t = total passenger capacity per vehicle
c_a = vehicle seating capacity
c_b = vehicle standing capacity
α = fraction of c_b allowed

Hence, capacity R_c of a bus routing during any time period is

$$R_c = \frac{60c_t}{h_m} = \frac{60(c_a + \alpha c_b)}{h_m} \tag{21}$$

The fleet size, or the number of vehicles needed to serve a particular route, can be determined, based on the time it takes a bus to complete a round trip. Thus,

$$t_R = \frac{d}{v_c} \tag{22}$$

where

t_R = round-trip travel (hr)
d = distance of a round trip (miles or km)
v_c = average vehicle speed or commercial speed (mph or km/hr)

A minimum layover and recovery time (say, 10 minutes) is provided at the end of each round trip. The number of vehicles needed (fleet size) can be determined from

$$N_f = St_R = \frac{t_R}{h} \tag{23}$$

where N_f is the fleet size, and S is the service frequency = $1/h$.

Example 5

A bus system needs to be set up between the Washington State University Campus and the University of Idaho, a distance of 8.5 miles. The operating time is 30 minutes. It has been estimated that the peak-hour demand is 400 passengers/hr and 45-seater buses are available, which can safely accommodate 20 standees. Design the basic system and determine the fleet size, assuming that the policy headway is 30 minutes and that the minimum terminal time is 7.5 minutes, which may be revised if necessary.

Solution

Operating speed, $v_0 = 60L/t_o = 60 \times 8.5/30 = 17.0$ mph; t_o = operating time

Policy headway = 30 min (which is arbitrary)

Terminal time = 7.5 min

Headway, $h_{min} = 60c_t/R_c = \dfrac{60(45 + 20)}{400} = 9.75$ min (adopt 10 min)

Cycle time, $T = 2(t_0 + t_t) = 2(30 + 7.5) = 75$ min

Fleet size, $N_f = T/h = 75/10 = 7.5 = 8$ vehicles

Revised cycle time, $T' = N_f h = 8 \times 10 = 80$ min

Revised terminal time, $t_t' = (T' - 2t_0)/2$

$$= [80 - (2 \times 30)]/2 = 10 \text{ min}$$

Commercial speed, $v_c = d/t_R = 120L/T' = 120(8.5)/80$

$$= 12.75 \text{ mph}$$

In summary,

Headway, $h = 10$ min

Cycle time, $T = 80$ min

Terminal time, $t_t = 10$ min

Fleet size, $N_f = 8$ vehicles

Commercial speeds, $v_c = 12.75$ mph

9. SCREENING OF TRANSPORTATION MODAL OPTIONS

9.1 Decision-Making Process

Before considering any improvements or additions to the existing transportation system, it is best to determine both its deficiencies and the possible opportunities for improvement. Any investment in a new system should be considered only after these potential opportunities have been thoroughly explored. In many cases, a range of highway and transit alternatives may have to be considered. Transportation engineers, planners, economists, and sociologists, along with potential decision makers and community leaders, are involved in choosing and making modal comparisons and identifying techniques for resolving the problems.

An indication that the existing system is deficient can be measured by the overcrowding on transit lines, excessive travel times by automobile and transit, excessive fares, high accident rates, inadequate transit equipment to handle the existing demand, poor travel operation, and severe surface street and freeway congestion.

A conceptual view of the decision-making process for transit facilities is shown in Figure 10-6. One of the most important factors is the resolution of issues. Tied to these issues is a set of objectives for evaluating the investment likely to be used for transit improvements and expansions. A typical list of these objectives is given in Table 10-8.

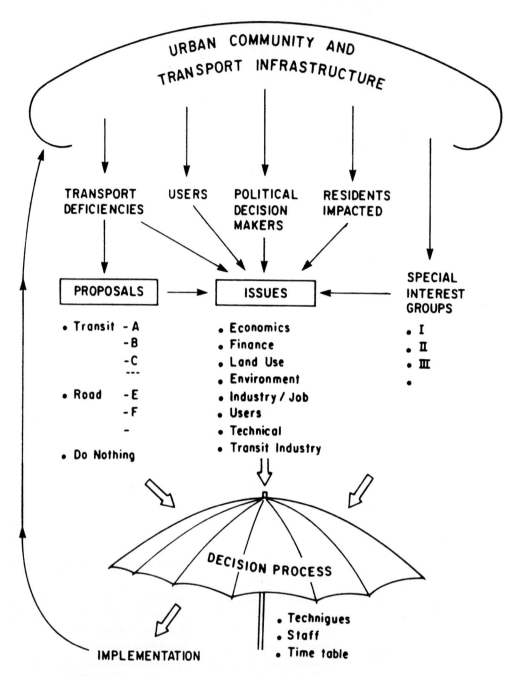

Figure 10-6   Conceptional View of the Decision-Making Process (CUTA/RTAC, 1985).

TABLE 10-8 TYPICAL LIST OF OBJECTIVES
FOR EVALUATING TRANSIT INVESTMENTS

Measures of community objectives:
  Air pollution
  Land consumption
  Differential accessibility
  Differential unit cost of travel
  Transportation expenditure
Measures of nonuser objectives:
  Market accessibility
  Residential dislocation
  Job dislocation
  Noise impacts
  Neighborhood disruption
  Transportation facility compatibility
Measures of user objectives:
  Personal accessibility
  Total travel time
  Excess time
  Congestion
  Total travel cost
  Direct travel cost
  Coverage
  Traveler satisfaction
  Accident rate
  Criminal activity
Measures of operator objectives:
  Gross revenues
  Operating costs
  Capital costs
  Productivity
  Facility utilization

Source: Peat, Marwick, Mitchell & Co., 1973.

9.2 Screening Process

The main purpose in screening options is to determine which of the options may be worthy of a detailed feasibility study and at the same time to determine the appropriateness of systems that already exist or are now being proposed and recommended. As a first step, it is necessary to gain some indication of the level of demand that can be expected in the future. For cities where a transportation study has been done, it is easy to predict the probable share of demand to be satisfied by the proposed transit system. Once the range of demand along the main corridors has been determined, it is possible to consider the systems that are likely to be able to cope with this demand. Combinations of alternatives for a transit system and alternative modes for various population ranges are shown in Figure 10-7 (CUTA/RTAC, 1985).

COMBINATIONS OF ALTERNATIVES FOR A TRANSIT SYSTEM

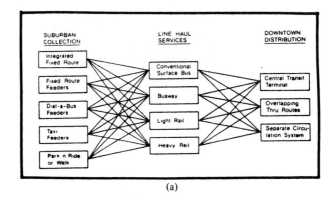

(a)

ALTERNATIVE TRANSIT MODES FOR VARIOUS POPULATION RANGES

		10 - 50,000	50 - 250,000	250 - 750,000	750 - 2,000,000	2,000,000+
SUBURBAN COLLECTION	Satellite Areas	Vanpooling	Subscription Buses	Subscription & Express Buses	Express Buses	Commuter Rail
	Low Density Areas	Taxis and Vanpooling	Dial-a-Bus Vans and Taxis	Subscription Dial-a-Bus Small Buses	Dial-a-Bus or Fixed Route Feeders	Dial-a-Bus or Fixed Route Feeders
	Medium Density Areas		Dial-a-Bus, Fixed Route Feeders or Integrated Rtes	Dial-a-Bus, Fixed Route Feeders or Integrated Rtes	Fixed Route Feeders or Integrated Routes	Mainly Integrated Routes
LINE HAUL SERVICES		Dial-a-Bus or Fixed Route Services using Buses, Vans or Taxis	Fixed Route Buses Express Service in the peak	Fixed Route Buses. Busways. Light Rail in heavy corridors	Fixed Route Buses. Busways, Light Rail, Heavy Rail, PRT or ICTS	Fixed Route Buses. Busways, Light Rail, Heavy Rail, PRT or ICTS
DOWNTOWN DISTRIBUTION			Overlapping Routes	Overlapping Routes or Separate System	Overlapping Routes or Separate System	Central Terminal. Thru Routes or Separate System

(b)

Figure 10-7 Combinations of Alternative Modes for a Transit System (CUTA/RTAC, 1985).

9.3 Costing Process

The cost characteristics of transit systems vary considerably. For example, the operating costs of bus systems are high in comparison to capital costs, in the ratio 5:1, whereas the opposite is true for, say, underground rail systems (1:3). Operating costs, in turn, are greatly influenced by labor, energy, and material costs. Capital costs are related to the useful lives of vehicles and the infrastructure: 8 to 15 years for buses, 30 to 40 years for rail cars, and 100 years for tunnels. In calculating comparative costs, the operating costs of each system and the capital cost in terms of annual depreciation and interest changes are examined. The cost effectiveness of the options under consideration can then be compared by expressing total costs in terms of passenger-miles or passenger-kilometers. To sum up, the following procedure provides a quick way of determining transit costs in just enough detail to permit broad comparison of the options.

1. The characteristics of each system are examined: type, demand, capacity, performance.
2. Approximation of operating costs can be calculated by applying unit rates for the following:
   (a) Distance-related costs (energy, maintenance, servicing of vehicles, and so on, in terms of vehicle-miles)
   (b) Time-related costs (operating staff wages, and other costs in terms of total number of hours run by the fleet of vehicles in vehicle-hours)
   (c) Route-related costs (maintenance of roadway, track, signals, stations, in terms of cost per mile per day)
   Distance- and time-related costs are called *variable costs*. Approximate operating unit costs are given in Table 10-9.
3. Approximation of capital costs are annualized and represent depreciation and interest changes, and are calculated on the following basis:
   (a) Each category of capital costs is assumed to be financed by a loan for a term equal to its useful life
   (b) An interest rate is assumed (say, 6%)
   (c) Constant annual payments are made on the loan and are calculated by annualizing the cost of each element, using conventional tables
4. The cost effectiveness of various systems is compared by expressing total costs in terms of cost per passenger-mile.

Example 6

A light-rail system is proposed for a linear city about 1 million population. The track will run 75% grade-separated and 25% through regular street intersections. The following details apply:

Route length = 15 miles
Spacing of stop/station = 0.3 mile
Operating hours/day = 18

TABLE 10-9 APPROXIMATE OPERATING UNIT COSTS (DOLLARS)

Cost	Bus	LRT	RRT
Distance cost (car-mile)	0.70	2.00	2.00
Time cost (car-hour)	16	10	11
Route cost (per mile of route/day)	25	320	1200

Operating days/years = 365

Average trip length = 3.75 miles

Journey speed = 12.5 mph

Peak-period operation = 3 hours

Passengers/day = 500,000

Average hourly boardings (peak = 12%) = 60,000

Heaviest flow in one direction (peak) = 24,000 on busiest section

Headway = 120 sec (peak); 240 sec (off-peak)

Capacity/train = 900 passengers

Capacity/car = 225 passengers

Load factor = 90%

Capital costs:

Segregated ROW: 9.16 million/mi; life 40 years

Track (double): 3.33 million/mi; life 30 years

Signals: 1.67 million/mile; life 30 years

Power: 5.00 million/mile; life 30 years

Stations (stops): 0.15 million each; life 40 years

Yards (2): 12.5 million each; life 40 years

Workshops (1): 25.0 million each; life 40 years

Rolling stock: 0.80 million each; life 25 years

Interest rate = 6% per annum

Operating costs:

Distance cost per car-mile = $2

Time cost per car-hour = $9

Route costs per mile per day = $375

What is the cost per passenger-mile?

Solution

Demand:

1. Daily passenger boardings:

$$\text{Peak: 3 hours} \times 60{,}000 = 180{,}000 \text{ in 3 hours}$$

$$= 60{,}000 \text{ passengers/hr}$$

$$\text{Off-peak: } 500{,}000 - 180{,}000 = 320{,}000 \text{ in 15 hours}$$

$$= 21{,}333 \text{ passengers/hr}$$

2. Heaviest flow:

Peak: 24,000; off-peak: 12,000 (assuming 50% peak-hour)

Vehicle requirements:

3. Hourly capacity (90% loading) = line 2 ÷ 90%

$$\text{Peak: } \frac{24{,}000}{0.90} = 27{,}000$$

$$\text{Off-peak: } \frac{12{,}000}{0.90} = 13{,}000$$

4. Headways: peak = 120 sec; off-peak = 240 sec

5. Frequency (trains/hr): peak: 30; off-peak: 15

6. Capacity/train: 900

7. Capacity/car: 225

8. Cars/train: 4

9. Cars/hr: peak = 120; off-peak = 60

10. Round-trip time (including stopover of 21 min):

$$2 \text{ hr, } 24 \text{ min} + 21 \text{ min} = 2 \text{ hr, } 45 \text{ min} = 2.75 \text{ hours}$$

11. Fleet size: 120 cars/hr × 2.75 hours ÷ 90% = 367 cars (90% availability)

12. Car-miles/day:

$$\text{Peak: } 120 \text{ cars/hr} \times 3 \text{ hours/day} \times 30 \text{ mi} = 10{,}800$$

$$\text{Off-peak: } 60 \text{ cars/hr} \times 15 \text{ hours/day} \times 30 \text{ mi} = 27{,}000$$

$$\text{Total} = 37{,}800 \text{ car-miles/day}$$

13. Train operating hours/day:

$$\text{Peak: } 30 \text{ trains/hr} \times 3 \text{ hours/day} = 90 \text{ trains/day}$$

$$\text{Off-peak: } 15 \text{ trains/hr} \times 15 \text{ hours/day} = 225 \text{ trains/day}$$

$$\text{Total} = 315 \text{ trains/day} \times 2.75 \text{ hours} = 866 \text{ train-hours/day}$$

14. Car operating hours/day:

$$866 \text{ train-hours/day} \times 4 \text{ cars/train} = 3464 \text{ car-hours/day}$$

Costs:

15. Total capital cost (in millions) = 637.3

$$\text{Annual cost} = \frac{\text{principal} \times i\%}{1 - (1 + i\%)^{-n}} \qquad \text{where } i = 6\%$$

16. Annualized capital cost = \$46.8 million

17. Annual operating costs:

$$\text{Daily distance cost: } 37{,}800 \text{ car-miles} \times \$2.00 = \$75{,}600$$

$$\text{Daily time cost: } 3464 \text{ car-hours} \times \$9.00 = \$31{,}176$$

$$\text{Daily route cost: } 15 \text{ route-miles} \times \$375.00 = \$5{,}625$$

$$\text{Total} = \$112{,}401$$

$$\text{Annual operating cost} = \$112{,}400 \times 365 = \$41.03 \text{ million}$$

Element	Unit cost	Cost (millions of dollars)	Life	Annual cost (millions of dollars)
Segregated ROW (15 mi)	$9.16/mi	137.5	40	9.14
Track (15 mi)	3.33/mi	50.0	30	3.63
Signals (15 mi)	1.67/mi	25.0	30	1.82
Power (15 mi)	5.00/mi	75.0	30	5.45
Stations/stops (45 + 5)	@0.15	7.5	40	0.50
Yards (2)	@12.5	25.0	40	1.66
Workshop (1)	@25.0	25.0	40	1.66
Rolling stock (367)	@0.80	293.6	25	22.96
Total		638.6		46.8

18. Total annual cost = 46.8 + 41.03 = \$87.83 million
19. Annual passenger-miles = 500,000 × 3.75 × 365
    = 685 million
20. Cost/passenger-mile = 87.83/685 = \$0.128

9.4 Impact Estimation

A fundamental element of the decision-making process revolves around the estimation of impacts. Figure 10-8 shows a flowchart of the subcomponents of the impact analysis. This is quite a major undertaking, consuming a lot of effort, time, and money. The minimum level of analysis for a transit investment would consist of ridership and costs. These would then be combined with an economic analysis. Transit proposals are compared to highway/street improvement alternatives in which consideration of user and nonuser impacts are included for a fair comparison. A variety of ranking methods are available for selecting the "best" alternative.

10. ROUTES AND NETWORKS

Public transportation is a collection and distribution process and the shape and form of transit networks reflect the demand patterns prevalent. Transit users can be classified into two groups: captive riders and choice riders. The *captive rider* market consists of people who must operate without the use of a car. Even the choice of their home and work location is dictated because of the existence of a transit line. *Choice riders* are those who use transit because it is cheaper, faster, and more convenient as compared to their automobiles. This group uses transit almost exclusively for CBD-oriented work trips, because car licensing, maintenance, and parking, coupled with congestion, can add up to frustration and expense. Transit routes using buses can be classified in many ways: service type, network type, and hierarchical type, and these are described next.

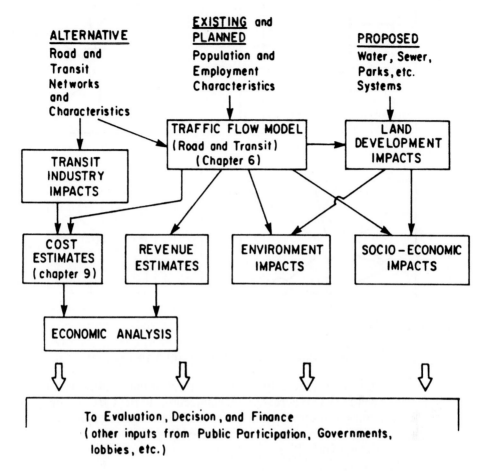

Figure 10-8   Components of Impact Estimation (CUTA/RTAC, 1985).

10.1 Service Type

Fixed routes (no deviation) are used where service is fixed in both time and space and where high capacity is required. Fixed routes with special-purpose deviation are used where the driver is allowed to deviate for certain passenger groups, such as the elderly and handicapped. Deviation may not be allowed during rush hours.

Demand-responsive routes (complete deviation) are those where no formal routes are designated, drivers responding to requests along their paths. Low vehicle utilization, high costs, and large delays to passengers have been observed in such routing.

10.2 Network Type

Grid networks [Figure 10-9(a)] follow the existing street network, with routes passing through the central business district being dominant. Some trips require transfers from one route to another. Most routes are straightforward.

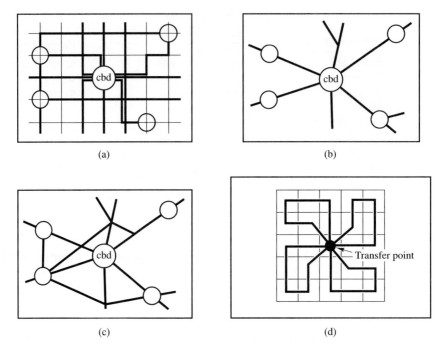

Figure 10-9   Network Types: (a) Grid Networks; (b) Radial Networks; (c) Circumferential Connections; (d) Territorial Networks (CUTA/RTAC, 1985).

Radial networks [Figure 10-9(b)] are typical of many cities with routes fanning out from the center city. Whereas direct connections to the CBD are obvious from the suburbs, the connections from suburb to suburb may be problematic unless circumferential or cross-town connections are established [Figure 10-9(c)].

Territorial networks [Figure 10-9(d)] are used in small and medium-sized towns. Routes are laid out to give good coverage to every territory. In larger cities, such networks can be used in the suburban areas with a timed transfer focal point. Trip generators such as community centers and shopping and recreation centers are good focal points (Carter and Homburger, 1978).

10.3 Network Evaluation

The basic objectives in designing a transit network are to provide a level of personal mobility for captive and choice riders, to encourage an efficient utilization of the transit system, and to ensure that transit and highway improvements reflect the overall community policies concerning mobility and environmental qualities. Evaluations of alternative networks depend heavily on usage of the system and this is measured by the overall patronage and the annual rides per capita.

11. PLANNING CONCERNS AND PLANNING GUIDELINES

Much effort has been expended by several researchers and practitioners to synthesize data, information, and attributes connected with public transportation to decide which modal options should be considered, and those that are trivial and unwarranted. In the final analysis, meaningful answers to some basic questions must be available to decision makers as well as to the public. A sample of such questions follow:

1. What are the transportation problems currently existing in the city, and can these be mitigated by some form of public transportation system, in case one does not currently exist? What types of transit are suitable: buses, rail, light rail, and so on?

2. If a public transportation system does exist, can improvements be made to the system to mitigate problems, without requiring major transit investments? Or is there a need to consider introducing another transit mode to supplement the one in existence? What would be the likely benefits and adverse impacts associated with each alternative?

3. Considering the anticipated growth of the city, is it practical from a transit point of view to introduce a mode that may prove to be underutilized?

4. Is intermodalism being thoroughly considered with the various transit options proposed?

Because the majority of public transportation options are highly capital-intensive, answers to these and other pertinent questions require thorough investigation, requiring realistic assessment of demands, costs, benefits, and revenues. Dean and Pratt (1992) indicate three primary factors that ought to be carefully identified and studied before embarking on a transit project.

1. Financial and institutional factors that dictate the constraints within which the system is to be planned, financed, designed, built, and made operational. Broad-based political and public support is essential if the transit project is to be a success.

2. Attitudinal factors that exist in the community independent of the plan and/or the planning process for the transit system. It also makes economic sense when planning a multimillion dollar system to get the communities together, knowing that financial assistance is most likely to be sought for from different levels of government.

3. Physical and analytical factors that involve the physical layout of the system and the ridership it is supposed to serve. Interaction with other modes of transportation must also be critically examined. Guidelines and criteria related to the boundary conditions under which different modes of transit can operate efficiently must be used with great caution.

Planning of public transportation options for a city should reflect specific local circumstances: needs, goals, resources, finances, area, population, density, structure, CBD size, activity and intensity, and system development costs. Levinson and his colleagues (1975) provide a number of city attributes that can help to quickly identify crucial measurements of potential passenger demand that can justify rapid transit, as shown in Table 10-10. Pushkarev and Zupan (1977) have also set forth guidelines shown in Table 10-11, based on past experience, connecting transit modes to residential densities. They have worked out the corresponding minimum floor space in the central business district for four modes, and this is shown in Table 10-12.

TABLE 10-10 GENERAL CONDITIONS CONDUCIVE TO URBAN RAPID TRANSIT DEVELOPMENT

Primary determinants	Rail — Desired conditions for rail system development	Minimum conditions — Rail or busway[a]	Minimum conditions — Busway
1. Urban area population	2,000,000	1,000,000	750,000
2. Central city population[b]	700,000	500,000	400,000
3. Central city population[b] density, in people per square mile	14,000	10,000	5000
4. High-density corridor development	Extensive and clearly defined	Limited but defined	Limited but defined
5. CBD function	Regional	Regional or subregional	Regional or subregional
6. CBD floor space, in square feet	50,000,000	25,000,000	20,000,000
7. CBD employment	100,000	70,000	50,000
8. Daily CBD destinations per square mile	300,000	150,000	100,000
9. Daily CBD destinations per corridor	70,000	40,000	30,000
10. Peak hour cordon person movements leaving the CBD (four quadrants)	100,000	70,000	35,000

[a] May also apply to light-rail transit.
[b] "Effective central city"—central city and contiguously developed areas of comparable population density.
Source: Adapted from Levinson, Adams, and Hoey, 1975.

TABLE 10-11   TRANSIT MODES RELATED TO RESIDENTIAL DENSITY

Mode	Service	Minimum necessary residential density (dwelling units/acre)	Remarks
Dial-a-bus	Many origins to many destinations	6	Only if labor costs are not more than twice those of taxis
	Fixed destinations or subscription service	3.5 to 5	Lower figure if labor costs are twice those of taxis; higher if thrice those of taxis
Local bus	Minimum, $\frac{1}{2}$-mi route spacing, 20 buses/day	4	
	Intermediate, $\frac{1}{2}$-mi route spacing, 40 buses/day	7	Average, varies as a function of downtown size and distance from residential area to downtown
	Frequent, $\frac{1}{2}$-mi route spacing, 120 buses/day	15	
Express bus reached on foot	5 buses during 2-hr peak period	15 (average density over 2-mi^2 tributary area)	From 10 to 15 mi away to largest downtowns only
reached by auto	5 to 10 buses during 2-hr peak period	3 (average density over 20-mi^2 tributary area)	From 10 to 20 mi away to downtowns larger than 20 million ft^2 of nonresidential floor space
Light rail	5-min headways or better during peak hour	9 (average density for a corridor of 25 to 100 mi^2)	To downtown of 20 to 50 million ft^2 of nonresidential floor space
Rapid transit	5-min headways or better during peak hour	12 (average density for a corridor of 100 to 150 mi^2)	To downtown of larger than 50 million ft^2 of nonresidential floor space
Commuter rail	20 trains a day	1 to 2	Only to largest downtowns, if rail line exists

Source: Pushkarev and Zupan, 1977.

TABLE 10-12 MINIMUM FLOOR SPACE IN CENTRAL
BUSINESS DISTRICT

Mode	Millions of square feet
Commuter rail and	
rail transit	75
Light rail transit	35
Express bus	20
Local bus: 10-min service	18
30-min service	5–7

Source: Pushkarev and Zupan, 1977.

The number of riders needed to justify a major transit project depends on the expected daily ridership threshold volumes. An estimate of such passenger volumes can be obtained as follows:

$$V = pM/r \qquad (24)$$

where

V = passenger volume on line

M = length of line (miles)

r = average length of ride

and p = passenger-mile/mile of line

Table 10-13 provides daily ridership threshold volumes suggested for various types of construction and modes. Note that values are shown in terms of passenger miles of line (Pushkarev, Zupan, and Cumella, 1982).

As has been indicated before, bus transit service in North America has been one of the popular modes of public transportation. Table 10-14 shows suggested bus service planning guidelines for service patterns, service levels, new routes, and factors covering passenger comfort and safety (TRB, 1980).

TABLE 10-13 THRESHOLD VOLUMES FOR RAPID TRANSIT DEVELOPMENT[a]

Mode	Type of construction	Daily pass.-mi/mi of route
Rail rapid	Above ground	14,000
	One-third tunnel	17,000–24,000[b]
	All tunnel	24,000–42,000[b]
LRT	Low capital	4000
	Considerable grade separation	7000
	One-fifth in tunnel	13,500
	All tunnel	40,000
Downtown people mover	Above ground	12,000
	All tunnel	30,000

[a] Keyed to type of structure. Minimum service frequency, 8 min.
[b] Range reflects varying criteria for cost/weekday passenger-mi of travel.
Source: Adapted from Pushkarev, Zupan, and Cumella, 1982, p. 116.

TABLE 10-14 SUGGESTED BUS SERVICE PLANNING GUIDELINES

1. Service pattern
    1.1 Service area and route coverage
       a. Service area is defined by operating authority or agency.
       b. Provide $\frac{1}{4}$-mile coverage where population density exceeds 4000 persons per square mile or three dwelling units per acre. Serve at least 90% of residents.
       c. Provide $\frac{1}{2}$-mile coverage where population density ranges from 2000 to 4000 persons per square mile (less than three dwelling units per acre). Serve 50 to 75% of the population.
       d. Serve major employment concentrations, schools, hospitals.
       e. Two-mile radius from park-and-ride lot.
    1.2 Route structure and spacing
       a. Fit routes to major street and land use patterns. Provide basic or modified grid system where streets form grid and land uses permit; provide radial or radial-circumferential system where irregular or radial street patterns exist.
       b. Space routes at about $\frac{1}{2}$ mile in urban areas, 1 mile in low-density suburban areas, and closer where terrain inhibits walking.
    1.3 Route directness—simplicity
       a. Routes should be direct and avoid circuity. Routes should be not more than 20% longer in distance than comparative trips by car.
       b. Route deviation shall not exceed 8 minutes per round trip, based on at least 10 customers per round trip.
       c. Generally, there should be not more than two branches per trunk-line route.
    1.4 Route length
       a. Routes should be as short as possible to serve their markets; excessively long routes should be avoided. Long routes require more liberal travel times because of the difficulty in maintaining reliable schedules.
       b. Route length generally shall not exceed 25 miles round-trip or 2 hours.
       c. Two routes with a common terminal may become a through route if they have more than 20% transfers and similar service requirements, subject to (b). This usually results in substantial cost savings and reduces bus movements in the central business district.
    1.5 Route duplication
       a. There should be one route per arterial except on approaches to the CBD or a major transit terminal. A maximum of two routes per street (or two branches per route) is desirable.
       b. Express service should utilize freeways or expressways to the maximum extent possible.
       c. Express and local services should be provided on separate roadways, except where frequent local service is provided.
2. Service levels
    2.1 Service period
       a. Regular service: 6 A.M. to 11 P.M./midnight, Mon.–Fri.
       b. Owl service: selected routes, large cities—24 hours.
       c. Suburban feeder service: weekdays, 6–9 A.M., 4–7 P.M. (Some services 6 A.M. to 7 P.M.)
       d. Provide Saturday and Sunday service over principal routes except in smaller communities, where Sunday service is optional.
    2.2 Policy headways—desirable minimum service frequency
       a. Peak: 20 minutes—urban; 20–30 minutes—suburban.
       b. Midday: 20 minutes—urban; 30 minutes—suburban.
       c. Evening: 30 minutes—urban; 60 minutes—suburban.
       d. Owl: 60 minutes.
    2.3 Loading standards
       a. Peak 30 minutes: 150%.
       b. Peak hour: 125–150%.
       c. Transition period: 100–125%.
       d. Midday/evening: 75–100%.
       e. Express: 100–125%.
       f. Suburban: 100%.
       Note: Policy headways may result in considerably lower load factors.

TABLE 10-14 *(cont.)* SUGGESTED BUS SERVICE PLANNING GUIDELINES

2.4 Bus stops
   a. Stop frequency: central areas, 10–12 stops/mile; urban area, major signalized intersections, 6–8 stops/mile; suburban areas, 2–5 stops/mile; express or suburban service, 2–4 stops/mile in pickup zone.
   b. Stop location: depends on convenience and safety; no parking in curb lane all day or peak hours, far side; parking in curb lane, near side (except where conflicting with right turns).
   c. Stop length (one 40-ft bus): near side, 105–150 ft; midblock, 140–160 ft; far side, 80–100 ft; straight approach, 140 ft after right turn; add 20 ft for 60-ft articulated bus; add 45 ft for each additional 40-ft bus expected to stop simultaneously.

2.5 Route speeds (typical)
   a. Central area: 6–8 mph.
   b. Urban: 10–12 mph.
   c. Suburban: 14–20 mph.

2.6 Service reliability
   a. Peak: 80% of buses 0 to 3 minutes late.
   b. Off-peak: 90–95% of buses 0 to 3 minutes late.

3. New routes
3.1 Service evaluation
   a. Examine physical constraints/street patterns for reliability.
   b. Estimate ridership and costs.
   c. Compare with existing route performance.

3.2 Service criteria
   a. Minimum density of 2000 persons/square mile.
   b. Twenty to 25 passengers/bus-hour weekdays; 15 Saturday, 10 Sundays and holidays. Fewer riders if route provides continuity or transfers.
   c. Fares should cover 40 to 50% of direct costs of service.

3.3 Frequency of change
   a. Major changes not more than two or three times per year. Other changes may be at other times.

3.4 Length of trial period
   a. Minimum 6 months for experimental service.

4. Passenger comfort and safety
4.1 Passenger shelters
   a. Provide at all downtown stops.
   b. Provide at major inbound stops in residential neighborhoods.
   c. At stops that serve 200 to 300 or more boarding and transferring passengers daily.

4.2 Bus maintenance
   a. Spares should not exceed 10 to 12% of scheduled fleet.
   b. Five thousand-mile preventive maintenance inspection.

4.3 Bus route and destination signs
   a. Provide front- and side-mounted signs.
   b. Front sign should give at least route number and general destination; side sign should give route number and name (front sign may give all three types of information).

4.4 Passenger information service
   a. Provide telephone information service for period that system operates.
   b. Ninety-five percent of all calls should be answered in 5 minutes.

4.5 Route maps and schedules
   a. Provide dated route maps annually.
   b. Provide printed schedules on a quarterly basis or when service is changed.
   c. Schedules should provide route map (for line).

4.6 Driver courtesy, efficiency, appearance
   a. Carefully select, train, supervise, and discipline drivers.
   b. Avoid extremes in personal appearances.
   c. Provide driver incentive programs.

4.7 Passenger and revenue security system
   a. Exact fare procedures—generally usable.
   b. Provide radio communication for driver with secret emergency alarm.
   c. Provide transit police (large properties).

Source: Adapted from TRB, 1980.

SUMMARY

In an era when energy prices are high, streets and highways are congested, the air-pollution levels are rising, and the overall costs of travel in metropolitan areas are escalating, the role of public transportation is bound to play an important part in our lives. In this chapter, we explain briefly the technical, operational, and cost characteristics of bus, light-rail, and rail rapid transit.

It is essential for the reader to get a flavor for urban public transportation in the total scheme of urban planning because the shaping, growth, and decay of urbanized areas has been a complex interaction between public and private transportation on the one hand and the distribution and densities of land use on the other. In recent years, the impact of transportation technology has been a tremendous force in shaping the lives of the urban dweller and the reintroduction of light-rail systems in several urban areas of this country and abroad is a testimony to the general importance of public transport.

REFERENCES

ARMSTRONG-WRIGHT, A. (1986). *Urban Transit Systems, Guidelines for Examining Options*, World Bank Technical Paper 52, The World Bank, Washington, DC.

CANADIAN URBAN TRANSIT ASSOCIATION and THE ROADS AND TRANSPORTATION ASSOCIATION OF CANADA (CUTA/RTAC) (1985). *Canadian Transit Handbook*, 2nd Ed., CUTA/RTAC, Toronto.

CARTER, E. C., and W. S. HOMBURGER (1978). *Introduction to Transportation Engineering*, Reston Publishing, Reston, VA.

DEEN, THOMAS B., and RICHARD H. PRATT (1992). Evaluating Rapid Transit in *Public Transportation*, 2nd Ed., eds. George E. Gray and Lester A. Hoel, Prentice-Hall, Englewood Cliffs, NJ, pp. 293–332.

GRAY, GEORGE E., and LESTER A. HOEL (1979). *Public Transportation: Planning, Operations and Management*, Prentice-Hall, Englewood Cliffs, NJ. Reproduced by permission of the publisher.

GRAY, GEORGE E., and LESTER A. HOEL (Eds.) (1992). *Public Transportation*, 2nd Ed., Prentice Hall, Englewood Cliffs, NJ.

GUENTHER, K. (1971). *Incremental Implementation of Dial-a-Ride Systems*, Special Report 124, Highway Research Board, National Research Council, Washington, DC.

HUTCHINSON, B. G. (1974). *Principles of Urban Transport Systems Planning*, Scripta Publishing, Silver Spring, MD.

ISMAIL, M., and C. J. KHISTY (1974). Mass Transit in the Mid-Ohio Region, *Traffic Engineering Journal*, I.T.E., July, pp. 16–19.

KHISTY, C. J., and P. J. KAFTANSKI (1981). *Demand Analysis of the New York City Subway System*, Transportation Research Record 797, National Academy of Science, Washington, DC.

KIRBY, R. F., ET AL. (1974). *Paratransit—Neglected Options for Urban Mobility*, The Urban Institute, Washington, DC.

LEVINSON, H. S., C. ADAMS, and W. F. HOEY (1975). *Bus Use of Highways, Planning and Design Guidelines*, NCHRP Report 155, Transportation Research Board, National Research Council, Washington DC.

MEYER, J. R., and J. A. GÓMEZ-IBÁÑEZ (1981). *Autos, Transit and Cities*, Harvard University Press, Cambridge, MA.

MEYER, J., J. KAIN, and M. WOHL (1965). *The Urban Transportation Problem*, Harvard University Press, Cambridge, MA.

PEAT, MARWICK, MITCHELL, & CO. (1973). *Transportation Systems Evaluation Indicators*, U.S. DOT, Washington, DC.

PUSHKAREV, B. S., and J. M. ZUPAN (1977). *Public Transportation and Land Use Policy*, Indiana University Press, Bloomington.

PUSHKAREV, B. S., J. M. ZUPAN, and R. S. CUMELLA (1982). *Urban Rail in America*, Indiana University Press, Bloomington.

TRANSPORTATION RESEARCH BOARD (TRB) (1980). *Bus Route and Schedule Planning Guidelines*, NCHRP Synthesis No. 69, Transportation Research Board, National Research Council, Washington, DC.

TRANSPORTATION RESEARCH BOARD (TRB) (1985). *Highway Capacity Manual*, Special Report 209, National Research Council, Washington, DC.

U.S. DEPARTMENT OF TRANSPORTATION (U.S. DOT) (1977). *Alternatives for Improving Urban Transportation: A Management Overview*, U.S. DOT, Washington, DC.

U.S. DEPARTMENT OF TRANSPORTATION (U.S. DOT) (1980). *Public Transportation, An Element of the Urban Transportation System*, U.S. DOT, Washington, DC.

VUCHIC, V. R. (1981). *Urban Public Transportation—Systems and Technology*, Prentice-Hall, Englewood Cliffs, NJ.

YU, JASON (1982). *Transportation Engineering*, Elsevier, Amsterdam.

EXERCISES

1. Select a city of your choice that has a public transportation system. Research the major developments that have happened in the system's growth and character, taking into consideration its population, area, form, economic development, and politics. Has the urban form been influenced and shaped because of its transport system over the years?

2. Now that you have acquired a basic appreciation of public transportation, what is your opinion as to how the "transportation gap" should be filled? Can government intervention help? If necessary, refer to Chapter 1 regarding the meaning of transportation gaps.

3. What is meant by the term "balanced transportation system"? How can the government (state or federal) or the city provide this balance?

4. A city with a population of 550,000 has a labor force of 100,000. In a recent survey done by the city, it was found that the overall traffic characteristics of the city could be categorized as follows:

Category	Average length of one-way trip (mi)	Percent of workers	Average travel speed (mph)
A	5	50	10
B	10	30	15
C	15	20	18

The present average car occupancy is 1.10 persons/car. It is proposed to start a campaign to implement a limited carpool program to attract 15% of the labor force (across the board) to get into this program, thereby increasing the speeds of all categories as follows:

Category	Average travel speed (mph)
A	15
B	20
C	25

Calculate the total time saved per day by the commuters of the city, and reduction in vehicles, assuming the revised car occupancy = 2.0 for carpoolers.

5. An observer at a busy downtown bus stop did a 30-minute study of nine buses.

Bus	Number of doors	Front door Alight	Front door Board	Back door alight
1	1	7	12	—
2	2	10	7	8
3	2	12	10	7
4	1	8	9	—
5	1	7	14	—
6	2	7	8	22
7	2	6	4	18
8	1	12	10	—
9	2	7	8	9

Determine (a) the busiest door; (b) the maximum number of buses that can use this bus stop. (Assume $a = 2$ sec; $b = 3$ sec; $c = 5$ sec.)

6. Small town USA has a bus configuration as shown in Figure 10-P6. E is the downtown bus stop. The data pertains to a 2-hour peak.

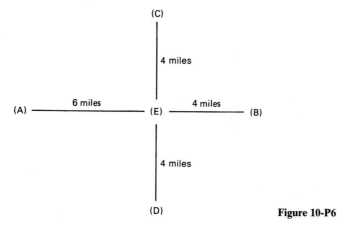

Figure 10-P6

	Route AB	Route CD
Length	10 miles	8 miles
Operating time	60 minutes	50 minutes
Maximum load	500 passengers	350 passengers
Vehicle capacity	45 seats	45 seats
Policy headway	45 minutes	45 minutes
Load factor	0.8	0.8
Minimum terminal time	10 minutes	10 minutes

Develop an operational plan so that good coordination is established for passengers on the AB route to transfer to the CD route. Summarize your results and provide the following information: (1) headways, (2) cycle time, (3) terminal time, (4) fleet size, and (5) commercial speed.

7. The bus system described in Example 5 needs to be worked out for off-peak periods when the demand is 250 passengers/hr. All other factors are the same. Design the basic system.

8. The bus system in Example 5 has a possibility of being extended to 1 mile on either side of the 8.5-mile run on the WSU and U of I campuses to cater to the demands of the cities of Pullman and Moscow, respectively. In other words, a bus will pick up passengers from Pullman to WSU (1 mile), continue to U of I (8.5 miles), and then terminate its run to Moscow (1 mile). The peak demand is 500 passengers/hr. Design the system.

9. Refer to Example 4 on rail operation design. The transit authority has revised the peak-hour demand to 15,000 passengers/hr with a required speed of 30 mph. Other factors remain the same. How many cars should be provided per train, and what will be the corresponding headway?

10. A rail transit system has prescribed the following factors: deceleration, $d = 3$ ft/sec^2; safety factor, $K = 1.4$; peak-hour demand = 35,000 passengers/hr; required speed between 40 and 45 mph; car length, $L = 80$ ft; car capacity = 180 passengers/vehicle; station platform length to accommodate a maximum of 12 vehicles/train; load factor = 0.8; reaction time, $t_r = 6$ sec; and guideway utilization factor, $\alpha = 0.75$. Design the system, assuming minimum headway = 100 sec.

11. A light-rail system similar to Example 6 needs to be constructed between the downtown and the main suburb. All information remains the same as given in the example except the following: route length = 10 miles; spacing of stations = 0.4 mile; operating hours/day = 20; average trip length = 2.5 miles; journey speed = 15 mph; peak period operation = 2.5 hours; and interest rate = 8%. What is the cost per passenger-mile?

12. A rail rapid transit line operates 15 five-car trains per track per hour. Schedule loadings average 2.0 passengers per seat. How many people can the line carry? Cars have 75 seats each.

13. A bus route having seven schedule stops has the following number of passengers alighting and boarding the bus:

Stop	1	2	3	4	5	6	7
Alighting	0	3	5	6	8	7	12
Boarding	20	5	7	3	4	2	0

Passengers board and alight through the front door. Alighting time/passenger = 1.8 sec and boarding time/passenger = 3 sec. Compute the dwell time at each stop and the total dwell time. (Assume the clearance time = 5 sec).

14. What are the major inherent characteristics of the automobile that have made it the predominant mode of transportation in North America? Is there any chance that this predominance could be corrected by intervention on the part of the government through policy changes?

15. What lessons have we, as a nation, learned in the last three decades with respect to our transportation system? Can these lessons be put to use, and how?

16. Taxi service, in general, is one of the more complicated forms of transportation. From a passenger's point of view, what positive and negative features does this mode have with respect to the total transportation system?

17. A transit system needs to be planned for a metropolitan region currently served by an old bus system. The present CBD employment is 40,000, and the population, area, and density of the region are as follows:

	CITY	URBANIZED AREA	RURAL AREA
Population	1,000,000	1,600,000	500,000
Land area (sq. mi)	420	750	1500
Density (p/sq. mi)	3000	1800	300

Discuss the various transit options you would suggest for this metropolitan region, providing proper justification for your options.

18. A close-knit community of about 8000 inhabitants is located 10 miles north of the central business district of a medium-sized city of 300,000 population, and 8 miles east of a large shopping and commercial center. The community has an average density of 10 persons/acre. A recent survey indicates that 80% of this community would regularly use some form of bus transportation. Using guidelines provided in this chapter, work out basic requirements for such a system, indicating service levels, bus stops, collection points, and so on. Make suitable assumptions stating them up front.

19. A city is considering a light-rail system in three corridors of the city. Ridership/day, length of the lines, and average trip length are as follows:

CORRIDOR	A	B	C
Ridership/day	20,000	17,000	10,000
Length of line (mi)	16	12	9
Trip length (mi)	10	9	5

Comment on the feasibility of these corridors to support LRT lines, based on criteria set forth in the guidelines in this chapter.

20. A proposed bus line is being planned for a city, with 40-seater buses. An acceptable load factor is 1.20. The estimated line ridership derived from a recent survey for the morning peak hour in passengers/hour is as follows:

STATION	DISTANCE (MI)	PASSENGERS		
		ON	OFF	ON BUS
1	0.0	700	0	—
2	0.5	100	80	720
3	1.0	50	80	690
4	1.5	50	100	640
5	2.0	50	100	590
6	2.5	20	100	510
7	3.0	10	140	380
8	4.0	10	100	290
9	5.0	10	100	200
10	6.0	0	200	

(a) Develop an operational plan for the morning peak hours for this line, making suitable assumptions.

(b) Develop a preliminary service pattern and schedule. The morning and afternoon peaks are 2 hours each.

Chapter 11

Urban Transportation Planning

1. INTRODUCTION

Planning is done by human beings for human beings. It is future-oriented and optimistic. City and regional planning involves the arrangement of spatial patterns over time. However, the spatial arrangement is not planning: It is the object of a process, and this process is the planning process. Transportation planning is a part of this process.

Transportation of people and goods is a medium of activity, and is the joint consequence of the land-use potential and the traffic-carrying capabilities of various modes of transportation. Engineers and planners prepare plans for the future to guide the city or community to control and govern a course of action. The methodological process of planning leads to the eventual preparation of a transportation plan for, say, the year 2020.

During the last 30 years a great mass of information has been collected and scores of books and papers have been written on the transportation planning process. Several of these books are cited in the references, and readers interested in furthering their knowledge are well advised to consult them. One of these references is *An Introduction to Urban Travel Demand Forecasting—A Self-Instructional Text*, from which some of the material contained in this chapter has been derived, with due modifications. Another source is *PLANPAC/BACKPAC General Information Manual* (FHWA, 1977).

Urban transportation planning is the process that leads to decisions on transportation policies and programs. The objective of the transportation planning process is to provide the information necessary for making decisions on when and where improvements should be made in the transportation system, thus promoting travel and land development patterns that are in keeping with community goals and objectives. Figure 11-1 shows the major activities in the transportation planning process. Each step is discussed briefly in this chapter.

449

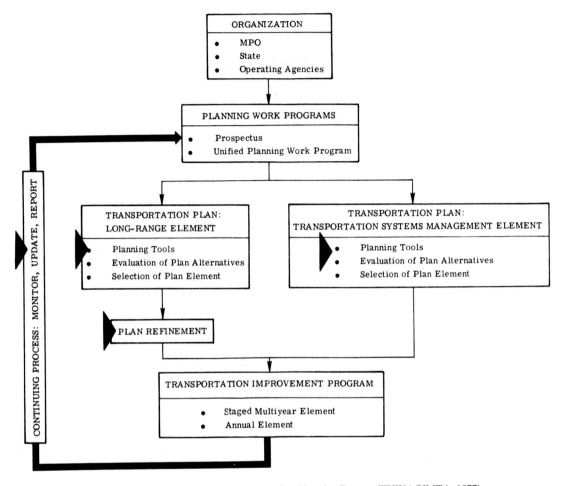

Figure 11-1   Urban Transportation Planning Process (FHWA/UMTA, 1977).

2. ORGANIZATION

Transportation planning is an extensive undertaking, and the need for an effective organization is obvious. The planning process must operate within the framework of the goals and objectives of the study area. Early in the process, ways to promote inter-action with public officials, public agencies, and the citizens of the area must be defined to make sure that the goals and objectives reflect current community values.

The governor of each state designates a *metropolitan planning organization* (MPO) to be responsible, with the state, for urban transportation planning. The MPO's planning activities are carried out in cooperation with state and local agencies.

3. PLANNING WORK PROGRAM

Work programs connected with planning are developed to make sure that the planning is conducted efficiently and comprehensively and that it addresses all pertinent issues. Figure 11-1 indicates that there are two parts to this phase: the prospectus and the unified planning work program.

The *prospectus* establishes a multiyear framework for the planning process. It summarizes the planning procedures, discusses the important issues that will be addressed during planning, describes the responsibilities of each agency that is participating in planning, and describes the status of all elements in the planning process.

The *unified planning work program* has two functions. It describes all urban transportation and transportation-related planning activities that are anticipated over the next year or two, and it documents work to be performed with federal planning assistance.

4. TRANSPORTATION PLAN

There are two elements in the preparation of a comprehensive transportation plan to guide improvements to the transportation system. The transportation systems management element and the long-range element work together to formulate a transportation improvement program.

4.1 The Transportation Systems Management Element

The key word here is management. *The transportation systems management (TSM) element* is concerned with making existing systems as efficient as possible and with making provisions for an area's short-range transportation needs. Automobiles, taxis, trucks, terminals, public transit, pedestrians, and bicycles are all parts of the urban transportation system. The four basic strategies applied to increase the efficiency of the system are: actions to ensure the efficient use of existing road space, actions to reduce vehicle use in congested areas, actions to improve transit service, and actions to improve internal management efficiency.

Planning tools have been developed for use of transportation planners to work out details of alternatives and to provide sound advice to decision makers. Because of the importance of this element a separate chapter has been devoted to this topic.

4.2 The Long-Range Element

To make provision for the long-range transportation needs of the urban area, the *long-range element* identifies facilities to be constructed, major changes to be made to existing facilities, and long-range policy actions. This element of the transportation plan might, for example, consider future land development policies by adding a highway link or by installing a busway system.

As in the TSM element, many long-range alternatives must be evaluated before decisions can be made. Thus planning tools for analysis of long-range alternatives must be developed. These tools also provide information to help decision makers select the most promising alternatives. Once again, travel-demand forecasting plays an important role as a major contributor to the planning tools used in evaluating alternatives.

5. PLAN REFINEMENT

After the long-range plan elements have been selected, the plan is refined by further detailed studies. For instance, the corridor in which improvements are planned could be studied in greater detail, along with studies of various types of technology—for example, buses versus rapid transit—and also studies to determine the proper staging (scheduling) of the planned projects. Travel-demand forecasting plays a role here in refining estimates of such things as patronage, market areas, congestion, and turning movements.

6. TRANSPORTATION IMPROVEMENT PROGRAM

After selections have been made in the transportation systems management and long-range elements, and after plan refinement, a *transportation improvement program* (TIP) is developed. This program ensures that the transportation plan will be implemented in an orderly, efficient manner and represents a statement as to how the transportation system will improve in the next few years. The program has two major elements: the staged multiyear element and the annual element.

The *staged multiyear element* describes the general aspects of the program over the next 3 to 5 years. This element indicates priorities among the projects identified for implementation, groups projects into appropriate staging periods, and makes estimates of cost and revenues for the program period.

The *annual element* identifies the details of projects that will be implemented within the next year. For each project, the annual element contains:

1. A basic description
2. Cost involved
3. Revenue sources
4. The local agency that is responsible for implementation

7. THE CONTINUING PROCESS

The continuing process (the stage that most urban areas are in now) consists of monitoring changes that could make it necessary to modify the transportation plan, updating the data that serve as a base for planning, updating the methods used in transportation planning—including those for travel forecasting—and reporting on activities and findings.

Transportation studies have been made in all areas with populations of over 50,000. Planners have made inventories of the characteristics that affect travel in urban areas. These inventories, which make up the database, include

1. Population
2. Land use
3. Economic activity
4. Transportation systems
5. Travel
6. Laws and ordinances
7. Financial resources
8. Community values

In the continuing process, one of the tasks is to update these inventories as necessary to make sure that the database for planning is complete and accurate. This update is done using secondary sources, such as the census or through small sample surveys. Complete large-scale surveys are no longer feasible, due to their high cost.

Monitoring changes is extremely important. For example, a decision to install a major shopping center or sports arena in or near a city would affect traffic patterns and eventually the transportation plan. In the continuing process, we monitor both the transportation system and its performance.

Through careful studies of data on people's travel behavior, relationships have been developed to predict how many trips people will make, where they will go, by which mode of transportation, and by which specific route. These relationships are the basis for travel demand forecasting and must be reviewed and reevaluated, if necessary, in the continuing process.

The continuing process ensures that the transportation plan will respond to the area's transportation needs—needs that are constantly changing. The level of effort involved in the continuing process will depend on the size of the urban area and the complexity of its problems.

8. OVERVIEW OF INFORMATION NEEDS

Prior to embarking on the forecasting process a significant amount of work is necessary to accumulate the information that is needed in the forecasting process. These information needs include broad items such as defining the area for which forecasting will be done and specific items such as identifying detailed information about streets and bus routes. To understand the travel forecasting process and its terminology, a brief exposure to its information requirements is essential. For the purpose of this discussion, information needs have been divided into four broad categories, as follows:

1. The study area
2. Urban activities

3. Transportation system

4. Travel

With knowledge of these four categories of information, the transportation planner has the data necessary to begin the travel-demand forecasting process.

8.1 The Study Area

Defining the Boundaries. Obviously, before forecasting travel for an urban area, the planner must clearly define the exact area to be considered. This planning area generally includes all the developed land plus the undeveloped land that the urban area will encompass in the next 20 to 30 years.

The boundary of the planning area is demarcated by the cordon line. In addition to considering future growth, the establishment of the cordon line might take into account political jurisdictions, census area boundaries, and natural boundaries. The cordon should intersect a minimum number of roads to save on subsequent interview requirements.

Subdividing the Area for Forecasting. The study area must be divided into analysis units to enable the planner to link information about activities, travel, and transportation to physical locations in the study area. The transportation analysis units are known as *zones*. The zones vary in size depending on the density or nature of urban development. In the central business district (CBD), zones may be as small as a single block, and in the undeveloped area they may be as large as 10 or more square miles. An area with a million people might have 600 to 800 zones, and an area of 200,000 people might have 150 to 200 zones. The zones attempt to bound homogeneous urban activities; that is, a zone may be all residential, all commercial, all industrial, and so on. Zones should also consider natural boundaries and census designations (see Figure 11-5).

An important consideration in establishing zones is their compatibility with the transportation network to be used. As a general rule, the network should form the boundaries of the zones. Zones are usually grouped into larger units known as *districts*. Districts might contain 5 to 10 zones, and a city of 1 million might have about 100 districts. Districts often follow travel corridors, political jurisdictions, and natural boundaries such as rivers.

The downtown area is usually defined as one district. Groups of districts moving radially from the CBD are known as *sectors*. Districts in circumferential groups are known as *rings*. Dividing the area into zones, districts, sectors, and rings is a great help to the transportation planner in organizing information and interpreting results.

8.2 Urban Activities

Once the study area has been divided into appropriate analysis units (zones and districts), information about activities in these areas can be gathered. Knowledge about the forecasting procedure is essential at this point, because only data relevant to the calibration and forecasting process need be gathered. Collecting data that are not eventually used is a wasteful practice.

Urban activity forecasting is the source of information regarding activities that might influence travel in the urban area. These activity forecasts are done on a zonal basis. The results of a typical activity analysis provide the planner with present levels of activities in zones to help in predicting future levels that provide a basis for forecasting.

8.3 Network Geometry

The transportation system consists of networks that represent the available modes (auto, bus, etc.). The network description is an abstraction of what is actually on the ground, and as such does not include every local street or collector street in the area. A network description is developed to describe auto and truck travel, with a separate description for transit if transit is a consideration. These descriptions could include the geometry of the transportation system. Network geometry includes numbering the intersections (called *nodes* for assignment purposes). Numbering the nodes allows us to identify the segments between them (called *links*). In transit networks we also identify the segments between them (called *lines*). This geometric description of the transportation network shows all possible ways that travel can take place between points in the area.

In the network description, zone *centroids* (center of activity) are identified; they are connected to nodes by imaginary links called *centroid connectors*. Centroids are used as the points at which trips are "loaded" onto the network. Sample vehicle and transit network maps are shown in Figure 11-2 for a small city.

9. TRAVEL FORECASTING

Travel-demand forecasting is used in at least four elements of planning: TSM, long-range, plan refinement, and updating. There are several different techniques for demand forecasting from which to choose, depending on the requirements of the analysis. These techniques differ in complexity, cost, level of effort, sophistication, and accuracy, but each has its place in travel forecasting. Figure 11-3 shows the general relationship of these planning tools to their most likely applications (Khisty, 1985). Each tool is explained briefly in what follows.

Sketch Planning Tools.   Sketch planning is the preliminary screening of possible configurations or concepts. It is used to compare a large number of proposed policies in sufficient analytical detail to support broad policy decisions. Useful in both long- and short-range regional planning and in preliminary corridor analysis, sketch planning with minimum data yields aggregate estimates of capital and operating costs, patronage, corridor traffic flows, service levels, energy consumption, and air pollution. The planner usually uses sketch planning tools until the need arises to examine a strategic plan alternative at a finer level of detail.

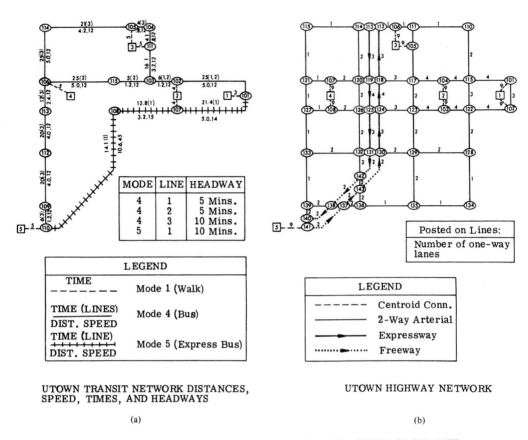

LEGEND (transit)

MODE	LINE	HEADWAY
4	1	5 Mins.
4	2	5 Mins.
4	3	10 Mins.
5	1	10 Mins.

LEGEND	
TIME — — — — —	Mode 1 (Walk)
TIME (LINES) / DIST. SPEED	Mode 4 (Bus)
TIME (LINE) / DIST. SPEED	Mode 5 (Express Bus)

UTOWN TRANSIT NETWORK DISTANCES,
SPEED, TIMES, AND HEADWAYS

(a)

Posted on Lines:
Number of one-way lanes

LEGEND	
— — — — —	Centroid Conn.
——————	2-Way Arterial
——————►	Expressway
·······►·····	Freeway

UTOWN HIGHWAY NETWORK

(b)

Figure 11-2   Transportation Networks for a Small City (FHWA/UMTA, 1977).

Traditional Tools.   Traditional tools treat the kind of detail appropriate to tactical planning; they deal with fewer alternatives than sketch tools, but in much greater detail. Examples include the location of principal highway facilities and delineated transit routes. At this level of analysis, the outputs are detailed estimates of transit fleet size and operating requirements for specific service areas, refined cost and patronage forecasts, and level-of-service measures for specific geographical areas. Household displacements, noise, and aesthetic factors can also be evaluated.

The cost of examining an alternative at the traditional level is about 10 to 20 times its cost at the sketch planning level, although default models, which dispense with many data requirements, can be used for a less expensive first look. Apparently promising plans can be analyzed in detail, and problems uncovered at this stage may suggest a return to sketch planning to accommodate new constraints. Typical traditional tools are the focus of this chapter.

Microanalysis Tools.   They are the most detailed of all planning tools. At this level of analysis, one may wish, for example, to make a detailed evaluation of the extension, rescheduling, or repricing of existing bus service; to analyze passenger and

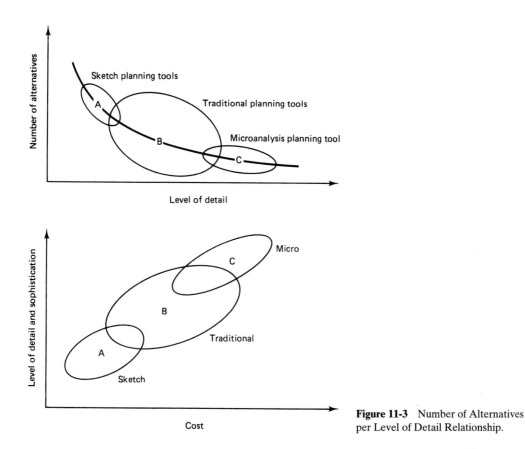

Figure 11-3 Number of Alternatives per Level of Detail Relationship.

vehicle flows through a transportation terminal or activity center; or to compare possible routing and shuttling strategies for a demand-activated system. Final analysis at this level is prohibitively expensive except for subsystems whose implementation is very likely, and whose design refinements would bring substantial increase in service or significant reductions in cost. It is most effective in near-term planning, when a great many outside variables can be accurately observed or estimated.

10. OVERVIEW OF THE FORECASTING PROCESS

In general, travel-demand forecasting attempts to quantify the amount of travel on the transportation system. Demand for transportation is created by the separation of urban activities. The supply of transportation is represented by the service characteristics of highway and transit networks.

The *traditional four-step process* will be discussed. This process has been developing over the past 25 years for forecasts of urban travel. In terms of the planning process, this discussion will focus on planning tools for the long-range element of the transportation plan, as modified and updated in the continuing process.

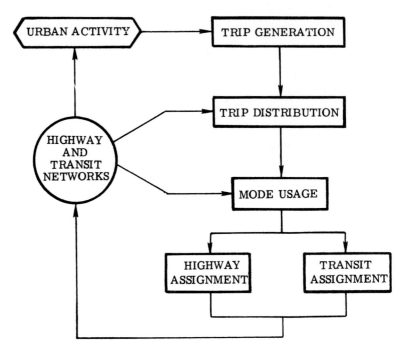

Figure 11-4   Four Basic Models Used in Transportation Planning (FHWA/UMTA, 1977).

There are four basic phases in the traditional travel-demand forecasting process.

1. *Trip generation* forecasts the number of trips that will be made.
2. *Trip distribution* determines where the trips will go.
3. *Mode usage* predicts how the trips will be divided among the available modes of travel.
4. *Trip assignment* predicts the routes that the trips will take, resulting in traffic forecasts for the highway system and ridership forecasts for the transit system.

Figure 11-4 shows how these phases fit together into the travel forecasting process.

Urban activity forecasts provide information on the location and intensity of future activity in an urban area and provide primary input to trip generation. Descriptions of the highway and transit networks provide the information necessary to define the "supply" of transportation in the area; the four phases predict the travel "demand." The feedback arrows shown represent checks of earlier assumptions made on travel times and determine if any adjustments are necessary.

11. URBAN ACTIVITY FORECASTS

Urban activity forecasts provide estimates of where people will live and where businesses will be located in the future. These forecasts also include the intensity of activity, such as the number of households and the number of employees in businesses.

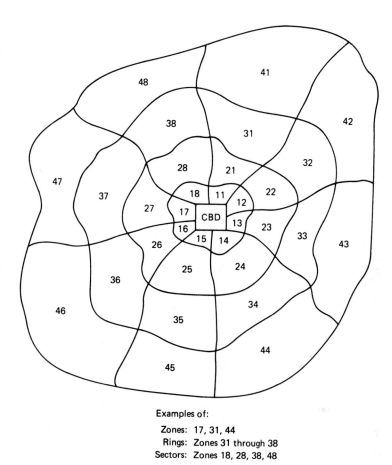

Examples of:

Zones: 17, 31, 44
Rings: Zones 31 through 38
Sectors: Zones 18, 28, 38, 48

Figure 11-5 Structures of Zones, Rings, and Sectors.

Figure 11-5 shows the structure of zones, rings, and sectors for a small city. They are a great help in organizing information and forecasting. Zone activity might appear as, say, residential population 1400, average income of $13,000, employment 752 (302 white collar, 450 blue collar). Several additional factors, such as car ownership, residential density, and amount of vacant land, may also be known.

These forecasts are done for small parcels of land called *zones*. Zones vary in size, with the smallest about the size of a block in the downtown area, whereas the largest on the urban fringe may be several square miles in area. An area with a million people could have 700 to 800 zones. Zonal urban activity forecasts are based on the following:

1. Total urban area population and employment estimates
2. Location behavior of people and businesses
3. Local policies regarding land development, transportation, zoning, sewers, and so on.

Once the study area has been divided into appropriate analysis units, such as zones and sectors, information about activities in these areas can be gathered and aggregated. The results of a typical activity analysis provide the planner with present levels of activities in zones to help in predicting future levels. These activity forecasts are direct inputs to the next stage of the process, trip-generation analysis.

12. TRIP GENERATION

Trip generation is the process by which measures of urban activity are converted into numbers of trips. For example, the number of trips that are generated by a shopping center is quite different from the number of trips generated by an industrial complex that takes up about the same amount of space. In trip generation, the planner attempts to quantify the relationship between urban activity and travel.

The inventory data discussed earlier provide the analyst's input for trip-generation analysis. Surveys of travelers in the study area show the numbers and types made; by relating these trips to land-use patterns, the analyst is able to forecast the number of trips that will be made in the future, given forecasts of population and other urban activity. Here's a simplified example. A small city's survey data show that zone 11 has an employment of 900 people and attracts 4511 trips. By dividing the trips by employees, we find that about five trips are attracted per employee. This rate can then be used to predict attractions for future employment levels. The output of trip-generation analysis is a table of trip ends—the number of trips produced and the number that are attracted.

As mentioned earlier, the study area is divided into zones for analysis purposes. After trip-generation analysis, the planner knows how many trips are produced by each zone and how many are attracted by each zone. In addition, the planner knows the purposes for the trips—the trips are put into several categories, like trips from home to work, or home to shop, or home to school. This categorization is necessary because each trip purpose reflects the behavior of the trip maker. For example, school trips and work trips are pretty regular; shopping and recreation trips are less so. There are basically two tools for trip-generation analysis, multiple linear regression and cross-classification, and these methods are explained in the following sections.

12.1 Multiple Linear Regression Technique

Transportation studies have shown that residential land use is an important trip generator. Also, nonresidential land use in many cases is a good attractor of trips. A typical equation connecting estimated trips generated by a residential zone could be

$$Y = A + B_1 X_1 + B_2 X_2 + B_3 X_3 \qquad (1)$$

where

$Y = $ trips/household
$X_1 = $ car ownership
$X_2 = $ family income

X_3 = family size

A, B_i = parameters determined through a calibration process

Model parameters and variables vary from one study area to another and are established by using base-year information. Once the equations are calibrated, they are used to estimate future travel for a target year. In developing regression equations it is assumed that

1. All the independent variables are independent of each other.
2. All the independent variables are normally distributed.
3. The independent variables are continuous.

It is not possible for the planner to conform to these specifications, and because of this departure, regression analysis has come in for considerable criticism. A typical regression equation may appear as follows:

$$Y = 0.0649X_1 - 0.0034X_2 + 0.0066X_3 + 0.9486X_4 + 12$$

where

Y = total trips/household

X_1 = family size

X_2 = residential density

X_3 = total family income

X_4 = cars/household

To derive an estimate of Y for a zone for a future year, appropriate estimates of X_1, X_2, X_3, and X_4 are substituted and the equation solved. The quality of fit of a regression line determined by multiple linear regression analysis is indicated by the multiple correlation coefficient (goodness of fit) represented by R, the value being between 0 and 1. The closer R is to 1, the better is the linear relationship between the variables.

When making preliminary investigations into trip making, it is useful to compute correlation coefficients between trip making and separate independent variables. The result of a typical computation is shown in Table 11-1, from which an appreciation of

TABLE 11-1 MATRIX OF LINEAR CORRELATION COEFFICIENTS

	Dependent variable	All home-based trips
X_1	Family size	0.41
X_2	Residential density	−0.76
X_3	Total family income	0.73
X_4	Cars(s) per household	0.86
X_5	Travel time to CBD	0.32
X_6	Proportion of school-going children	0.27

the pattern of linear relationships between pairs of variables may be obtained. It can be observed that the dependent variable, Y, is highly correlated with independent variable X_1, through X_4, but somewhat weakly correlated with X_5 and X_6.

The multiple regression technique is appealing to transportation analysts because it is easy to determine the degree of relationship between the dependent and independent variables. Also, it is possible to determine the ability of the equation to predict accurately. Apart from the coefficient of multiple correlation referred to before, it is the practice to determine the standard error of estimate (s_y), which is a measure of the deviation of, say, observed trips from values predicted by the equation. In addition, the partial correlation coefficients (r_j) of each of the independent variables is calculated. It provides the relationship between the dependent variable (Y) and the particular independent variable (X_i) under investigation. The t-test may be used to determine whether an estimated regression coefficient is significant.

In summary, if

$$Y_e = A + B_1 X_1 + B_2 X_2 + B_3 X_3 + \cdots + B_n X_n \tag{2}$$

Then $B_1, \ldots, B_n$ are referred to as partial regression coefficients, and

$$R^2 = \frac{\Sigma (Y_e - \mu_Y)^2}{\Sigma (Y_i - \mu_Y)^2} \tag{3}$$

$$s_e = \frac{\Sigma Y_e^2}{N - (n + 1)} \tag{4}$$

$$s_{B_n} = \frac{s_e^2}{s_{x_i} N (1 - R_{x_i}^2)} \tag{5}$$

$$t = \frac{B_n}{s_n} \tag{6}$$

where

Y_e = estimated magnitude of the dependent variable
μ_Y = mean of the dependent variable
R^2 = coefficient of multiple determination
R = multiple correlation coefficient
s_e = standard error of estimate
s_{B_i} = standard deviation of the independent variable X_i
R_{x_i} = coefficient of multiple correlation between X_i and all other independent variables

References (such as Gutman et al., 1982) on linear regression should be consulted for working with trip-generation analysis.

Example 1

An analyst came up with the following regression equations and a simple correlation matrix for 20 zones, as given:

$$Y = 50.5 + 0.80X_1$$
$$s_e = 210, \quad R^2 = 0.95, \quad t = 34$$
$$Y = 308 + 0.79X_2$$
$$s_e = 844, \quad R^2 = 0.88, \quad t = 29$$
$$Y = 52.7 + 0.85X_2 + 1.75X_3$$
$$s_e = 205, \quad R^2 = 0.98, \quad t = 60; 22$$
$$Y = -105 + 1.38X_2 - 0.4X_3 + 0.1X_4$$
$$s_e = 155, \quad R^2 = 0.97, \quad t = 3; 2; 0.5$$

where

Y = trips produced
X_1 = total population
X_2 = blue-collar population
X_3 = white-collar population
X_4 = school-going children

	Y	X_1	X_2	X_3	X_4
Y	1.0	0.95	0.85	0.42	0.23
X_1		1.00	0.92	0.53	0.22
X_2			1.00	0.35	0.09
X_3				1.00	0.12
X_4					1.00

Comment on the suitability of these equations for use in a transportation study.

Solution

Equation 1 appears logical and satisfactory, and the t value is significant at the 1% level of significance. $t_{0.01, \text{df} = 18} = 2.88 < 34$. R^2 is very high.

Equation 2 is also logical and quite satisfactory, although $A = 308$ is not reasonable. In Eq. 3, with two independent variables, X_2 and X_3, but with about the same standard error, there is not much of an improvement over Eq. 1. Both partial regression coefficients are statistically significant ($t = 2.898$, df = 17). Note that although R^2 is very high, this is not the best equation to work with.

Equation 4 has a very high R^2, but it appears to be unreasonable in two respects: first, the negative coefficient (-105) is not logical; also, for df = 16, $t_{0.01} = 2.92$ is greater than 2.0 and 0.5 for X_3 and X_4. An examination of the matrix reveals that X_1 and X_2 are highly correlated with Y, and X_3 and X_4 are not as highly correlated (0.42 and 0.23). However, X_1 is highly correlated with X_2 and therefore X_2 could be easily eliminated.

Of all the equations, Eq. 1 seems the most logical to use. Two points may be noted: (1) a high R^2 by itself means little if the t-test is marginal or poor, and (2) just having a large number of independent variables does not mean very much. A large number of variables is expensive from the data-gathering point of view.

TABLE 11-2   SAMPLE TRIP-GENERATION RATES

Land-use category	Land area (ft^2)	Person-trips	Trips/1000 ft^2 per day
Residential			
Single family	3020	11,838	3.92
Apartments	4900	17,101	3.49
Commercial			
Retail	9540	88,054	9.23
Wholesale	5325	25,613	4.81

Source: TRB, 1978.

12.2 Trip-Rate Analysis Technique

In Chapter 3 it was demonstrated that the trips produced by different land-use categories (trip generator) and the trips attracted by different land uses, such as shopping centers and recreational areas, can be aggregated resulting in trip-generation rates grouped by generalized land-use categories. For example, in a recent survey of a small city, it was found that the shopping trips to the CBD and the shopping center could be aggregated to be 2.75 and 9.34 shopping trips per employee on weekdays, respectively. Table 11-2 shows the land area or floor area trip rates for a medium-sized city. Trip rates and characteristics of generators are listed in Table 11-3. Tables 11-4 and 11-5 list trip-generation characteristics and parameters, respectively. The material presented in Tables 11-2 to 11-5, derived from Sosslau et al. (1978), is representative of a wide range of values for each generator.

12.3 Category Analysis

One of the techniques that is widely used to determine the number of trips generated is called category analysis or cross-classification analysis. The approach is based on a control of total trips at the home end. The amount of home-end travel generated is a function of number of households, the characteristics of households, the income level, and car ownership. The density of households could also be considered. At the nonhome end, a distribution index is developed based on land-use characteristics, such as the number of employees by employment category, land-use type, and school enrollment. Some of the advantages of this approach are as follows:

- Ease of understanding by decision makers and the public.
- Efficient use of data. If no current origin–destination data are available, a small stratified sample survey will be enough.
- Easily monitored and updated.
- Validity. The process is valid in forecasting as well as in the base-year accuracy check.
- Sensitivity. The process is policy-sensitive: for example, the impact of differing population densities or car ownership can be used for assessing needed changes in a district or neighborhood.

TABLE 11-3 AVERAGE VEHICLE TRIP RATES AND OTHER CHARACTERISTICS OF GENERATIONS

Generator[a]	Vehicle-trips[b] to and from per day per		Percent trips in hour shown			Typical auto occupancy	Typical percent transit of total person-trips[c]
	Dwelling unit	Acre	A.M. peak	P.M. peak	Peak hour of generation		
Residential							
Single family							
1 DU/acre	9.3	9.3	8.0	10.8	10.8	1.62	3.2
2 DU/acre	9.3	18.6	8.0	10.8	10.8	1.62	3.2
3 DU/acre	10.2	30.6	8.0	10.8	10.8	1.67	3.2
4 DU/acre	10.2	40.8	8.0	10.8	10.8	1.67	3.2
5 DU/acre	9.1	45.5	8.0	10.8	10.8	1.62	3.2
Medium density							
(duplex, townhouses, etc.)							
5 DU/acre	7.0	35.0	8.0	10.8	10.8	1.57	5.6
10 DU/acre	7.0	70.0	8.0	10.8	10.8	1.57	5.6
15 DU/acre	7.0	105.0	8.0	10.8	10.8	1.57	5.6
Apartments							
15 DU/acre	6.0	90.0	7.9	10.8	10.8	1.56	12.4
25 DU/acre	6.0	150.0	7.9	10.8	10.8	1.56	12.4
35 DU/acre	6.0	210.0	7.9	10.8	10.8	1.56	12.4
50 DU/acre	6.0	300.0	7.9	10.8	10.8	1.56	12.4
60 DU/acre	6.0	360.0	7.9	10.8	10.8	1.56	12.4
Mobile home park							
5 DU/acre	5.5	27.5	8.3	10.8	12.5	1.54	1.0
10 DU/acre	5.5	55.0	8.3	10.8	12.5	1.54	1.0
15 DU/acre	5.5	82.5	8.3	10.8	12.5	1.54	1.0
Retirement community							
10 DU/acre	3.5	35.0	12.1	12.1	12.1	1.48	6.0
15 DU/acre	3.5	52.5	12.1	12.1	12.1	1.48	6.0
20 DU/acre	3.5	70.0	12.1	12.1	12.1	1.48	6.0
Condominiums							
10 DU/acre	5.9	59.0	7.1	7.1	7.1	1.56	9.0
20 DU/acre	5.9	118.0	7.1	7.1	7.1	1.56	9.0
30 DU/acre	5.9	177.0	7.1	7.1	7.1	1.56	9.0
Planned unit development							
5 DU/acre	7.9	39.5	10.1	10.1	10.1	1.58	7.1
15 DU/acre	7.9	118.5	10.1	10.1	10.1	1.58	7.1
25 DU/acre	7.9	197.5	10.1	10.1	10.1	1.58	7.1

(continued)

TABLE 11-3 *(continued)*

Generator[a]	Vehicle-trips[b] to and from per day per: Dwelling unit	Acre		Percent trips in hour shown: A.M. peak	P.M. peak	Peak hour of generation	Typical auto occupancy	Typical percent transit of total person-trips[c]
Miscellaneous								
Service station	*Station* 748	*Pump* 133		1.5	3.0	4.0	1.55	—
	See individual generator below							
Race track	*Seat* 0.61	*Attendee* 1.08		—	—	—	2.05	—
Pro baseball	0.16	1.18		—	—	—	2.05	—
Military base	*Military personnel* 2.2	*Civilian employees* 7.1 / *Total employees* 1.8		—	—	—	1.42	—

Generator[a]	1000 ft² GFA	Employee	Acre	A.M. peak	P.M. peak	Peak hour of generation	Typical auto occupancy	Typical percent transit of total person-trips[c]
Retail								
Freestanding								
Supermarket	135.3	—	1000	0	8.7	12.6	1.64	1
Discount store	50.2	57.2	—	0	5.1	9.7	1.64	1
Discount store with supermarket	81.2	30.3	—	0	6.9	11.1	1.64	1
Department store	36.1	32.8	900	—	—	—	1.64	2
Auto supply	88.8	—	—	—	—	—	1.64	1
New car dealer	44.3	—	—	—	—	—	1.64	1
Convenience								
24 hr	577.0	—	—	—	—	—	1.64	1
15-16 hr	322.0	—	—	—	—	—	1.64	1
Shopping center regional								
(over 1 million ft²)	33.5	30.9	580	1.9	9.7	11.5	1.64	3
½–1 million ft²	34.7	20.4	370	2.8	9.6	—	1.64	3
Community 100,000-500,000 ft²	45.9	20.6	330	—	11.2	11.3	1.64	3
Neighborhood under 100,000 ft	97.0	—	—	3.3	11.5	12.4	1.64	3
Central area (high density)	40.0	—	900	—	—	—	1.64	12

	Vehicle-trips[b] to and from per day per			Percent trips in hour shown			Typical auto occupancy	Typical percent transit of total person-trips[c]
	1000 ft² GFA	Employee	Acre	A.M. peak	P.M. peak	Peak hour of generation		
Industrial/Manufacturing								
Freestanding general manufacturing	4.2	2.3	40.5	18.4	19.3	32.2	1.33	5
Warehouse	5.3	4.4	67.5	12.7	32.2	—	1.25	5
Research/development	5.1	2.4	60.8	21.1	20.4	—	1.33	5
Industrial park	8.8	3.9	71.9	13.2	14.7	—	1.33	5
General light industrial	5.5	3.2	52.4	21.1	20.4	21.2	1.33	5
All industry average	5.5	3.0	59.9	15.8	19.4	—	1.40	5
Offices								
General	11.7	3.5	145	20.7	19.1	—	1.35	5
Medical	63.5	25.0	426	—	—	8.5	1.45	5
Governmental	48.3	12.0	66	8.5	16.0	—	1.35	5
Engineering	23.0	3.5	282	16.9	14.6	—	1.35	5
Civic center	25.0	6.1	33	9.0	11.4	—	1.35	5
Office park	21.0	3.3	277	16.9	14.6	—	1.35	5
Research center	9.3	3.1	37	16.0	18.5	20.2	1.35	5
Restaurants								
High-quality restaurant	56.3	—	200	1.8	6.0	12.5	1.93	3
Other sitdown	198.5	—	932	29.0	6.4	—	1.93	3
Fast food	533.0	—	1825	16.0	5.7	—	1.93	1
Banks	388	75	—	—	—	—	1.45	—
Parks and recreation								
Marina	—	259.0	18.5	—	—	—	2.05	—
Golf course	—	34.2	7.4	—	—	—	2.05	—
Bowling	—	—	296.3	—	—	—	2.05	—
Participant sports	—	—	26.5	—	—	—	2.05	—
City park	—	—	60.0	—	—	—	2.05	—
County park	—	26.5	5.1	—	—	—	2.05	—
State park	—	61.1	0.6	—	—	—	2.05	—
Wilderness park	—	—	0.7	—	—	—	2.05	—

[a] Most of the generators are located outside the CBD. DU stands for dwelling unit.
[b] The vehicle trip rates are actually volumes into and out of the site. As such, they may include some trips that would be passing the site. Their percentage is not established.
[c] The typical transit percentage shows wide variation and, therefore, gross approximation.

467

TABLE 11-4 DETAILED TRIP-GENERATION CHARACTERISTICS[a]

Income range 1970 (thousands of dollars)	Average autos per HH[b]	Average daily person trips per HH[c]	Percent HH by autos owned[d]				Average daily person trips per HH by no. of autos/HH[e]				Percent average daily person-trips by purpose[f]		
			0	1	2	3+	0	1	2	3+	HBW	HBNW	NHB
Urbanized area population: 50,000–100,000													
0–3	0.56	4.5	53	39	7	1	2.0	6.5	11.5	12.5	21	57	22
3–4	0.81	6.8	32	58	10	1	2.2	8.0	13.0	15.0	21	57	22
4–5	0.88	8.4	26	61	12	1	2.6	9.5	14.5	16.5	21	57	22
5–6	0.99	10.2	20	62	17	1	3.0	11.0	15.5	18.0	18	59	23
6–7	1.07	11.9	15	64	20	1	3.0	12.5	16.5	19.5	18	59	23
7–8	1.17	13.2	11	64	23	2	3.5	13.3	17.0	21.5	16	61	23
8–9	1.25	14.4	8	62	28	2	4.8	14.0	17.5	22.5	16	61	23
9–10	1.31	15.1	6	60	32	2	5.5	14.3	17.5	24.0	16	61	23
10–12.5	1.47	16.4	3	49	44	3	6.2	15.0	18.5	25.5	15	62	23
12.5–15	1.69	17.7	2	38	52	8	6.1	15.0	19.0	25.5	14	62	24
15–20	1.85	18.0	2	28	57	13	6.0	13.5	19.5	23.0	13	62	25
20–25	2.03	19.0	1	21	58	20	6.0	13.0	20.0	23.0	13	62	25
25+	2.07	19.2	1	19	59	21	6.0	12.5	20.0	23.0	13	62	25
Weighted average	1.55	14.1	12	47	35	6	4.6	12.6	17.2	21.4	16	61	23
Urbanized area population: 100,000–250,000													
0–3	0.49	4.0	57	37	6	0	1.0	7.5	10.5	13.8	20	63	17
3–4	0.72	6.8	36	56	8	0	1.7	9.2	13.3	16.4	22	60	18
4–5	0.81	8.4	29	61	10	0	2.5	10.2	14.5	17.6	22	58	20
5–6	0.94	10.2	21	65	13	1	3.5	11.4	14.5	19.0	22	58	20
6–7	1.01	11.7	17	66	16	1	4.5	12.5	15.6	20.5	20	58	22
7–8	1.14	13.6	12	65	21	2	5.4	13.8	17.0	22.2	20	57	23
8–9	1.25	15.3	9	61	28	2	5.8	15.0	17.5	23.0	20	57	23
9–10	1.34	16.2	6	58	33	3	6.3	15.8	18.0	23.5	19	57	24
10–12.5	1.50	17.3	4	50	40	6	6.8	16.0	19.0	24.5	19	57	24
12.5–15	1.65	18.7	2	40	51	7	7.0	16.0	20.4	25.0	19	56	25
15–20	1.85	19.6	2	28	57	13	7.2	15.0	21.0	25.5	18	56	26
20–25	2.01	20.4	1	20	61	18	7.5	15.0	21.0	25.5	18	55	27
25+	2.07	20.6	1	19	59	21	7.5	15.0	21.0	25.2	18	55	27
Weighted average	1.55	14.5	14	48	33	6	5.4	13.7	18.4	22.4	20	57	23

[a] Total of internal and external trips generated by area residents.
[b] Source: 1970 Census. HH stands for household.
[c] Origin–destination surveys.
[d] Calculated using b.
[e] Calculated using b and c.
[f] National figures (1966 and 1976).
Source: Sosslau et al., 1978.

TABLE 11-5 TRIP-GENERATION PARAMETERS

(a) Trip production estimates

Urbanized area population	Average daily person trips per HH[a]	Average daily person-trips by mode[b] (%)			Average daily person-trips by purpose[a] (%)			Auto-person trips as a percent of total person-trips[c]			Auto-driver trips as a percent of total person-trips[d]		
		Public transit	Auto passenger	Auto driver	HBW	HBNW	NHB	NBW	HBNW	NHB	HBW	HBNW	NHB
50,000–100,000	14	2	40	58	16	61	23	96	99	98	70	54	68
100,000–250,000	14	6	30	64	20	57	23	88	97	94	64	54	66
250,000–750,000	12	8	31	61	20	55	25	84	96	92	62	54	64
750,000–2,000,000	8	13	30	57	25	54	21	74	93	86	56	53	60

(b) Useful characteristics for trip estimation

	External travel characteristics		Total areawide truck trips as a percent of areawide auto-driver trips[f]
Urbanized area population	Percent of total external trips passing through area[c]	Percent of total external trips to the CBD[e]	
50,000–100,000	21	22	27
100,000–250,000	15	22	17
250,000–750,000	10	18	16
750,000–2,000,000	4	12	16

(continued)

469

TABLE 11-5 TRIP-GENERATION PARAMETERS (*continued*)

(c) Trip attraction estimating relationships (all population groupings for either vehicle- or person-trips)[g]

To estimate trip attractions for an analysis area, use:

HBW trip attractions = F_1[1.7 (analysis area total employment)]

$$\text{HBNW trip attractions} = F_2\left[10.0\left(\begin{array}{c}\text{analysis area}\\\text{retail}\\\text{employment}\end{array}\right) + 0.5\left(\begin{array}{c}\text{analysis area}\\\text{nonretail}\\\text{employment}\end{array}\right) + 1.0\left(\begin{array}{c}\text{analysis area}\\\text{dwelling}\\\text{units}\end{array}\right)\right]$$

$$\text{NHB trip attractions} = F_3\left[2.0\left(\begin{array}{c}\text{analysis area}\\\text{retail}\\\text{employment}\end{array}\right) + 2.5\left(\begin{array}{c}\text{analysis area}\\\text{nonretail}\\\text{employment}\end{array}\right) + 0.5\left(\begin{array}{c}\text{analysis area}\\\text{dwelling}\\\text{units}\end{array}\right)\right]$$

where F_1, F_2, and F_3 are areawide control factors.

To develop areawide control factors, use:

$$F_1 = \frac{\text{areawide productions for HBW trips}}{1.7 \ (\text{areawide total employment})}$$

$$F_2 = \frac{\text{areawide productions for HBNW trips}}{10.0\left(\begin{array}{c}\text{areawide}\\\text{retail}\\\text{employment}\end{array}\right) + 0.5\left(\begin{array}{c}\text{areawide}\\\text{nonretail}\\\text{employment}\end{array}\right) + 1.0\left(\begin{array}{c}\text{areawide}\\\text{dwelling}\\\text{units}\end{array}\right)}$$

$$F_3 = \frac{\text{areawide productions for NHB trips}}{2.0\left(\begin{array}{c}\text{areawide}\\\text{retail}\\\text{employment}\end{array}\right) + 2.5\left(\begin{array}{c}\text{areawide}\\\text{nonretail}\\\text{employment}\end{array}\right) + 0.5\left(\begin{array}{c}\text{areawide}\\\text{dwelling}\\\text{units}\end{array}\right)}$$

a From Table 11-4.
b Source: Baerwald, 1976.
c Source: Origin–destination surveys.
d Calculated using c and average daily auto-occupancy rates for urbanized areas.
e Source: Public Roads, 1961.
f Source: Baerwald, 1976.
g Source: Office of Planning Methodology and Technical Support, UMTA.
Source: Sosslau et al., 1978.

- Flexibility. Application at different study levels is possible: district, zonal, regional, corridor, and so on.
- Easy transferability of analysis between cities or parts of study areas of the same size and character.
- Wide use of data. Census data can be used extensively in these analyses, particularly the socioeconomic data.

Example 2

Twenty households in a city were sampled for household income, autos per household, and trips produced.

Household	Trips	Income (dollars)	Autos
1	2	4000	0
2	4	6000	0
3	10	17,000	2
4	5	11,000	0
5	5	4500	1
6	15	17,000	3
7	7	9500	1
8	4	9000	0
9	6	7000	1
10	13	19,000	3
11	8	18,000	1
12	9	21,000	1
13	9	7000	2
14	11	11,000	2
15	10	11,000	2
16	11	13,000	2
17	12	15,000	2
18	8	11,000	1
19	8	13,000	1
20	9	15,000	1

Develop matrices connecting income to automobiles available, and also draw a graph connecting trips per household to income. How many trips will a household with an income of $10,000 owning one auto make per day?

Solution

1. A matrix based on family income and auto availability is set up with cells and numbers representing the appropriate household sample numbers.

Income (thousands of dollars)	Autos available		
	0	1	2 or more
≤ 6	1, 2	5	—
6–9	8	9	13
9–12	4	7, 18	14, 15
12–15	—	19, 20	16, 17
> 15	—	11, 12	3, 6, 10

2. The average number of trips the household generates in each cell is calculated. For example, the average trip rate for households with two or more autos and an income between $12,000 and $15,000 is 11.5, because households 16 and 17 together make a total of 33 trips. These average rates are shown.

Income (thousands of dollars)	Autos available		
	0	1	2 or more
≤ 6	3.0	5.0	—
6–9	4.0	6.0	9.0
9–12	5.0	7.5	10.5
12–15	—	8.5	11.5
> 15	—	8.5	12.7

3. The data from the second matrix can be plotted on a graph connecting trips per household with income, and smooth curves drawn through the points.

4. A household with $10,000 income and one auto per household will make 7 trips per day. (See Figure 11-E2.)

In urban areas it is possible that one could use survey data to set up graphs showing the relationship between percent of trips by categories: home-based work (HBW), home-based other (HBO), and nonhome-based (NHB). It is also possible to borrow data from other urban areas, if necessary. An example of this relationship is shown in Figure 11-6.

Just as trip production rates were developed based on household characteristics, one can establish trip attraction rates by analyzing the urban activities that attract trips. Trips are attracted to various locations, depending on the character, location, and amount of activities taking place in a zone. Trip attraction rates can be borrowed from other urban areas or developed from survey data relating the number of trip attractions to activity character, location, and amount. Table 11-6 is a typical example.

TABLE 11-6 ATTRACTION RATES

	Attractions per household	Attractions per nonretail employee	Attractions per downtown retail employee	Attractions per other retail employee
Home-based work	Negligible	1.7	1.7	1.7
Home-based other	1.0	2.0	5.0	10.0
Nonhome-based	1.0	1.0	3.0	5.0

Source: FHWA/UMTA, 1977.

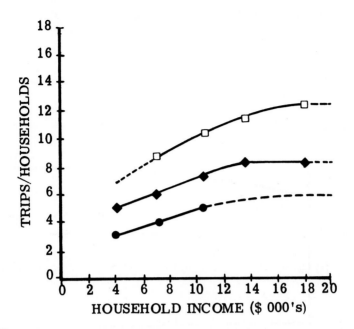

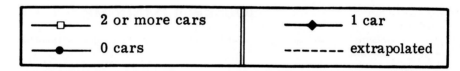

Figure 11-E2 Trips per Household Based on Household Income and Auto Ownership.

Example 3

A number of suburban zones have a total of 1000 dwelling units (DU). The average income per DU is $12,000. Using the curves *a*, *b*, and *c* provided, estimate the number of trips produced by the zones (Figure 11-E3).

Solution

Refer to Figure 11-E3.

1. Enter curve *a* with zonal income per dwelling unit to determine car ownership level by household:

2%	0 auto households =	20 dwelling units
32%	1 auto households =	320 dwelling units
52%	2 auto households =	520 dwelling units
14%	3 auto households =	140 dwelling units

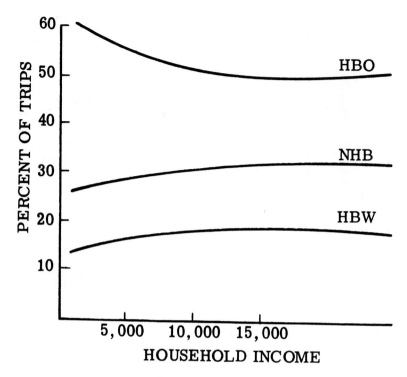

Figure 11-6   Relationship between Percent of Trips by Categories and Household Income.

2. Enter curve *b* with income, to determine the total production (person-trips) from each household:

$$\text{Trips from 0 auto household} = 5.5 \text{ trips/DU} \times 20 \text{ DU}$$

$$= 110 \text{ trips}$$

$$\text{Trips from 1 auto household} = 12.0 \text{ trips/DU} \times 320 \text{ DU}$$

$$= 3840 \text{ trips}$$

$$\text{Trips from 2 auto household} = 15.5 \text{ trips/DU} \times 520 \text{ DU}$$

$$= 8060 \text{ trips}$$

$$\text{Trips from 3 auto household} = 17.2 \text{ trips/DU} \times 140 \text{ DU}$$

$$= 2408 \text{ trips}$$

Total trips = 14,418

Average trips/dwelling unit = 14.4

3. Enter curve *c* with income to determine the trips produced by purpose:

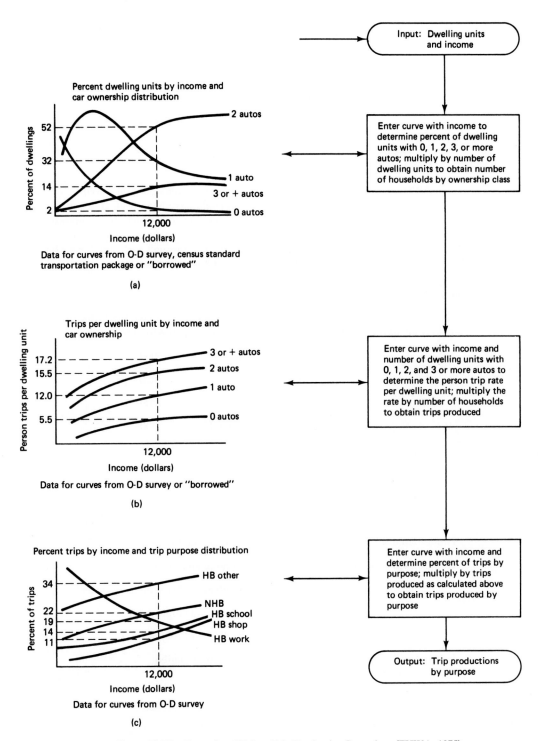

Figure 11-E3 Example of Urban Trip Production Procedure (FHWA, 1975).

Home-to-work trips $= 19\% \times 14{,}418 = 2739$ trips

Home-to-shop trips $= 11\% \times 14{,}418 = 1586$ trips

Home-to-school trips $= 14\% \times 14{,}418 = 2018$ trips

Home-to-other trips $= 34\% \times 14{,}418 = 4903$ trips

Nonhome-based $= 22\% \times 14{,}418 = 3172$ trips

Total $= 14{,}418$ trips

Discussion

Although the problem states that the average income per household $= \$ 12{,}000$, it would have been better if these characteristics could have been broken down by percent households by income category (e.g., low, medium, high), in which case, somewhat more accurate estimation would have resulted. Of course, in such a case, each step in this example would then involve multiple calculations, one pass through each step for each income category. Published census data are a good source for this information.

Example 4

A large suburban zone on the outskirts of a city is likely to have the following activities and housing development in the next 10 years. Calculate the total trip attractions.

Number of dwelling units $= 3000$

High school students $=\ \ 800$

Elementary school students $= 1800$

Shopping center retail employees $=\ \ 200$

Other retail employees $=\ \ 100$

Nonretail employees $=\ \ \ 50$

Solution

The trip attractions are obtained as follows, based on sample attraction rates indicated:

Home-based work attractions $= 1.7$ (total zonal employees) $= 1.7 \times 350 = 595$

Home-based shop attractions $= 2.0$ (CBD retail employees)

$+ 9.0$ (shopping center retail employees)

$+ 4.0$ (other retail employees)

$= 2(0) + 9(200) + 4(100) = 2200$

Home-based school attractions $= 0.90$ (university students)

$+ 1.60$ (high school students)

$+ 1.20$ (other school students)

$= 0.9(0) + 1.6(800) + 1.2(1800) = 3440$

Home-based other attractions $\quad = 0.70$ (number of households)

$\qquad\qquad + 0.60$ (nonretail employees)

$\qquad\qquad + 1.10$ (CBD retail employees)

$\qquad\qquad + 4.00$ (shopping center retail employees)

$\qquad\qquad + 2.30$ (other retail employees)

$\qquad\qquad = 0.7 (3000) + 0.6 (50) + 1.1 (0) + 4.0 (200) + 2.3 (100)$

$\qquad\qquad = 3160$

Nonhome-based attractions $\quad = 0.30$ (number of households)

$\qquad\qquad + 0.4$ (nonretail employees)

$\qquad\qquad + 1.00$ (CBD retail employees)

$\qquad\qquad + 4.60$ (shopping center retail employees)

$\qquad\qquad + 2.30$ (other retail employees)

$\qquad\qquad = 0.3 (3000) + 0.4 (50) + 1.0 (0) + 4.6 (200) + 2.3 (100)$

$\qquad\qquad = 2070$

Total attractions $= 11{,}465$

13. TRIP DISTRIBUTION

After the trip-generation stage, the analyst knows the numbers of trip productions and trip attractions that each zone shown in Figure 11-5 will have. But where do the attractions in zone 1 come from and where do the productions go? What are the zone-to-zone travel volumes?

Trip-distribution procedures determine where the trips produced in each zone will go—how they will be divided among all other zones in the study area. The output is a set of tables that show the travel flow between each pair of zones. In a hypothetical five-zone city, zone 1 may produce 2000 trips, and zones 1, 2, 3, 4, and 5 may attract 300, 600, 200, 800, and 100 trips, respectively. The decision on where the trips go is represented by comparing the relative attractiveness and accessibility of all zones in the area.

There are several methods of trip distribution analysis: the Fratar method, the intervening opportunity model, and the gravity model. The Fratar method and the gravity model are discussed here because they are the most widely used.

13.1 The Fratar Method

While working with the Cleveland, Ohio Metropolitan Region, T. J. Fratar used a simple method to distribute trips in a study area. He made the following assumptions: (1) the distribution of future trips from a given origin zone is proportional to the present trip distribution, and (2) this future distribution is modified by the growth factor of the zone to which these trips are attached.

The Fratar formula can be written as

$$t_{ij}^f = t_{ij}^0 \frac{O_i^f}{O_i^0} \frac{D_j^f}{D_j^0} \frac{\sum\limits_{k=1}^{n} t_{ik}^0}{\sum\limits_{k=1}^{n} (D_k^f/D_k^0)(t_{ik}^0)} \tag{7}$$

where

O_i^f and O_i^o = future and base-year origin trips from zone i
D_j^f and D_j^0 = future and base-year destination trips to zone j
t_{ij}^f and t_{ij}^0 = future and base-year trips from i to j

This model has been used extensively in several metropolitan study areas, particularly for estimating external trips coming from outside the study areas to zones located within the study area. As can be detected, the Fratar model is not capable of being applied in a situation where a new zone (origin or destination) is created after the base-year volume has been determined. Also, the model does not in any shape or form take account of impedance between zone traffic movement.

Example 5

An origin zone i with 20 base-year trips going to j, k, and l numbering 4, 6, and 10, respectively, has growth rates of 2, 3, 4, and 5 for i, j, k, and l, respectively in 25 years. Determine the future trips from i to j, k, and l in the future year.

Solution

Refer to Figure 11-E5. $t_{ij}^0 = 4$; $t_{ik}^0 = 6$; $t_{il}^0 = 10$. Growth: $i = 2$; $j = 3$; $k = 4$; $l = 5$.

$$O_i^f = 20 \times 2 = 40$$

$$D_j^f = 4 \times 3 = 12$$

$$D_k^f = 6 \times 4 = 24$$

$$D_l^f = 10 \times 5 = 50$$

$$\Sigma D^f = 12 + 24 + 50 = 86$$

$$t_{ij}^f = (4 \times 2 \times 3)\frac{4 + 6 + 10}{86} = 6$$

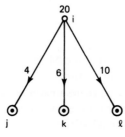

Figure 11-E5   Trip Diagram for Example 5.

$$t_{ik}^f = (6 \times 2 \times 4) \frac{4 + 6 + 10}{86} = 11$$

$$t_{il}^f = (10 \times 2 \times 5) \frac{4 + 6 + 10}{86} = 23$$

Total = 40

Notice that in this problem, the future origin ($20 \times 2 = 40$) and the future destinations $[(4 \times 3) + (6 \times 4) + (10 \times 5)] = 86$, are not balanced ($40 < 86$). Although the future trips produced by zone i meet the requirement that $O_u^f/O_i^o = 2$, the requirement that $D_j^f/D_j^o = 3$, $D_R^f/D_R^o = 4$, and $D_e^f/D_i^o = 5$ are not met. Example 6 shows how such a balance can be achieved, through iterations.

Example 6

A 3-by-3 trip table representing a total of 2500 trips is shown in the following table, which is for the base year.

Orig. \ Dest.	1	2	3	Total
1	1	4	2	7
2	6	2	3	11
3	4	1	2	7
Total	11	7	7	25

The next table indicates the origin and destination growth factors for the horizon year.

Zone	1	2	3
Origin factor (production)	2.0	3.0	4.0
Destination factor (attraction)	3.0	4.0	2.0

Use the Fratar technique to distribute the trips in the horizon year.

Solution

In the horizon year, the desired trip table should resemble the following matrix, where the row and column total equal the corresponding base-year totals multiplied by the origin and destination growth factors.

Orig. \ Dest.	1	2	3	Total
1	x	x	x	14
2	x	x	x	33
3	x	x	x	28
Total	33	28	14	75

Finding the missing x values (ij) is the main purpose of the Fratar technique. The next step is to multiply the destination growth factors (DGF) by the cell numbers, giving the following matrix.

Orig. \ Dest.	1	2	3	Actual total	Desired total	Row factors
1	3.00	16.00	4.00	23.00	14.00	0.61
2	18.00	8.00	6.00	32.00	33.00	1.03
3	12.00	4.00	4.00	20.00	28.00	1.40
Total	33.00	28.00	14.00	75.00		

However, the actual row totals and the desired row totals do not match and a set of row factors to correct them is calculated. Now we multiply the row factors by the cell figures in the preceding matrix and obtain cell values as follows:

Orig. \ Dest.	1	2	3	Total
1	2	10	2	14
2	19	8	6	33
3	16	6	6	28
Actual total	37	24	14	75
Desired total	33	28	14	
Column factors	0.89	1.17	1.00	

Notice that, again, the column totals do not match the desired column totals, and therefore a set of column factors are derived that will possibly correct the situation. The column factors are multiplied by the cell figures of matrix, giving us a new matrix:

Orig. \ Dest.	1	2	3	Actual total	Desired total	Row factors
1	1.78	11.70	2.00	15.48	14	0.90
2	16.90	9.36	6.00	32.26	33	1.02
3	14.24	7.02	6.00	27.26	28	1.03
Total	32.92	28.08	14.00	75.00		

Once again, the row totals and column totals are calculated and the process goes through for a second time, producing a matrix that is good enough for planning purposes.

Orig. \ Dest.	1	2	3	Total
1	2	10	2	14
2	17	10	6	33
3	15	7	6	28
Actual total	34	27	14	75
Desired total	33	28	14	
Column factors	0.97	1.04	1.00	

It is possible to continue for one more iteration and get the following result:

Orig. \ Dest.	1	2	3	Total
1	2	10	2	14
2	16	11	6	33
3	15	7	6	28
Total	33	28	14	75

In summary, the column factor convergence is as follows:

Column factor	1	2	3
First iteration	0.89	1.17	1.00
Second iteration	0.97	1.04	1.00
Third iteration	1.00	1.00	1.00

and the row factor convergence is as follows:

Row factor	First	Second	Third
1	0.61	0.90	0.98
2	1.03	1.02	1.01
3	1.04	1.03	1.01

Discussion

The planner can continue to iterate until satisfied that the destinations calculated almost match the destinations desired. Generally, two iterations will do the job.

13.2 The Gravity Model

The gravity model derives its base from Newton's law of gravity, which states that the attractive force between any two bodies is directly related to their masses and inversely related to the distance between them. Similarly, in the *gravity model*, the number of trips between two zones is directly related to activities in the two zones, and inversely related to the separation between the zones as a function of the travel time.

The gravity model formula appears as follows:

$$T_{ij} = \frac{P_i \, A_j \, F(t)_{ij} K_{ij}}{\displaystyle\sum_{j=1}^{n} A_j F(t)_{ij}} \tag{8}$$

where

T_{ij} = number of trips produced in zone i and attracted to zone j
P_i = trips produced from zone i
A_j = trips attracted to zone j
$F(t)_{ij}$ = friction factor for interchange ij (based on travel time between i and j)
i = origin zone
j = destination zone
n = number of zones in the study area
K_{ij} = socio-economic characteristics of zones.

The gravity model states that the trips (P_i) produced in zone i will be distributed to each other zone j (T_{ij}) according to the relative attractiveness of each zone j $(A_j/\Sigma A_j)$ and the relative accessibility of each zone $j [F(t)_{ij}/\Sigma F(t)_{ij}]$; this means that

$$\text{Trips between } i \text{ and } j = \text{trips produced at } i \times \frac{\substack{\text{attractiveness and accessibility} \\ \text{characteristics of } j}}{\substack{\text{attractiveness and accessibility} \\ \text{characteristics of all zones} \\ \text{in the area}}} \tag{9}$$

Thus, zone j gets a portion of zone i's trip productions according to its characteristics as compared to the characteristics of all other zones in the study area. This leads to the term *share model*, often applied to the gravity model and other models having these characteristics.

In practice, a separate gravity model is developed for each trip purpose, because different trip purposes have different distribution characteristics. Before the number of trip interchanges can be computed, several parameters must be defined. The travel time between each pair of zones in the study area is determined by the trip assignment process.

A more general term used to represent travel time (for the separation between zones) is *impedance*. Impedance can represent travel time, cost, distance, or a combination of factors. Generally, impedance is a weighted sum of various types of times (walking, waiting, riding) and types of cost (fares, operating cost, tolls, parking cost). In the past, travel time was used in the gravity model to measure separation and to develop friction factors. We now have the capability to include other factors, such as tolls and operating cost, in the impedance function, which more accurately represent the separation between zones.

Another parameter (K_{ij}) that can be used reflects the unique socioeconomic characteristics of the various zones—characteristics that are not otherwise accounted for—and how these characteristics affect the travel patterns in the study area. These adjustment factors affect the number of trip interchanges determined by the model. Like the friction factors, socioeconomic factors are determined in the calibration process. They should be used with caution and only when justified to account for an area's unique characteristics.

Example 7

We need to distribute 602 work-trip productions from zone 3 to zones 1, 2, 4, and 5. The numbers of work-trip productions and attractions were determined in the trip-generation phase (Figure 11-E7). Values of $F(t)_{ij}$ were obtained from past records.

Solution

Socioeconomic (K_{ij}) factors are not used in this problem. T_{ij} is calculated as follows:

From zone:	To zone:	Attractions, A_j	Impedance (min)	$F(t)_{ij}$	$A_j F(t)_{ij}$	T_{ij}
3	1	1080	20	6	6480	147
3	2	531	7	29	15,399	350
3	3	76	5	45	3420	78
3	4	47	10	18	846	19
3	5	82	25	4	328	8
Total					26,473	602

Source: FHWA/UMTA, 1977, pp. 4–16.

As can be seen, the computations in even a very simple situation can be complex and cumbersome. In an actual situation, several more steps are necessary, such as balancing attractions to assure that the gravity model does not distribute more trips to a zone than it attracts according to trip-generation analysis.

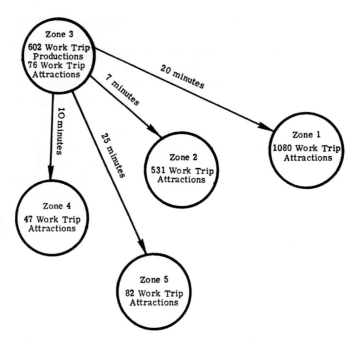

Figure 11-E7   Trip Distribution for Small City USA, for Example 7 (FHWA/UMTA, 1977).

13.3 Calibrating a Gravity Model

Calibration of the gravity model is accomplished by developing friction factors and developing socioeconomic adjustment factors. As noted before, friction factors reflect the effect travel time of impedance has on trip making. A trial-and-error adjustment process is generally adopted. One way, of course, is to use the factors from a past study in a similar urban area.

Three items are used as input to the gravity model for calibration:

1. Production-attraction trip table for each purpose
2. Travel times for all zone pairs, including intrazonal times
3. Initial friction factors for each increment of travel time

Essentially, then, the calibration process involves adjusting the friction factor parameter until the planner is satisfied that the model adequately reproduces the trip distribution as represented by the input trip table—until the model's trip table agrees substantially with the table from the survey data, using indications such as the trip-time frequency distribution and the average trip time. The process is as follows:

1. Use the gravity model to distribute trips based on initial inputs.
2. Total trip attractions at all zones j, as calculated by the model, are compared to those obtained from the input "observed" trip table.

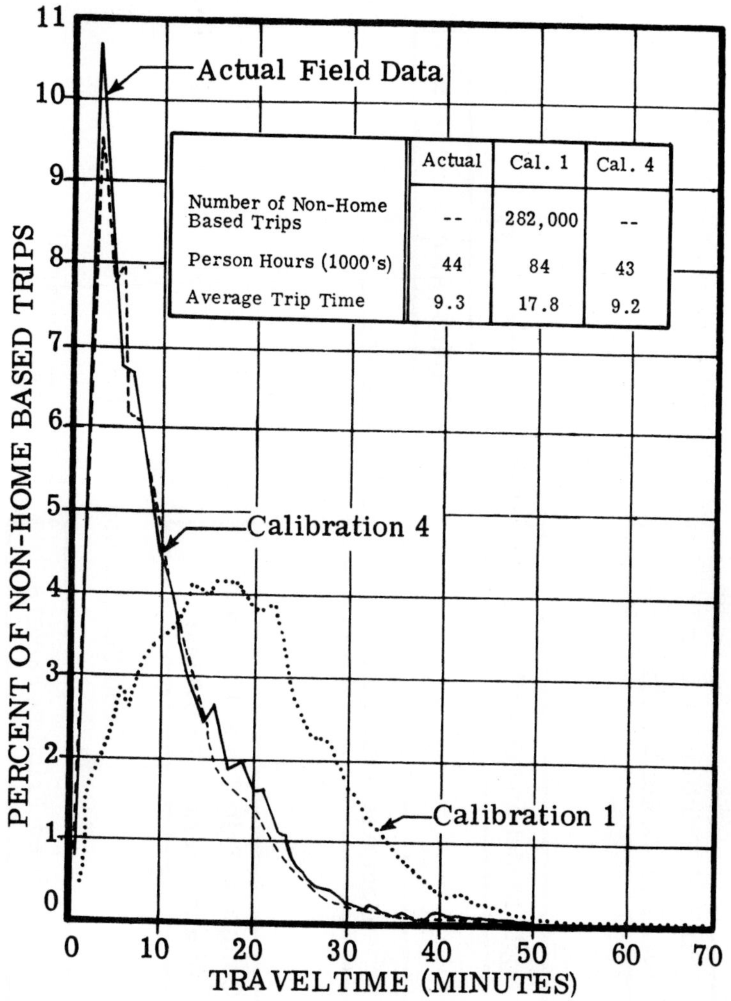
Actual Field Data
Actual
Cal. 1
Cal. 4
Number of Non-Home Based Trips
--
282,000
--
Person Hours (1000's)
44
84
43
Average Trip Time
9.3
17.8
9.2
Calibration 4
Calibration 1
PERCENT OF NON-HOME BASED TRIPS
0 1 2 3 4 5 6 7 8 9 10 11
0 10 20 30 40 50 60 70
TRAVELTIME (MINUTES)

Figure 11-7 Travel Times versus Percent of Nonhome-Based Trips (FHWA/UMTA, 1977).

3. If this comparison shows significant differences, the attraction A_j is adjusted for each zone, where a difference is observed.
4. The model is rerun until the calculated and observed attractions are reasonably balanced.
5. The model's trip table and the input travel time table can be used for two comparisons: the trip-time frequency distribution and the average trip time. If there are significant differences, the process begins again.

Figure 11-7 shows the results of four iterations comparing travel-time frequency. The flowchart (Figure 11-8) is intended to illustrate the sequence of steps needed to calibrate the gravity model. An example of smoothed values of F factors is illustrated in

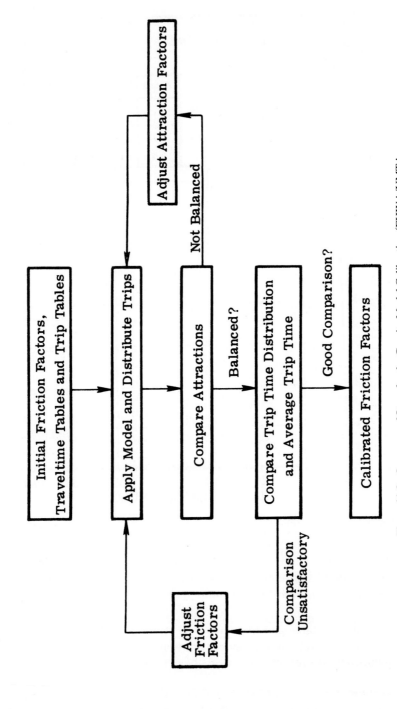

Figure 11-8 Sequence of Steps for the Gravity Model Calibration (FHWA/UMTA, 1977).

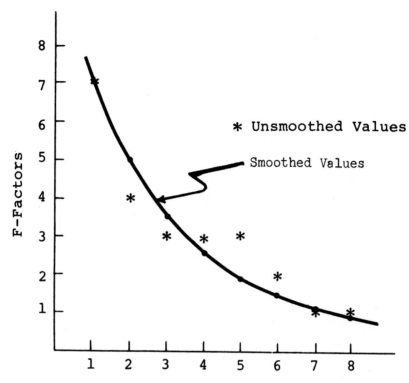

Figure 11-9   Smoothed Adjusted Factors, Calibration 2 (FHWA, 1977).

Figure 11-9. Note that, in general, values of F decreases as travel times increase, and may take the form F varies as t^{-1}, t^{-2}, or e^{-t}. As noted earlier, a more general term used for representing travel time (or a measure of separation between zones) is impedance, and the relationship between a set of impedances (W) and friction factors (F) can be written as:

$$F_{ij} = 1/W_{ij}^c \qquad (10)$$

Example 8

A gravity model was calibrated with the following results:

Impedances (travel times, min)	$W =$ 4	6	8	11	15
Friction factors	$F =$ 0.035	0.029	0.025	0.021	0.019

Using F as the dependent variable, calculate parameters A and c of the equation

$$F = A/W^c$$

Solution

This equation can be written as

$$\ln F = \ln A - c \ln W$$

and

$$\ln W = \quad 1.39 \qquad 1.79 \qquad 2.08 \qquad 2.40 \qquad 2.71$$
$$\ln F = -3.35 \qquad -3.54 \qquad -3.69 \qquad -3.86 \qquad -3.96$$

These figures yield the following values of $A = 0.07$ and $c = 0.48$. Hence $F = 0.07 \, W^{-0.48}$.

Example 9

(a) A four-zone city has two residential zones, A and B, generating 725 and 575 trips, respectively. These trips go to two employment zones C and D, attracting 875 and 425 trips, respectively. The travel time, in minutes, between zones is $AC = 8$, $BC = 10$, $BD = 13$, and $AD = 15$. Friction factors (F_{ij}) corresponding to these travel times are 90, 60, 50, and 10, respectively, taken from the gravity model (GM). The socioeconomic factor $K_{ij} = 1$. What is the distribution of trips?

Solution

Productions should be equal to attractions:

From $A + B = 725 + 575 = 1300$ trips produced

To $C + D = 875 + 425 = 1300$ trips attracted

Further,

$$T_{ij} = \frac{T_i A_j F_{ij} K_{ij}}{\Sigma A_j F_{ij} K_{ij}}$$

$$T_{AC} = \frac{725 \times 875 \times 90 \times 1}{(875 \times 90 \times 1) + (425 \times 10 \times 1)} = 688 \left.\begin{array}{c} \\ \\ \end{array}\right\}$$

$$T_{AD} = \frac{725 \times 425 \times 10 \times 1}{(875 \times 90 \times 1) + (425 \times 10 \times 1)} = 37 \qquad \left.\begin{array}{c} \\ \\ \end{array}\right\} = 725$$

$$T_{BC} = \frac{575 \times 875 \times 60 \times 1}{(875 \times 60 \times 1) + (425 \times 50 \times 1)} = 409 \left.\begin{array}{c} \\ \\ \end{array}\right\}$$

$$T_{BD} = \frac{575 \times 425 \times 50 \times 1}{(875 \times 60 \times 1) + (425 \times 50 \times 1)} = 166 \qquad \left.\begin{array}{c} \\ \\ \end{array}\right\} = 575$$

Route	Travel time (min)	F_{ij}	K_{ij}	Trips
AC	8	90	1	688
AD	15	10	1	37
BC	10	60	1	409
BD	13	50	1	166

(b) The origin–destination (O–D) survey that was performed for this city indicates that the number of trips on each route was as follows: $AC = 650$; $AD = 75$; $BC = 400$; $BD = 175$. Determine the new F_{ij}'s in order to replicate the actual trip movements.

Route	Percent O–D trips		Percent computed by GM		O–D% / GM%	Old F_{ij}	New F_{ij}
AC	650	50.0	688	52.9	0.94	90	85
AD	75	5.8	37	2.8	2.07	10	21
BC	400	30.7	409	31.5	0.98	60	59
BD	175	13.5	166	12.8	1.05	50	53
Total	1300	100.0	1300	100.0			

Check the results from the gravity model with new F_{ij}.

$$T_{AC} = \frac{725 \times 875 \times 85 \times 1}{(875 \times 85 \times 1) + (425 \times 21 \times 1)} = 647$$

$$T_{AD} = \frac{725 \times 425 \times 21 \times 1}{(875 \times 85 \times 1) + (425 \times 21 \times 1)} = 78$$

$\left. \right\} = 725$

$$T_{BC} = \frac{575 \times 875 \times 59 \times 1}{(875 \times 59 \times 1) + (425 \times 53 \times 1)} = 400$$

$$T_{BD} = \frac{575 \times 425 \times 53 \times 1}{(875 \times 59 \times 1) + (425 \times 53 \times 1)} = 175$$

$\left. \right\} = 575$

The result derived by using the new F_{ij}'s appears satisfactory.

Example 10

A three-zone city has the following trips produced in and attracted to the three zones as follows.

ZONE	1	2	3	TOTAL
Trips produced (P_i)	700	200	0	900
Trips attracted (A_j)	0	400	500	900

The impedance and corresponding friction factors have been calibrated as follows:

Impedance	2	4	6	8	(Travel times in min)
Friction factors (F_{ij})	10	7	6	5	

Travel times (min)

Orig. \ Dest.	1	2	3
1	2	4	6
2	4	2	8
3	6	8	2

Distribute the trips between the zones, assuming $k_{ij} = 1$.

Solution

First iteration ($m = 1$):

$$T_{ij} = P_i(A_j.F_{ij})/(\Sigma A_j.F_{ij})$$
$$T_{1\text{-}1} = (700)(0)(10)/[(0)(10) + (400)(7) + (500)(6)] = 0$$
$$T_{1\text{-}2} = [(700)(400)(7)]/5800 = 338$$
$$T_{1\text{-}3} = [(700)(500)(6)]/5800 = 362$$
$$T_{2\text{-}1} = [(200)(0)(7)]/[(0)(7) + (400)(10) + (500)(5)] = 0$$
$$T_{2\text{-}2} = [(200)(400)(10)]/6500 = 123$$
$$T_{2\text{-}3} = [(200)(500)(5)]/6500 = 77$$

$T_{3\text{-}1}$, $T_{3\text{-}2}$, and $T_{3\text{-}3}$ are all equal to zero.

Zone-to-zone trips: first iteration:

Orig. \ Dest.	1	2	3	Total
1	0	338	362	700
2	0	123	77	200
3	0	0	0	0
Total	0	461	439	900

Computed A_1	0	461	439	900
Given A_j	0	400	500	900

Notice that although the sum of the trip productions and the sum of the trip attractions add up to 900, the total trip attractions (A_j) do not equal the desired trip attractions. Therefore, further iterations are necessary. Adjusted attraction factors according to the following expression can be calculated.

$$A_{jk} = [A_j.A_{j(k-1)}]/C_{j(k-1)}$$

A_{jk} = adjusted attraction factor for attraction zone (column) j, iteration k
A_{jk} = A_j, when $k = 1$
C_{jk} = actual attraction (column) total for zone j, iteration k
A_j = desired attraction total for attraction zone (column) j
j = attraction zone number, $j = 1, 2, 3, \ldots, n$
n = number of zones
k = iteration number, $k = 1, 2, 3, \ldots, m$
m = number of iterations

Calculate the adjusted attraction factors according to the following formula:

$$A_{jk} = [A_j.A_{j(k-1)}]/C_{j(k-1)}$$

Zone 1: no adjustment needed; $A_1 = 0$

Zone 2: $A_2 = 400 \times 400/461 = 347$

Zone 3: $A_3 = 500 \times 500/439 = 569$

Second iteration ($m = 2$):

$$T_{1-1} = [(700)(0)(10)]/[(0)(10) + (347)(7) + (569)(6)] = \quad 0$$

$$T_{1-2} = [(700)(347)(7)]/5843 \qquad\qquad\qquad = 291$$

$$T_{1-3} = [(700)(569)(6)]/5843 \qquad\qquad\qquad = 409$$

$$T_{2-1} = [(200)(0)(7)]/[(0)(7) + (347)(10) + (569)(5)] = \quad 0$$

$$T_{2-2} = [(200)(347)(10)]/6315 \qquad\qquad\qquad = 110$$

$$T_{2-3} = [(200)(569)(5)]/6315 \qquad\qquad\qquad = \quad 90$$

Zone-to-zone trips: second iteration:

Orig. \ Dest.	1	2	3	Total
1	0	291	409	700
2	0	110	90	200
3	0	0	0	0
Total	0	401	499	900

Computed A_2	0	401	499	900
Given A_j	0	400	500	900

Discussion

The convergence of the computed attractions is very close to given attractions. This process is usually continued until there is reasonable agreement between estimated values using the gravity model and the values furnished.

14. MODE USAGE

In this phase of travel-demand forecasting, we analyze people's decisions regarding mode of travel; auto, bus, train, and so on, are analyzed. In the flowchart of the travel-demand forecasting process, mode usage comes after trip distribution. However, mode usage analysis can be done at various points in the forecasting process. Mode usage analyses are also commonly done within trip-generation analyses. The most common point is after trip distribution, because the information on where trips are going allows the mode usage relationship to compare the alternative transportation services competing for users.

Before we can predict how travel will be split among the modes available to the travelers, we must analyze the factors that affect the choices that people make. Three broad categories of factors are considered in mode usage:

1. The characteristics of the trip maker (e.g., family income, number of autos available, family size, residential density)
2. The characteristics of the trip (e.g., trip distance, time of day)
3. The characteristics of the transportation system (e.g., riding time, excess time)

The planner looks at how these characteristics interact to affect the trip maker's choice of mode. When the relationships have been discovered, the planner can predict how the population of the future will choose from among the modes that will be available.

Generally, at this point in the forecasting process, some consideration is given to predicting the number of occupants in autos for those choosing that mode. This consideration of auto occupancy either can be included in the mode usage relationship with each level of occupancy being considered a separate mode, or a separate relationship might be developed.

14.1 Direct-Generation Usage Modes

Figure 11-10(a) shows how direct generation works in a two-mode situation: generation of auto trips and transit trips. The trips generated by mode are distributed to their destinations and are assigned to highway and transit networks. This approach is generally appropriate to smaller urban areas without major transit service. Equations connecting trips to population (POP), income (INC), and automobiles (AUTO), may take the form

$$P(\text{transit}) = A + B(\text{POP}) - C(\text{INC})$$

$$P(\text{auto}) = A + B(\text{POP}) + C(\text{AUTO})$$

Another way to develop direct-generation models is to use trip-generation models that generate total trips, similar to what was explained under category analysis, and then develop cross-classification tables for the areas that are served by transit. One of the tables developed before is repeated below, along with a table that estimates transit trips (Table 11-7). Notice that when income is high and automobile availability is 2+, transit usage is almost zero, which is logical. These tables can be converted to graphs, for general use.

14.2 Trip-Interchange Mode Usage Models

Figure 11-10(b) shows trip-interchange models in the travel forecasting process. Trip-interchange mode usage models are used after the trip distribution phase. Income, auto availability, and trip purpose are widely used in these models; hence, the analyst has a strong measure of characteristics of the transportation system.

Consider, for example, a model in which income is used as the trip-making characteristic. Trip purpose and trip orientation (to CBD or elsewhere) are used as trip

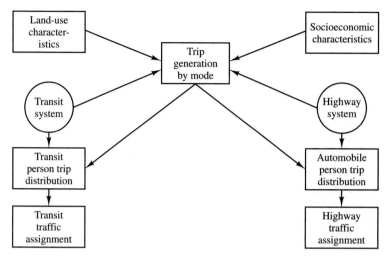

(a) Direct-generation usage mode

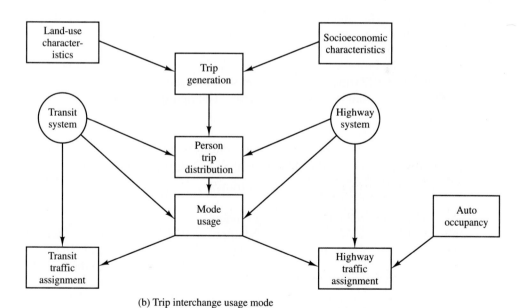

(b) Trip interchange usage mode

Figure 11-10   Modal Split Model: (a) Direct-Generation Usage Mode; (b) Trip Interchange Usage Mode (FHWA/UMTA, 1977).

TABLE 11-7 TRANSIT TRIP ESTIMATION WITH
CATEGORY ANALYSIS

Income (thousands of dollars)	Autos available		
	0	1	2 or more
Total person-trips per household			
≤ 6	3.0	5.0	—
6–9	4.0	6.0	9.0
9–12	5.0	7.5	10.5
12–15	—	8.5	11.5
> 15	—	8.5	12.7
Total transit trips produced per household in areas with transit service			
≤ 6	2.1	0.7	0.4
6–9	1.3	0.4	0.3
9–12	0.5	0.2	0.1
12–15	0.2	0.1	0.0
> 15	0.1	0.0	0.0

characteristics. The characteristics of the transportation system are described in terms of the difference in cost, in-vehicle time, and excess time for each pair of zones in the study area.

Figure 11-11 is for CBD work trips made by people with a household income of $10,000. Non-CBD trips, trips with other purposes, and trips made by people with different incomes would have different curves. The figure is read as follows:

1. Auto in-vehicle time is 15 min less than transit in-vehicle time (difference − 15).
2. Auto usage cost 25 cents more than transit usage (difference + $0.25).
3. Excess time for auto is 3 min more than transit (difference + 3).
4. Therefore, 37% of trips will be by transit.

The trip-interchange mode usage model is appropriate for any size urban area with any level of transit usage. It is most appropriate in larger urbanized areas with an appreciable level of transit usage, at least in some areas. This model is an example of a family of models using a logit formulation. The term *logit* refers to the S-shaped logit curve shown in Figure 11-12, used to fit the model data.

The logit formulation is a share model (as was the gravity model) that divides the persons between the various modes depending on each mode's relative desirability for any given trip. Modes are said to be relatively more desirable if they are faster, cheaper, or have other more favorable features than competitive modes.

The better a mode is, the more utility it has for the potential traveler. The logit model takes the following form to trade off the relative utilities of various modes: Probability of using mode i, P_i, is given by

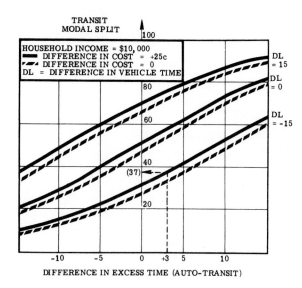

Figure 11-11   Difference in Excess Time (Auto–Transit)—Modal Split Model (FHWA/UMTA, 1977).

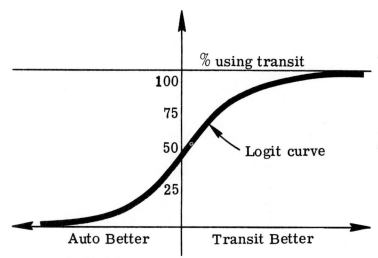

Figure 11-12   Logit Curve (FHWA/UMTA, 1977).

$$P_i = \frac{e^{V(i)}}{\sum_{r=1}^{n} e^{V(r)}} \qquad (11)$$

where

$V(i)$ = utility of mode i

$V(r)$ = utility of mode r

n = number of modes in consideration

Example 11

The calibrated utility functions for auto and transit travel are

Auto: $V_a = -0.3 - 0.04X - 0.1Y - 0.03C$
Transit: $V_t = -0.04X - 0.1Y - 0.03C$

where

V_i = utility function of mode i
X = in-vehicle travel time
Y = out-of-vehicle travel time
C = cost of travel/income

A traffic zone has the following characteristics:

	Auto travel	Transit travel
In vehicle time (min)	15	20
Out-of-vehicle time (min)	5	10
Travel cost (cents)	300	75

What is the probability that a person with an income of $10,000 will travel by transit?

Solution

$$V_a = -0.3 - 0.04\,(15) - 0.1\,(5) - 0.03\left(\frac{300}{10,000}\right) = -1.4$$

$$V_t = -0.04\,(15) - 0.1\,(10) - 0.03\left(\frac{75}{10,000}\right) = -2.6$$

The probability of the trip maker taking transit is

$$P_t = \frac{e^{V_t}}{e^{V_t} + e^{V_a}} = \frac{e^{-2.6}}{e^{-2.6} + e^{-1.4}} = 0.23 \quad \text{or} \quad 23\%$$

Example 12

A calibrated utility function for travel in a medium-sized city by automobile, bus, and light rail is

$$U = a - 0.002X_1 - 0.05X_2$$

where X_1 is the cost of travel (cents), and X_2 is the travel time (min). Calculate the modal split for the given values.

Mode	a	X_1	X_2
Automobile	−0.30	130	25
Bus	−0.35	75	35
Light rail	−0.40	90	40

If a parking fee of $1.00 per trip is imposed, what would be the split to the other two modes?

Solution

Automobile: $U_a = -0.30 - 0.002\,(130) - 0.05\,(25) = -1.81$

Bus: $U_b = -0.35 - 0.002\,(75)\ \ - 0.05\,(35) = -2.25$

Light rail: $U_1 = -0.40 - 0.002\,(90)\ \ - 0.05\,(40) = -2.58$

Mode	U	e^u	P	Percent
Automobile	−1.81	0.164	0.475	48
Bus	−2.25	0.105	0.304	30
Light rail	−2.58	0.076	0.221	22
Total		0.345	1.000	100

If a parking fee of $1.00 per trip is imposed, U_a would be $-0.3 - 0.002\,(230) - 1.25 = -2.01$.

Mode	U	e^u	P	Percent
Automobile	−2.01	0.134	0.425	43
Bus	−2.25	0.105	0.333	33
Light rail	−2.58	0.076	0.242	24
Total		0.315	1.000	100

Discussion

Even a flat parking of $1.00 makes a 5% difference in automobile ridership.

15. TRIP ASSIGNMENT

Trip assignment is the procedure by which the planner predicts the paths the trips will take. For example, if a trip goes from a suburb to downtown, the model predicts the specific streets or transit routes to be used. The trip-assignment process begins by constructing a map representing the vehicle and transit networks in the study area. The network maps show the possible paths that trips can take.

The intersections (called *nodes*) on the network map are identified, so that the sections between them (called *links*) can be identified. After the links are identified by nodes, the length, type of facility, location in the area, number of lanes, speed, and travel time are identified for each link. If transit is available, additional information, which identifies fares, headways (time between vehicles), and route descriptions, is included on a separate network. This information allows the computer to determine the paths that the traveler might take between any two points on the network and to assign trips between zones to these paths. The output of trip-assignment analysis shows the paths that all trips will take, and therefore the number of cars on each roadway and the number of passengers on each transit route.

Using these analyses of trip generation, trip distribution, mode usage, and trip assignment, the planner can obtain realistic estimates of the effects of policies and programs on travel demand. Once travel demand is known, the planner can assess the performance of alternative transportation systems and identify various impacts that the system will have on the urban area, such as energy use, pollution, and accidents. With information on how transportation systems perform and the magnitude of their impacts, planners can provide decision makers with some of the information they need to evaluate alternative methods of supplying a community with transportation services.

15.1 Trip-Assignment Procedures

Several techniques are available to determine which paths through a network are to be assigned trips between zones. Two techniques are discussed: minimum path and minimum path with capacity restraint.

Minimum-Path Techniques. Minimum-path techniques are based on the assumption that travelers want to use the minimum impedance route between two points. Efficient methods of determining minimum paths had to be developed, because manual determinations would be nearly impossible. In Figure 11-13(a), 40 different paths must be tested to determine the minimum between *A* and *B*. You can imagine the problem of finding the shortest path in a network with thousands of links and nodes.

Work that was undertaken to determine the minimum paths for long-distance telephone calls provided the help that planners needed. Rather than simply testing each path, these new *algorithms* allowed planners to find minimum paths to complete networks. The algorithm used most commonly is Moore's algorithm. Details of algorithms are given in standard texts on operations research.

In using Moore's algorithm, minimum paths are developed by fanning out from the origin to all other nodes. Determining the minimum paths from node 1 to each of the other nodes results in a "skimtree" from node 1 to all other nodes [Figures 11-13(b) and 11-13(c)].

Once the minimum paths are found, the trips between zones are loaded onto the links making up the minimum path. This technique is sometimes referred to as "all-or-nothing," because all trips between a given origin and destination are loaded on the links comprising the minimum path and nothing is loaded on the other links. After all possible interchanges are considered, the result is an estimate of the volume on each link in the network. This method can cause some links to be assigned more travel volume than the link has capacity at the original assumed speed. This volume/capacity problem led to the development of trip-assignment procedures taking capacity restraints into consideration.

Minimum Path with Capacity Restraints. Capacity-restraint techniques are based on the finding that as the traffic flow increases, the speed decreases. There is a relationship between impedance and flow for all types of highways. This relationship is shown graphically in Figure 11-14. The trip-assignment process assigns trips according

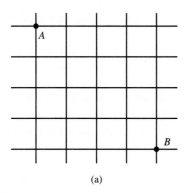

(a)

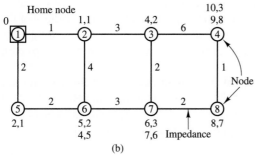

(b)

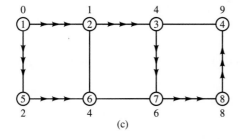

(c)

Figure 11-13 Minimum Path Technique: (a) Small Network: 24 links, 16 nodes; (b) Minimum Path through a Network; (c) Skim Tree from Node One to All Other Nodes (FHWA/UMTA, 1977).

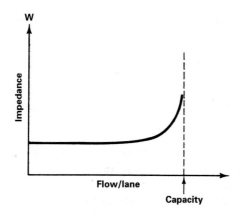

Figure 11-14 Impedance/Flow Relationship.

to the impedances coded on the links of the network. The result of this process is the traffic flow on each link of the network.

Because there is a direct relationship between travel time (or speed) on a link and the volume on the link, a process was developed to allow for consideration of this relationship. This process is called *capacity restraint*. Capacity restraint attempts to balance the assigned volume, the capacity of a facility, and the related speed.

There are several methods of utilizing capacity restraint in a trip assignment. The most common method is simply to load the network and adjust assumed link speeds after each loading to reflect volume/capacity restraints. These loadings and adjustments are done incrementally until a balance is obtained between speed, volume, and capacity. Experience has shown that a reasonable balance can be obtained after three or four loadings. This assignment should result in a more realistic representation of traffic on the network and is now in fairly widespread use. Regardless of the technique used, the highway trip-assignment procedure results in estimates of the volume of traffic on each link in the network. Figure 11-15 shows a graphical representation of a loaded network.

Capacity-restraint assignment deals with overloaded links in the network. Several methods have been suggested (Black, 1981; Blunden and Black, 1984). The Bureau of Public Roads (BPR) method is often used. This traffic-flow-dependent travel-time relationship is represented by the general polynomial function:

$$T_Q = T_0\left[1 + \alpha\left(\frac{Q}{Q_{max}}\right)^\beta\right] \tag{12}$$

where

T_Q = travel time at traffic flow Q

T_0 = "zero-flow" travel time

   = travel time at practical capacity $\times$ 0.87

Q = traffic flow (veh/hr)

Q_{max} = practical capacity = $\frac{3}{4}\times$ saturation flow

α, β = parameters

Assume that a link, 1 mile long, has a practical capacity = 40,000 veh/day and a speed at that capacity of 40 mph.

$$\text{Travel time at that volume} = 1.5 \text{ min}$$

$$\text{Travel time, } T_0 = 1.5 \times 0.87 = 1.31 \text{ min}$$

But after the link is loaded, it is found that 60,000 veh/day are assigned to it. Assume that $\alpha = 0.15$ and $\beta = 4$:

$$T = 1.31\left[1 + 0.15\left(\frac{60,000}{40,000}\right)^4\right] = 2.3 \text{ min}$$

This works out to 26 mph at which 60,000 vehicles travel per day.

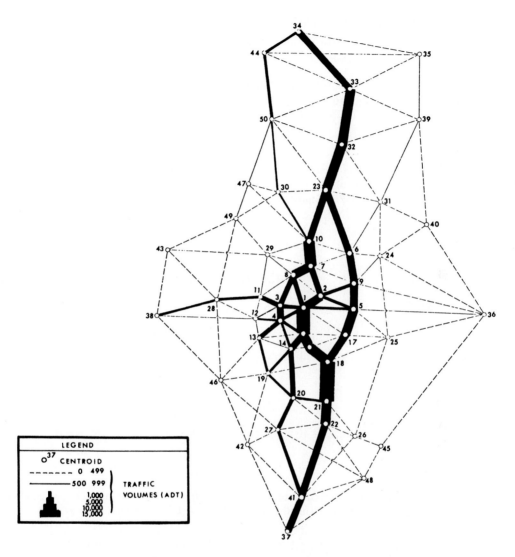

Figure 11-15   Trip Assignment to Network (FHWA/UMTA, 1977).

Davidson (1966) has suggested the use of an expression giving travel time relationship similar to the BPR formula:

$$T_Q = T_0 \frac{1 - (1 - \tau) Q/Q_{max}}{1 - Q/Q_{max}} \tag{13}$$

where

T_Q = travel time at traffic flow Q
T_0 = "zero-flow" travel time

Q = traffic flow (veh/hr)

Q_{max} = saturation flow (veh/hr)

τ = level-of-service (LOS) parameter

The LOS parameter is related to the type of road, road width, frequency of signals, pedestrian crossing, and parked vehicles. Blunden and Black (1984) suggest $\tau = 0$ to 0.2 for freeways, 0.4 to 0.6 for urban arterials, and 1 to 1.5 for collector roads.

Example 13

A freeway section 10 miles long has a free speed of 60 mph, Q_{max} = 2000 veh/hr, Q = 1000 veh/hr, τ = 0.1, α = 0.474, β = 4, and T_0 = 10 min. Apply the three methods—(a) Davidson's, (b) BPR's, and (c) Greenshields'—to find T_Q.

Solution

(a) *Davidson*:

$$T_Q = T_0 \frac{1 - (1 - \tau)Q/Q_{max}}{1 - Q/Q_{max}}$$

where T_0 = 10 min, Q = 1000 veh/hr, Q_{max} = 2000 veh/hr, and τ = 0.1.

$$T_Q = 10 \frac{1 - (1 - 0.1)1000/2000}{1 - 1000/2000} = 11 \text{ min}$$

(b) *BPR*:

$$T_Q = T_0 \left[1 + \alpha \left(\frac{Q}{Q_{max}} \right)^{\beta} \right]$$

where T_0 = 10 min, Q = 1000 veh/hr, Q_{max} = 2000 veh/hr, α = 0.474, and β = 4.

$$T_Q = 10 \left[1 + 0.474 \left(\frac{1000}{2000} \right)^4 \right] = 10.30 \text{ min}$$

(c) *Greenshields (q–k–v)*: d = 10 miles, free speed = 60 mph, Q_{max} = 2000 veh/hr, and Q = 1000 veh/hr. Using the relationship $v = A - Bk$, we have A = 60 and

$$q = \frac{A^2}{4B} = 2000 \quad \Rightarrow \quad B = \frac{60 \times 60}{2000 \times 4} = 0.45$$

Also

$$q = \left(\frac{A}{B} \right)v - \left(\frac{1}{B} \right)v^2$$

$$\Rightarrow \quad 1000 = \left(\frac{60}{0.45} \right)v - \left(\frac{1}{0.45} \right)v^2$$

Therefore, v = 51.2 mph and t = 11.7 min.

Figure 11-16 illustrates the application of the all-or-nothing assignment technique. The network is shown in Figure 11-16(a). The minimum-path trees are shown in

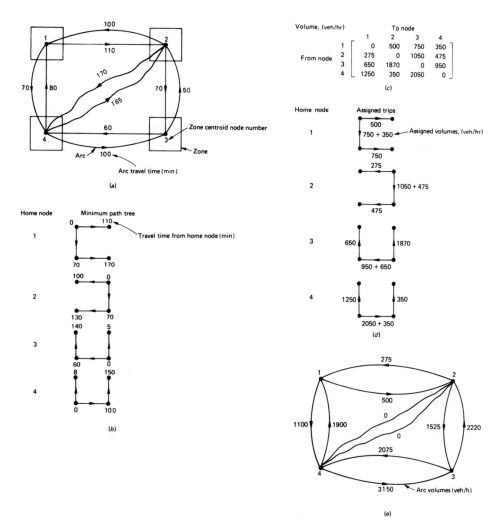

Figure 11-16 Example of All-or-Nothing Traffic Assignment: (a) Network; (b) Minimum-Path Trees; (c) Origin–Destination Trip Table; (d) Assignment of Trips to Minimum-Path Trees; (e) Assigned Traffic Volumes (Morlock, 1978).

Figure 11-16(b). The origin-to-destination flows corresponding to each node of the trip table [Figure 11-16(c)] is assigned with the links that make up the minimum path, and are shown in Figure 11-16(d). The aggregate of volumes on each link is finally shown in Figure 11-16(e).

Example 14

A highway network consisting of five nodes and eight links is shown in Figure 11-E14. The cost of transportation is also shown. A trip table showing the numbers of vehicles per hour wanting to go from one node to another is also provided. Assign the trips to the network using the all-or-nothing method. All the links are two-way. Find the total volume on each individual link and the total cost of all the trips.

Solution

See Figure 11-E14.

Flows in Networks. From a theoretical standpoint, it is interesting to know how traffic is distributed in a network. Several cases are illustrated through worked examples.

Two Links in Series. Two links are connected in series, as shown in Figure 11-17(a), having cost functions of $C_{AB} = 3 + f_{AB}$ and $C_{AB} = 2 + 2f_{AB}$. We want to find the cost function for the combination. In general, C_{ij} is the cost of travel on link ij and f_{ij} is the flow along link ij. For continuity,

$$f_{AC} = f_{AB} = f_{BC}$$

$$C_{AC} = C_{AB} = C_{BC}$$

$$= (3 + f_{AB}) + (2 + 2f_{BC})$$

$$= 5 + 3f_{AC} \quad \text{(because } f_{AC} = f_{AB} = f_{BC})$$

For example, if $f_{AC} = 100$,

$$C_{AC} = 5 + (3 \times 100) = 305 \text{ (total cost)}$$

$$C_{AB} = 3 + 100 = 103$$

$$C_{BC} = 2 + 100 = \frac{202}{305}$$

Note that when links are in series, we add the cost of each link to obtain the total cost.

Two Links in Parallel. Two links are connected in parallel, as shown in Figure 11-17(b), having cost functions of $C_1 = 2 + f_1$ and $C_2 = 2 + f_2$. We want to find the cost of the travel from A to B. Referring to Figure 11-17(b) gives

$$f_{AB} = f_1 + f_2$$

$$f_1 = C_1 - 2$$

$$f_2 = \frac{C_2}{2} - 1$$

Because $C_1 = C_2$ (Wardrop's principle) and $f_{AB} = f_1 + f_2 = 3C_1/2 - 3$,

$$C_1 = 2 + \frac{2(f_1 + f_2)}{3}$$

$$C_{AB} = C_1 + C_2 \quad \text{and} \quad C_1 = C_2$$

$$C_{AB} = C_1 + C_2 = 2C_1 = 2\left[2 + \frac{2(f_1 + f_2)}{3}\right]$$

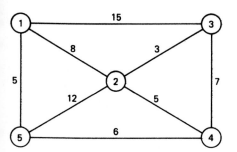

Trip Table

From \ To	1	2	3	4	5	
1	0	50	60	70	30	210
2	40	0	30	60	80	210
3	90	40	0	20	50	200
4	80	70	90	0	30	270
5	30	40	50	60	0	180
	240	200	230	210	190	1070

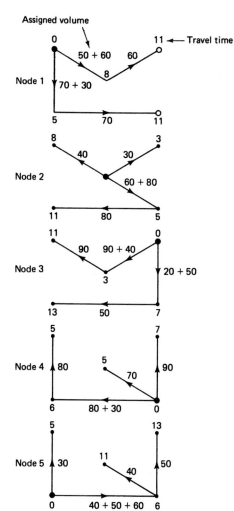

Link	Cost m	Vol hr	Total Vol	Total Cost
1 – 2	8	50 + 60	110	880
2 – 1	8	40 + 90	130	1040
1 – 3	15	0	0	0
3 – 1	15	0	0	0
1 – 5	5	70 + 30	100	500
5 – 1	5	80 + 30	110	550
2 – 3	3	60 + 30	90	270
3 – 2	3	90 + 40	130	390
2 – 4	5	60 + 80	140	700
4 – 2	5	70 + 40	110	550
2 – 5	12	0	0	0
5 – 2	12	0	0	0
3 – 4	7	20 + 50	70	490
4 – 3	7	90 + 50	140	980
4 – 5	6	80 + 50 + 80 + 30	260	1560
5 – 4	6	70 + 40 + 50 + 60	220	1320

Figure 11-E14   Solution to Example 14, Using the All-or-Nothing Method.

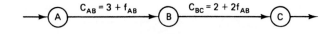

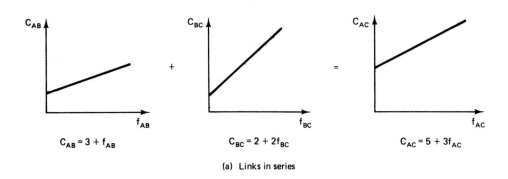

(a) Links in series

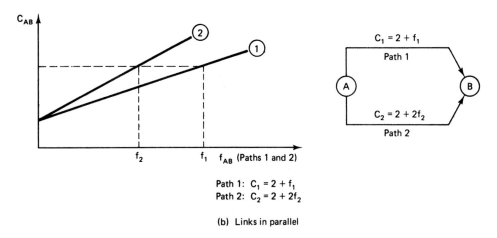

Path 1: $C_1 = 2 + f_1$
Path 2: $C_2 = 2 + 2f_2$

(b) Links in parallel

Figure 11-17   Flows in Networks.

For a cost of C_{AB}, there will be f_2 vehicles on path 2 and f_1 vehicles on path 1. For example, if $f_{AB} = 99$, $C_{AB} = 2[2 + (\frac{2}{3} \times 99)] = 136$, $C_1 = 68$, $C_2 = 68$; $f_1 = 68 - 2 = 66$ and $f_2 = 34 - 1 = 33$.

Transport Networks with a Demand Function

Case 1: Single Road (Two Links in Series).   As before, we have two links having cost functions as noted in what follows. By referring to Figure 11-18(a), the supply functions are as follows:

Link *AB:* $C_{AB} = 4 + 2f_{AB}$
Link *AB:* $C_{BC} = 4 + 4f_{BC}$

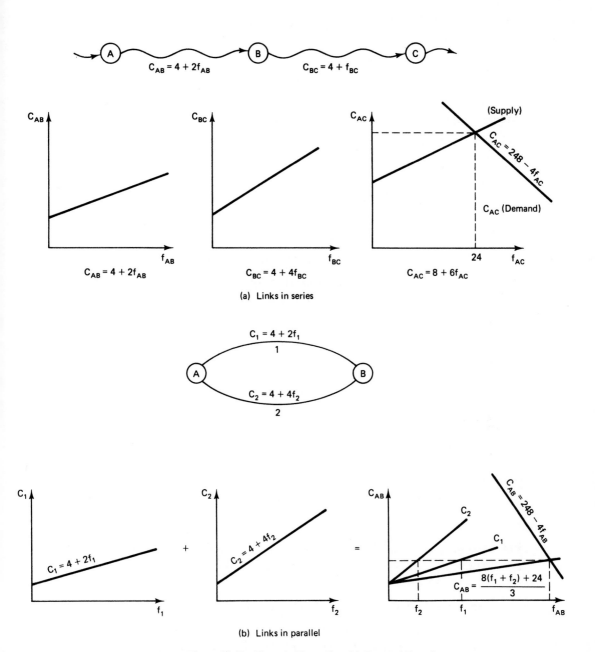

(a) Links in series

(b) Links in parallel

Figure 11-18   Flows in Networks with Demand Function.

For continuity, $f_{AC} = f_{AB} = f_{BC}$. Therefore,

$$C_{AC} = C_{AB} + C_{BC} = (4 + 2f_{AB}) + (4 + 4f_{BC})$$
$$= (8 + 6f_{AC})$$

Demand function:

$$C_{AC} = 248 - 4f_{AC}$$

Equating supply and demand functions, we have

$$8 + 6f_{AC} = 248 - 4f_{AC}$$

$$10f_{AC} = 240 \quad \Rightarrow \quad f_{AC} = 24$$

$$C_{AB} = 4 + 2f_{AB} = 4 + 2(24) = 52 \qquad (\text{because } f_{AC} = f_{AB})$$

$$C_{BC} = 4 + 4f_{BC} = 4 + 4(24) = 100 \qquad (\text{because } f_{AC} = f_{BC})$$

Total cost $C_{AC} = 152$

$$C_{AC} = 8 + 6f_{AC} = 8 + 6(24) = 152 \qquad (\text{check})$$

Case 2: Parallel Roads (Two Links).   Again, this example is similar to the one solved before. Referring to Figure 11-18(b), we find that

$$f_1 = \frac{c_1}{2} - 2$$

$$f_2 = \frac{c_2}{4} - 1$$

Therefore,

$$f_1 + f_2 = \left(\frac{c_1}{2} - 2\right) + \left(\frac{c_2}{4} - 1\right)$$

According to Wardrop's principle, $C_1 = C_2$. Therefore,

$$f_1 + f_2 = \frac{3C_1 - 12}{4}$$

and

$$C_1 = \frac{4(f_1 + f_2) + 12}{3}$$

Now

$$C_{AB} = C_1 + C_2 = 2C_1 = 2\left[\frac{4(f_1 + f_2) + 12}{3}\right]$$

$$= \frac{8(f_1 + f_2) + 24}{3}$$

so

$$\text{Supply function} = \frac{8(f_1 + f_2) + 24}{3} = C_{AB} = \frac{8(f_{AB} + 24)}{3}$$

and

$$\text{Demand function} = 248 - 4f_{AB} = C_{AB}$$

Therefore,

$$\frac{8(f_{AB} + 24)}{3} = 248 - 4f_{AB}$$

and

$$f_{AB} = 36.00$$

$$C_{AB} = \frac{8(f_1 + f_2) + 24}{3} = 104$$

$$C_{AB} = C_1 + C_2 \qquad \text{and} \qquad C_1 = C_2 = 52$$

$$f_1 = \frac{C_1}{2} - 2 = 52 - 2 = 24$$

$$f_2 = \frac{C_2}{4} - 1 = \frac{52}{4} - 1 = 12$$

$$f_{AB} = f_1 + f_2 = 36$$

Example 15

A highway connecting two small cities has the following characteristics: The time (t, min) to travel on a certain stretch of a highway is $t_1 = 12 + 0.01q_1$, where q_1 is the flow of vehicles (veh/hr). The demand function is $q = 4800 - 100t$.

(a) Estimate the equilibrium flow and travel time.

(b) The traffic department wants to close the existing highway and replace it with a better highway with a supply function of $t_2 = 12 + 0.006\,q_2$, with the same demand function. How much additional traffic will be induced by this new highway?

(c) Citizens currently using the existing highway want to continue using it, and, in addition, demand the new highway as well. What will be the equilibrium flow and travel time for this scenario, assuming the demand for travel remains unchanged (Wardrop's principle applies)?

(d) If the new road is built with a supply function of $t_3 = 10 + 0.005t_3$, and the existing highway is used as well, what would be the new equilibrium flow and time?

Solution

(a) $q_1 = 4800 - 100t_1$ and $t_1 = 12 + 0.01q_1$

Therefore, $q_1 = 4800 - 100(12 + 0.01q_1)$

$q_1 = 1800$ veh/hr and $t_1 = 30$ min

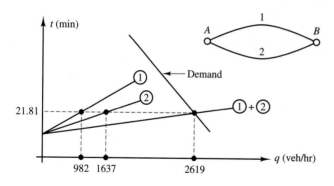

Figure 11-E15(c).

(b) $q_2 = 4800 - 100\,t_2$ and $t_2 = 12 + 0.006\,q_2$

Therefore, $q_2 = 2250$ veh/hr and $t_2 = 25.5$ min

Additional traffic induced $= 2250 - 1800 = 450$ veh/hr

(c) See Figure 11-E15(c).

$$t_1 = 12 + 0.01\,q_1;$$

Therefore, $q_1 = (t_1 - 12)/0.01$

$$t_2 = 12 + 0.006\,q_2;$$

Therefore, $q_2 = (t_2 - 12)/0.006$
and

$$q_1 + q_2 = (t_1 - 12)/0.01 + (t_2 - 12)/0.006$$

but $t_1 = t_2$ (Wardrop's principle). Therefore,

$$q_1 + q_2 = q_{AB} = 266.67 t_1 - 3200$$

and because demand $q_1 = 4800 - 100 t_1$,

$$t_1 = t_2 = 21.82 \text{ min}$$

$$q_1 = (21.82 - 12)/0.01 = 982 \text{ veh/hr}$$

$$q_2 = (21.82 - 12)/0.006 = 1637 \text{ veh/hr}$$

Combined,

$$q_1 + q_2 = 2619 \text{ veh/hr}$$

Additional induced traffic $= 2619 - 1800 = 619$ veh/hr.

(d) See Figure 11-E15(d).

Regime 1: Up to $q = 400$ veh/hr, all the flow will be on Route 2.

Regime 2: When $q > 400$ veh/hr, flow will be on Routes 1 and 2.

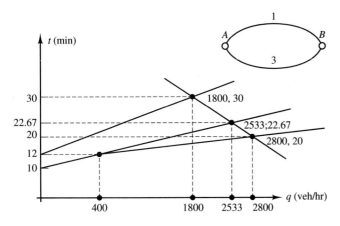

Figure 11-E15(d).

As before,

$$q_1 = (t_1 - 12)/0.01 \quad \text{and} \quad q_2 = (t_2 - 10)/0.005$$

and

$$q_1 + q_2 = 100t_1 - 1200 + 200t_2 - 2000 = Q$$

But $t_1 = t_2$, as per Wardrop's principle. Therefore,

$$300t_1 - 3200 = 4800 - 100t_1$$

$$t_1 = t_2 = 20 \text{ min}$$

and, as before,

$$q_1 = 800 \quad \text{and} \quad q_2 = 2000 \text{ veh/hr}$$

16. BACK-OF-THE-ENVELOPE–TYPE CALCULATIONS

A common local area problem encountered by planners is to estimate approximate land-use activities and derive infrastructure needs for the community. The following worked out example serves to demonstrate how such needs can be estimated approximately.

Example 16

A city planner has provided the following data for a new residential expansion of the city.

1. Residential area; single-family dwelling units (DUs) at 4 persons/DU (average DU occupancy)
2. Trip generation: 3 trips/person/day (average)
3. Trip length: 5 miles (average) on arterials
4. Residential density: 6 DUs/acre (gross)
5. Auto occupancy: 1.25
6. Modal split: 70% automobile; 30% transit

7. Peak-hour factor: 0.10

8. Lane capacity: 800 veh/hr

Determine (a) lane-miles of arterials per square mile and (b) spacing of four-lane arterials in a grid pattern. (c) Derive an expression connecting the length of arterials in terms of area of zone (A) and spacing of arterials (S).

Solution

6 DUs/acre $\times$ 640 acres/mi^2 = 3840 DUs/mi^2

3840 DUs/mi^2 $\times$ 4 persons/DU = 15,360 persons/mi^2

15,360 persons/mi^2 $\times$ 3 trips/person/day = 46,080 person-trips/day/mi^2

46,080 person-trips/day/mi^2 $\times$ 0.10 = 4608 person-trips/peak hr/mi^2

4608 person-trips/peak hr/mi^2 $\times$ 0.7 (modal split for auto use) = 3226 person-trips/peak hr by auto/mi^2

3226 person-trips/peak hr by auto/mi^2 $\div$ 1.25 (auto occupancy) = 2580 auto-vehicle trips/peak hr/mi^2

2580 auto-vehicle trips/peak hr/mi^2 $\times$ 5 mi = 12,902 veh-mi auto-veh trips/peak hr/mi^2

(a) 12,902 veh-mi of auto-vehicle trips/hr/mi^2 $\div$ 800 veh/hr/lane = 16 lane-mi/mi^2

(b) If the arterials are laid in a grid pattern, $\frac{1}{2}$ N–S and $\frac{1}{2}$ E–W, we would have 8 lane-miles/mi^2 running N–S, and 8 lane-miles/mi^2 running E–W. A four-lane arterial, therefore, would have a spacing of $\frac{4}{8} = \frac{1}{2}$ mile.

(c) See Figure 11-E16.

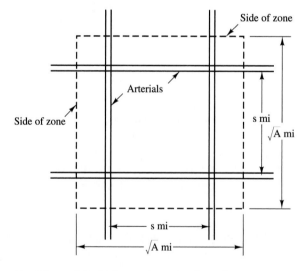

Area of zone (assumed square)	= A sq mi
Spacing of arterials	= s mi
Length of arterials	= L mi
Length of side of zone	= $\sqrt{A}$ mi
Number of arterials in one direction	= $\sqrt{A}$/s
Length of arterials in one direction	= $\{\sqrt{A}/s\}\{\sqrt{A}\}$ = A/s mi
Length of arterials in both directions	= 2A/s mi

Figure 11-E16 Zone and Arterial Grid.

17. SPECIFICATION, CALIBRATION, AND VALIDATION

Before forecasting travel, considerable effort must be expended to analyze inventory data and establish relationships among travel choices and several other variables. Discovering the reasons for making travel decisions, such as where or how to travel, is done in two steps. First, the types of models to be used and their variables are specified; and second, those models are calibrated to reproduce observed travel. Refer to Figure 11-19 to see how these steps relate to one another.

In model specification, a choice must be made among several mathematical formulations and many possible variables. Research has shown which formulations and variables will probably yield the best results; therefore, the task is somewhat simplified in testing a few options. During this step, the level of analysis for the models must be specified; that is, a decision must be made whether to model individual travel behavior or that of a larger group, such as a zone.

The calibration process is basically an attempt to duplicate travel for the year in which calibration data are available. The year for which data are available is called the *base year*. Surveys are made to see how people travel in the study area; base-year data are used to calibrate the trip-generation, trip-distribution, mode usage, and trip-assignment relationships separately. The calibration process also includes intuitive tests of models to see if the variables and their coefficients are reasonable.

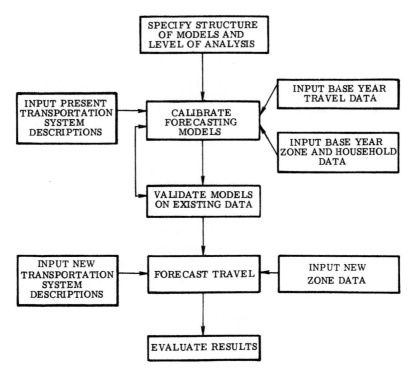

Figure 11-19 Specification, Calibration, and Validation Procedure (FHWA/UMTA, 1977).

Once the models are calibrated, they should go through a validation process by applying them sequentially to the base year in exactly the same way that they would be applied in future years. This is necessary to see if the procedure produces reasonable comparisons to the base-year observed data. For example, if models are calibrated using household information (income, family size, access time to bus, parking costs, etc.), they should be validated at the same level of aggregation to be used in forecasting (average zone income, average zone access time to bus, etc.). This validation will show how well the entire chain of calibrated models can forecast observed travel in the same way that they will be used to forecast future travel. If the series of models cannot produce traffic volumes and transit riderships similar to what is observed on roadways and bus lines, the models must be reevaluated and appropriate adjustments made. The validation process can continue in future years by comparing measured traffic volumes with model estimates. This continued reappraisal of the travel forecasting models assures that forecasts will remain as accurate as possible.

Before proceeding, it should be noted that the travel-demand forecasting process is not without problems. The forecasting process has in recent years come under criticism because of its sequential nature; first, a decision to make a trip is modeled, then a destination, then a mode, and finally, a path. Is this the way people make travel decisions? Some analysts have thought not, and have attempted with varying degrees of success to place the entire process within a single model.

Forecasting techniques are being improved continually in terms of their theoretical basis and ability to respond to changing requirements. Recent research is aimed at overcoming the problems mentioned and enhancing the capabilities to represent behavior accurately (Hutchinson, 1974; Meyer and Miller, 1984).

Despite some criticisms, the current forecasting process has been shown to work quite well over extensive periods of time with major changes in the amount of activity in the area, rising affluence of households, and changes to the transportation system. Extensive research work has been done in simplifying transportation planning and this effort continues unabated (Khisty and Alzahrani, 1984). A similar thrust to adopt inexpensive land-use models is also evident (Khisty, 1982, 1985).

SUMMARY

The transportation planning process is an extensive and expensive endeavor consuming a great deal of effort and time. One of the most important tasks involved in this process is the estimation of travel demand for the purpose of providing facilities and services. In this chapter, an overview of the transportation planning process has been described. This process consists of the following major steps:

1. Establishing goals and objectives
2. Defining the boundaries of the planning area
3. Subdividing the study area into districts and zones
4. Collecting appropriate data and forecasting the variables, such as population and economic activities

5. Running the four-step models: trip generation, trip distribution, modal split, and traffic assignment

6. Validating and refining the process

7. Evaluating the resulting and recommending actions

8. Implementing the actions

Although the planning process has matured over the last 30 years, planners are still working at simplifying the state of the art. In recent years the thrust has been to use manual methods coupled with methods that make extensive use of the microcomputer.

REFERENCES

BAERWALD, J. E., (Ed.) (1976). *ITE Transportation and Traffic Engineering Handbook,* Prentice-Hall, Englewood Cliffs, NJ.

BLACK, J. (1981). *Urban Transportation Planning,* Johns Hopkins University Press, Baltimore.

BLUNDEN, W. R., and J. A. BLACK (1984). *The Land Use/Transportation System,* 2nd Ed., Pergamon Press, Elmsford, NY.

DAVIDSON, K. B. (1966). *A Flow Traveltime Relationship for Use in Transport Planning,* Proceedings, Australian Road Research Board 3.

FEDERAL HIGHWAY ADMINISTRATION (FHWA) (1975). *Trip Generation Analysis,* U.S. Department of Transportation, Washington, DC.

FEDERAL HIGHWAY ADMINISTRATION (FHWA) (1977), *PLANPAC/BACKPAC General Information,* U.S. Department of Transportation, Washington, DC.

FEDERAL HIGHWAY ADMINISTRATION and URBAN MASS TRANSPORTATION ADMINISTRATION (FHWA/UMTA) (1977). *An Introduction to Urban Travel Demand Forecasting: A Self-Instructional Text,* U.S. Department of Transportation, Washington, DC.

GUTTMAN, I., S. S. WILKS, and J. S. HUNTER. *Introductory Engineering Statistics,* 3rd Ed. (1982). John Wiley, New York.

HUTCHINSON, B. G. (1974). *Principles of Urban Transportation Systems Planning,* McGraw-Hill, New York.

KHISTY, C. J. (1982). *Land-Use Forecasting Sensitivity Analysis: Civil Engineering for Practicing Design Engineers,* Pergamon Press, New York.

KHISTY, C. J. (1985). *Research on Appropriate Planning Methodology in Developing Countries,* Transportation Research Record 1028, National Academy of Sciences, Washington, DC.

KHISTY, C. J., and A. ALZAHRANI (1984). *Inexpensive Travel Demand Model for Small and Medium-sized Cities,* Transportation Research Record 980, National Academy of Sciences, Washington, DC.

MEYER, M. D., and E. J. MILLER (1984). *Urban Transportation Planning: A Decision-Oriented Approach,* McGraw-Hill, New York.

MORLOK, E. K. (1978), *Introduction to Transportation Engineering and Planning,* McGraw-Hill, New York.

PUBLIC ROADS (1961). *Traffic Approaching Cities,* FHWA, U.S. Department of Transportation, Washington, DC.

SOSSLAU, A. B., ET AL. (1978). *Quick-Response Urban Travel Estimation Techniques and Transferable Parameters: User's Guide,* Report 187, Transportation Research Board, Washington, DC.

TRANSPORTATION RESEARCH BOARD (TRB) (1978). *Quick Response Urban Travel Estimation Technique,* NCHRP Report 187, National Research Council, Washington, DC.

EXERCISES

1. A small study area represented by six traffic zones has the following characteristics:

Zone	1	2	3	4	5	6
Trip production	600	450	900	850	750	290
Car ownership	250	200	710	615	280	130

Set up a linear regression equation, illustrate the data, calculate R^2, and apply the t-test for significance.

2. The following equations were established for a study area.

$$T = 4.0 + 4.89X_1 - 0.004X_2 - 0.2L_3 - 0.01L_4$$

$$(t_1 = 5, t_2 = 3, t_3 = 9, t_4 = 1.2, R^2 = 0.69)$$

$$T = 30 + 5.9X_1 \quad (t = 25, R^2 = 0.45)$$

$$T = 300 + 0.25X_2 \quad (t = 41, R^2 = 0.23)$$

$$T = 25 + 2.4X_1 + 0.15X_2 \quad (t_1 = 35, t_2 = 95, R^2 = 0.85)$$

where

$T = $ trips/household
$X_1 = $ vehicle ownership (cars/household)
$X_2 = $ population density (persons/acre)
$X_3 = $ distance from the CBD (miles)
$X_4 = $ family income ($1000)

The number of observations used in setting up these equations is 53. (R^2 and t-tests results are also given.) Comment on the reasonableness and logic of these equations for use in this study.

3. An analysis of the four independent variables given in Example 1 are shown in the simple correlation matrix. Specify which ones should be considered for inclusion in regression equations, with reasons.

	T	X_1	X_2	X_3	X_4
T	1.0	0.75	0.95	0.44	0.75
X_1		1.00	0.43	0.25	0.95
X_2			1.00	0.29	0.45
X_3				1.00	0.28
X_4					1.00

4. An urban area consisting of four zones has the base-year trip matrix shown. The growth rates for the origin and destination trips have been projected for a 25-year period. Using Fratar's techniques, calculate the number of trip interchanges in the horizon year. Do just two iterations.

Origin	Destination				Total	Origin growth factors
	1	2	3	4		
1	3	5	8	12	28	2
2	4	1	9	10	24	1
3	2	4	2	7	15	4
4	9	12	8	4	33	2
Total	18	22	27	33	100	
Destination growth factors:						
	3	0.5	4	1		

5. A four-zone city has the following productions and attractions:

Zone	Production	Attraction
1	1000	3000
2	2000	3000
3	3000	2000
4	4000	2000

The travel time matrix is

Zone	Travel time (min)			
	1	2	3	4
1	2	5	7	10
2	5	3	8	12
3	7	8	2	11
4	10	12	11	3

Travel time (min)	F_{ij}
2	3.0
3	2.5
5	2.3
7	1.5
8	1.2
11	0.95
12	0.90

Apply the gravity model to distribute the trips ($K_{ij} = 1$).

6. A small town has two residential zones, A and B, producing 900 and 600 work trips, respectively. Zones C, D, and E are work-opportunity zones attracting 900, 400, and 200 trips. The travel times between zones and the actual observed trips are as follows:

Links	Time (min)	Actual trips
AC	7	425 ⎫
AD	10	325 ⎬ 900
AE	12	150 ⎭
BC	8	475 ⎫
BD	15	75 ⎬ 600
BE	9	50 ⎭

Develop a set of travel time factors (F_{ij}) for use in a gravity model that will produce a reasonable approximation of the observed trips. Start with an initial value of $F_{ij} = 1$ for all links. Assume that $K_{ij} = 1$.

7. A three-zone city has the following characteristics:

	Production	Attraction
Zone 1	500	400
Zone 2	600	800
Zone 3	200	100

Initial travel time factors are

Travel time (min)	10	15	20
F_{ij}	80	60	40

Distances (time):

   1–2: 10 min
   2–3: 15 min
   3–1: 20 min

Distribute the trips using the gravity model.

8. In Example 12, the cost of travel is now 200, 100, and 150 cents for the automobile, bus, and light rail, respectively. In addition, all automobiles have to pay a parking fee of $2. The travel time has also altered to 30, 40, and 45 minutes, respectively. What is the modal split?

9. A freeway section 20 miles long has a free speed of 70 mph, $Q_{max} = 3000$ veh/hr, $Q = 1500$ veh/hr, $\tau = 0.15$, $\alpha = 0.474$, $\beta = 4$, and $T_0 = 17.14$ minutes. Apply the Davidson, BPR, and Greenshields methods to find T_Q. Make suitable assumptions as appropriate.

10. A simple network shown in Figure 11-P10 has two-way links. The time cost is also shown. Find the shortest path from nodes A, B, C, and D to all other nodes and intersections.

11. A trip table (veh/hr) needs to be loaded on the network shown for Exercise 10. Find the total volume on each link assuming an all-or-nothing assignment.

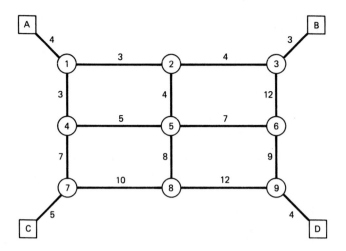

Figure 11-P10   Network for Exercise 10.

From	To	A	B	C	D
A		—	50	40	20
B		30	—	80	10
C		90	80	—	20
D		60	70	50	—

12. A network (Figure 11-P12) connected to four centroids is loaded with trips, as shown in the trip table. Assign the trips (using the all-or-nothing technique) assuming the following: (1) figures on links indicate travel cost; (2) a left turn, a right turn, and going straight through an intersection carries a penalty of 3, 2, and 1 units, respectively; and (3) all links are two-way.

From	To	A	B	C	D
A		—	900	400	700
B		200	—	700	300
C		600	800	—	400
D		100	200	500	—

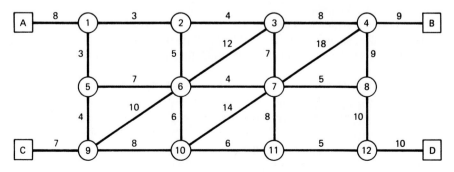

Figure 11-P12   Network for Exercise 12.

13. Examine the land use and transportation study done for your city (Metropolitan Planning Organization, or MPO) and write a brief report on the planning process. You may refer to the eight steps enumerated in the summary of this chapter.

14. **(a)** The following information [connecting family size, household (HH), trips, and auto ownership] was collected for an MPO:

| Family size | Auto ownership | | | | | |
| | 0 | | 1 | | 2 or more | |
	HH	Trips	HH	Trips	HH	Trips
1	900	1100	2000	5000	120	200
2	1500	2000	2000	6000	700	1500
3	1300	1900	3000	14,000	5000	20,000
4 or more	800	1500	4000	18,000	5000	26,000

What is the number of trips per household with respect to family size and auto ownership? Develop appropriate curves representing your results.

(b) The forecasted number of households in a typical zone is as follows:

| Family size | Auto ownership | | |
	0	1	2 or more
1	36	50	10
2	12	65	200
3	18	45	250
4 or more	10	25	410

Using the curves (or your calculations), determine the number of trips produced by this zone.

15. If you are living in a metropolitan area, visit the office of the Metropolitan Planning Organization. Examine the original transportation planning document (which probably used the 4-step planning process). Find out how this document has been revised from time to time, and then write a report.

16. A mode-choice model for a city includes the following modes: autos (A), light rail (L), buses (B), and rapid rail (R), with utility functions (U), as shown in the table:

FUNCTION	COST (C)	TIME (T)
$U(A) = 3.2 - 0.3C - 0.04T$	5	30
$U(L) = 1.0 - 0.2C - 0.04T$	3	25
$U(B) = \quad\; - 0.1C - 0.01T$	2.5	40
$U(R) = 1.5 - 0.3C - 0.05T$	6	20

where C is the cost in dollars, and T is the time in minutes.

(a) Based on an estimate that 12,000 workers will head for downtown each morning, how many workers will choose to take a particular mode?

(b) If the government subsidizes light rail by 30%, buses by 20%, and rail rapid by 10%, and at the same time increases automobile costs by 15%, what will be the new modal distribution?

17. A city has a utility function for use in a logit model of the form

$$U = -0.075A - 0.5W - 0.04R - 0.02C$$

where A is the access time in minutes, W is the waiting time in minutes, R is the riding time in minutes, and C is the out-of-pocket cost in cents.

(a) What modal distribution would you expect, using the following values for A, W, R, and C, for the four modes used in the city.

MODE	A	W	R	C
Auto	6	1	25	300
Rail	7	10	15	75
Bus	10	15	35	60
Bike	1	0	45	10

(b) The city is seriously thinking of subsidizing rail and bus by 50%, encouraging biking by constructing bike paths and thus reducing biking time by 20%, and increasing auto costs (through higher parking charges) by 10%. What is likely to be the new modal distribution with these changes?

18. A busy travel corridor connecting a suburb with the city center is served by two routes having a typical travel time function, $t = a + b(q/c)$, where t is the time in minutes, q is the vehicular flow in veh/hr, and c is the capacity of the route in veh/hr. The existing characteristics of the two routes is as follows:

ROUTE	a	b	c
1	3	4	3000
2	4	2	4000

(a) If the existing peak-hour demand is 5000 veh/hr, what is the traffic distribution on the two routes?

(b) If repair work on Route 1 reduces its capacity to 2000 veh/hr, what is likely to be the traffic distribution on the two routes for the duration of the repairs?

(c) It is anticipated that after the repairs are completed on Route 1, its capacity will be 4200 veh/hr. How will this affect the distribution?

(d) Sketch the three cases, marking important values on the same.

19. A new residential expansion in Seattle needs to be planned in a grid pattern with arterials and local streets, with the following specifications:

Single-family dwelling units (DUs) housing: 3 persons/DU (average)
Average trips generated per DU: 4 trips/person/day
Average trip length: 4.5 mi on arterials

Average residential density: 5 DUs/acre (gross)
Average auto occupancy: 1.6
Average modal split: 70% auto, 20% bus, 10% walk and bike
PHF: 0.15
Capacity per lane of arterial: 900 veh/hr
Assume bus occupancy   40 persons/bus
Assume 1 bus takes up the road space of 3 cars

Determine (a) lane-miles of arterials/sq. mile, running north–south and east–west in a grid pattern and (b) spacing of four-lane arterials to satisfy the specifications given.

20. Refer to Example 16 in the text and Exercise 19. If the specifications for a new housing development include adopting a mix of 70% of what is described in Example 16 and 30% as per Exercise 19, what would you suggest would be needed (a) in terms of lane-miles of arterials and (b) their spacing in a grid pattern?

21. A developer has constructed 4-lane arterials with a lane capacity of 750 veh/hr, in a grid pattern, 1 mile N–S and 1 mile E–W, in a new development. The traffic engineer of the county has specified the following characteristics for this development:

Average trips generated/DU: 3.5 trips/person/day
Average trip length: 4.0 miles on arterials
Single family dwelling units (DUs):   3.5 persons/DU
Auto occupancy: 1.5
Modal split: 65% auto, 15% transit, 20% bike and walk
PHF: 0.18
Assume bus occupancy of 40 persons/bus, and that 1 bus takes up the same road space as 3 cars.

What maximum residential density would satisfy this development?

22. (a) Determine the minimum-path tree for the network shown in Figure 11-P22, rooted in centroid A (home node), using the costs shown against each link.

(b) If, in addition, penalties are introduced for right-turn, left-turn, and straight through, equivalent to 3, 5, and 2 units, respectively, determine the minimum-path tree rooted in centroid A.

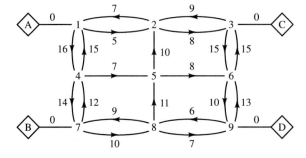

Figure 11-P22   Network for Exercise 22.

23. In the simple highway network shown in Figure 11-P23, the cost function (C) depends on the link traffic flow (f).

  (a) If $f_1 = 1$, compute the value of f_4.

  (b) If $f_1 = 9$, compute the value of C_4 (Wardrop's principle applies).

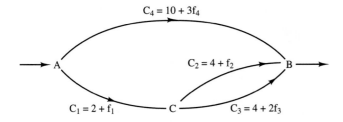

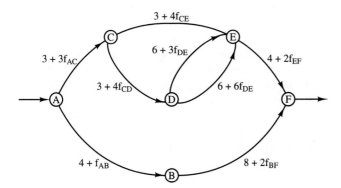

Figure 11-P23   Network for Exercise 23.

24. Compute the cost function C_{AF} from the highway network shown in Figure 11-P24, where f_{AF} represents the traffic flow (Wardrop's principle applies). Note that the cost function of each link is given in terms of flow.

Figure 11-P24   Network for Exercise 24.

Chapter 12

Local Area Traffic Management

1. INTRODUCTION

With the introduction of transportation system management regulations in 1975, there has been a consistent interest at both the metropolitan scale and at the local level to reduce energy consumption, relieve traffic congestion, enhance the mobility of central business areas, promote pedestrian and bicycle use, and improve parking management. Over the years, an enormous amount of experience relevant to local area planning has been acquired. Although a separate chapter is devoted to transportation system management, a few selected topics, such as pedestrian movement, bicycle planning, traffic management, and parking design, are discussed in this chapter that are particularly applicable to local area planning and design.

Local area planning embraces solutions for existing neighborhoods and their expansions or renovations. New developments can also be included in this category. Local area solutions must, of course, be dovetailed with planning solutions at the wider metropolitan scale, through community involvement and negotiation (Black, 1981; Blunden and Black, 1984; Ogden and Bennett, 1984).

Although the general principles, planning methodologies, and design practices of traffic engineering are applicable to local area planning, there are several areas where the emphasis lies in microplanning. This emphasis is pointed out in subsections of this chapter.

2. PEDESTRIAN FACILITIES

It is only within the last decade that pedestrian facility planning and programming has attained some importance. Local agencies are attempting to include pedestrian facilities such as sidewalks and footbridges, in new construction with major funding commitments. "Retrofit" programs are also being undertaken to provide facilities for pedestrian movement.

In this section, we describe the basic pedestrian characteristics and their application to planning and designing of pedestrian facilities. The level-of-service concept, used so successfully in highway design, is also applied to pedestrian design. Two excellent references on pedestrian planning and design are by Fruin (1971) and Pushkarev and Zupan (1975).

2.1 Terminology

Pedestrian design analysis makes use of terms commonly used in traffic engineering. These include the following:

Pedestrian speed: average walking speed, expressed in ft/sec.

Pedestrian flow rate: number of pedestrians passing a point per unit time, expressed as pedestrians/minute or pedestrians/15 minutes. "Point" refers to a perpendicular line of sight across the width of a walkway.

Unit width flow: pedestrians per minute per foot.

Platoon: a number of pedestrians walking together in a group, usually involuntarily.

Pedestrian density: average number of pedestrians per unit area within a walkway or queueing area, expressed in pedestrians/per square foot.

Pedestrian space: average area provided for each pedestrian, which is the inverse of density, expressed in square feet per pedestrian.

2.2 Human Space Requirements

Individual persons, on an average, require a minimum area when standing, which is known as the *body ellipse*. It measures 18 in. by 24 in., as shown in Figure 12-1. This minimum space requirement of 2.3 ft^2/person is not sufficient if human beings are carrying luggage or backpacks. For personal comfort, Fruin suggests about 7 to 10 ft^2/person. Note that these space requirements are for persons standing without motion (Fruin, 1971; FHWA, 1980).

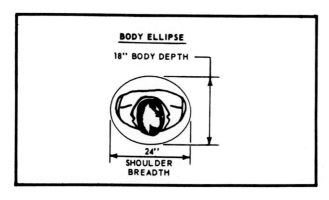

Figure 12-1 Body Ellipse Configuration (FHWA, 1980).

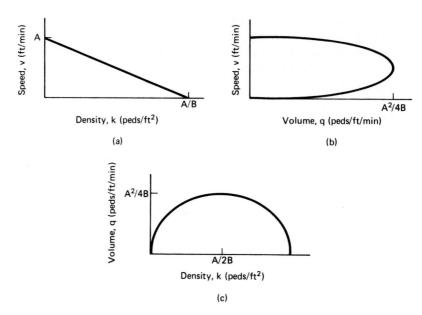

Figure 12-2   Theoretical Model of Pedestrian Flow in Single Channels: (a) Speed versus Density; (b) Volume versus Density; (c) Speed versus Volume.

Pedestrian Speed–Density–Flow Relationship.   Pedestrian flow is described in terms of speed and flow, which can be approximated by a parabolic curve that is similar to motor vehicle flow (Greenshields, 1934; TRB, 1985). A theoretical speed–density–flow relationship is shown in Figure 12-2. As density increases, the speed of pedestrians in the traffic stream decreases. However, density (D) is an inconvenient concept, particularly when one deals with fractions of pedestrians; therefore, the reciprocal of density or available space per pedestrian is often used (Greenshields, 1934; Khisty, 1985; TRB, 1985).

$$v = S \times D \tag{1}$$

where

v = pedestrian flow in ped/min/ft
S = pedestrian speed in ft/min
D = pedestrian density in ped/ft^2

or

$$v = \frac{S}{M} \tag{2}$$

where M = pedestrian module in ft^2/ped.

Also, pedestrian demand is expressed as ped/15 min, using a peak 15-min period of flow as the basis for analysis. The average pedestrian flow (v) is then computed as:

$$v = \frac{V}{15\,W_E} \tag{3}$$

where

V = peak pedestrian flow (ped/15 min)

W_E = effective walkway width (ft)

Relationships between pedestrian speed and density, speed and flow, and flow and space are shown in Figures 12-3, 12-4, and 12-5, respectively. Also, the relationship between walking speed and available space is shown in Figure 12-6. It shows that at an average space of about 15 ft^2/ped, even the slowest pedestrian cannot achieve his or her desired walking speed. Pedestrians wanting to walk at say 350 ft/min are not able to achieve such speed freely until the average space is about 40 ft^2/ped. This fact gives us a clue regarding level-of-service boundaries, described later.

Effective Walkway Width.   The concept of a defined pedestrian "lane" as applied in motor vehicle traffic is not applicable in pedestrian flow analysis. Pedestrians passing each other generally require 2.5 ft width each. Pedestrians walking together may require a width of 2.2 ft each, in which case, there is every likelihood of body contact due to sway.

The width of a walkway that can effectively be used by pedestrians is called the *clear walkway width*. Poles, signs, and benches, for example, reduce the effective walkway width. Typical obstructions and the estimated width of walkways that they preempt are provided in Table 12-1, and Figure 12-7 shows an example of walkway widths curtailed by curbs and buildings.

Walking Speeds.   Many researchers have contributed to measuring walking speeds. Figures 12-8 and 12-9 show that there is a wide range of speeds among pedestrians. Trip purpose, land use, age, and other environmental factors all affect walking speeds. Designers must adjust numerical analysis to reflect this large variation in speeds.

Level-of-Service Criteria for Walkways.   In pedestrian facility design the basic measure of effectiveness is space. Capacity is taken to be 25 ped/min/ft. Table 12-2 shows the criteria for pedestrian level of service. Figure 12-10 illustrates walking level of service.

Short-term fluctuations can occur in most pedestrian traffic flow as pedestrians arrive and depart randomly, such as at sidewalks. When sidewalks and other facilities show signs of platooning effects, it is recommended that these surges should be timed and counted. An expression relating maximum platoon flow rates to average flow rate is

$$\text{Platoon flow} = \text{average flow} + 4 \tag{4}$$

where both flows are expressed in ped/min/ft.

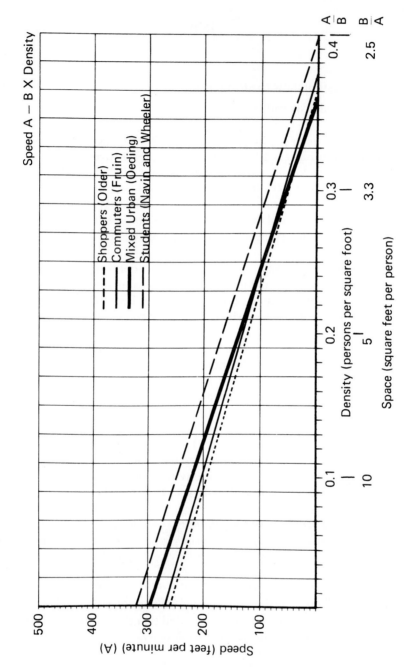

Figure 12-3 Relationships between Pedestrian Speed and Density (Pushkarev and Zupan, 1975).

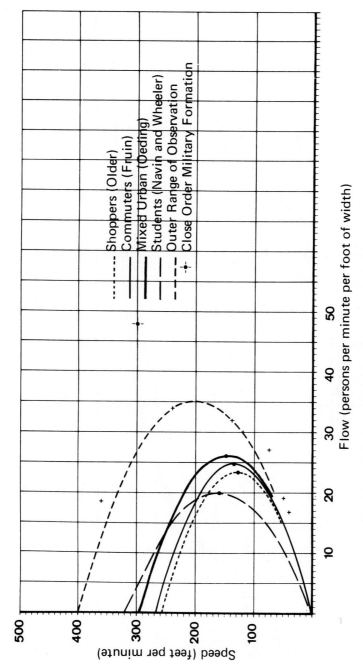

Figure 12-4 Relationships between Pedestrian Speed and Flow (Pushkarev and Zupan, 1975).

529

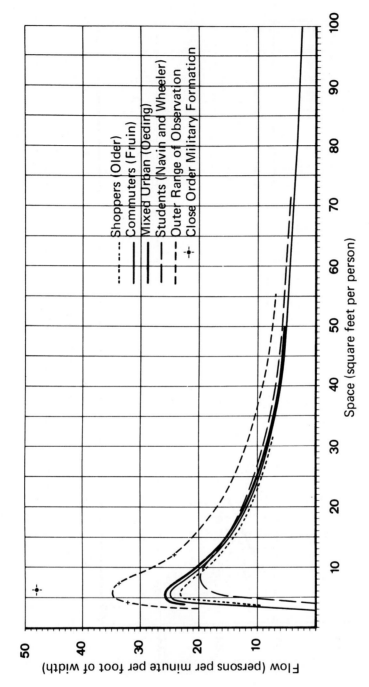

Figure 12-5 Relationships between Pedestrian Speed and Flow and Space (Pushkarev and Zupan, 1975).

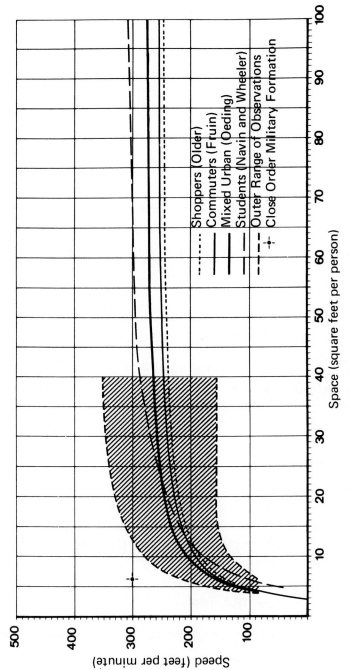

Figure 12-6 Relationships between Pedestrian Speed and Space (Pushkarev and Zupan, 1975).

TABLE 12-1   FIXED-OBSTACLE-WIDTH ADJUSTMENT FACTOR FOR WALKWAYS[a]

Obstacle	Approx. width preempted[b] (ft)
Street furniture	
Light poles	2.5–3.5
Traffic signal poles and boxes	3.0–4.0
Fire alarm boxes	2.5–3.5
Fire hydrants	2.5–3.0
Traffic signs	2.0–2.5
Parking meters	2.0
Mailboxes (1.7 ft by 1.7 ft)	3.2–3.7
Telephone booths (2.7 ft by 2.7 ft)	4.0
Wastebaskets	3.0
Benches	5.0
Public underground access	
Subway stairs	5.5–7.0
Subway ventilation gratings (raised)	6.0+
Transformer vault ventilation gratings (raised)	5.0+
Landscaping	
Trees	2.0–4.0
Planting boxes	5.0
Commercial uses	
Newsstands	4.0–13.0
Vending stands	Variable
Advertising displays	Variable
Store displays	Variable
Sidewalk cafes (two rows of tables)	Variable, try 7.0
Building protrusions	
Columns	2.5–3.0
Stoops	2.0–6.0
Cellar doors	5.0–7.0
Standpipe connections	1.0
Awning poles	2.5
Truck docks (trucks protruding)	Variable
Garage entrance/exit	Variable
Driveways	Variable

[a] To account for the avoidance distance normally occurring between pedestrians and obstacles, an additional 1.0 to 1.5 ft must be added to the preemption width for individual obstacles.
[b] Curb to edge of object, or building face to edge of object.
Source: TRB, 1985.

2.3 Environmental Factors (TRB, 1985)

The following factors should be considered in designing pedestrian facilities:

1. *Comfort:* such as weather protection, climate control, transit shelter, skywalks
2. *Convenience:* walking distances, directness, grade on ramps, stair suitable for elderly, directory maps, and other factors that contribute to the ease of pedestrian movement

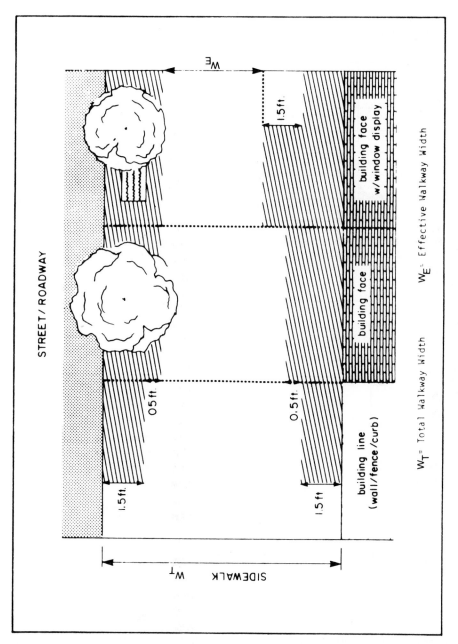

Figure 12-7 Preemption of Walkway Width (TRB, 1985).

Unimpeded Free-Flow

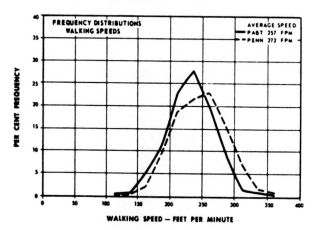

PABT = Port Authority Bus Terminal
PENN = Pennsylvania Station

Figure 12-8   Pedestrian Walking Speeds (FHWA, 1980).

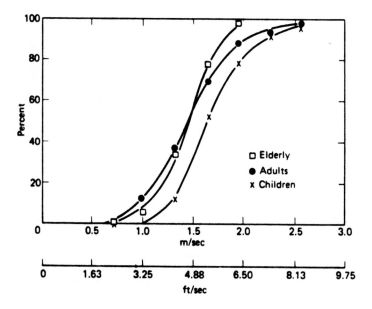

Figure 12-9   Pedestrian Types and Walking Speeds (FHWA, 1980).

3. *Safety:* separation of pedestrian traffic from vehicular traffic, malls meant only for pedestrians, traffic control devices that safeguard the lives of pedestrians

4. *Security:* lighting, line of sight, crime-free environment

5. *Economy:* minimization of travel delays

TABLE 12-2   PEDESTRIAN LEVEL OF SERVICE OF WALKWAYS[a]

Level of service	Space (ft^2/ped)	Expected flows and speeds		
		Ave. speed, S (ft/min)	Flow rate, v (ped/min/ft)	Vol./cap. ratio, V/c
A	≥ 130	≥ 260	≤ 2	≤ 0.08
B	≥ 40	≥ 250	≤ 7	≤ 0.28
C	≥ 24	≥ 240	≤ 10	≤ 0.40
D	≥ 15	≥ 225	≤ 15	≤ 0.60
E	≥ 6	≥ 150	≤ 25	≤ 1.00
F	< 6	< 150	Variable	Variable

[a] Average conditions for 15 min.
Source: TRB, 1985.

Engineers and planners involved with complex pedestrian facilities design are advised to work in collaboration with environmental psychologists (Khisty, 1979).

2.4 Planning for Pedestrians

Careful assessment of demand, design standards, functional elements, and space design is called for in pedestrian facility planning. It is best to begin a study by defining the goals and objectives of the project. For instance, one of the principal objectives in planning a major shopping center may be to cope with crowds by designing adequate circulation routes for pedestrians only. Safety may be the predominant objective when pedestrian-oriented malls permit access to buses, taxis, and delivery vehicles. A flowchart showing the elements of the pedestrian facility planning process is given in Figure 12-11.

Pedestrian demand consists of estimating traffic volumes, traffic patterns, and composition. Land-use patterns and building types will provide information regarding trip generation. For instance, sports stadiums and theaters will produce pedestrian movements depending on seats provided. Some generators will produce high seasonal peak demands.

Optimal space design may be considered as the best functional space envelope that most economically, effectively, and safely accommodates the movements of pedestrians. It is common to use the "system" approach in understanding and in applying the principles of pedestrian design to all problems. A systems diagram for the design of a subway transit platform from the point of view of pedestrians is shown in Figure 12-12. There are particularly good references giving details for designing pedestrian facilities and these may be consulted for details (Fruin, 1971; Pushkarev and Zupan, 1975).

A more recent analytical approach proposed by Fruin (1992) and Benz (1987) uses a time–space (T–S) concept, taking into account the balance between time and space. T–S analysis is particularly applicable in dealing with complicated cases of personal space occupancies related to pedestrian activities. T–S supply is the product of the time of the analysis and the area of the space being analyzed. Likewise, the T–S demand is the product of the total number of pedestrians using the analysis space and

LEVEL OF SERVICE A

Pedestrian Space: ≥ 130 sq ft/ped   Flow Rate: ≤ 2 ped/min/ft

At walkway LOS A, pedestrians basically move in desired paths without altering
their movements in response to other pedestrians. Walking speeds are freely
selected, and conflicts between pedestrians are unlikely.

LEVEL OF SERVICE B

Pedestrian Space: ≥ 40 sq ft/ped   Flow Rate: ≤ 7 ped/min/ft

At LOS B, sufficient area is provided to allow pedestrians to freely select
walking speeds, to bypass other pedestrians, and to avoid crossing conflicts with
others. At this level, pedestrians begin to be aware of other pedestrians, and to
respond to their presence in the selection of walking path.

LEVEL OF SERVICE C

Pedestrian Space: ≥ 24 sq ft/ped   Flow Rate: ≤ 10 ped/min/ft

At LOS C, sufficient space is available to select normal walking speeds, and to
bypass other pedestrians in primarily unidirectional streams. Where reverse-
direction or crossing movements exist, minor conflicts will occur, and speeds
and volume will be somewhat lower.

LEVEL OF SERVICE D

Pedestrian Space: ≥ 15 sq ft/ped   Flow Rate: ≤ 15 ped/min/ft

At LOS D, freedom to select individual walking speed and to bypass other
pedestrians is restricted. Where crossing or reverse-flow movements exist, the
probability of conflict is high, and its avoidance requires frequent changes in
speed and position. The LOS provides reasonably fluid flow; however,
considerable friction and interaction between pedestrians is likely to occur.

LEVEL OF SERVICE E

Pedestrian Space: ≥ 6 sq ft/ped   Flow Rate: ≤ 25 ped/min/ft

At LOS E, virtually all pedestrians would have their normal walking speed
restricted, requiring frequent adjustment of gait. At the lower range of this LOS,
forward movement is possible only by "shuffling." Insufficient space is provided
for passing of slower pedestrians. Cross- or reverse-flow movements are
possible only with extreme difficulties. Design volumes approach the limit of
walkway capacity, with resulting stoppages and interruptions to flow.

LEVEL OF SERVICE F

Pedestrian Space: ≤ 6 sq ft/ped   Flow Rate: variable

At LOS F, all walking speeds are severely restricted, and forward progress is
made only by "shuffling." There is frequent, unavoidable contact with other
pedestrians. Cross- and reverse-flow movements are virtually impossible. Flow is
sporadic and unstable. Space is more characteristic of queued pedestrians than
of moving pedestrian streams.

Figure 12-10   Illustration of Walkway Level of Service (TRB, 1985).

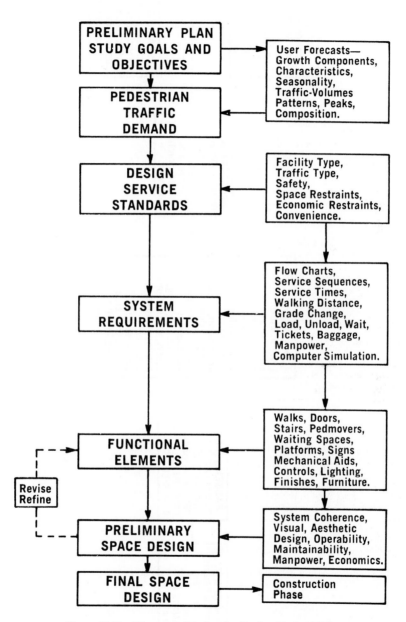

Figure 12-11   Elements of Pedestrian Design (Fruin, 1971).

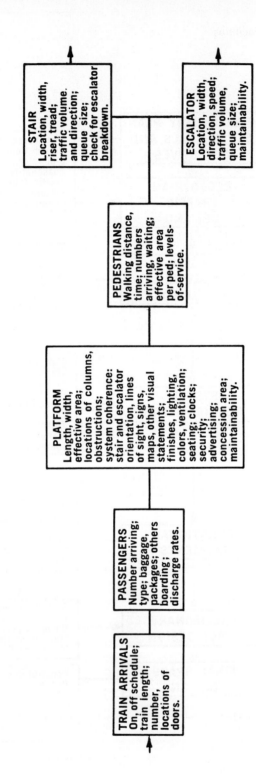

Figure 12-12 System Description: Transit Platform (Fruin, 1971).

their time of occupancy. The objective is to evaluate the adequacy of a given space for a forecasted peak-period demand and the occupancy time of pedestrians walking or waiting in this space. By dividing the *T–S* supply by the *T–S* demand, the average area occupied per pedestrian and the corresponding LOS can be determined, and this is expressed as follows:

$$a = \frac{T\text{–}S\text{ supply}}{T\text{–}S\text{ demand}} = \frac{TS}{nt} \tag{4}$$

where

$a =$ average area per pedestrian (ft^2/ped) within the analysis space, during the analysis period

$T =$ time of analysis period (min)

$S =$ net effective area of analysis space (ft^2)

$n =$ number of pedestrians occupying the space (walking, waiting, standing, etc.)

$t =$ predicted occupancy times of pedestrians for functions performed during the analysis period

2.5 Applications

It is best to illustrate the application of design principles through worked examples.

Example 1

A sidewalk has a peak 15-minute pedestrian flow of 1000 ped/min. The sidewalk is 12 ft wide, has a curb on one side, and windows with shopping displays on the other. At what LOS does the sidewalk operate (a) on the average; (b) within platoons?

Solution

$$\text{Sidewalk width} = 12 \text{ ft}$$

$$\text{Effective sidewalk width} = 12 - 1.5 - 3.0 = 7.5 \text{ ft}$$

$$\text{Average unit width flow} = v = \frac{V}{15W_E}$$

$$= \frac{1000}{15 \times 7.5}$$

$$= 8.9 \text{ ped/min/ft}$$

$$\text{Platoon unit flow} = v_p = v + 4$$

$$= 8.9 + 4$$

$$= 12.9 \text{ ped/min/ft}$$

Table 12-2 indicates that these flows correspond to LOS C and D, respectively.

Example 2

The expected 5-minute peak flow through the main entrance of an office building is 650 persons. If a LOS C is desired, find the number of doors and the approach corridor width required.

Solution

LOS C corresponds to a space ≥ 24 ft^2/ped, average speed ≥ 250 ft/min, and flow rate ≤ 10 ped/min/ft. Assuming that doors are 7 ft wide, representing two pedestrian lanes,

$$\text{Pedestrian flow per door at LOS C} = 7 \text{ ft} \times 10 \text{ ped/min/ft}$$

$$= 70 \text{ ped/min}$$

Further,

Number of doors required in peak direction

$$= \frac{650 \text{ ped}}{5 \text{ min} \times 70 \text{ ped/min}} = 2 \text{ doors of 7 ft width}$$

Add a 3.5-ft door for reverse flow.

$$\text{Corridor width} = (2 \times 7) + 3.5 = 17.5 \text{ ft}$$

Adding 2-ft wide standing and door opening lane each side of the corridor, we have $17.5 + 4 = 21.5$ ft.

Example 3

A 2400-ft^2 covered corridor connects two high-rise buildings. During a 15-min peak period, 2000 persons use the corridor taking an average walking time of 20 sec to traverse this corridor. However, 10% of these persons spend an average of an additional 15 sec to look at the exhibits in the show windows on either side of the corridor. Determine the theoretical average space provided per person.

Solution

Because we do not have any information regarding the actual length and width of this corridor, no allowance will be made for side-width clearance adjustments.

$$\text{TS supply} = TS = (15)(2400) = 36,000 \text{ ft}^2\text{-min}$$

$$\text{TS demand} = nt = [(2000)(20) + (2000)(0.1)(15)]/60$$

$$= 716 \text{ person-min}$$

Hence,

$$a = 36,000 \text{ ft}^2\text{-min}/716 \text{ person-min}$$

$$= 50.28 \text{ ft}^2/\text{person}$$

Discussion

This is the theoretical square foot/person available. If the length of the corridor as well as other details were known, then one could determine the required corridor width, given the LOS desired.

Example 4

A 150-ft-long corridor is being planned to accommodate a pedestrian flow of 300 persons/min with a walking speed of 4.2 ft/sec, and an average space of 30 ft^2/person. What should be the theoretical width of this corridor?

Solution

Here $S = wl$, where $w = $ the width and $l = $ length of the corridor in feet. Also $t = 150/4.2 = 35.71$ sec, or 0.60 min, and $T = 1$ min.

Using Eq. 4 and rearranging terms,

$$w = ant/Tl = (30)(300)(0.6)/(1)(150) = 36 \text{ ft}$$

Discussion

Note that 36 ft is the theoretical width. The actual width would be 36 ft + 2 ft = 38 ft to account for clearance from the walls of the corridor.

Example 5

A subway pedestrian corridor 300 ft long is being planned to connect two buildings in a transit station, with a two-way peak-hour pedestrian flow of 12,000 persons per hour. There will be a retail shopping development along both sides of the corridor. What should be the minimum width of this corridor? Assume an average of 12.5 ped/min/ft.

Solution

Assume that the typical 15-min peak would be about 40% of the peak-hour flow.

$$\text{Net effective width} = (12,000)(0.4)/(15)(12.5) = 25.6 \text{ ft}$$

$$\text{Add edge effect of shopping doors } (2 \times 3) = \quad 6.0 \text{ ft}$$

$$\text{Total} = 31.6 \text{ ft}$$

3. BICYCLE FACILITIES

Although bicycle traffic composes only a small percentage of the total traffic stream, it is sufficient enough to have an impact on street planning and design. Recent accident studies have indicated that the bicyclist has been increasingly involved in motor car/bicycle collisions. Local officials are therefore taking cognizance of this fact, and many cities have initiated extensive programs to provide bicycle facilities designated as bikeways as well as bicycle lanes on streets and highways. Bicycling is no longer a recreational pastime, but is considered a feasible alternative to motoring particularly in milder climates.

3.1 Definitions

A bicycle is defined as a vehicle having two tandem wheels propelled solely by human power on which any person or persons may ride. A bikeway is a trail, path, part of a highway or shoulder, sidewalk, or any other means specifically marked and assigned for bicycle use. Bikeways are generally classified as follows:

> *Class I bikeway:* completely separated from vehicular traffic and within an independent right-of-way or the right-of-way of another facility. Bikeways separated from vehicles but shared by both bicycles and pedestrians are included in this classification, as shown in Figure 12-13(a).
>
> *Class II bikeway:* part of the roadway or shoulder is marked by pavement markings or barriers. Vehicle parking, crossing, or turning movements are permitted within the bikeway. This class of bikeway is shown in Figure 12-13(b).
>
> *Class III bikeway:* shares right-of-way with motor vehicles; are designated by signing only [Figure 12-13(c)]. There is hardly any protection from motor vehicles, although the signing helps to make the motorist aware of the presence of bicyclists.

The *Highway Capacity Manual* (TRB, 1985) classifies bicycle facilities in two basic forms. When a portion of a roadway is striped, signed, and marked for exclusive or preferential use of the bicyclists, it is called a *bike lane.* On the other hand, when a bikeway is physically separated from motorized vehicular traffic, either within the highway right-of-way or within an independent right-of-way, it is called a *bike path.*

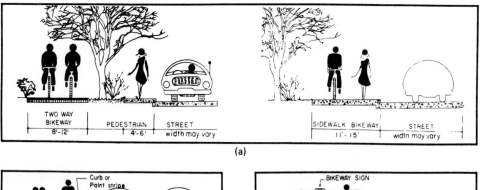

(a)

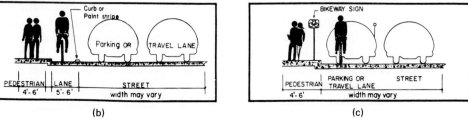

(b)   (c)

Figure 12-13   Bikeway Classification (FHWA, 1980).

3.2 Capacity Implications

Comparatively little is known regarding impacts of bicycles on the capacity and LOS of highway facilities. The *Highway Capacity Manual* (TRB, 1985) addresses the following three aspects of bicycle capacity:

1. The effect of bicycles sharing a lane with other vehicles can be accounted for by assigning an appropriate passenger-car equivalent (PCE) for each bicycle.
2. The turning movements of vehicles affect bicycle streams in addition to conflicts normally presented by opposing streams of pedestrians and motor vehicles, as presented in Chapters 7 and 8.
3. Bicycles moving straight may be considered as unopposed. Left-turning bicycles are considered as opposing vehicular flow on two-way streets. Right-turning bicycles may or may not encounter significant pedestrian interference, and therefore can be considered as either opposed or unopposed, depending on pedestrian traffic conditions.

Table 12-3 gives the values of passenger-car equivalents for bicycles. Consider a signalized intersection with a vehicular volume of 700 veh/hr that shares a 12-ft lane with a vehicular volume of 150 bicycles/hour, one-half of which are opposed, then

Equivalent volume = 700 + (150 × 0.5 × 0.5) + (150 × 0.5 × 0.2) = 753 veh/hr

The capacity of one-way and two-way bicycle facilities is given in Table 12-4.

TABLE 12-3 PASSENGER-CAR EQUIVALENT FOR BICYCLES

Bicycle movement	Lane width (ft)		
	< 11	11–14	> 14
Opposed	1.2	0.5	0.0
Unopposed	1.0	0.2	0.0

Source: TRB, 1985.

TABLE 12-4 REPORTED ONE-WAY AND TWO-WAY HIGH VOLUMES OF BICYCLE FACILITIES

Type of facility	Number of lanes[a]	Range of reported capacities (bicycles/hr)
One-way bike lane or path	1	1700–2530
Two-way bike path	1	850–1000
	2	500–2000

[a] Lane widths: 3–4 ft/lane.
Source: TRB, 1985.

3.3 Bikeway Location

Usually, bikeway facilities or networks have to be planned and located to integrate with the existing street and highway network. Barriers and obstacles then can be identified along each corridor and specific travel routes assessed for their suitability. Corridor studies provide a rough draft for preliminary cost estimates. Because bikeways are environmentally favorable there is little trouble in going through an environmental impact analysis.

Some important criteria used in evaluating feasible bikeway routes are as follows:

1. Determine the potential demand for the use of the route.
2. Provide the basic width needed for safe operation. It is best to consider one-way operation.
3. Continuity and directness of route without much detour is essential, connecting points of importance.
4. Safety is of prime importance. Attempts to minimize vehicular/pedestrian conflicts should be given the highest priority.
5. Grades should be within tolerable range. A maximum grade of 5% is desirable.
6. The selection of pavement is of greater importance to bicyclists as compared to motorists. Ride quality of pavement and even safety are affected by poor pavement surfaces.
7. Among the automobile emissions, carbon monoxide is particularly hazardous to pedestrians and bicyclists.
8. Motor vehicle traffic, particularly trucks, moving at speeds of 50 mph can upset the balance of a bicyclist.

3.4 Bikeway Design (HRB, 1971; ITTE, 1975; FHWA, 1980; Homburger, 1982)

Highlights of important design elements are as follows.

Design speed: Although speeds approaching 30 mph are possible, normal speeds from 7 to 20 mph are common. A working speed of 20 mph may be adopted.

Bikeway width: The bikeway pavement width depends on the bicycle width, maneuvering allowance, clearance between oncoming and passing bicycles, and edge conditions. There must be at least 2.5 ft of horizontal separation between bicycles and pedestrians. A sidewalk between 11 and 15 ft wide would be required to accommodate pedestrians and bicyclists. Figure 12-14 shows accepted clearances. The AASHTO guidelines and FHWA criteria (FHWA, 1976) should be consulted for desirable widths.

Stopping sight distance: Assuming a perception-reaction time ranging between 1 and 2.5 sec, Figures 12.15 and 12.16 show distances on crest vertical curves and around horizontal curves, respectively.

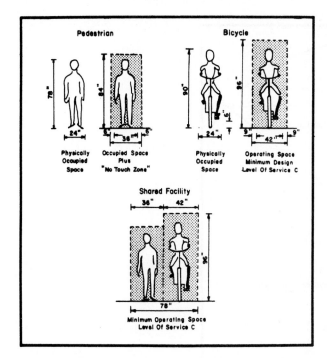

Figure 12-14 Bikeway Clearance Requirements (FHWA, 1980).

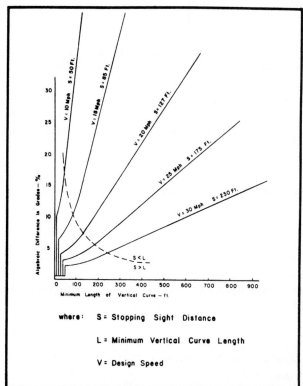

Figure 12-15 Bicycle Stopping Sight Distance on Crest Vertical Curves (FHWA, 1980).

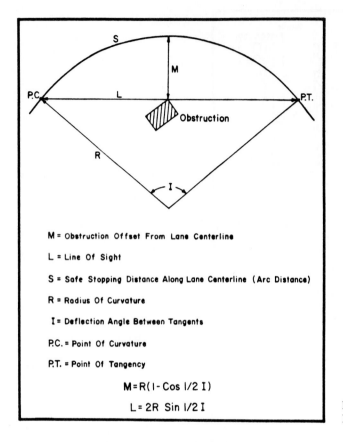

M = Obstruction Offset From Lane Centerline

L = Line Of Sight

S = Safe Stopping Distance Along Lane Centerline (Arc Distance)

R = Radius Of Curvature

I = Deflection Angle Between Tangents

P.C. = Point Of Curvature

P.T. = Point Of Tangency

$$M = R(1 - \cos 1/2\ I)$$

$$L = 2R\ \sin 1/2\ I$$

Figure 12-16   Sight Distance Around Horizontal Curves (FHWA, 1980).

Grades: Bicyclists are sensitive to grades. Over short distances of say 100 ft or less, a 10% grade is tolerable, although grades of 4 to 5% represent the maximum desired. Figure 12-17 shows desirable and acceptable standards for bikeway grades.

3.5 Application

Once again, one of the best ways of illustrating the use of design principles is via a worked example.

Example 6

An intersection approach has two traffic lanes and a right-hand-curb bike lane. The approach volumes are: median lane 230 veh/hr; right lane 200 veh/hr. There are 60 bikes/hr making left turns. What is the impact of this bike volume on motor vehicular traffic? Lanes are 12 ft wide.

Solution

From Table 12-3, each bike has an equivalent of 0.5 PCE (opposed, 12-ft lane). Sixty left-turn bikes/hr are equivalent to $60 \times 0.5 = 30$ veh/hr. These PCEs should be added to the volume in the approach lanes.

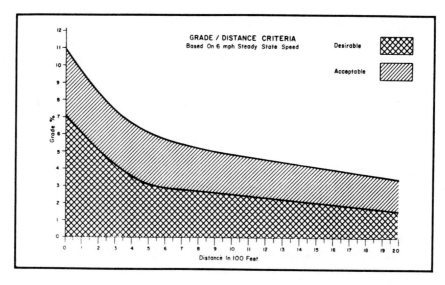

Figure 12-17   Bikeway Grade Criteria (FHWA, 1980).

Curb lane: 230 + 30 = 260 veh/hr
Right lane: 200 + 30 = 230 veh/hr

In effect, the PCEs are added to each lane that is crossed by the bikes, transferring from the bike lane to the leftmost traffic lane.

One of the objectives of transportation planning is to provide for the safe and efficient movement of people and goods among diverse land uses. Although two of the objectives of land-use planning are to provide compatibility and to achieve functional efficiency, these objectives are not always congruent (Stover and Koepke, 1988).

4. TRAFFIC PLANNING AND MANAGEMENT AT THE LOCAL LEVEL

There are a wide range of land-use and traffic planning problems that call for quantitative and qualitative examination ranging from siting a new shopping center to aligning the best approach to a townhouse. Development applications call for meaningful traffic appraisals by local traffic engineers on a day-to-day basis. Examples of traffic generation appraisals can take various forms, such as the following:

- Assessing the acceptability of a proposed hospital or hotel in a suburb of a large city
- Estimating the site's future effects on the traffic services of existing adjacent roads
- Planning a major housing development with an internal circulation system, parking arrangements, and connections with existing surface roads
- Assuming various alternatives for proposed changes to land-use zoning ordinances and subdivision regulations over extensive areas of the city

Generally, the use of traffic generation analysis to assess traffic impacts is what is required in most local area planning. For performing this analysis, the traffic activity associated with a site or a particular land-use activity is needed. From this information, the number of one-directional vehicular movements arriving at or departing from the studied area per unit of time is determined.

4.1 Data Requirements

The basic data requirements are the following:

1. Expected daily traffic generation rates for each land use or activity encountered
2. Expected hourly traffic generation rates
3. Expected traffic generation rates at peak-hour traffic conditions

In Chapter 11, there were several tables providing information on average trip rates and other characteristics of generators. These tables should be considered for site-specific estimates of traffic generation, and the input required for performing traffic analysis follows:

1. *Residential area:* type of residence and number of dwelling units or acres of development
2. *Industrial/manufacturing, offices:* gross floor area (GFA), or employees, or acres of development
3. *Restaurants:* gross floor area or acres of development
4. *Banks:* gross floor area or employees
5. *Parks and recreation areas:* acres (or employees for a few types)
6. *Hospitals:* staff or beds
7. *Educational facilities:* students or staff
8. *Airports:* takeoffs and landings or employees or acres
9. *Hotels/motels:* rooms or employees
10. *Retail stores:* gross floor area of employees or acres
11. *Military bases:* military personnel and civilian employees or total employees
12. *Race tracks, baseball stadiums:* seats or attendees

Definitions of units used in Table 11-2 are as follows:

Dwelling unit (DU): a place of domicile such as a single-family home, apartment, and the like
Acre: 43,560 ft^2; as used here it includes all developed land area connected to a site, including parking lots and the like
1000 GFA: the gross floor area of a site under roof in terms of 1000 ft^2
Employee: a person who works at the location, generally in the employ of a business located at the site

Staff: for hospitals, a doctor, nurse, or other employee; for schools, a faculty member, teacher, or other employee

Student: a person enrolled full- or part-time in a course at a school or other educational facility

Bed: for hospitals, the number of beds available for patient care

Takeoff/landing: for airports, the sum of aircraft takeoffs plus landings

Room: for hotels and motels, a room or suite of rooms available for overnight stay of a guest(s)

Military personnel: a member of the armed forces assigned to work or train at a military installation

Civilian employee: a nonmilitary worker whose place of employment is at a military establishment

Total employees: for a military base, the sum of civilian and military personnel

Seat, attendee, station, pump, and the like: self-explanatory

4.2 Auto Restricted Zones

The principal objective of the auto restricted zone (ARZ) project is to create a shift in travel patterns that will reduce auto usage in congested activity centers, especially in downtown areas. By reducing automobile traffic in congested zones, greater attention can be paid to pedestrian movements and to facilitation of transit operations. Also, a shift to higher occupancy vehicles, both transit and car- and vanpooling, is expected when the ARZ projects are implemented.

Design Aspects.   Many of the design concepts used for partial or full conversion of vehicular streets to pedestrian or transit streets are taken from European experience. The *traffic cell* technique has been used in several cities. This concept involves the creation of small areas through which only local traffic can circulate. All movement from one cell to another and all through movements must use circumferential routes. In some cases, conversion of all streets within a cell to one-way operation has added to the flow and safety benefits obtained.

The most important considerations for the street design team to consider when analyzing street closures for automobiles are as follows:

1. Parking facilities along the street must be serviced by allowing the street to operate as a driveway only, or by rerouting access to the facility. The UMTA demonstration program also may involve relocation of parking.
2. Structural design of the surface is as important as the curb design at the access point.
3. Amenities for pedestrians (if the street is to be used for pedestrians) should be included in the design. Benches, some landscaping, and convenient access to transit stops are factors to be considered.
4. Signing, striping, and signalization must conform to the closure and the approaching motorist must be given adequate information on proper actions to take.

5. Part-time closures during the most congested hours can often be a more acceptable approach to take, allowing truck deliveries to retain a normal schedule.

Pedestrian Malls.   There are many design arrangements for providing streets for pedestrians only or for pedestrians and transit.

Modified sidewalks: a conventional street, allowing for both pedestrian and vehicular movement but with modifications particularly designed to facilitate pedestrian movement

Transitway: a "street" dedicated to pedestrians and transit riders, but from which all private vehicles are excluded except for emergencies or temporary construction work, with "transit lanes" set apart from pedestrian areas.

Plazas or interrupted mall: blocks of the retail street that are given over to exclusive pedestrian use, with cross streets left open to vehicular traffic.

Continuous mall: a pedestrian "street" from which all but emergency vehicles are excluded and that extends the full length of the shopping area without interruption.

Concourses: intersectional connections at a second level (either below or above grade) in which pedestrians may pass, from store to store at important intersections without conflict with vehicular traffic.

Multilevel traffic separation: pedestrian level above or below street grade, usually combined with mass transit, or traffic level above or below the mall, for automobiles, trucks, or mass-transit vehicles.

4.3 Traffic Diverters

Most new residential street patterns recognize that through traffic having no destination in the immediate neighborhood should be restricted to collector and arterial streets. Residential streets themselves should be planned and designed for local traffic. New designs generally incorporate curvilinear patterns, cul-de-sacs, and T intersections to a large degree.

Many cities must contend with older residential areas laid out in a grid pattern. There are positive aspects to such patterns, such as use during closures of parallel collectors or arterials. However, in many instances the negative aspects outweigh the positive. Negative factors include the following:

- As traffic volumes build up on adjacent parallel collectors and arterials, many drivers begin using local streets to save time and/or avoid congestion.
- The increase of through trips on local streets is usually accompanied by an increase in average speed, which is undesirable along residential streets.
- Noise and air pollution created by through traffic brings an undesirable consequence to residents of the area.

In attempting to reduce traffic loading, many cities have begun using diverters at residential intersections. The concept is simply that when through trips become numerous enough to create undesirable neighborhood effects, making through travel difficult will eliminate these trips. Most typical residential streets operate with from 100 to 500 veh/day. When volumes reach 500 or 1000 veh/day, the "local" and "residential" aspects of the street are perceived by most observers to be diminished. It is at this traffic level and any higher level that diversion techniques might be considered.

Traffic diversion is intended to accomplish one or more of the following effects:

- Reduction in total vehicular traffic
- Reduction in average speed of traffic
- Corresponding reductions in nuisance factors such as noise and air pollution
- Greater protection for bicyclists and pedestrians
- Increase in safety
- Greater cohesiveness of the residential area

4.4 Applications

Example 7

A shopping center is proposed in a suburb with a floor space of 18,000 ft^2. Surface streets are two-lane, two-way, and the frontage road to the site is expected to carry up to 20,000 veh/day in the next 15 to 20 years. An estimated 150 parking spaces are provided. Analyze this situation for adequacy, making suitable assumptions as appropriate.

Solution

1. Estimate the opposing traffic volume in the busiest hour (PHF = 10%; directional split 35/65).

$$\text{Opposing volume, } q = 20{,}000 \times 0.10 \times 0.35 = 700 \text{ veh/hr}$$

2. For a left turn across two lanes of opposing traffic, critical gap, $T = 5$ sec; follow-up headway, $T_0 = 3$ sec.

$$N = \frac{q \exp(-qT/3600)}{1 - \exp(-qT_0/3600)} = \frac{700 \exp(-700 \times 5/3600)}{1 - \exp(-700 \times 3/3600)} = 600 \text{ veh/hr}$$

For saturation to occur, the left-hand turning volume would have to exceed 600 veh/hr.

3. The types of stores in this shopping center produce 3 trips/hr/100 ft^2 of gross shopping space collectively.

$$\text{Total site generation} = 3 \times \frac{18{,}000}{100} = 540 \text{ veh/hr at street peak}$$

If it is assumed that traffic movements are 35/65 during peak period, it would still be within the 600-veh/hr limit, even if all 540 vehicles were making left-hand turns, which is highly unlikely.

4. Assuming an average of 3.0 veh/hr is the average turnover per parking space, a total of $3 \times 150 = 450$ vehicles would enter the site during the peak hour.

5. As calculated in step 4, 450 veh/hr enter the site and assuming that 50% enter from either direction, the highest estimated turning volume would be 225 veh/hr. If the

absorption capacity equation is used in reverse to find q given $N = 225$, we find $q = 1600$ veh/hr, which is well in excess of the planning year figure of 1000 veh/hr.

6. The calculations in steps 3 to 5 really amount to estimates of traffic generation.

7. It might be a good idea to check if left-hand turning vehicles do not form an excessive queue to enter the site. The volume of 225 veh/hr across a flow of 700 veh/hr produces a utilization factor (u) of

$$u = \frac{\sigma}{\mu} \tag{5}$$

where

$u =$ utilization factor

$\sigma =$ arrival rate

$\mu =$ practical absorption rate $= 0.8$ (theoretical absorption rate)

$$u = \frac{225}{600 \times 0.80} = 0.469$$

$$\text{Average queue length, } L = \frac{u}{1-u} = \frac{0.469}{1 - 0.469}$$

$$= 0.8828 = \text{less than 1 vehicle}$$

Had this queue length turned out to be too long, it could have been possible to put in a refuge lane or a continuous left-turn lane if several entrances were adopted.

Example 8

A hospital has 1200 beds, 3200 staff, and about 1500 outpatients per day. Estimate the peak-hour traffic generated by this facility, making suitable assumptions where necessary.

Solution

1. Say that 85% of the staff attend on an average day and that 90% of the staff drive to work. Also, each staff makes 2.5 vehicle movements per day.

    Total daily staff traffic generation $= 3200 \times 0.85 \times 0.90 \times 2.50$

    $= 6120$ vehicle movements per day

2. Outpatient traffic generation: Percentage driving car $= 60\%$

    Total daily outpatient traffic generation $= 1500 \times 0.60 \times 2$

    $= 1800$ vehicle movements per day

3. Visitor traffic generation: Number of visitors (at 2/bed) $= 2 \times 1200 = 2400/\text{day}$; proportion arriving by car $= 90\%$; car occupancy, 1.5/car.

    Total daily visitor traffic generation $= 2400 \times 0.90 \times \frac{2}{1.5}$

    $= 2880$ vehicle movements per day

4. Total daily traffic generation $= 6120 + 1800 + 2880$

    $= 10,800$ vehicle movements per day

5. Assuming that staff movements occur at the usual peak hours between 7 to 9 A.M. and 4 to 6 P.M. and that visitor movements take place between 6 and 8 P.M., one can calculate peak traffic generation as follows:

Staff: $6120 \times 10\% = 612$

Visitor: $2280 \times 50\% = 1440$

Outpatient: $1800 \times 10\% = 180$

Total = 2232 vehicle movements per peak hour

5. PARKING AND TERMINAL FACILITIES

Parking and terminal facilities are an essential part of the total transportation system. The planning and designing of these facilities demands an understanding of the characteristics of vehicles, the behavior of the drivers, the parking operation, and the parking generating characteristics of different land uses served.

As one of the activities of the urban complex, parking is competing for space, both on-street and off-street. Ideally, a motorist would like to be able to park right in front of his or her door, to avoid the need for walking, but this luxury is not always possible. Street space is more profitably used for moving traffic.

Of the many references available on parking, one of the best is *Parking Principles* (HRB, 1971). More recent data and information are contained in the *Transportation and Traffic Engineering Handbook* (Homburger, 1982).

5.1 Parking Policies

The formulation of parking policies is one of the more difficult tasks with which a planner has to contend. The difficulty lies in coordinating parking policies with several other planning objectives. The following considerations may be taken into account:

1. To strike a compromise between the amount of curb space devoted to parking spaces and that devoted to moving vehicles.
2. To make provision for parking of delivery vehicles and for short- and long-term parkers.
3. To design parking lots and their approaches so that street traffic is not adversely affected by the ingress and egress of vehicles.
4. To ensure that the interest of business establishments along the street is enhanced by good parking arrangements.
5. To ensure that parking policies and public transit policies are complementary; for example, car parking facilities adjacent to express bus routes would enhance bus ridership.
6. To preserve the character of the neighborhood by restricting parking and enforcing land-use controls.

TABLE 12-5 PARKING CONTROL

Control	Parking	
	On-street	Off-street

Price	Install meters
	Increase meter rates
	Street parking
	  permit with fee
		Parking tax rate
		  structure to
		  discourage long-
		  term parkers

Supply	Ban parking,
	  partially or totally
	Ban parking, except
	  specified groups,
	  such as residents
	Adjust meter rates
		Freeze new parking
		Reduce existing
		  parking
		Control future parking
		Vary time of opening
		  and closing

Source: Ogden and Bennett, 1984.

7. To control parking supply and demand through the pricing mechanism; encouraging short-term parking and discouraging long-term parking may serve to enhance the central business district (CBD).

Pricing mechanism or supply mechanism strategies for on-street and off-street parking could be implemented as shown in Table 12-5.

5.2 Parking Design for Automobiles

There are generally two types of parking facilities available for the automobile: curb parking and off-street parking. Engineers usually use a set of design standards and operating criteria to determine the best arrangements possible under specific site conditions. Some of the terms commonly used in connection with parking studies and design are as follows:

Parking accumulation: the number of parked vehicles in a parking facility (or study area) at a specified time (Figure 12-18)

Parking load: the area under the accumulation curve between two specific times (Figure 12-18)

Parking duration: the length of time a particular vehicle is parked in a specified parking space (generally specified as an average)

Parking efficiency: theoretical duration/actual duration

Parking turnover: the rate at which a parking space is used (e.g., the number of vehicles that park in a given space during an average day)

Parking volume: the number of vehicles using a parking facility during a specified time (e.g., 24 hours)

Parking space hour: the use of a single parking space during a unit time (e.g., 1 hour)

Parking deficiency: the extent to which parking demand exceeds supply, expressed in number of parking spaces

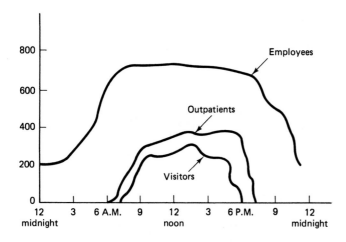

Figure 12-18   Accumulation Curve for a Local Hospital.

Example 9

An office has the following characteristics: hours of operation 6 A.M. to 8 P.M. (14 hours), number of parking spaces in the garage are 500, 80% are commuters with average parking duration of 8 hours, 10% are visitors parking for an average of 2 hours, and the balance are shoppers parking for an average of 3 hours. However, observations made at the garage indicate that 15% visitors during peak hours (10 to 12 noon and 1 to 3 P.M.) do not find parking. How many additional spaces should be added to the garage to meet the demand?

Solution

Space hours demanded:

Commuters:	80% × 500 × 8 = 3200
Visitors:	10% × 500 × 2 = 100
Shoppers:	10% × 500 × 3 = 150
Demand not served:	15% × 500 × 4 = 300
Total space-hours demanded	= 3750
Total space-hours served	= 3450
Balance space-hours	= 300

Because the additional demand is during 4 hours, the number of additional spaces needed is $300/4 = 75$ spaces.

Discussion

Although 75 additional spaces are needed only during the 4 peak hours, it is open to debate whether the garage owner should expend money to meet this demand, because for the rest of the operating time ($14 - 4 = 10$ hours), it is likely that the 75 spaces may not be occupied.

Curb Parking.   Depending on parking duration, turnover, space occupancy, and the distribution of vehicle size, it is possible to determine curb parking geometry. Figure 12-19 shows the geometric requirements for parking stalls. Although angle parking may allow more spaces per linear foot of curb, it restricts traffic movement on the street more than parallel parking. Tandem parallel parking reduces parking maneuver

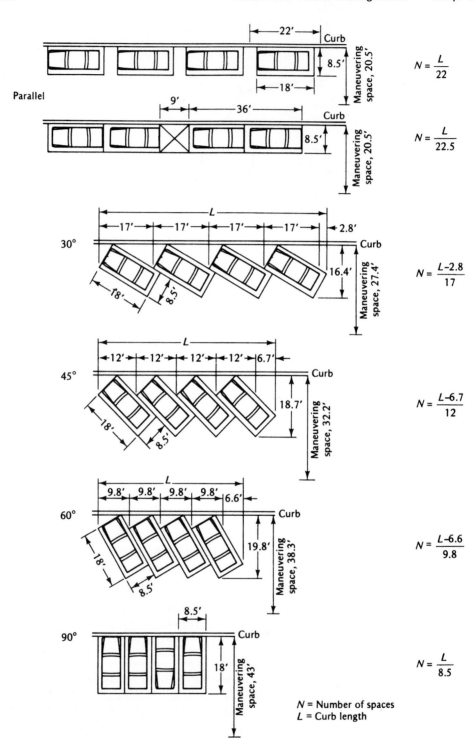

Figure 12-19 Curb Parking Geometry (Carter and Homburger, 1978).

and is recommended for major streets with heavy traffic. Safety considerations should be considered in all curb parking arrangements, and this factor is closely tied in with volume and speed of traffic on the street concerned (Homburger, 1982).

Off-Street Parking.   Many cities and their suburbs have off-street parking that is open to public free of charge. A significant proportion of off-street parking is either self-parking or attendant parking. Self-parking facilities are fast becoming by far the most common method of parking.

Typical layouts of self-parking lots are shown in Figure 12-20. Also, typical garage layouts are shown in Figure 12-21. The maximum amount of storage capacity from the

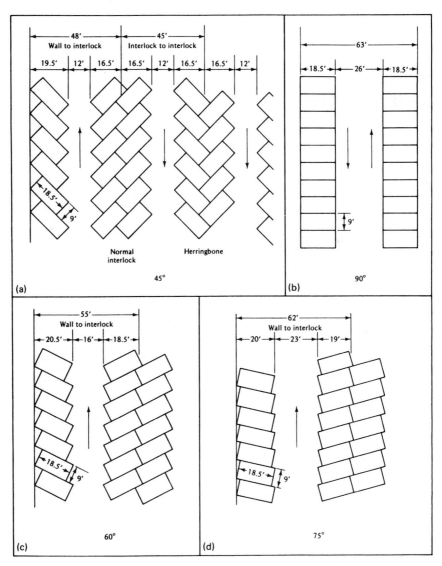

Figure 12-20   Typical Self-Parking Dimensions (Carter and Homburger, 1978).

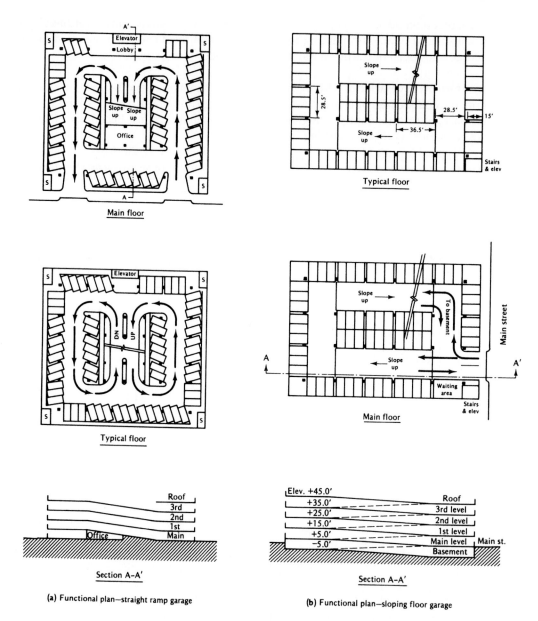

Figure 12-21   Typical Garage Layouts (Whiteside, 1961).

given working area is what the engineer is shooting for, consistent with the distribution of automobile sizes and dimensions. The capacity and spacing of access points to parking facilities must be sufficient to accommodate incoming vehicles without a backup on the surface streets. In the same context, the discharge of outgoing vehicles should pose no problems. This last concern is particularly important in cases where the approaches are close to street intersections (Homburger, 1982).

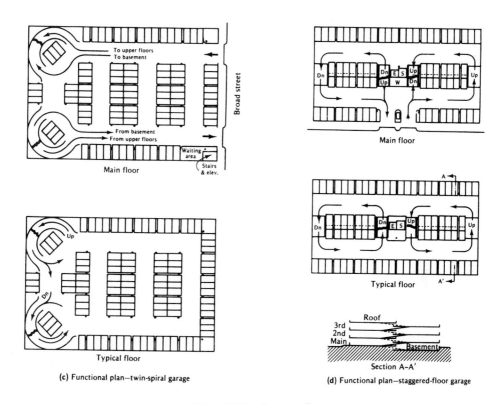

(c) Functional plan—twin-spiral garage

(d) Functional plan—staggered-floor garage

Figure 12-21   (*continued*)

5.3 Characteristics of Parking Supply and Demand

Free parking is part of the American way of life. It is said that 3 out of 4 cars driven to work are parked free in employer-provided spaces. If to these figures, one adds other free parking available on streets and lots, 93% of all commuters park free. If free parking were to be banned by 25%, commuters who currently drive to work alone would seriously reconsider changing their car occupancy.

Based on a national comprehensive inventory of parking facilities for urban areas ranging in size from 10,000 to over 1 million, the following was found:

1. The percentage of total spaces supplied at the curb decreases from 43% to 14%.
2. The percentage of spaces in lots ranges between 55% and 64%.
3. The percentage of spaces in garages steadily increases from 0% to 31%.
4. The proportion of off-street spaces provided in public facilities increases and the proportion provided in private facilities decreases for lots (HRB, 1971).

Tables 12-6, 12-7, and 12-8 provide information regarding the use of different types of parking, classification by trip purpose, and duration of parking, all as a function of city size, respectively (HRB, 1971).

TABLE 12-6 USE OF FACILITIES CLASSIFIED
BY TYPE

Population group of urbanized area	Location of parking spaces (%)		
	Curb	Lot	Garage
10,000–25,000	79	21	0
25,000–50,000	74	24	2
50,000–100,000	68	31	1
100,000–250,000	52	42	6
250,000–500,000	54	34	12
500,000–1,000,000	33	39	28
Over 1,000,000	30	54	16

Source: HRB, 1971.

TABLE 12-7 PARKING CLASSIFIED BY TRIP PURPOSE

Population group of urbanized area	Trip purpose of parkers (%)			
	Shopping	Personal business	Work	Other
10,000–25,000	38	23	21	18
25,000–50,000	27	35	21	17
50,000–100,000	24	31	20	25
100,000–250,000	21	34	26	19
250,000–500,000	19	33	30	18
500,000–1,000,000	13	25	47	15
Over 1,000,000	10	30	41	19

Source: HRB, 1971.

TABLE 12-8 LENGTH OF TIME PARKED CLASSIFIED BY TRIP PURPOSE

Population group of urbanized area	Trip purpose (hr)			Average all trips (hr)
	Shopping	Personal business	Work	
10,000–25,000	0.5	0.4	3.5	1.3
25,000–50,000	0.6	0.5	3.7	1.2
50,000–100,000	0.6	0.8	3.3	1.2
100,000–250,000	1.3	0.9	4.3	2.1
250,000–500,000	1.3	1.0	5.0	2.7
500,000–1,000,000	1.5	1.7	5.9	3.0
Over 1,000,000	1.1	1.1	5.6	3.0

Source: HRB, 1971.

5.4 Forecasting Parking Demands

Parking-demand analysis is a difficult task because the various factors affecting demand are interrelated. Some significant factors are the following:

1. Growth in population and motor vehicle registration
2. Trends in CBD growth such as floor space and retail sales
3. Public policies regarding parking supply, public transit, and parking pricing structure

Figure 12-22 provides a simplified flowchart of parking analysis for a small shopping area parking project. The supply of parking is derived by assessing potential sites for parking facilities, including the possibilities of utilizing any onstreet parking. The demand for parking can be conveniently derived from trip ends to relevant zones derived from the urban travel demand model. A detailed account of parking demand analysis is given in the *Transportation and Traffic Engineering Handbook* (Homburger, 1982).

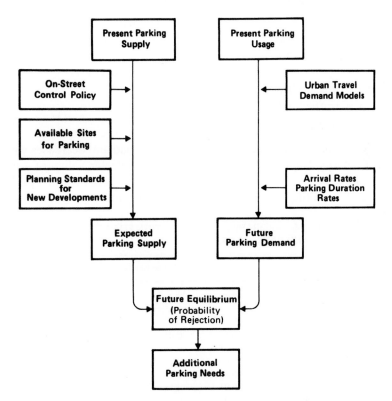

Figure 12-22 Main Steps in Forecasting Parking Requirements (Black, 1981. Reprinted with the permission of Croom Helm Ltd., Kent, UK).

5.5 Analysis of Parking Facilities

The average duration that a vehicle is parked in a facility and the number of arriving vehicles per unit time are indicators of the traffic load placed on the facility. If

A = traffic load
Q = number of vehicles arriving per unit time
T_d = mean parking duration in time units

then

$$A = QT_d \tag{6}$$

For example, if T_d = 50 min and Q = 120 veh/hr,

$$A = 50 \times \frac{120}{60} = 100$$

The Erlang formula is the basic rationale for parking analysis. The probability that a patron will find a parking space is given by the expression

$$P_L = \frac{A^M/M!}{1 + A + A^2/2 + \cdots + A^M/M!} \tag{7}$$

where

P_L = probability of rejection
A = traffic load
M = number of bays

The value of P_L increases with the traffic load and decreases with the number of parking spaces. Table 12-9 gives values of P_L for selected values of A and M. Use of the Erlang formula is a good method of evaluating the adequacy of a parking facility. Say, for example, in the calculation above that the number of parking spaces is 50, then

TABLE 12-9 PROBABILITY OF REJECTION FOR SELECTED TRAFFIC LOADS AND PARKING SPACES

Traffic load, A	Number of parking spaces				
	$M = 1$	$M = 5$	$M = 10$	$M = 50$	$M = 100$
1	0.50	0.00	0	0	0
2	0.67	0.04	0	0	0
3	0.75	0.11	0	0	0
4	0.80	0.20	0	0	0
5	0.83	0.28	0.02	0	0
10	0.91	0.56	0.21	0	0
50	0.98	0.90	0.80	0.10	0
100	0.99	0.95	0.90	0.51	0.08

Source: Black, 1981. Reprinted with the permission of Croom Helm, Ltd., Kent, UK.

$A = 100$ and $M = 50$, and therefore the probability of a randomly arrived patron finding no parking would be 51%. If the number of spaces were increased to 100, this rejection probability would reduce to 8%.

5.6 Parking Accumulation Analysis

Especially in small urban areas, one of the feasible solutions to relieve congestion is the removal of curb parking to provide additional lanes. Because curb parking is an element of parking supply, it is often necessary to estimate the demand for parking and to check the balance between supply and demand. If demand exceeds supply, additional off-street parking may have to be provided. The method consists of several basic steps:

1. Determine the number of auto-driver trip destinations to the study area.
   (a) The study area may be a single parking facility (shopping center) or a group of parking facilities; a traffic analysis zone (TAZ) or group of TAZs; or an entire study area.
   (b) Auto-driver trip destinations are stratified as long-term (work trips) or short-term (home-based shop, home-based other, and nonhome-based).
   (c) The number of auto-driver trip destinations is usually determined as the output of the four-step transportation planning procedure. The results are usually expressed as 24-hour values.
2. Determine the proportion of 24-hour trips that arrive during the daytime hours. A default value of 0.70 can be used, or a more precise value can be estimated by future analysis.
3. Determine the total daytime auto-driver trip destinations = parking demand = (24-hour auto-driver trip destination) $\times$ (proportion of 24 hours during daytime hours).
4. Determine the number of parkers accumulated during different periods of the day.
   (a) Long-term parkers (working trips) tend to arrive early and have a long parking duration.
   (b) Short-term trips tend to have dispersed arrivals and relatively short duration.
   (c) Accumulation factors are used to estimate the parking accumulation by period. The nature of these factors is illustrated in the following hypothetical example.
   (d) Default values for accumulation factors are given on the computation form shown in Table 12-10.

   The total number of daytime long-term parkers is 200, the same as the number of daytime short-term parkers. Seventy-five percent of the long-term parkers arrive the first hour, and the remaining 25% the second hour. All stay for a 9-hour period. Short-term parkers arrive more uniformly throughout the day, remaining for a 2-hour period.
5. Adjust the accumulation to determine practical capacity space requirements. The resulting value is the number of parking spaces that are required to accommodate each type of anticipated parking demand. The actual capacity requirements are greater than the estimated demand because of the following:

TABLE 12-10 PARKING ACCUMULATION WORKSHEET

Daily study area
Auto trip destinations

Daytime/total daily
Travel proportions

Daytime study area
Auto trip destinations

Time period		Home-based work		Home-based shop		Home-based other		Nonhome-based		Short-term parkers accumulated col. 4 + col. 6 + col. 8 (9)	Total parkers accumulated col. 2 + col. 9 (10)
		Accommodation factor (1)	Parkers accommodated (long-term parkers) (2)	Accommodation factor (3)	Parkers accommodated (4)	Accommodation factor (5)	Parkers accommodated (6)	Accommodation factor (7)	Parkers accommodated (8)		
9–10	1	(0.728)		(0.041)		(0.121)		(0.098)			
10–10:30	2	(0.754)		(0.067)		(0.131)		(0.112)			
10:30–11	3	(0.752)		(0.073)		(0.122)		(0.108)			
11–11:30	4	(0.754)		(0.125)		(0.136)		(0.132)			
11:30–12	5	(0.746)		(0.103)		(0.159)		(0.142)			
12–12:30	6	(0.754)		(0.109)		(0.195)		(0.170)			
12:30–1	7	(0.778)		(0.113)		(0.193)		(0.170)			
1–1:30	8	(0.762)		(0.132)		(0.175)		(0.163)			
1:30–2	9	(0.762)		(0.154)		(0.179)		(0.171)			
2–2:30	10	(0.788)		(0.183)		(0.173)		(0.175)			
2:30–3	11	(0.765)		(0.145)		(0.147)		(0.146)			
3–3:30	12	(0.750)		(0.133)		(0.160)		(0.152)			
3:30–4	13	(0.736)		(0.103)		(0.149)		(0.138)			
4:00–6:00	14	0		0		0		0			

LONG-TERM TRIPS

Period	Arrivals	Departures	Accumulation	Accumulation factor[a]
7–8	150	0	150	0.750
8–9	50	0	200	1.000
9–10	0	0	200	1.000
10–11	0	0	200	1.000
11–12	0	0	200	1.000
12–1	0	0	200	1.000
1–2	0	0	200	1.000
2–3	0	0	200	1.000
3–4	0	0	200	1.000
4–5	0	150	50	0.250
5–6	0	50	0	0.000
Total	200	200		

[a] Accumulation factor = accumulation/total number of parkers.

SHORT-TERM TRIPS

Period	Arrivals	Departures	Accumulation	Accumulation factor
8–9	10	0	10	0.005
9–10	15	0	25	0.125
10–11	20	10	35	0.175
11–12	20	15	40	0.200
12–1	25	20	45	0.225
1–2	25	20	50	0.250
2–3	30	25	55	0.275
3–4	20	25	50	0.250
4–5	20	30	40	0.200
5–6	15	20	35	0.175
6–7	0	20	15	0.125
7–8	0	15	0	0.000
Total	200	200		

(a) Time is wasted in parking and unparking maneuvers as the available unused supply of spaces approaches zero.

(b) Parkers do not have complete knowledge of space availability.

Practical capacity is normally defined as 85% of the available supply of parking spaces.

6. Compare space requirements to available parking supply.

5.7 Benefits of Good Parking Management

Parking is an economic commodity that is subject to the basic laws of economics. Thus, if a parking policy is enforced that reduces the parking spaces in downtown, a new equilibrium point will be established, resulting in a higher price to make a trip because of a higher parking price, the probability of making a trip by automobile to downtown would decrease. In short, parking policy affects energy consumption, traffic congestion, and transit usage.

Good parking management has been tried in many cities with startling results. Some of these are an increase in automobile occupation levels; a decrease in person-trips; faster travel times; an increase in transit usage; and most important, a decrease in street congestion.

It has been suggested over the years that street congestion cannot be solved by simply adding highway lanes. On the contrary, it has been argued that if motorists were subject to paying for a congestion price, the problem of congestion could possibly be mitigated. Parking pricing can also be considered as congestion pricing, and this idea has been used in Singapore in recent years (Khisty, 1980).

SUMMARY

This chapter covered a number of topics that are of much importance to engineers and planners at the local level. Pedestrian and bicycle facilities, parking and terminal facilities, and management of traffic are dealt with in an introductory way because these are the topics that are affecting the work of designers. Although planning at the macro level is important, the topics discussed in this chapter emphasize the microlevel factors.

REFERENCES

BENZ, GREGORY B. (1987). *Transit Platform Analysis Using the Time-Space Concept*, Transportation Research Record 1152, National Research Council, Washington, DC.

BLACK, JOHN (1981). *Urban Transport Planning*, Johns Hopkins University Press, Baltimore.

BLUNDEN, W. R., and J. A. BLACK (1984). *The Land-Use/Transport System*, 2nd Ed., Pergamon Press, Elmsford, NY.

CARTER, EVERETT C., and WOLFGANG S. HOMBURGER (1978). *Introduction to Transportation Engineering*, Reston Publishing, Reston, VA.

FEDERAL HIGHWAY ADMINISTRATION (FHWA) (1976). *Safety and Locational Criteria for Bicycle Facilities: Users Manual*, Vol. 2, U.S. Department of Transportation, Washington, DC.

FEDERAL HIGHWAY ADMINISTRATION (FHWA) (1980). *Design of Urban Streets*, U.S. Department of Transportation, Washington, DC.

FRUIN, JOHN J. (1971). *Pedestrian Planning and Design*, Metropolitan Association of Urban Designers and Environmental Planners, Inc., New York.

FRUIN, JOHN J. (1992). *Designing for Pedestrians*, Chapter 8, in George Gray and Lester Hoel (Eds.), Prentice Hall, Englewood Cliffs, NJ.

GREENSHIELDS, B. D. (1934). "A Study of Traffic Capacity," *Highway Research Board Proceedings*, Vol. 14, pp. 448–477.

HIGHWAY RESEARCH BOARD (HRB) (1971). *Parking Principles*, Special Report 125, National Research Council, Washington, DC.

HOMBURGER, WOLFGANG S. (Ed.) (1982). *Transportation and Traffic Engineering*

Handbook, 2nd Ed., Prentice-Hall, Englewood Cliffs, NJ.

INSTITUTE OF TRAFFIC AND TRANSPORTATION ENGINEERING (ITTE) (1975). *Bikeway Planning Criteria and Guidelines*, University of California at Los Angeles, Los Angeles.

KHISTY, C. J. (1979). "Use of Behavior Circuits for Pedestrian Safety," First International Symposium on Transportation Safety, San Diego.

KHISTY, C. J. (1980). "Some Views on Traffic Management Strategies—With Emphasis on Parking and Energy Use," *Traffic Quarterly*, Vol. 34, No. 10, pp. 511–522.

KHISTY, C. J. (1985). "Pedestrian Cross Flow Characteristics and Performance," *Environment and Behavior*, Vol. 17, No. 6, pp. 579–595.

OGDEN, K. W., and D. W. BENNETT (1984). *Traffic Engineering Practice*, 3rd Ed., Monach University, Department of Civil Engineering, Clayton, Victoria, Australia.

PUSHKAREV, BORIS S. and J. M. ZUPAN (1975). *Urban Space for Pedestrians*, The MIT Press, Cambridge, MA.

STOVER, V. G., and F. J. KOEPKE (1988). *Transportation and Land Development*, Prentice Hall, Englewood Cliffs, NJ.

TRANSPORTATION RESEARCH BOARD (TRB) (1985). *Highway Capacity Manual*, Special Report 209, National Research Council, Washington, DC.

WHITESIDE, ROBERT E. (1961). *Parking Garage Operations*, Eno Foundation for Transportation, Westport, CT.

EXERCISES

1. An observer counts 4900 pedestrians using a 50-ft-long skywalk in a 1-hour period. About 1500 of these took 15 seconds to cover the distance, and the balance took 20 seconds. At what LOS was the skywalk operating?

2. Solve Example 2 assuming that an LOS of B is desired.

3. A commercial building is expected to generate about 4700 pedestrians in a 15-minute period, in addition to the flow of about 500 pedestrians currently using the same sidewalk. With reference to the Figure 12-P3 and the following information, find the pedestrian LOS at points *A*, *B*, *C*, and *D*. The effective width of sidewalk is the actual width minus 3 ft. The perception-reaction time for pedestrians using the crosswalk is 4 seconds. Use a queueing standard of 5 ft²/ped, the effective width of the crosswalk is 15 ft, and the signal timing is 35 seconds walk/35 seconds don't walk.

4. An intersection approach has two traffic lanes and a right-hand-curb bike lane. The approach volumes are median lane 340 veh/hr and right lane 320 veh/hr. There are

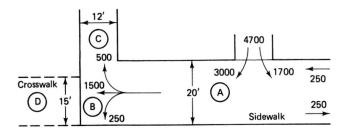

Figure 12-P3 Sketch for Exercise 3.

50 bikes/hr making left turns. What is the impact of this bike lane on motor vehicle traffic? Lanes are 11.5 ft wide.

5. During a 15-minute peak period, 1000 passengers in a transit station pass through a 1575-ft^2 enclosed ticketing area. The average walk time through this space is 12 seconds. However, 15% of these passengers spend an additional 20 seconds waiting in line to buy tickets and making inquiries. What is the average area per person in this area during this 15-minute period?

6. Determine the recommended width of a 450-ft-long subway corridor connecting two subway lines. The peak-hour pedestrian flow is 16,000 ped/hr in both directions. This corridor is planned to have small retail shops on one side and display windows on the other. An LOS D is desired and it is assumed that the 15-minute peak will be 35% of the peak-hour flow.

7. A new development consists of the following:

> Residential units:
>> Single family   20
>> Townhouses   30
>> Apartments   50
>
> Commercial:
>> Convenience shopping $27,000\text{ ft}^2$
>> Offices 5000 ft^2
>
> Park: Active recreation, 10 acres

Using appropriate tables and data referred to in the text, determine (a) person-trips by purpose generated by the residential development; (b) person-trips by purpose attached by the commercial and park development. (Assume that the city population is 100,000.)

8. What is the possible traffic impact of a newly established shopping mall on a street given the following information: ADT = 18,000 veh/day; PHF = 10%; directional split 40/60; for LT, critical gap T = 5 sec; follow-up headway, t = 3 sec; parking spaces = 240; estimated average turn over rate/parking space = 3 veh/hr; shopping mall floor space = $20,000\text{ ft}^2$; trip attraction rate = $4/\text{hr}/100\text{ ft}^2$; car occupancy = 1.2 persons/car. Make other assumptions, as needed.

9. Rework Example 8 with the following data: number of beds = 800, each staff makes 2 vehicle-trips to the hospital per day, car occupancy for outpatient trips = 3.0 going to the hospital and 2.0 leaving the hospital. Calculate the peak-hour traffic entering and leaving the hospital.

10. A quick-service restaurant serving 400 customers during the 2-hour lunch period serves 75% of its customers who arrive in self-driven cars. The average parking duration is 20 minutes. Calculate the probability of not finding a parking spot if the number of spaces is 10, 50, and 75.

11. A parking facility attached to a manufacturing plant has accumulation factors as given in tables on page 565. There are 500 workers, of which 90% use the parking facility (car occupancy = 1.25). Work out the cumulative arrivals and departures on a typical day. Assume 20% workers are short term.

12. An urban street has a curb length of 1000 ft available for parking on both sides of the street. The width of the street is 62 ft and a width of 24 ft is needed for two lanes of traffic. Based

on Figure 12-19, calculate the number of cars that can be parked on each side of the street. What are the advantages and disadvantages of using the five arrangements suggested?

13. Based on Exercise 11, plot the cumulative arrival and departure curves.
   (a) What is the maximum number of cars in the parking lot?
   (b) How well is the parking lot being utilized?

14. Visually examine a neighborhood in your city that has complaints of excessive through traffic. Determine ways of combating this problem.

15. Examine the CBD (downtown) in a city with which you are familiar. What are the chances of introducing an auto-restricted zone in part of the CBD to enhance the quality of life and ease of movement? Examine strategies that could be applied, and prioritize them.

Chapter 13

Energy Issues Connected with Transportation

1. INTRODUCTION

1.1 Framework and Objectives

Despite the glut of information available regarding energy, we do not know enough about how we can best become an energy-efficient nation. The best we can do is to draw on the results of completed and ongoing research about the relationship of energy to various facets of American planning. This chapter has been written primarily for the transportation engineer or planner who is interested in understanding the energy issues connected with transportation. It provides an introduction to techniques for energy planning and energy conservation.

1.2 Energy and Society

"Everything is based on energy. Energy is the source and control of all things, all value, and all the actions of human beings and nature" (Odum and Odum, 1976). However, we are at the beginning of an energy crisis, because certain fuels on which we place great reliance, such as petroleum, are nonrenewable and are being depleted rapidly. To add to our difficulties, energy supplies are inadequate, unpredictable, and vulnerable to be cut off, because a substantial proportion comes from politically volatile sources. For a nation that has become accustomed to cheap energy, this is a disconcerting realization. There is immediate need for society to conserve energy, to develop alternative energy technologies, to increase the efficiencies of various components of our infrastructure (particularly transportation), to develop policies to conserve energy in every sphere, and to educate society regarding energy issues.

Energy is a social as well as a technological issue. Yet, in many discussions of the factual basis of energy, the cultural and social contexts tend to be left implicit. This problem arises chiefly from our growing dependence on large institutions and a decreased dependence on both individual and group self-reliance. Such dependence has led to increased governmental planning and reliance on expertise. Major choices in energy planning are often made in the context of conflicting perceptions and beliefs (NAS, 1980).

1.3 The Energy Crisis

A crisis is defined as a turning point. The oil crisis of 1973–1974 constituted a turning point in postwar history, delivering a powerful economic and political shock to the entire world. "The term energy crisis refers either to a shortage, or a catastrophic price rise for one or more forms of useful energy, or to a situation in which energy use is so great that the resulting pollution and environmental disruption threaten human health and welfare" (Miller, 1980).

The key issues and related problems in dealing with the energy crisis are dependency on foreign oil sources, heavy outflow of U.S. dollars, political instability and attitudes of many foreign governments supplying oil, the competing demand of other countries for petroleum, and the relatively high price of oil.

2. ENERGY ISSUES IN TRANSPORTATION

2.1 Spatial Form and Structure

Urban spatial structure refers to the order and relationship among physical elements and land uses in urban and regional areas as they evolve from interactions among the key systems—individuals and households, firms, and institutions—and pass through transformations in time and space. Land-use planning generally utilizes the normative (prescriptive) approach in deciding what the future ought to be (Chapin and Kaiser, 1979). A land-use arrangement that is most efficient and least costly to the city and its citizens is the basic concern. Elements such as health and safety, convenience, environmental quality, social equity, and social choice have, in the past, been taken into account. In recent years, studies relating to energy-efficient patterns of land development have assumed importance. Energy efficiency is a special case of cost efficiency, and in view of the nature of the energy problem and the long-range implications of the built environment, spatial form and structure have emerged as interests to many disciplines. Also, because the transportation section is such a heavy consumer of fuel, it can be concluded that land-use alternatives that involve the least amount of aggregate travel are generally considered as the most energy-efficient solutions. Another consideration is the development intensity with which the land is put to use. For example, a city that suffers from urban sprawl and ribbon development would have more miles of streets, water pipes, and sewer lines than would a more compact city. Not only would the initial cost of developing and constructing the infrastructure be high, but also the cost of maintaining it, as compared to a compact city with the same population. In reality, the crucial issue is the costs the citizens are willing to pay in order to satisfy their wants. This willingness to pay is a function of a society's values, attitudes, and preferences.

2.2 Energy Demand

In general, Americans use energy in four basic sectors: transportation, residential, commercial, and industrial. The breakdown of energy uses in each sector is shown in Figure 13-1. Transportation accounts for about 40% of the energy used, with about 23% (direct energy) used to move people and goods, as indicated in Figure 13-1, and another 17% (indirect energy) used to build and maintain vehicles, highways, and

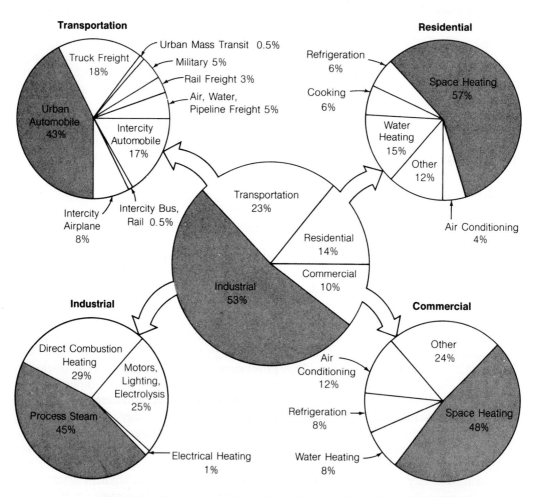

Figure 13-1   Use of Energy in the United States, in 1976 (Miller, 1980. Reprinted by permission of the publisher.).

TABLE 13-1   COMPARISON OF ENERGY CONSUMPTION/GDP

Country	Per capita		Energy/GDP ratio (thousand Btu/dollar)	Index numbers (U.S. = 100)		
	GDP (dollars)	Energy (million Btu)		Per capita		Energy/GDP ratio
				GDP	Energy	
United States	5643	335.5	59.5	100	100	100
Canada	4728	336.6	71.1	84	100	120
France	4168	133.1	31.9	74	40	54
West Germany	3991	165.6	41.5	71	49	70
Italy	2612	95.7	36.6	46	29	62
Netherlands	3678	188.1	51.1	65	56	86
United Kingdom	3401	152.9	45.0	60	46	76
Sweden	5000	213.4	42.7	89	64	72
Japan	3423	116.6	34.1	61	35	57

other vehicle support services (Miller, 1980). The indirect energy is accounted for in the industrial and commercial sectors.

A comparison of energy and productivity of the United States with some of the other eight industrialized countries of the world is given in Table 13-1. Public policy in the United States has been to control the price of energy; consequently, social, community, and industrial development and processes in the past were built around the assumption that energy was cheap and abundant. In Europe and Japan, on the other hand, public policy has been at the other end of the scale—one of resource conservation. This conservation is fostered in part by taxing energy and machines (including vehicles) using energy at a high level as compared to the United States. It is evident that the U.S. energy consumption per unit of gross domestic product ranges from 35% to 85% higher than in other industrial countries. It is possible for the United States to reduce energy consumption without fundamental changes in life-style.

A sketch of energy flows through the U.S. economy is shown in Figure 13-2. On the left-hand side are the inputs from coal, petroleum, natural gas, water, and nuclear power in units of 10^{17} calories. Note the extent of energy wasted. More energy is ultimately wasted than is used productively. The picture has not changed significantly today.

A fundamental factor in metropolitan expansion, particularly in the United States during the years 1945 to 1970, was the availability of cheap energy. Other elements that contributed to this expansion were the rapid increase in real per capita income, rapid diffusion of the automobile, development of beltways and freeways at the expense of public transportation, and planning policies that encourage low-density residential development. Since 1973, five major factors have changed patterns of metropolitan development: (1) recession; (2) high inflation and interest rates; (3) continuing decline in the size of households; (4) new environmental and land-use controls; and (5) energy costs and uncertain energy futures (William et al., 1979).

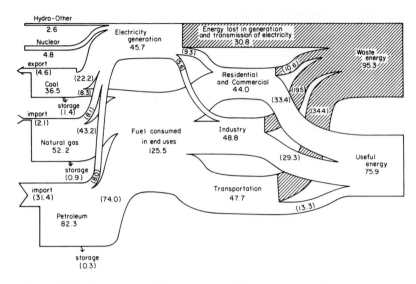

Figure 13-2   Production and Consumption of Energy in the United States, 1975 (10^{17} Calories) (Reprinted from *Environmental Science in Perspective* by Spiro and Stigliani, 1980, by permission of the State University of New York Press.).

3. ENERGY CONSERVATION

There are at least three basic strategies for meeting energy needs: (1) develop new sources of energy, (2) reduce energy waste, and (3) adopt new life-styles that use less energy. Although some combination of all three strategies is called for, the one that holds out the most promise, keeping in mind actual and potential environmental impacts, is energy conservation. Energy conservation is indeed energy efficiency. If the United States were to make a collective effort to conserve energy, it could save between 30% and 40% of the 1980 consumption level and still enjoy the same or even a higher standard of living (Stobaugh and Yergin, 1979).

Conservation can take several forms: out-and-out curtailment is one way if we are faced with emergencies; a second category is overhauling, where cities reduce sub-urbanization, or adopt certain minimum and maximum standards of urban density in order to encourage substantial amount of mass transit; a third category is a form of adjustment, where energy efficiency is greatly improved.

3.1 Transportation Energy Savings

Conservation of energy used for transportation is of vital concern to the nation. A major area for potential energy savings lies in the transportation sector, which accounts for one-fourth of the total energy and about one-half of the petroleum used in the United States. This is called the *total direct transportation energy consumption*. If, however, indirect energy consumption is included, transportation accounts for more than 40% of total energy consumption. Indirect energy consumption attributed to transportation is in the shape of refining and distribution losses of transportation fuels; manufacture and maintenance of vehicles and equipment; and construction, operation, and maintenance of fixed transportation-related facilities, such as highways, airports, truck terminals, railroad tracks, and ports. Accordingly, a large share of the savings required in the total national conservation effort must come from the transportation sector, directly and indirectly, especially from the automobile, which represents the largest fuel consumer.

The alternatives for reducing transportation energy consumption can be put under five categories (TRB, 1977):

1. Shift traffic to more efficient modes by lowering the Btu per seat-mile.
2. Increase the load factor by raising the passenger-miles per seat-mile.
3. Reduce the demand by reducing the passenger-miles.
4. Increase the energy conversion efficiency by lowering the Btu per seat-mile.
5. Improve the use pattern by lowering the seat-miles.

Figure 13-3 shows the alternatives under each category. Increasing the energy conversion efficiency of highway vehicles is the most important option in the short term, for the following reasons:

1. The savings potential of improving vehicle efficiency is much larger than that of any of the other approaches because motor vehicles now consume the major share of transportation energy.
2. Gains in vehicle efficiency will have little adverse impact on service quality.
3. Implementing improvements in vehicle efficiency will reduce total cost of transportation. Load-factor (occupancy) improvements are also important. Although inconveniences might make them unattractive for some users, such improvements could be implemented quickly with little or no capital costs and could add significantly to energy efficiency. Operational improvements in use patterns and declines in growth rates will reduce energy consumption. Modal shifts offer theoretical savings, although they are not likely to be induced by fuel price increase of the magnitude experienced over the past decade (TRB, 1977).

Transportation energy solutions are quite varied. Long-term solutions include increased supplies through synthetic fuels and land-use development patterns that reduce the need to travel. In the short term, outside of converting stationary liquid fuel users to natural gas or coal, the solutions are focused in the conservation arena. Rationing, taxes, decontrol of prices, and vehicle fuel efficiency improvements all have the highest potential fuel saving but require government regulations or politically sensitive actions. Improved driving habits, vehicle maintenance, and ride sharing all have potential but require changes in social behavior. A summary of highway energy conservation strategies is presented in Table 13-2, indicating estimated savings ranging from 0 to 50%.

Energy-efficient cost-effectiveness of urban transportation actions is a good indicator for making decisions regarding strategies likely to be adopted by a planner. For

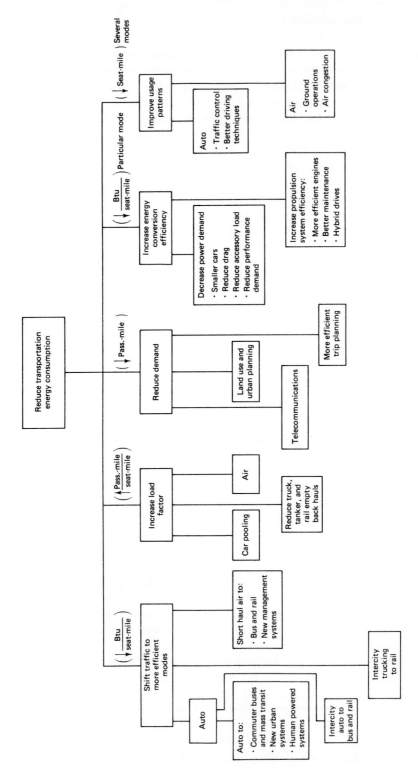

Figure 13-3 Alternatives for Reducing Transportation Energy Consumption (TRB, 1977).

TABLE 13-2 SUMMARY OF HIGHWAY ENERGY CONSERVATION STRATEGIES

Program area	Elements included	Estimated saving of TDTE[a,b] (%)

1. Vehicle technology improvements	Downsizing model lines
	Design improvements
	  Reduce weight
	  Reduce drag
	  Improve transmissions and drive trains
		10–20

2. Ride sharing	Ride sharing matching program Ride sharing marketing Employer programs HOV incentives	2–5
3. Traffic flow improvements	Traffic signal improvements One-way streets Reversible lanes Intersection widening Ramp metering Freeway surveillance and control	1–4
4. Other transportation system management strategies	Fringe parking Alternative work schedule Priority lanes for HOVs Pedestrian and bicycle improvements Pricing parking and highway facilities	1–4
5. Goods movement efficiency improvements	Improved routing and scheduling of urban goods delivery Truck size and weight changes Truck deregulation TOFC	1–4

6. Transit improvements	Modal shifts to transit through:
	  Park and ride
	  Improved service
	  Marketing
	  Preferential highway lanes
	  Fare reduction
	Improved routing and scheduling
	Improved maintenance
	Vehicle rehabilitation
		1–3

7. Construction and maintenance	Improved highway maintenance RRR Substitute sulfur-based materials for asphalt Pavement recycling	1–3

8. Fifty-five-mph speed limit	Better enforcement and compliance to
	  achieve fuel saving and reduced
	  fatalities
		0–2

9. Improved driving habits and vehicle maintenance	Radial tires Higher tire inflation Improved maintenance Travel planning trip linking	5–20
10. Rationing	Private autos Taxis/trucks	15–50
11. Pricing, decontrol	Gas tax Parking fees/policies Road pricing Vehicle registration	5–25

[a] TDTE = total direct transportation energy.
[b] Not additive.
Source: Cannon, 1980.

TABLE 13-3   COST-EFFECTIVENESS OF URBAN TRANSPORTATION ENERGY EFFICIENCY ACTIONS[a]

Urban transportation energy efficiency action	Estimated project expenditures (dollars) per gallon saved	Estimated gallons saved per project dollar expended
Ride sharing (carpooling and vanpooling)	0.06–0.26	4–17
Compressed work weeks	0.043	23
Flexible work hours	0.28	4
HOV priority treatments	1.20–5.40	0.2–0.8
Signal optimization	0.035–0.047	21–29
Signal interconnection/coordination	0.042–0.15	7–24
Advanced computer control of signals	0.19	5
Signal removal and flashing	Negative cost	
Freeway traffic management	1.29–1.58	0.6–0.7
Areawide express bus services	4.57	0.2
Broad transit expansion programs	7.62	0.1

[a] Single values should be interpreted as *midpoints of ranges*. All estimates are *approximate orders of magnitude* based on generalized cost analyses.
Source: ITE, 1980.

instance, signal optimization is estimated to cost a transportation agency only 3 to 4 cents per gallon of gas saved. In other words, each dollar invested in this project would result in about 25 gallons of fuel saved. Table 13-3 provides the cost-effectiveness of some selected actions applicable in an urban area. In general, the strategies are listed in increasing order of cost and difficulty.

The cost-effectiveness of different actions is highly variable. Planners should therefore carefully assess each action at the planning stage to ensure that the greatest impacts are achieved from expenditures on urban transportation energy efficiency programs.

Energy conservation strategies concerned with substitution of communication for transportation is a feasible area for investigation which may lead to a replacement of up to 50% of the current face-to-face business meetings in the future (Khisty, 1981). Other strategies, such as those connected with "congestion pricing" and parking, are also promising (Khisty, 1979, 1980). Good parking management, for example, can lead to the following results: (1) an increase in automobile occupancy levels; (2) a decrease in vehicle trips; (3) faster travel times and a decrease in travel delays; (4) an increase in transit usage; (5) a reduction of air pollutants; (6) a lower ambient noise level; and (7) a decrease in congestion. The first four points directly reduce energy consumption.

3.2 Land-Use Patterns

A long-term perspective is essential if the urban transportation planning process is to deal with problems of providing transportation in an environment characterized by fuel shortages. Land-use plans provide a pattern or arrangement of land uses which are adopted by a city to achieve the goals and objectives of that city. The cost of public services, such as transportation, water supply, sewers, telephones, gas, and electricity is almost directly dependent on land form. Urban sprawl tends to increase public service cost as well as energy consumption, whereas multicenter plans have generally lower infrastructure costs and lower energy consumption.

In a study made by Edwards (1978), it has been shown that structural changes in transportation and land-use patterns can produce significant reductions in energy consumption for urban passenger travel (Figure 13-4). Four dimensions of urban form were examined:

1. Shape of city
   (a) Concentric ring or grid shape
   (b) Pure linear shape
   (c) Polynucleated shape
   (d) Pure cruciform shape
2. Extent to which the city is compact or sprawling (its geographic extent)
3. Population concentration
4. Employment concentration

Some important conclusions drawn from this study are as follows: From the average work trip length, one can determine the total amount of energy required for transport in that city; energy consumption in concentric ring cities rises fairly rapidly with

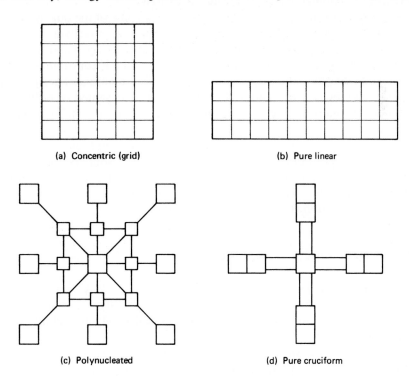

(a) Concentric (grid)   (b) Pure linear

(c) Polynucleated   (d) Pure cruciform

Figure 13-4   Urban Shapes.

increasing average work-trip length, whereas the rate of increase is much lower in polynucleated cities; a city with most employment concentrated in the downtown area will consume energy quite differently from one in which most business and industry is located along a beltway; transportation level of service is an important factor because traffic congestion, with its inefficient fuel use, can swallow up any advantage to an urban area's shape. The research suggests the desirability of controlling the spread of cities and of channeling development into higher-density, nucleated forms. This may also serve in the short term as a policy on rezoning requests and building permits and as a criterion for incremental construction to the urban infrastructure. There is need to improve traffic operations to reduce congestion without building new freeway-type facilities. Such facilities can be self-defeating, because they encourage horizontal spread of cities unless strict land-use controls are applied. Moving more people by transit is a promising energy-minimizing strategy. Because transit solutions reduce accessibility, better ways of providing service must be found if people are to use transit by choice (U.S. DOT, 1980a and 1980b).

Peskin and Schofer (1977) have expanded and defined Edwards and Schofer's (1976) work and come to the conclusion that polynucleated urban structures hold more promise for energy conservation than do other spatial arrangements. A U.S. DOT study suggests that it would be advisable to encourage (1) filling of city space, thus utilizing existing infrastructure; and (2) contiguous development, thereby avoiding scattered site development remote to the urban area.

In general, there is a growing concern over some of the effects of land use/transportation interaction. Not only have transportation systems covered large areas of land (in many cases very expansive land), but the increased accessibility created by trucks and autos has encouraged activities to disperse and has contributed to the decline of population densities and the economy of some central areas. Transportation and general development policy adopted by a city can be designed to reduce land-use consumption, to improve accessibility, and to reduce environmental damage in inner cities. Where new development is being undertaken it may be designed to reduce dependency on cars. Economic techniques, such as value capture and joint development, may be used to transfer to public transportation authorities some of the profits accruing to property owners from improvements in accessibility. The principles of value capture, joint development, and mixed use can be used to intensify these benefits, including energy conservation (U.S. DOT, 1980a, 1980b).

4. ENERGY CONTINGENCY STRATEGIES

Among the greatest vulnerabilities of the U.S. energy system is its reliance on oil imported from a region with a history of instability, halfway around the world. Central to the supply–demand balance underlying this universal vulnerability is our voracious appetite for oil.

The principles of resilient design—dispersion, diversity, redundancy, interconnection—are widely applied in industry, such as in aircraft and power plants. They are now being applied to our energy system. It is only by spreading awareness of the threat, knowledge of local energy potential, and readiness to act that we can prepare ourselves for energy contingency (Lowins and Lowins, 1981).

In recent years, a number of national actions have been taken to reduce the shock of another oil embargo. These actions include the gradual decontrol of fuel prices to stimulate greater domestic production, acceleration of synfuels development, and establishment of the Strategic Petroleum Reserve. In the short run, local and regional approaches can contribute to energy security much more quickly than the actions the federal government is likely to take (Clark, 1981).

Conservation planning is generally thought of as a continuing, long-range strategy to encourage reduction in energy consumption. Contingency planning is considered to be those actions implemented quickly in response to an unexpected but possibly emergency situation. In our energy situation, conservation and contingency planning are so interwoven that the two must occur as part of the same process. Thus, it is true that the best contingency plan is to have in place an aggressive conservation strategy, and an efficient conservation plan must include an effective contingency strategy.

The objective of contingency planning is twofold: first, to ease short-term crisis and help people cope with the problems generated by such situations, and second, to do so in a manner that helps to solve the long-term problem of reducing fuel demand. Actions that satisfy both objectives are the most desirable contingency actions (U.S. DOT, 1980b).

4.1 Energy Contingency Strategy for Transportation

To be adequately prepared for the energy climate of the future, all levels of government need to prepare a variety of conservation and contingency strategies that can be employed as necessary in response to a wide variety of energy supply scenarios. Inconsistent actions by different levels of government will increase confusion and anxiety when a major reason for the plans is to reduce anticipated confusion and anxiety.

The contingency planning process should produce three principal products. First, the process itself: The interaction among many levels of government and private institutions can generate conflicts over roles, responsibilities, and finance. To avoid confusion, this conflict should be resolved before energy shortages occur. Second, the process should lead to ongoing working relationships among various levels of government, private businesses, and institutions for implementation of strategies. Third, the process should produce contingency plans and implementation strategies that are designed to cover several different types of shortages, for example:

- Expected shortages over a long period
- A local, 3-week-long, 8% to 12% shortfall resulting from local misallocation of fuel
- A national, 6-month-long, 8% to 12% shortfall that results from an international event
- A more severe 12% to 20% shortfall
- A shortfall of over 20% accompanied by rationing

All relevant actors need to know what actions they take under the various contingencies, what actions they can rely on others to take, how their actions will be financed, where their trained personnel will come from, and what the limits of their effectiveness will be. This information should be specific enough to serve as a basis for estimating real potential of transit, paratransit, and ridesharing as well as the costs of providing this potential. In short, the "plans" should clearly delineate implementation responsibilities, expected results, and appropriate timetables (U.S. DOT, 1980b).

Almost all metropolitan planning organizations (MPOs) have prepared transportation energy contingency plans in response to federal directives. Recommendations of one MPO, based on an exhaustive study, are as follows (U.S. DOT, 1980b):

1. Modify state and federal fuel contingency regulations to provide priority fuel allocations to public transportation providers.
2. Maintain the present metropolitan carpool programs.
3. Expand or develop fuel storage reserves.
4. Designate a local energy coordinator (LEC) in counties and major cities.
5. Encourage flexible-work-hour programs.
6. Increase transit system bus availability.
7. Modify state laws to permit the use of school buses for the general public under emergency conditions.
8. Investigate the impact of an energy shortage on taxicabs and their possible role in local mobility during an emergency.
9. Develop regional park-and-ride programs and an exclusive-lane bus plan.
10. Draft contingency agreements to be used between local governments, transit operators, and taxicab operators for mutual assistance.
11. Begin intergovernmental dialogue regarding possible energy contingencies and local solutions.

Naturally, each MPO will evolve its own priorities and mechanisms for implementing contingency plans, depending on such factors as population, economics, city configuration, level of service of transit and its extent, employment characteristics, and so on (Barker and Cooper, 1978; U.S. DOT, 1980b).

5. ENERGY ANALYSIS INFORMATION AND METHODS

5.1 Energy Efficiency of Modes and Travel Characteristics

Although several sources provide conflicting figures regarding the energy efficiencies of various modes, the information given in Table 13-4 is sufficiently authentic for rough comparisons. Note that bicycling and walking do not consume any fossil fuel. Also, the general average for the automobile depends on several factors, discussed later.

Passenger cars, which compose about 80% of all U.S. motor vehicles, are driven an average of 10,000 miles annually, and this figure has remained surprisingly constant

TABLE 13-4   ENERGY INPUTS FOR VARIOUS
TRANSPORT MODES

Transport mode	Energy input
Freight transport	Btu/ton-mi
Oil pipeline	450–660
Waterway	500–750
Railroad	650–750
Truck–trailer combination	1,600–3,400
Aircraft	27,000–63,000
Intercity passenger transport	Btu/passenger-mi
Motorcycle	2276
Bus	1100–1600
Railroad	2730–3530
Automobile	2900–3400
Aircraft	7700–9700
Urban passenger transport	Btu/passenger-mi
Bicycle	100–200
Walking	300–500
Rail transit	1650
Commuter railroad	2500
Bus transit	2680–3700
Automobile (general average)	5580–8100
Small car (all trip purposes)	2620
Work and related business	5770
Shop and family business	3020
Social and recreation	1670
Standard car (all trip purposes)	5100
Work and related business	7970
Shop and family business	6040
Social and recreation	2970

Source: Homburger and Kell, 1980.

for the past 35 years. Total travel is therefore increasing at about the same rate as vehicle registration. Table 13-5 lists average annual travel and fuel consumption for different types of vehicles. The trip-length distribution and the corresponding percentages are given in Table 13-6.

5.2 General Considerations

It must be apparent by now that motor vehicles consume a large proportion of transportation-related fuel, and therefore it is not surprising that considerable attention has been paid to making these vehicles energy efficient. Vehicle design and operating factors connected with fuel economy are two major factors considered here.

Vehicle Design.   The energy needed to overcome inertia when accelerating and climbing grades are dependent on the weight of the vehicle and the load it is carrying. In the last decade, the weight of cars has been substantially reduced and the attempt to reduce weights further is continuing (e.g., the average weight dropped from about 4000 lb in 1975 to about 3000 lb in 1980). Rolling resistance of vehicles is a function of vehicle weight plus the weight it is carrying, tire characteristics, and drive train bear-

TABLE 13-5 ANNUAL MOTOR VEHICLE USE AND FUEL CONSUMPTION, 1985

Type of vehicle	Travel				Fuel consumption	
	Rural (veh-mi × 10⁹)	Urban (veh-mi × 10⁹)	Total (veh-mi × 10⁹)	Average (mi/veh)	Amount (gal/veh)	Rate (mi/gal)
Automobiles[a] ⎱	474.0	795.6	⎰ 1260.6	9560	525	18.2
Motorcycles ⎰	200.9	219.2	⎱ 9.1	1669	33	50.6
Single-unit trucks	200.9	219.2	420.1	11,115	946	11.7
Truck combinations	52.6	27.0	79.6	56,725	10,889	5.2
Buses[b]	2.7	2.2	4.9	8276	1407	5.8
All vehicles	730.2	1044.0	1774.2	10,018	685	14.6
	(41.2%)	(58.8%)	(100.0%)			

[a] Includes taxis.
[b] Commercial and school buses.
Source: FHWA, 1986.

TABLE 13-6 DISTRIBUTION OF AUTOMOBILE TRIP LENGTHS AND VEHICLE MILES OF TRAVEL, 1983

Trip length (mi)	All trips Number (%)	All trips Travel (%)	Commute trips (%)	Trip length (mi)	All trips Number (%)	All trips Travel (%)	Commute trips (%)
≤ 5	62.4	17.9	54.1	31–50	2.2	11.2	⎫
6–10	18.7	18.9	20.3	51–75	0.7	5.3	⎬ 3.6
11–15	7.7	13.1	10.7	76–100	0.3	3.8	⎭
16–20	4.3	10.1	6.1	≥ 100	0.4	8.8	—
21–30	3.3	10.9	5.2	Total	100.0	100.0	100.0

Source: FHWA, 1985.

ings. It is relatively independent of speed. Power-operated accessories such as power brakes and steering consume fuel.

Vehicle Operating Factors.   Aerodynamics drag is negligible at low speeds, but becomes significant at speeds above 50 mph. Acceleration requires more power than cruising. Additional fuel consumption in case of speed changes can be calculated from Figure 13-5. If measures can be taken to minimize speed changes through traffic engineering, one can save fuel. Even for vehicles traveling at uniform speed, fuel consumption per unit distance is very high at low speeds and decreases as speed increases to about 35 mph (Figure 13-6).

Manufacturers of motor vehicles have been challenged by federal regulations to produce motor vehicles that meet certain overall fuel consumption standards. This challenge has met with fair compliance. Whereas the 1974 average passenger fuel consumption for new cars was merely 14.2 mi/gal, the corresponding figure for 1986 models is 27.9. Table 13-5 shows the annual motor vehicle use and fuel consumption for the year 1985. However, it should be noted that these figures reflect an average consumption rate of 18.9 mi/gal because the automobile fleet represented here is for all automobiles on the highway.

Intersection Control.   Traffic engineers have been concerned about the possibilities of fuel conservation gained through choosing proper intersection controls. Although the choice of such controls is dictated by warrants, the effects of energy savings should also be given consideration. Figure 13-7 shows the average fuel consumption as a function of major/minor volume ratios. Note that flashing red/red is a four-way stop intersection and a flashing yellow/red is a two-way stop.

5.3 Energy Analysis Methods

Several concepts and pieces of information can be used in transportation energy analysis (such as direct and indirect energy consumption) and energy effectiveness in the modal comparison of trips. Direct energy is energy used to propel a vehicle and to support vehicle-related auxiliaries (e.g., air conditioning). Indirect energy is that which is

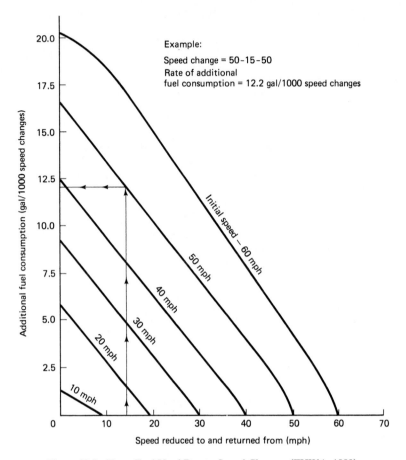

Figure 13-5   Extra Fuel Used Due to Speed Changes (FHWA, 1980).

consumed during manufacture, construction of the infrastructure, and maintenance. Energy effectiveness can be analyzed through energy efficiency or energy intensity. The former is equal to the distance (miles) divided by the energy consumed (gallons), and the latter is equal to energy consumed (gallons) divided by the distance (miles). For consistency, it is preferable to express energy consumption in terms of joules (1 Btu = 1055 J). Table 13-7 is an energy conversion table, and Table 13-8 lists the heat content for various fuels.

     A typical energy analysis procedure is outlined. Some of the steps shown may be eliminated depending on the nature of the analysis (CUTA/RTAC, 1985):

1. Define the population group and the mode(s) affected, such as the auto or transit mode, the type of trip, the time of the day, and vehicle speeds.
2. Choose the appropriate measures of effectiveness (MOEs), such as changes in vehicle efficiency, changes in vehicle-miles of travel, and changes in load factors.

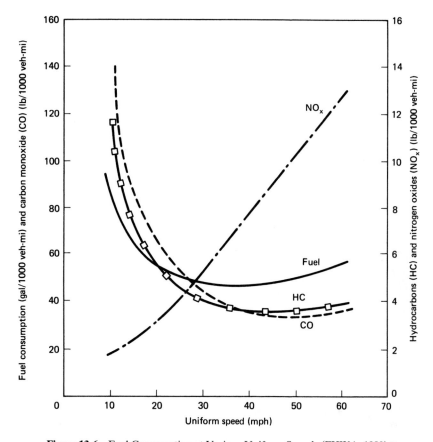

Figure 13-6 Fuel Consumption at Various Uniform Speeds (FHWA, 1980).

3. Estimate "before" and "after" conditions such as changes in trip length and transit ridership. Although general information may be readily available (or transferable from other studies), it is possible that a special survey may be required for a site-specific case.

4. Calculate the energy consumption by applying the data collected in step 3 to each MOE identified in step 2.

5. Calculate the net change in energy consumption. This will show an overall reduction.

6. Perform a cost–benefit analysis.

5.4 Fuel Consumption Models (OECD, 1985)

Fuel consumption models are mathematical relationships giving fuel consumption as a function of other measurable or calculable quantities that are usually aggregated over the portion of the transportation system being analyzed. The quantities measured or calculated may be vehicle-miles traveled, number of stops per unit of time, total vehicular

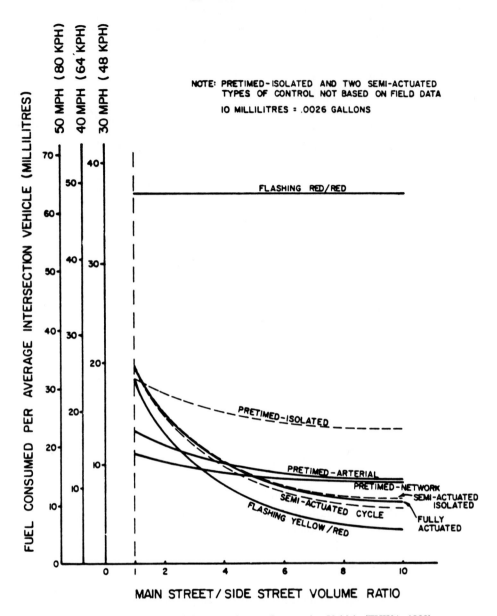

Figure 13-7   Average Fuel Consumption per Intersection Vehicle (FHWA, 1980).

delay per unit of time, average speed, and so on. Two forms that have found the most use in practice to date are drive-mode elemental models and average-speed models.

Drive-Mode Elemental Models.   As the name implies, drive-mode elemental models are made up of the elements that contribute to fuel consumption while driving (i.e., fuel used in cruising, idling, and accelerating). The basic assumptions in an ele-

TABLE 13-7 ENERGY CONVERSIONS

Unit converted from: (1)	Units converted to:						
	Foot-pound (2)	Kilogram-meter (3)	Horsepower-hour (4)	Metric horsepower-hour (5)	British thermal unit (6)	Kilowatt-hour (7)	Joule (8)
ft-lb	1	0.1383	5.0505×10^{-7}	5.12×10^{-7}	1.285×10^{-3}	3.766×10^{-7}	1.356
kg-m	7.233	1	3.655×10^{-6}	3.704×10^{-6}	9.295×10^{-3}	2.724×10^{-6}	9.80655
hp-hr	1.98×10^{6}	2.7375×10^{5}	1	1.0139	2544	0.7457	2.6845×10^{6}
metric hp-hr	1.953×10^{5}	270,000	0.9865	1	2510	0.75555	2.648×10^{5}
Btu	778.2	107.6	3.93×10^{-4}	3.985×10^{-4}	1	2.931×10^{-4}	1055
kWhr	2.655×10^{5}	3.671×10^{5}	1.341	1.3596	3412	1	3.6×10^{6}
joule	0.7376	0.10197	0.3725×10^{-6}	0.3777×10^{-6}	0.9478×10^{-3}	0.2778×10^{-6}	1

1 quad Btu = 0.4724 million barrels crude per day = 0.1724 billion barrels crude per year

TABLE 13-8 HEAT CONTENT FOR VARIOUS FUELS

Type of fuel (1)	Heat content (2)	Unit[a] (3)
Fuel oils		
Crude	138,100	Btu/gal
Residual	149,700	Btu/gal
Distillate	138,700	Btu/gal
Automotive gasoline	125,000	Btu/gal
AVGAS	124,000	Btu/gal
Jet fuel		
Kerosene	135,000	Btu/gal
Naphtha	127,500	Btu/gal
Diesel oil (No. 2)	138,700	Btu/gal

[a] 1 Btu/gal = 278.7 J/liter = 2.787×10^5 J/m^3; 1 Btu/short ton = 942.0 J/metric ton.

mental model are that the elements are independent and their sum equals the total fuel consumed.

The simplest form of the drive-mode elemental model is

$$G = f_1 L + f_2 D + f_3 S \qquad (1)$$

where

G = fuel consumed per vehicle over a measured distance (total section distance)

L = total section distance traveled

D = stopped delay per vehicle (i.e., time spent in idling)

S = number of stops

f_1 = fuel consumption rate per unit distance while cruising

f_2 = fuel consumption rate per unit time while idling

f_3 = excess fuel used in decelerating to stop and accelerating back to cruise speed

A procedure used in the United States involves the use of graphs to obtain fuel consumption values for uniform speed and for speed changes. This procedure is essentially the same as an elemental model.

Average-Speed Models. The basic model relating fuel consumption to trip time or its inverse, average speed, is

$$F = k_1 + k_2 T \qquad (2)$$

$$= k_1 + \frac{k_2}{v} \qquad \text{for } 10 \leq v \leq 56 \text{ km/hr}$$

where

F = fuel consumed per vehicle per unit distance (e.g., liters/km or gal/mi)

T = travel time per unit distance, including stops and speed changes (e.g., min/km or min/mi)

v = average speed measured over a distance, including stops and speed changes

k_1 = parameter associated with fuel consumed to overcome rolling resistance, approximately proportional to vehicle weight (gal/veh-mi)

k_2 = parameter approximately proportional to fuel consumption while idling (gal/hr)

This form of the average speed model is not valid at speeds higher than 56 km/hr (35 mph) because at higher speeds the effects of air resistance become increasingly stronger. For speeds lower than 15 km/hr (9 mph), fuel consumption increases rapidly and a better fit to the data is

$$F = F_R + \frac{800F_0}{v(v + 8)} \tag{3}$$

where

F_R = fuel consumed while in motion (liters/100 km)

F_0 = fuel consumed during stopped time while idling (liters/hr)

v = average speed (km/hr)

The validity of the average-speed model was extended to suburban areas for speeds up to 88 km/hr (55 mph) by the introduction of a term involving v^2:

$$F = k_1 + \frac{k_2}{v} + k_3 v^2 \tag{4}$$

The parameters k_1 and k_2 in the basic average-speed model have been estimated from regression analyses of fuel consumption data in several countries. Data on 37 U.S. passenger cars of the 1976 model year were found to be closely correlated with vehicle weight and fuel flow rate at idle, giving the relationships

$$k_1 = 9.58 \times (10^{-6})\,W \qquad \text{(gal/mi)} \tag{5}$$

$$= 49.57 \times (10^{-6})\,Wm \qquad \text{(liters/km)} \tag{6}$$

$$k_2 = 0.998I = I \times f_2 \tag{7}$$

where

W = vehicle weight (lb)

Wm = vehicle weight (kg)

I = fuel consumption rate while idling

Depending on vehicle weight, engine size, and technological features, the values of k_1 and k_2 can vary widely. Data from field tests on nine 1973–1976 passenger cars ranging in weight from 2270 to 5474 lb (1035 to 2488 kg) gave values of k_1 ranging from 0.0194 to 0.0518 gal/mi (0.0456 to 0.1218 liter/km) and values of k_2 ranging from 0.418 to 1.031 gal/hr. (For use in a metric version of the model this would be equivalent to k_2 ranging from 0.983 to 2.42 in order to yield 1/km.) The average values were incorporated into a version of the model published by FHWA:

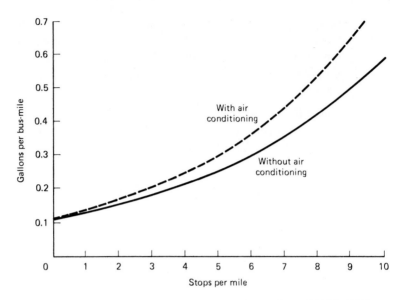

Figure 13-8   Bus Fuel Consumption as a Function of Stops (OECD, 1985).

$$F = 0.0362 + \frac{0.746}{v} \qquad \text{(gal/veh-mi)} \qquad (8)$$

By using the relationship $k_1 = 9.58 \times (10^{-6})\,W$, the value of $k_1 = 0.0362$ represents a vehicle weighing 3800 lb (1718 kg).

Buses operating in an urban environment may present a unique case because of their schedule of frequent stops. In this case, bus fuel consumption may be calculated from graphs such as Figure 13-8 calibrated with average number of stops per unit distance rather than average speed. The estimate for stops should include both the scheduled stops and the stops resulting from traffic congestion and traffic controls.

Example 1

A "before-and-after" study of a transit system revealed the following information:

> 1986 transit (bus) ridership (T_B): 50,000 per day
>
> 1988 transit (bus) ridership (T_A): 60,000 per day
>
> 1986 car passengers X_1: 85%
>
> 1986 car trip length X_2: 5 miles
>
> 1986 car fuel consumption rate X_3: 6 gal/100 miles
>
> 1986 car occupancy X_4: 1.25

Assuming that the city size remains the same, find the change in car direct energy.

Solution

$$\text{Change in car direct energy/day} = \frac{(T_B - T_A)(X_1)(X_2)(X_3)}{X_4}$$

$$= \frac{(-10{,}000)(0.85)(5)(6)}{(1.25)(100)}$$

$$= 2040 \text{ gal of gasoline saved/day}$$

If car gasoline is equivalent to 35 megajoules/liter and 1 liter = 0.264 gal,

$$\text{Energy savings/day} = 2040 \times \frac{35}{0.264} = 270{,}454 \text{ megajoules}$$

Example 2

Each day, 10,000 commuters in a bedroom suburb travel to the city center via an arterial at an average speed of 20 mph over a distance of 10 miles. Because of parking problems in the city center, 25% form carpools with a car occupancy of 2.25, and 10% arrange for subscription bus service (50 seaters). The rest are single-car drivers. The peak period congestion is somewhat better and the average speed is now 30 mph. What is the saving in fuel, in Btu? (Assume that the 1976 fleet composition applies and number of bus stops = 9.)

Solution

Before: 10,000 commuters (car occupancy = 1), 10 miles.

$$F = 0.0362 + 0.746/v \text{ gal/veh-mi}$$

$$= (0.0362)(10) + (0.746/20)(10) = 0.735 \text{ gal/veh}$$

$$\text{Total fuel consumption} = 0.735 \times 10{,}000 = 7350 \text{ gal}$$

$$= 7350 \text{ gal} \times 125{,}000 \text{ Btu/gal}$$

$$= 9.1875 \times 10^8 \text{ Btu}$$

After: 25% of 10,000 form carpools = 0.25 × 10,000 = 2500; car occupancy = 2.25.

$$\text{Carpool vehicles} = \frac{2500}{2.25} = 1111 \text{ vehicles}$$

$$10\% \text{ arrange bus (50 seater)} = 0.1 \times \frac{10{,}000}{50} = 20 \text{ buses}$$

$$\text{Balance} = (10{,}000 - 2500 - 1000) = 6500 \text{ commuters or } 6500 \text{ single-car drivers}$$

$$F = (0.0362)(10) + \left(\frac{0.746}{30}\right)(10) = 0.6107 \text{ gal/vehicle}$$

$$\text{Car fuel consumption} = 0.6107(6500 + 1111) = 4648 \text{ gal}$$

$$= 4648 \text{ gal} \times 125{,}000 \text{ Btu/gal}$$

$$= 5.81 \times 10^8 \text{ Btu}$$

$$\text{Bus fuel consumption (with nine stops)} = 0.5 \times 20 \times 10$$
$$= 100 \text{ gal (diesel)}$$
$$= 100 \text{ gal} \times 138{,}000 \text{ Btu/gal}$$
$$= 1.38 \times 10^7 \text{ Btu}$$

Btu before $= 9.1875 \times 10^8$ Btu

Btu after $= (5.81 \times 10^8) + (1.38 \times 10^7)$ Btu $= 5.948 \times 10^8$ Btu

Discussion

Apart from the obvious savings in fuel, there will be reductions in pollutions emitted by the vehicles and reductions in noise levels.

Example 3

Spokane highway lanes have the following characteristics:

$$v = 53.8 - 0.351k$$

where v is the speed in mph, and k is the density in veh/mi. If the average speed model is used for analysis, plot the fuel consumption rate of traffic density between 10 and 35 mph.

Solution

$$v = 53.8 - 0.351k \rightarrow k = \frac{53.8 - v}{0.351}$$

and

$$F = k_1 + \frac{k_2}{v} = 0.0362 + \frac{0.746}{v}$$

Speed, v	Density, k	F (gal/veh-mi)
5	139	0.1854
10	125	0.1108
20	96	0.0735
30	68	0.0611
35	54	0.0575

See Figure 13-E3.

SUMMARY

The consumption of energy on a per capita basis in the United States is very high compared to that in other developed countries having comparable standards of living. Transportation accounts for over 50% of all petroleum consumed in the United States and a large proportion of this percentage is imported. The long-range consequences of

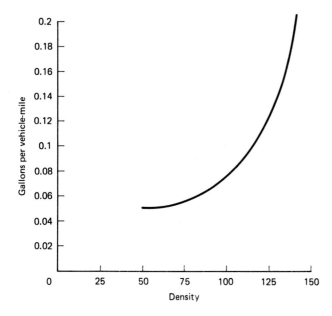

Figure 13-E3 *F* versus *k* Graph for Example 3.

such a heavy dependence on imported oil could be devastating in the event of an energy crisis.

Conserving energy in urban transportation lies in reducing the energy consumed by automobiles. Four possibilities, among others, hold out promise: by improving the fuel efficiency of the automobile fleet through mandatory fuel-economy standards for new cars; by substituting vanpools, carpools, and public transportation for cars; by highway and street pricing and increased gasoline taxes; and finally, through a concerted attempt in adopting land-use patterns that will produce the least amount of travel, consistent with the general welfare of communities.

Along with the conservation of energy it is necessary to be prepared for possible emergency situations where gasoline is in short supply. A contingency plan to deal with short-term crisis and long-term problems created by reduced fuel supply is described.

A major part of this chapter deals with methods of conducting transportation-related energy analysis. Such analyses are generally conducted in the context of transportation system management (a topic dealt with in detail in Chapter 14).

REFERENCES

BARKER, WILLIAM G., and LAWRENCE C. COOPER (1978). *An Approach to Local Transportation Planning for National Energy Contingencies*, North Central Texas Council of Governments, Arlington, TX.

CANADIAN URBAN TRANSIT ASSOCIATION and ROADS and TRANSPORTATION ASSOCIATION OF CANADA, CUTA/RTAC (1985). *Canadian Transit Handbook*, 2nd Ed., Toronto.

CANNON, BRUCE (1980). *Federal Highway Administration Program on Energy Conservation*, paper presented at the American Society of Civil Engineers' Annual Meeting, Spokane, WA, Oct. 3.

CHAPIN, F. STUART, and EDWARD J. KAISER (1979). *Urban Land-Use Planning*, 3rd Ed., University of Illinois Press, Champaign.

CLARK, WILSON (1981). "National Security and Community Energy Systems," *The Energy Consumer*, Dec./Jan, p. 8.

EDWARDS, JERRY (1978). *The Effect of Land-Use on Transportation Energy Consumption, Energy and the Community*, Ballinger Publishing, Cambridge, MA.

EDWARDS, JERRY, and JOSEPH SCHOFER (1976). *Relationships between Transportation Energy Consumption and Urban Structure*, Transportation Research Record 599, Transportation Research Board, Washington, DC.

FEDERAL HIGHWAY ADMINISTRATION (FHWA) (1980). *Procedure for Estimating Highway User Costs, Fuel Consumption, and Air Pollution*, U.S. Department of Transportation, Washington, DC.

FEDERAL HIGHWAY ADMINISTRATION (FHWA) (1985). *Personal Travel in the U.S. (1983–1984)*, U.S. Department of Transportation, Washington, DC.

FEDERAL HIGHWAY ADMINISTRATION (FHWA) (1986). *Highway Statistics 1985*, U.S. Department of Transportation, Washington, DC.

HOMBURGER, W. S., and J. H. KELL (1980). *Fundamentals of Traffic Engineering*, University of California, Berkeley, CA.

INSTITUTE OF TRANSPORTATION ENGINEERS (ITE) (1980). *Energy Impacts of Transportation Improvements*, Washington, DC.

KHISTY, C. J. (1979). "Energy Conservation through Parking Policy and Management," in *Compendium of Technical Papers*, Institute of Transportation Engineers, Washington, DC.

KHISTY, C. J. (1980). "Some Views on Traffic Management Strategies," *Traffic Quarterly*, Vol. 34, No. 10, pp. 511–522.

KHISTY, C. J. (1981). *Select Strategies for Energy Conservation*, paper presented at the ASCE Specialty Conference on Energy in the Man-Built Environment, Vail, CO.

LOWINS, AMORY, and HUNTER LOWINS (1981). "Getting Ready for a Surprise-Full Figure," *The Energy Consumer*, Dec./Jan., p. 6.

MILLER, G. TYLER (1980). *Energy and Environment*, 2nd Ed., Wadsworth, Belmont, CA.

NATIONAL ACADEMY OF SCIENCES (NAS) (1980). *Energy Choices in a Domestic Society*, Supporting Paper 7, NAS, Washington, DC.

ODUM, HOWARD T., and ELIZABETH C. ODUM (1976). *Energy Basis for Man and Nature*, McGraw-Hill, New York.

ORGANIZATION FOR ECONOMIC COOPERATION AND DEVELOPMENT (OECD) (1985). *Energy Savings and Road Traffic Management*, OECD, Paris.

PESKIN, ROBERT L., and JOSEPH L. SCHOFER (1977). *The Impacts of Urban Transportation and Land-Use Policies on Transportation Energy Consumption*, National Transportation Information Service, Springfield, VA.

SPIRO, THOMAS G., and WILLIAM M. STIGLIANI (1980). *Environmental Science in Perspective*, State University of New York Press, Albany.

STOBAUGH, R., and D. YERGIN (1979). *Energy Futures*, report of the energy project at the Harvard Business School, Random House, New York.

TRANSPORTATION RESEARCH BOARD (TRB) (1977). *Energy Effects, Efficiencies, and Prospects for Various Modes of Transportation*, NCHRP Synthesis of Highway Practice Report 43, National Research Council, Washington, DC.

U.S. DEPARTMENT OF TRANSPORTATION (U.S. DOT) (1980a). *Innovations in Urban Transportation in Europe and Their Transferability to the United States*, U.S. DOT, Washington, DC.

U.S. DEPARTMENT OF TRANSPORTATION (U.S. DOT) (1980b). *Transportation Energy Contingency Strategies*, U.S. DOT, Washington, DC.

WILLIAM, JOHN S., ET AL. (1979). "Metropolitan Impacts of Alternative Energy Futures," in *Sociopolitical Effects of Energy Use and Policy*, Supporting Paper 5, ed. C. T. Unseld, et al., National Academy of Sciences, Washington, DC, pp. 37–77.

EXERCISES

1. A city of 400,000 people has a labor force of 90,000. The overall traffic characteristics of this city for the journey to work are as follows:

Category	Average one-way trip (mi)	Percent of workers	Average speed (mph)
A	3	30	12
B	7	40	17
C	12	20	20

The existing car occupancy is 1.25 persons/car overall. It is proposed to encourage at least 10% of the labor force to improve their car occupancy to 3 persons/car across the board, resulting in higher speeds:

A: 15 mph

B: 20 mph

C: 25 mph

Make suitable assumptions (stating them clearly) and calculate the following:

(a) How much time (in hours) is saved per day for all commuters?

(b) How much gasoline is saved per day?

2. Many faculty, staff, and student groups have voiced their opinion that carpooling/vanpooling could easily be implemented for their university campus. As a consultant to your university, frame a "first-cut" proposal advising the president on what to do.

3. A two-lane highway 12 miles long connecting an industrial complex to a suburb has a speed–density relationship of $v = 53.8 - 0.351k$, where v = speed in mph, and k = density of vehicles in veh/mile. During the peak hour, the average speed of the vehicles is about 25 mph. In an attempt to improve the flow the industrial complex shifts 1200 employees from their cars to 40-seater buses. Assuming that this two-lane highway caters to 90% commuter traffic to the industrial complex during the peak hour, how much energy will be saved because of this change? How much time will be saved? (Assume that originally 3000 employees make the morning trip at a car occupancy of 1.5 during the peak hour.)

4. A small city has decided to take the following actions to improve traffic circulation: (1) convert 10 four-way-stop-controlled streets to two-way stops, (2) convert 20 two-way-stop-controlled streets to pretimed isolated signals, and (3) convert 10 pretimed-isolated controlled streets to semiactuated signals. The main-street/side street volume ratio in cases 1, 2, and 3 are 6, 4, and 8, respectively, and the vehicles affected are 1000, 5000, and 9000 veh/hr, respectively. What is the total fuel savings?

5. Make a broad transportation energy use study of the city with which you are most acquainted. List 20 ways by which transportation energy could possibly be conserved. Select 5 of the 20 ways for final implementation. Write a brief report based on this exercise.

6. Why does the average American citizen spend more energy than his or her counterpart in, say, Germany?

7. Does the amount of energy consumed per capita have any relationship to the quality of life?

8. Draw a simple flowchart connecting traffic, energy consumed, pollution (air and noise), quality of life, congestion, per capita expenditure on transportation, and any other factors

connected with these items. Discuss what would result from the use of, say, half the amount of energy per capita, enforced by the government.

9. Study the current parking rates, parking demand, and parking supply prevalent in the downtown (or other major activity center) of your city. Prepare a report showing how parking rates (and parking supply) could be used to save fuel consumption (and incidentally reduce congestion).

10. Assume the city you live in experiences a month-long shortfall of 15% resulting from local misallocation of transport fuel.

   (a) How much would it affect your city?

   (b) What advice would you provide the city to avoid undue disruption and inconvenience to the public?

Chapter 14

TSM Planning: Framework

1. INTRODUCTION

On September 17, 1975, the Urban Mass Transportation Administration and the Federal Highway Administration issued a series of joint regulations describing a concept known as *transportation system management* (TSM) (TIP, 1975). TSM covers a broad range of potential improvement strategies, focusing on nonfacility and low-capital-cost operations. It is the short-range element of the transportation work program and contributes directly to the makeup of the transportation improvement program. In Chapter 11, we described the TSM element briefly and indicated how TSM fits into the traditional urban transportation planning process. This chapter highlights, among other topics, the need for TSM, the strategies used in planning TSM, and the evaluation of TSM-type improvements.

2. WHAT IS TSM?

During the last three decades, the federal and state governments have jointly responded to the nation's need for increased accessibility and mobility. The effect has not been an unqualified success, particularly in the larger and more heavily populated urban areas of the country. Traffic congestion compounded with environmental pollution and high energy consumption have created increasingly vexing problems. Financial constraints, recognition of negative impacts, and changes in social priorities have indicated that the expansion of the transportation system is not the only solution for solving the nation's transportation problems.

Transportation system management is a process for planning and operating a unitary system of urban transportation. Its key objective is conservation of fiscal resources, of energy, of environmental quality, and enhancement of the quality of life. The federal regulations state that automobiles, public transportation, taxis, pedestrians, and bicycles should all be considered as elements of a single transportation system. The objective of TSM is to coordinate these individual elements through operating, regulating, and service policies so as to achieve maximum efficiency and productivity for the system as a whole (TIP, 1975).

Much literature on TSM has been generated during the last decade, and those interested in furthering their knowledge are urged to consult the references at the end of this chapter, especially TRB (1975, 1983, 1986), Remak and Rosenbloom (1976 and 1979), Rowan et al. (1977), Voorhees & Associates (1978), Cannon (1980), Dale (1980), and Roark (1981).

2.1 TSM Objectives, Scope, and Outline

The main objective of TSM is to maximize urban mobility within a given existing system through the development of specific applicable actions in the following four categories:

1. Actions to ensure efficient use of existing road space
2. Actions to reduce vehicle use in congested areas
3. Actions to improve transit service
4. Actions to improve internal transit management efficiency

Note that TSM actions should be consistent with efforts to conserve energy, improve air quality, and increase social, economic, and environmental amenity. Reference should be made to Chapter 13, which clearly shows the connection between traffic congestion and energy consumption.

TSM actions cover all categories of improvements appropriate to a region, and the level of analysis should be scaled to the size of the urbanized area and to the magnitude of the transportation problems. Assessment of candidate measures (i.e., evaluation criteria) or selection, programming, and implementation fall under the purview of local prerogatives.

2.2 TSM Process Outline

The following outline, applicable to a metropolitan area, serves as a "guide," and examples given under each major item are merely suggestions.

1. Organization
   (a) Metropolitan planning organizations (MPOs) assume the coordinating responsibility in cooperation with the state and publicly owned transit operators.
   (b) Input and participation is solicited from all relevant agencies (i.e., planning organizations, highway developments, traffic operators, and city and traffic engineers, with the possible inclusion of these agencies on the technical advisory committees).
2. Formulation of overall policy strategy and development of short-range objectives. For example:

     (a) Efficient use of existing facilities
     (b) Increase in urban mobility
     (c) Social, economic, and environmental considerations
     (d) Energy conservation
     (e) Mobility and accessibility of elderly and handicapped people
     (f) Improvement in air quality
     (g) Land-use impacts
     (h) Accessibility
     (i) Costs
     (j) Safety
     Note that short-range objectives should be consistent with long-range goals.

3. Collection of data to identify existing system characteristics. For example:
     (a) Physical characteristics
     (b) Operating characteristics
     (c) Management characteristics
     (d) Surveillance data from long-range planning element

4. Identification of problems and deficiencies resulting in inefficient use of the transportation system. For example:
     (a) Congestion
     (b) Transit service

5. Development of possible solutions for each problem or deficiency from actions to do the following:
     (a) Ensure efficient use of existing road space
     (b) Reduce vehicle use in congested area
     (c) Improve transit service
     (d) Increase internal transit management efficiency

6. Development of evaluation criteria consistent with short-range objectives. For example:
     (a) Accidents
     (b) Travel time
     (c) Level of service (highway capacity, transit frequency, coverage)

7. Evaluation of possible solutions relative to the foregoing criteria
8. Selection of specific solutions for each problem or deficiency identified
9. Listing of selected solutions for each problem or deficiency
10. Documentation of the process (i.e., items 1 to 9) in a TSM Report
11. Project initiation under TIP procedures

3. THE NEED FOR TSM

The United States has a mature transportation system and infrastructure. The overall feeling has been that this mature system is in need of good management. Several complex and interrelated factors have contributed to the development of a TSM "philosophy" to deal with the problems confronting this mature system. Some of the more important factors include the following:

1. The ever-increasing costs of construction and maintenance of highways, coupled with the high competition for public funds
2. Decline in population growth, resulting in a deflation of earlier forecasts
3. Continued public debate over new highway construction, community disruption, environmental impacts, efficiency of public transit systems, and equity issues
4. Cost-effectiveness of major projects

These interrelated factors are by no means new. They were recognized by the federal and state governments in the 1960s. As early as 1967, the Bureau of Public Roads (now the Federal Highway Administration) announced a program known as TOPICS (Traffic Operations Program to Increase Capacity and Safety) to provide federal funds on a matching basis for improvements to relieve bottlenecks and reduce hazards on roads in urban areas. TSM was therefore perceived by many professionals as having neo-TOPICS overtones. TSM, in summary, may be considered as a coherent, generic improvement strategy appropriately responsive to a new era for transportation. Its synergistic potential is now beginning to be realized.

4. LONG-RANGE VERSUS TSM PLANNING

The dovetailing of long-range and short-range planning (TSM) is not as simple as it may sound. Long- and short-range planning differ fundamentally and therefore may be at cross purposes if not connected through common goals and objectives. Lockwood (TRB, 1977) cites the following significant relationships:

1. Whereas long-range planning deals with capital-intensive improvements requiring several years to plan and implement, TSM embraces low-budget actions of short durations (possibly a year or two to plan and implement).
2. Recycling of TSM projects during the long-range time range is quite possible.
3. TSM strategies may work out to be alternatives to long-range components.
4. "Major" TSM projects may be synonymous to "minor" capital-intensive projects, particularly at middle-range levels.
5. There is an inherent flexibility in TSM projects. Proper monitoring and effective feedback may result in appropriate modifications to the original project and may lead to overall economy.

In summary, TSM is fundamentally different from long-range planning. TSM should reflect some of the key differences set out in Table 14-1.

5. TSM PLANNING CYCLE

TSM development is a cyclical normative process embedded in an annual program consisting of the following steps: initiation, diagnosis of existing performance, project selection, cost and impact estimation, consideration of priorities, and preparation of

TABLE 14-1   KEY DIFFERENCES BETWEEN TSM AND LONG-RANGE PLANNING

	TSM	Long-range
Problems	Clearly defined, observable	Dependent on growth scenarios and projected travel
Scale	Usually local, subarea, or corridor	Usually corridor or regional
Objectives	Problem-related	Broad, policy-related
Options	Few specific actions	Several modal, network, and alignment alternatives
Analysis procedures	Usually analogy or simple operational relationships	Based on trip and network models
Response time	Quick response essential	Not critical
Product	Design for implementation	Preferred alternative for further study or detailed design

Source: TRB, 1983.

the recommended program of projects. The sequence feedback of this planning cycle is shown in Figure 14-1 (Gray and Hoel, 1979).

Because the TSM concept is so broad, it is necessary that the set of objectives established be clear, measurable, and objective. It is quite possible that objectives may conflict with one another. In a sense, objectives help to identify problems and assist in defining the appropriate TSM concepts. Furthermore, they provide the basis for the development of measures of effectiveness (MOE) used in evaluating the impact of various strategy combinations. It will seldom happen that objectives will favorably impact simultaneously. Thus, a TSM strategy could have a favorable impact on mobility and a negative impact on social costs. This is illustrated in Figure 14-2, where the four types of potential solution spaces with trade-offs are indicated. Performance measures must be developed for each objective of interest. Typical goals, objectives, and performance measures are shown in Tables 14-2 and 14-3.

Measures of effectiveness (MOEs) play a critical role in the evaluation of TSM plans. They must be clear and quantifiable. A good MOE has the following characteristics: It must define the target, it must be responsive to the incidence of impact, it must be characterized by geographic area of application and influence, it must be oriented to specific time periods, it must be directly or indirectly related to the objective, and it must be formulated at a proper level of detail for the type of analysis being performed.

TSM actions, singly or in combination, are established for an urban area for, say, corridors, central business districts, neighborhoods, and major activity centers on an individual or collective basis. Metropolitan planning organizations (MPOs) and their constituent entities, such as cities, counties, and townships, identify conflicts and potential trade-offs among the range of TSM actions proposed.

Design and cost of each TSM action estimates the determination of their impacts, and ultimately their evaluation is carried out locally and regionally, culminating in the integration of approved actions into an overall program. A cost-effectiveness index for a list of candidate actions can assist in determining priorities. Finding constraints limits

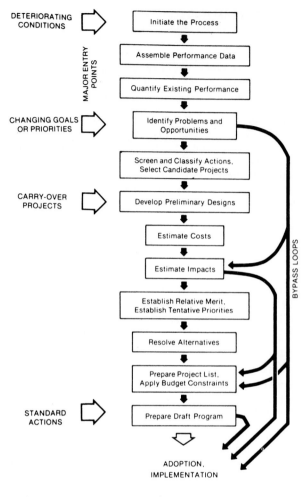

Figure 14-1   Overall TSM Planning Cycle (Normative) (Gray and Hoel, 1979).

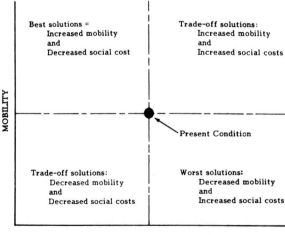

Figure 14-2   Potential Solution Spaces (TRB, 1977).

TABLE 14-2 EXAMPLES OF GOALS AND OBJECTIVES

Goals	Objectives
Improve personal mobility	Improve level of service of urban travel
	Improve reliability of travel
	Provide attractive alternatives to driving
	Private automobiles
	Provide good quality, affordable transportation services to elderly, handicapped, poor, and very young
	Improve facilities for nonmotorized travel modes (pedestrian and bicycle)
	Improve responsiveness to changing urban transportation needs
Improve public safety	Reduce occurrence of traffic accidents
	Reduce injuries and deaths resulting from traffic accidents
	Improve personal security of urban travelers
Enhance environmental and community quality	Reduce automotive emissions and impacts
	Reduce noise and vibration impacts
	Minimize adverse impacts on natural environment
	Minimize community disruption and relocation
	Enhance aesthetic qualities of urban environment
	Complement long-range urban land-use goals
Conserve energy resources	Reduce fuel consumed in urban travel
Improve the economic efficiency of transportation	Increase person and goods movement capacity of existing transportation facilities
	Reduce personal costs of urban travel
	Reduce public costs of urban transportation systems
	Achieve greater equity in payments for urban transportation
	Reduce costs of urban goods movement
	Minimize adverse economic impacts caused by urban transportation
	Maximize positive economic impacts caused by urban transportation

Source: TRB, 1977.

the number of TSM actions and those that are eventually chosen are integrated into the transportation improvement program (TIP).

6. TSM STRATEGIES

Over the years, transportation managers have identified 150 different TSM actions. Because of the synergistic characteristics of TSM, it has been strongly suggested that these actions should be grouped and applied collectively in order to derive the maximum benefits for the system (Roark, 1981). Four classification systems have emerged for grouping TSM measures:

1. Combining compatible individual techniques into action packages for critical problems, such as peak-period traffic congestion

TABLE 14-3   EXAMPLES OF OBJECTIVES AND PERFORMANCE MEASURES

Example TSM objective	Example performance measures
Improve level of service of urban travel	Total travel time or delay (person-hours) Weighted average speed for person travel
Provide attractive alternatives to driving private autos	Mode split percentages Percentage of population within × miles (walking distance) of scheduled transit service at home and at work Average occupancy per vehicle-trip
Provide good quality, affordable transportation services to the elderly, handicapped, poor, and very young	Percentage of special groups population to whom any specialized transportation services are available Percentage of special groups disposable income expended on transportation
Improve facilities for nonmotorized travel modes (pedestrian, bicycle)	Total miles of improved bicycle or pedestrian pathways Total person-miles of travel by pedestrians or bicyclists
Reduce the occurrence of traffic accidents	Total number of motor vehicle accidents Number of accidents per million vehicle-miles
Reduce automotive emissions and impacts	Grams of carbon monoxide, hydrocarbons, and nitrogen oxides emitted Grams of emittants per person-mile
Reduce noise impacts	Noise levels in decibels at different distances from transportation facilities Percentage of residents subjected to noise levels exceeding specified tolerance limits
Reduce fuel consumed in satisfying urban travel	Gallons of gasoline and diesel fuel consumed Average fuel economy in person-miles per gallon
Reduce public cost of urban transportation systems	Net annual cost of ownership and operation of transportation facilities by mode (total cost less direct fares, toll revenues, and parking charges) Net annual cost per capita for urban area

Source: TRB, 1977.

2. Combining strategies to overcome common institutional problems, such as congestion reduction techniques

3. Assessing each TSM action on the basis of its scope, complexity, design detail, and degree of coordination

4. Combining and grouping TSM actions to produce desired shifts in transportation system equilibrium, based on supply–demand impacts.

The last classification is of particular interest and is described in some detail in the next section.

7. TSM CLASSIFICATION: ASSESSMENT OF IMPACTS

Wagner and Gilbert's (1978) research on the classes of TSM actions is worthy of some elaboration. This conceptual approach is grounded in transportation supply–demand theory and is used to assess the effects of different actions in producing "shifts" in transportation system equilibrium. For example, a ride-sharing program would reduce demand for vehicle travel, whereas truck restrictions would enhance highway supply. Use of this concept has led to the conclusion that all TSM actions can be put into four general categories as a means of understanding their impacts and interrelationships.

Figure 14-3 illustrates four sets of strategies grouped according to their influence on the supply and demand sides of the travel equation. The transportation supply curve depicts the level of service provided by the transportation system as a function of the demand for personal travel. The demand curve depicts the quantity of travel demand that the public will generate at different levels of services.

The TSM actions in the right column reflect a wide range of possible actions. The impacts of different choices is shown in the figure, where the ordinate represents system disutility (minutes per mile spent in travel) and the abscissa represents travel demand (vehicle-miles of travel VMT). The four pie sections, labeled A through D, show the general effects of applying strategy groups.

As an example, Table 14-4 shows calculations of work-trip supply-demand equilibrium changes for each major class of action. Note that the application of combined strategies, A through D, could result in substantial reductions of 11.4% and 21.0% in vehicle-miles of travel and travel times, respectively.

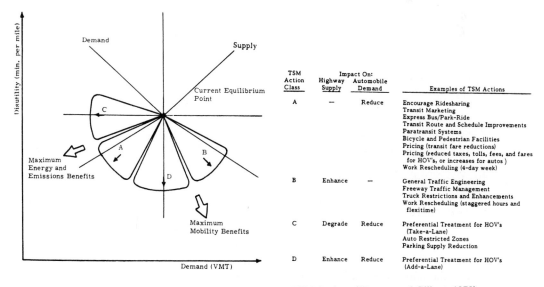

Figure 14-3 Equilibrium Shifts from TSM Actions (Wagner and Gilbert, 1978)

TABLE 14-4 IMPACT OF TSM ACTION CLASSES ON WORK-TRIP VMT
AND TRAVEL TIME FOR PROTOTYPICAL URBAN AREA OF 1 MILLION
PEOPLE

TSM strategy	Work-trip changes after equilibrium is reached (%)	
	VMT	Travel time
Class A (exclusive of pricing actions that reduce travel demand)	−5.3	−5.3
Pricing actions alone	−2.4	−2.4
Class B: actions that enhance highway supply	+0.6	−9.4
Class C: actions that reduce demand and degrade supply	−6.0	−4.9
Class D: actions that reduce demand and enhance supply	−0.6	−1.4
Combined strategies		
Classes A, B, and D	−5.4	−16.1
Classes A, C, and D	−11.9	−11.6
Classes A, B, C, and D	−11.4	−21.0

Source: Wagner and Gilbert, 1978.

1. A mobility dominant strategy, comprising classes A, B, and D.
2. An energy conservation/emissions reduction strategy, comprising A, C, and D.
3. All TSM actions combined.

The cost-effectiveness of applying various TSM strategies is shown in Table 14-5.

8. AIR QUALITY IMPACTS

It is quite appropriate at this stage to provide some insight into the relationship between traffic congestion and air pollution. The relationship between traffic congestion and energy consumption has been dealt with briefly in Chapter 13.

Air pollution is the contamination of the ambient air by chemical compounds or particulate solids at a concentration that adversely affects human health, materials, vegetation, or aesthetics. Carbon monoxide, oxides of sulfur, oxides of nitrogen, unburned hydrocarbons, and particulates make up the major manufactured pollutants. A large portion of the emissions in urban areas is generated by the automobile. Several factors, such as the type of engine, the mode of operation, the fuel composition, the emission control devices attached to the engine, and the atmospheric condition, contribute to the concentration of air pollution. The local meteorological condition, the topography, and the land-use activities, including their locations, are important elements influencing the air quality of an urban area. Researchers have suggested a num-

TABLE 14-5 COST-EFFECTIVENESS OF SELECTED TSM PROGRAMS[a]

| TSM action | Areawide reduction (%) | | Annual cost | Cost per: | |
	VMT	Travel time		VMT reduced	VHT reduced
Ridesharing					
Current program	0.2		$200,000	2 cents	
(Expanded program)	(1.0)		$400,000	1 cent	
Express bus	0.3		$5–7 million	40 cents	
Local bus (50% increase)	0.3		$5–6 million	43 cents	
Work rescheduling		0.4	$200,000		25 cents
Signal timing optimization		6.0	$250,000		2 cents
Computerized traffic control		1.5	$800,000		27 cents
Freeway surveillance and control		0.5	$1 million		$1.00
Truck restrictions/ enhancements		0.2	$200,000		50 cents
Comprehensive preferential treatments	−0.1	+0.4 (increase)	$800,000	16 cents (increase)	
Exclusive HOV lanes	0.1	0.4	$5–8 million	$1.30	$8.00

[a] Base conditions for urban area of 1 million: annual VMT ≃ 5000 million; annual VHT ≃ 200 million.
Source: Roark, 1981.

ber of emission models utilizing these factors. The main objective of these models is to estimate the pollution load that results from various combinations of traffic flows and vehicle mixes.

The relationship of carbon monoxide (CO), hydrocarbon (HC), and nitrous oxides (NO_x) to speed is shown in Figure 14-4. These curves are based on emissions from vehicles in use in 1975. The current trend is that the total emissions are being reduced annually, partially because the newer models of cars have less polluting engines (Dale, 1980).

Example 1

About 4000 vehicles enter the downtown area per hour in the $1\frac{1}{2}$-hour morning peak. Because of congestion problems, the city has turned the downtown into an auto restricted zone (ARZ) and only 500 essential vehicles are allowed to come into the area. Estimate the total reduction in CO in the $1\frac{1}{2}$-hour morning peak. Assume that the average vehicle speed is 15 mph and the average street network in the CBD used is 2 miles.

Solution

Reduction in vehicle-miles traveled in peak period

$$= 3500 \text{ veh/hr} \times 1.5 \text{ hr/peak} \times 2 \text{ miles}$$

$$= 10,500 \text{ veh-mile/peak period}$$

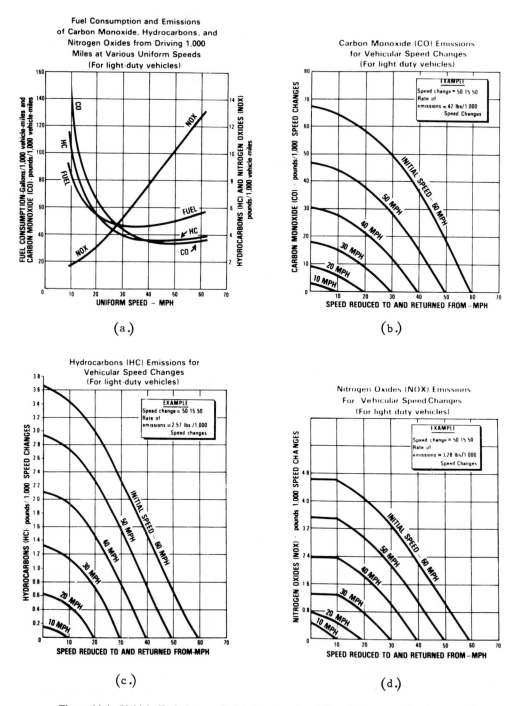

Figure 14-4 Vehicle Emissions as Related to Speed and Speed Changes (Homburger and Kell, 1988).

For a speed of 15 mph, the CO emission rate is 95 lb/1000 veh-mi. Therefore,

$$\text{Total load} = \frac{10,500 \times 95}{1000} = 9975 \text{ lb}$$

Discussion

A similar amount would possibly be reduced during the evening peak period.

9. STRATEGIC MANAGEMENT FRAMEWORK FOR TSM (TRB, 1983)

More than a decade of experience in applying the TSM philosophy has resulted in a growing body of experience. A new approach to implementing TSM has resulted in strategic management. Strategic management is a set of activities undertaken to provide an efficient, problem-oriented framework for planning, designing, evaluating, and implementing low-cost TSM projects. It covers not only the programming of recommended projects, but also the setting of realistic guidelines for design and development so that projects can be funded and implemented without excessive delays. A framework is shown in Figure 14-5.

Some of the key terms used are defined in the following:

Action: either a type of physical or operational change (e.g., "reversible lanes" or "employer vanpool program") or a specific change (e.g., ban parking along Main Street between Spruce and Elm from 7 to 9 A.M.). In general, the implied level of detail increases as planning progresses.

Package: a group of actions, either general or specific, proposed to solve a specific set of problems.

Program: a list of specific actions and packages to be implemented during the next fiscal period(s), usually 1 or 2 years, supplemented with funding and scheduling information.

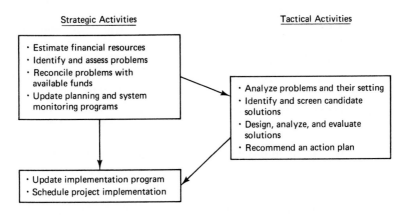

Figure 14-5   TSM Improvement Planning and Programming Process.

Project or tactical planning: activities that develop detailed solution to a specific set of problems.

Strategic management: activities that establish an agency's, community's, or region's needs, objectives, priorities, and improvement programs.

Strategy: a general approach that can be taken to solving specific problems (e.g., "improve traffic flow" or "encourage ride sharing").

Each of the activities in Figure 14-5 is described briefly in what follows.

9.1 Strategic Activities (TRB, 1983, 1986)

1. *Estimate financial resources.* These are funds to a transportation agency for planning, implementing, and operating low-cost improvements.

2. *Identify and assess problems.* Problems are identified through monitoring, and some are identified by the public and outside sources. The task of sorting and assessing problems can be addressed through a checklist of questions.

3. *Reconcile problems with available funds.* Agencies develop procedures that extend the concept of a traditional needs analysis to consider funding limitations. In some cases, impact estimation techniques have been used (such as speed–flow curves along with benefit–cost analysis) to examine the implications of different performance and design standards. What is generally needed is only a few, simple performance guidelines that provide a reasonable consistent basis for judging serious deficiencies and which allow engineers to design solutions that can be implemented with available funds.

4. *Update planning and system monitoring programs.* Questions should be raised and at least partially answered, such as:

    (a) Can a detected problem be treated in isolation or is it better addressed in combination with related problems identified in the same vicinity or along the same travel corridor? (For example, intersection delay and accident problems along an artery might best be solved with a package that includes signal coordination along the artery.)

    (b) Are problems expected to occur in the near future that might be efficiently addressed along with more immediate problems? (For example, improvements to an artery might allow for travel in the parking lanes, reversible lanes, and/or additional access point once adjacent sites are redeveloped.)

    (c) Is the solution of a given problem likely to introduce new problems or shift problems to nearby areas? (For example, closing a street to create a pedestrian mall may create traffic congestion on adjacent streets if bypass routes are not planned.)

    (d) Is a detected problem a symptom of a more fundamental problem? (For example, increase cross-town travel that creates traffic congestion on radial segments where dogleg movements are necessary.)

    (e) Are special data collection or analysis efforts required to resolve a problem? If so, how can data collection be tailored to staff and funding resources?

Detailed answers may not be available at this stage, but may be possible after additional data are assembled, or when tactical planning is under way.

9.2 Tactical Activities

Details of these activities are shown in Figure 14-5 and are self-explanatory. A word, however, should be said about key analysis and evaluation issues, which may occur throughout the four-phase project planning process. For example, an analysis in phase I may involve the estimation of near-term travel growth, a screening of possible actions in phase II may involve a rough estimate of their effectiveness in reducing travel delays, and the development of an implementable project in phase III may involve the estimation of travel speeds and implementation costs and an assessment of potential safety and air quality impacts.

The general relationships among performance measures is important. Figure 14-6, for example, shows the relationships among six categories of measures. The ones within the box are direct transportation measures (supply/capacity, service quality, volume/use), and the ones below the box are derived measures.

Once the four-phase tactical activities of a project have been completed, an agency may commit funds for implementation, defer a decision until funds are available, or resubmit the project (and problem) for further analysis. The actual scheduling of projects therefore depends on several factors, such as the availability of funds, the agency's contract procurement procedures, coordination with operating procedures, or the availability of staff to supervise or conduct the implementation.

9.3 Typical Examples of TSM (TRB, 1983, 1986)

It is intended to consider bus lanes that might be implemented to increase bus operating speeds and improve schedule adherence in a heavily trafficked corridor of a city. Options to be considered are (a) limited action, resulting in small savings in running time; (b) more extensive action, resulting in operator offering the same service with fewer buses, producing savings in labor requirements and other operating costs; and

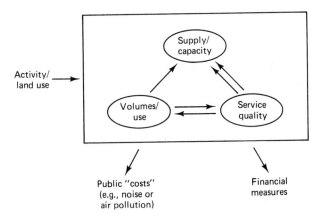

Figure 14-6 Basic Relationships among Performance Measures.

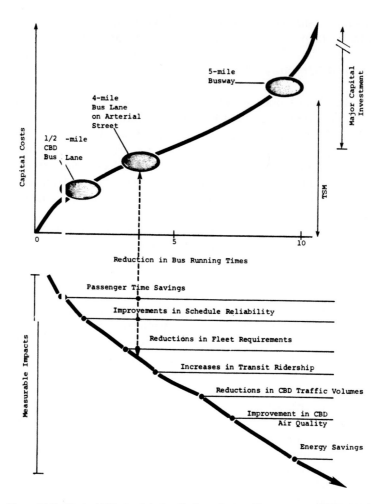

Figure 14-7 Typical Effects of Action Scale on Impact Measurement (TRB, 1986).

(c) even further expansion, producing measurable ridership increases and mode shifts. Figure 14-7 shows typical effects of actions.

A method of determining the measures required for analysis is to draw a simple flowchart of relevant evaluation factors and performance measures. An example of one such flowchart is shown in Figure 14-8. The proposal is to reduce transit operating costs by increasing operating speeds and reducing their variability. Notice the following:

1. Reduced travel times on buses and improved access to employment sites may encourage more commuters to use transit.
2. Improvement to traffic flow resulting from parking bans and turn prohibitions will therefore limit potential shifts in mode.
3. As a result, the primary evaluation of the lane can be limited to the measures shown in the left half of the flowchart.

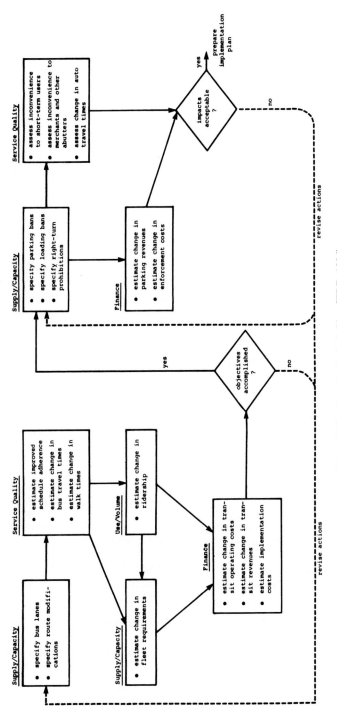

Figure 14-8 Sample Analysis Plan (TRB, 1986).

4. A ban on parking and loading is proposed to free lanes for buses, and right-turn restrictions are proposed at selected intersections to minimize conflicts between autos and buses. The right half of the flowchart shows these potentials.

In most cases, a qualitative estimate is enough as a first cut.

10. EXPERIENCE WITH FREEWAY CORRIDORS: AN EXAMPLE (ROARK, 1981)

Perhaps more work has been done on identifying and evaluating TSM measures for freeway corridors than for any of the other TSM operating environments. Because the urban freeway corridor includes parallel arterial roadways, transverse arterials that feed the freeway, and the express and fixed-route transit systems that operate on the freeway, the application of TSM strategies cover much more than the principles of traffic engineering on the freeway itself. For example, increased transit ridership through reduced fares or an intensive marketing campaign will benefit the freeway corridors.

Goals, objectives, and measures of effectiveness (MOEs) are given in Table 14-6. The synergistic effects of regionally applied TSM strategies have yet to be measured in an urban area, and little technology is available to project these impacts in advance of implementation. TSM actions for freeway corridors is motivated primarily by (1) freeway congestion and the demand for more efficient use of existing facilities, (2) federal

TABLE 14-6   GOALS, OBJECTIVES, AND MOEs FOR THE FREEWAY CORRIDOR OPERATING ENVIRONMENT

Goal	Objective	MOE	TSM measures for each goal
Maintain or improve the quality of transportation services on the existing transportation system	Minimize travel time	Person-hours of travel Vehicle delay	Ramp access control Preferential access for HOV
	Minimize travel costs	Point-to-point, out-of-pocket travel costs	Exclusive lanes for HOV
	Maximize safety	Accidents Accident rate	
Increase the efficiency of the existing transportation system	Minimize auto usage	Number of carpools Traffic volume	Ramp access control Preferential access for HOV
	Maximize transit usage	Transit passengers	Exclusive lanes for HOV
Minimize the undesirable impacts of existing transportation facilities	Minimize air pollution	Tons of emissions	Ramp access control Preferential access for HOV
	Minimize energy consumption	Energy consumption	Exclusive lanes for HOV

Source: Roark, 1981.

and state requirements to reduce air pollution and achieve air quality standards, and (3) concern for efficient use of motor fuels in transportation system operations.

Guidelines regarding implementation include such actions as ramp metering, exclusive bus facilities, contra-flow lanes, HOV lanes, and park-and-ride locations. Support activities include enforcement of TSM measures, introduction of bus collection and distribution, and introduction of ride-sharing programs.

10.1 Banfield Freeway HOV Lanes (Roark, 1981)

The Banfield Freeway (I-84) in Portland, Oregon, was initially constructed (1951–1958) as a six-lane facility from the Portland central area to 39th Avenue and as a four-lane facility from 39th Avenue to 74th Avenue. In 1975, HOV lanes were added to the freeway. This project illustrates the add-a-lane procedure for reserving a freeway lane for HOV.

The Banfield Freeway is the primary commuter route to Portland CBD from east Multnomah County. Between 1960 and 1970, the average weekday traffic on the freeway increased from 40,000 to 100,000 vehicles. By 1975, serious peak-hour congestion on the four-lane section reduced the average speed to less than 15 mph (24 km/hr). Also, the pavement had deteriorated to an unsafe condition.

A technical advisory committee consisting of representatives from the Oregon Department of Transportation (DOT), the Oregon State Police, the City of Portland, Multnomah County, Tri-MET (regional transit operator), and local civic groups was formed by the Oregon DOT to evaluate alternatives on the Banfield Freeway. The decision was made to replace the median barrier, overlay the roadway surface and the shoulder with asphaltic concrete, restripe the four-lane segment for six lanes, and operate the median lane as an HOV lane between 74th and 21st Avenues westbound and between 44th and 74th Avenues eastbound. Buses were to operate in mixed flow in all other segments. No shoulders were available for emergency parking, so seven parking bays were provided. When the construction was completed in December 1975, HOV lanes were reserved for buses and carpools (three or more persons) for 24 hours daily. After substantial public criticism, reservation of the HOV lane was reduced to hours between 6:00 and 10:00 A.M. westbound and 3:00 and 7:00 P.M. eastbound.

Objectives of this project were to (1) reduce air pollution by increasing the number of persons per vehicle; (2) reduce traffic congestion on the freeway and on parallel arterial facilities; (3) improve safety; (4) reduce travel time and fuel consumption; and (5) provide an interim, low-cost improvement to the freeway "until such time as a major revision can be accomplished." An evaluation study conducted after 18 months of HOV lane operation reported the following results:

- Increased average weekly traffic on the freeway (higher than projected for the freeway had the HOV lane not been added)
- Three percent decrease in traffic volumes on three parallel arterials
- Increased average peak-hour occupancy rate for all lanes from 1.217 persons per vehicle to 1.262 persons per vehicle
- Annual travel times savings of 62,500 person-hours

- Increased express bus ridership (two-way), from 300 persons per day to 633 persons per day
- Increased carpools during peak hours from 106 to 578
- Increased air pollution (2%)

In February 1979, the HOV lane was opened to carpools of two or more persons. This change resulted from the finding that in the A.M. peak hour, 3750 vehicles used the two freeway lanes and 200 vehicles used the HOV lane. As a result of the change, average speeds in the freeway lanes increased from 33 to 48 mph (53 to 77 km/hr) in April 1980. In January 1981, 1244 carpools used the HOV lane during the A.M. peak and 747 carpools during the P.M. peak. Violation rates decreased from 24% during the A.M. peak prior to the two-person carpools to 4% after; however, the violation rate has since increased to 9% in January 1981. On April 1980, enforcement on the Banfield Freeway was transferred from Oregon State Police to the City of Portland.

11. CASE STUDY: WHITE PLAINS CBD STUDY (TRB, 1983, 1986)

The White Plains CBD Study is a good example that illustrates the application of simplified procedures for preparing and evaluating low-cost TSM projects.

Purpose.   This study illustrates the use of simple, conventional analysis methods to develop integrated street, transit, and pedestrian improvements designed both to solve current problems and to accommodate forecast growth.

Site Description.   White Plains is a 10-square-mile city located in southern Westchester County, New York. The city's resident population is 47,000, and its daytime population (including those residents who remain in the city and others who come to the city to work) is 250,000. Over the past decade, White Plains has emerged as one of the largest suburban commercial and office hubs in the region. Transportation lines that intersect the city include the Cross Westchester Expressway (I-287), the Bronx River and Hutchinson River Parkways, NYS Routes 22, 119, and 127, and the Harlem River commuter rail line of the MTA. These facilities are shown on a map of the city and its CBD (Figure 14-9).

Step 1: Assemble and review information.

(a) Analyze existing and anticipated conditions. The study was prompted by the city's transportation officials, who recognized that continued economic growth depended heavily on convenient access to and circulation within the CBD by all modes.
(b) Study strategic management activities. A downtown comprehensive transportation plan was commissioned to improve mobility and reduce existing and anticipated congestion. Key steps in identifying and analyzing problems included:
   (1) Agency meetings held with traffic, parking, planning, transit, and development officials to identify problems, define goals, and denote opportunities

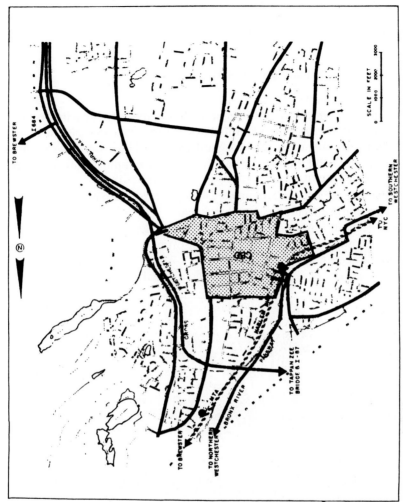

Figure 14-9 Site Map for White Plains (TRB, 1986).

(2) Field observations made to collect traffic data and to discover operational problems

(3) Transportation and development reports from the previous 15 years, reviewed for data, previous recommendations, and current relevancy

Step 2: Analyze problems and their setting.

(a) Collect performance data.

(1) Existing traffic volumes were analyzed to identify capacity problems. Seven of the 28 portals were identified as nearing or at capacity (see Figure 14-10).

(2) Bus transportation was assessed to determine route coverage (bus stop locations) and the system's ability to carry rail passengers from rail stations to the CBD.

(3) Parking capacity and occupancy were analyzed. A parking shortage was likely to occur in specific areas within 5 years.

(4) Pedestrian flows were also studied. The major shortcoming noted was a lack of sidewalks connecting the area of primary pedestrian activity with a large retail shopping complex.

(5) An analysis of taxi service indicated the major demand generators.

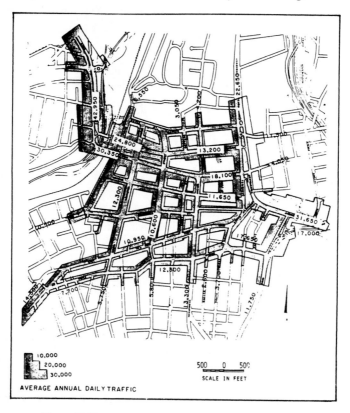

Figure 14-10   Average Annual Daily Traffic (TRB, 1986).

(b) Assess and identify problems.
   (1) Vehicle conflicts with parked cars, buses, and pedestrians
   (2) Traffic congestion
   (3) Limited street capacity
   (4) Limited short-term parking
   (5) Pedestrian circulation void
(c) Develop graphical representations. The locations of these problems were mapped, as shown in Figure 14-11.

Steps 3 and 4: Identify candidate strategies and screen actions.   Candidate strategies and actions were identified and screened using the following criteria:

(1) Coordinate transport facilities and capacities.
(2) Catalyze developments by improving existing facilities.
(3) Maintain and maximize accessibility of the city center.
(4) Provide travelers with choice of mode and route.
(5) Encourage public transport ridership.
(6) Provide a balance of long-term and short-term parking facilities.
(7) Expand capacity across the major barriers to movement.
(8) Develop a cohesive CBD by making intra-CBD movements easy.
(9) Refine bus routes.
(10) Coordinate rail, parking pedestrian, and transit proposals.
(11) Encourage and assist paratransit facilities.

Step 5: Specify initial packages.

(a) Street improvements:
   (1) Improve circulation throughout the city.
   (2) Expand capacity at major gateways without adding additional traffic to neighborhood streets.
   (3) Increase intersection and segment capacity to assess one-way street pairs and minor intersection improvements.
(b) Transit and pedestrian improvements:
   (1) Transform an existing street into a two-block "bus-only" street in the heart of a shopping district.
   (2) Close a street.
   (3) Construct new street segments.

Step 6: Plan the analysis.   To verify the feasibility of proposed actions, traffic flow and volume maps were prepared for the year 2005 using manual assignments of estimated traffic.

Steps 7 and 8: Analyze performance of solution package and secondary impacts.   The proposed CBD traffic flow patterns would accommodate a 30%

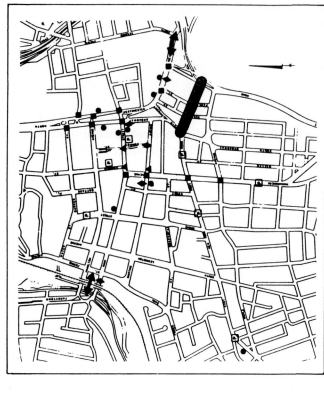

Figure 14-11 Locations of Problems for White Plains (TRB, 1986).

PARKING VEHICLE
CONFLICTS

LIMITED SHORT TERM
PARKING

CONGESTION

PEDESTRIAN-VEHICLE
CONFLICTS

BUS-VEHICLE CONFLICTS

PEDESTRIAN CIRCULATION
VOID

AREA OF LIMITED STREET
CAPACITY

500   0   500

SCALE IN FEET

increase in peak-hour traffic with only three new street segments. The increases in east-bound right turns from Main onto Court (resulting from transferring 250 cars to this intersection during the peak hour) and the impacts of these right turns on pedestrian flow were analyzed. Because most pedestrians were oriented to the midblock mall rather than the Main Street, the conflicts were believed manageable. Future transit operations and curb side usage were also forecast and mapped (Figure 14-12).

Step 9: Recommend an action plan.   Roadway and traffic improvements for the CBD were recommended to provide an integrated-system of street, bus, parking, and pedestrian facilities (Figure 14-13). A general three-stage construction plan was established as follows:

STAGE I (1982–1984)

   (1) Alleviate major capacity restrictions.

   (2) Improve public transit.

STAGE II (1985–1987)

   (1) Begin construction projects.

   (2) Make improvements to the downtown circulation system.

STAGE III (1988–1991)

   (1) Complete street extensions.

   (2) Complete the skywalk system.

Simplified procedures for evaluating low-cost TSM projects are given in NCHRP Reports 263 (TRB, 1983) and 283 (TRB, 1986) published by the Transportation Research Board. The examples given in this chapter are drawn from this literature.

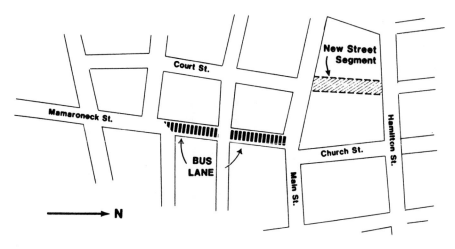

Figure 14-12   Bus Lanes (White Plains) (TRB, 1986).

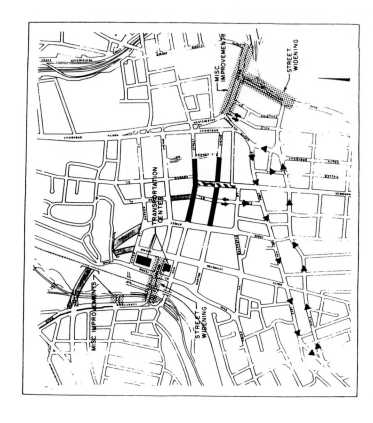

STREET IMPROVEMENT AS NOTED

TRANSIT MALL

TRANSIT PRIORITY CORRIDOR

NEW ONE WAY STREET

NEW TWO WAY STREET

NEW STREET

SCALE IN FEET

Figure 14-13 CBD Improvements (TRB, 1986).

Readers are encouraged to refer to these and other works cited in the references at the end of the chapter.

12. TRANSPORTATION DEMAND MANAGEMENT

There are generally two broad options to reduce traffic congestion and other transportation-related environmental problems. Either the road capacity can be increased or the existing capacity can be used more efficiently by reducing the overall travel demand. Obviously, the second option is often the cheaper, considering benefits and costs, and that is the reason this option has been widely adopted under the name *transportation-demand management* (TDM). Although congestion, air pollution, and parking capacity may be considered as the main justifications for TDM, other objectives such as user cost savings, energy conservation, reduced maintenance costs, and more efficient land use may induce city planners to adopt some form of TDM (Litman, 1994).

A growing number of cities and regions have adopted ordinances to create a legal obligation for employers and developers to alleviate transportation problems. Two types of ordinances have emerged. The first typically offers land developers reduced on-site parking requirements in return for specified traffic mitigation measures. The second type may be a ride-sharing ordinance to reduce trip reduction. Mandatory travel reduction requirements imposed by regulation at the point of employment have emerged as a policy to face the twin challenge of reducing traffic congestion and air pollution. For instance, employment sites in a city center with 100 or more employees are encouraged to consider alternatives to driving alone, say, by carpooling, vanpooling, or using public transit. Scores of localities have enacted trip reduction ordinances (TROs) implementing Employee Trip Reduction (ETR) programs to comply with the requirements of the Clean Air Act (Orski, 1992).

Although economists have long argued that the most efficient means of reducing traffic congestion is through road pricing, apparently due to the high costs of implementing such pricing schemes coupled with widespread political opposition, few road pricing programs have been implemented. Instead, we have witnessed hundreds of TSM and TDM strategies substituted for road pricing. Note that TSM usually involves low to moderate capital expenditures, and includes such strategies as high-occupancy-vehicle (HOV) lane use, ramp metering, and signal preemption, whereas TDM generally involves little or no capital expenditures, and includes such strategies as carpool matching, preferential parking, and bus-fare subsidies.

The institutional context in which TSM and TDM strategies are implemented have proven to be critical to their ultimate success. On the regulatory side, trip reduction ordinances (TROs), and on the voluntary side, transportation management associations (TMAs) have jointly helped many cities to derive the best results from TSM and TDM. It has been demonstrated that the involvement of the private sector in TSM and TDM can be highly beneficial. One of the critical links between public objectives and private sector actions is the local zoning ordinance. Such ordinances generally contain a set of off-street parking requirements that are intended to control the amount of parking supply created during private land development. When developers implement TSM actions by reducing the number of parking spaces, there is a proportionate reduction in vehicle trips. Thus, parking requirements are a potentially powerful tool in enabling-private sector and public-sector decisions to influence one another, with benefits to everybody.

A number of cities and counties are using trip reduction ordinances to establish incentives for developers, landowners, and employers to institute ride-sharing and transit programs in exchange for reductions in parking requirements, a major incentive for developers through cost savings. Table 14-7 illustrates typical cost savings based on increased vehicle occupancy levels.

In the last decade, the introduction of Intelligent Vehicle Highway Systems (IVHS) has been seriously researched, and instances of their initial use is gaining popularity worldwide. There are many reasons for using advanced technologies in TSM/TDM applications because IVHS technology can improve the efficiency and traffic-handling capacity of existing roadways, and can handle inherent safety limitations attributable to both human factors and the roadway system. The real-time information provided by IVHS technologies can include the following:

1. Location of reconstruction and maintenance activities
2. Location of underused or overused facilities and services
3. Identification of restricted or out-of-service facilities
4. Identification of alternative routes (or detours), modes, and services
5. Identification of ride-sharing and transit opportunities
6. Monitoring and routing of heavy and hazardous shipments

SUMMARY

Good engineering practice and astute management in urban transportation is necessary to maintain a certain level of productivity and efficiency. Although in a very real way we were using TSM-type actions to better our transportation system for a long time, we have now made deliberate efforts toward initial delineation of operating environments.

Successful TSM experiences in the United States seem to have common characteristics. Some of these include the following:

1. Coordinated team work among transportation entities is essential. Brainstorming among entities has resulted in synergistic strategies and creative management.
2. Problem identification is probably the single most difficult entity to sift out. Problems that apparently appear to be traffic-oriented may turn out to be economically or socially oriented. Elected officials and citizens are often the first to identify problems.
3. It is essential to do an adequate planning analysis to determine system impacts.

TABLE 14-7 SAMPLE COST SAVINGS FOR INCREASES IN AUTO OCCUPANCY

Auto occupancy increased to:	Parking reduction		Construction cost savings		Land cost savings		Commute cost savings (annual)	Annualized constr. & maint. cost savings	
	%	Spaces	Surface	Structured	Surface	Structured		Surface	Structured
1.2	4	24	$ 24,000	$ 120,000	$ 79,000	$ 26,000	$ 12,500	$ 2,600	$ 9,600
1.3	12	72	72,000	360,000	238,000	79,000	34,700	7,900	28,800
1.4	18	108	108,000	540,000	356,000	119,000	54,000	11,900	43,200
1.6	28	168	168,000	840,000	554,000	185,000	85,000	18,500	67,200
1.8	36	216	216,000	1,080,000	713,000	238,000	109,000	23,800	86,400
2.0	42	252	252,000	1,260,000	832,000	277,000	128,000	27,700	100,800
2.5	54	324	324,000	1,620,000	1,090,000	356,000	162,000	35,700	129,600

Assumptions:
Base auto occupancy = 1.15 (typical of a non-CBD site with no ride-sharing program).
Assumed average trip length = 8 miles one way.
Assume 200,000-square-foot office building with 800 employees with minimum parking requirement of 600 spaces.
Assume land cost = $10 per square foot.
Assume cost of parking construction per space = $1000 surface, $5000 structured, $10,000 underground.
Assume vehicle operating costs = $.15 vehicle-mile (approximate travel-related cost of vehicle operation, based on AASHTO's *Manual on User Benefit Analysis*).
Source: Smith and TenHoor, 1983.

4. Packaging of TSM strategies is probably the most difficult but rewarding task needed to be done. Sufficient literature has now been published to put together a proper package of TSM actions and support activities.

Both simplified and somewhat more complicated approaches to analyzing the effectiveness of TSM/TDM actions have been presented in this chapter to demonstrate the variety of actions to better manage traffic and increase mobility. The potential use of IVHS was also briefly mentioned. For additional information on TSM and TDM, Chapter 9 of the Transportation Planning Handbook contributed by Judycki and Berman (1992) is a good source.

REFERENCES

CANNON, B. (1980). *Energy Considerations in Highway Planning and Financing,* paper presented at the ASCE National Convention, Portland, OR.

DALE, CHARLES W. (1980). *Procedure for Estimating Highway User Cost, Fuel Consumption, and Air Pollution,* Federal Highway Administration, Washington, DC.

GRAY, GEORGE E., and LESTER A. HOEL (1979). *Public Transportation: Planning, Operations, and Management,* Prentice-Hall, Englewood Cliffs. NJ.

HOMBURGER, WOLFGANG S., and JAMES H. KELL (1988). *Fundamentals of Traffic Engineering,* 12th Ed., University of California, Berkeley.

JUDYCKI, D. C., and WAYNE BERMAN (1992). Transportation System Management, in *Transportation Planning Handbook*, ed. John D. Edwards, Prentice Hall, Englewood Cliffs, NJ.

LITMAN, T. (1994). Bicycling and Transportation Demand Management, in *Nonmotorized Transportation Around the World,* Transportation Research Record # 1441, National Research Council, Washington, DC.

ORSKI, C. K. (1992). Congestion Pricing: Promise and Limitations, *Transportation Quarterly,* Vol. 46, No.2, pp. 157–167.

REMAK, R., and S. ROSENBLOOM (1976). *Peak-Period Traffic Congestion—Options for Current Programs,* NCHRP Report 169,

Transportation Research Board, Washington, DC.

REMAK, R., and S. ROSENBLOOM (1979). *Implementation Packages of Congestion Reducing Techniques,* NCHRP Report 205, Transportation Research Board, Washington, DC.

ROARK, JOHN J. (1981). *Experiences in Transportation System Management,* NCHRP Synthesis of Highway Practice Report 81, Transportation Research Board, National Research Council, Washington, DC.

ROWAN, N., ET AL. (1977). *Alternatives for Improving Urban Transportation:—A Management Overview,* Technology Sharing Report 77-215, U.S. Department of Transportation, Washington, DC.

SMITH, S. A., and S. J. TENHOOR (1983). *Model Parking Code Provisions to Encourage Ridesharing and Transit Use,* Federal Highway Administration, Washington, DC.

TRANSPORTATION IMPROVEMENT PROGRAM (TIP) (1975), *Federal Register,* Vol. 40, No. 81, pp. 42,976–42,983.

TRANSPORTATION RESEARCH BOARD (TRB) (1975). *Better Use of Existing Transportation Facilities,* Special Report 153, National Research Council, Washington, DC.

TRANSPORTATION RESEARCH BOARD (TRB) (1977). *Transportation System Management,* Special Report 172, National Research Council, Washington, DC.

TRANSPORTATION RESEARCH BOARD (TRB) (1983), *Simplified Procedures for Evaluating Low-Cost TSM Projects—User's Manual,* NCHRP Report 263, National Research Council, Washington, DC.

TRANSPORTATION RESEARCH BOARD (TRB) (1986). *Training Aid for Applying NCHRP Report 263,* NCHRP Report 283, National Research Council, Washington, DC.

ALAN M. VOORHEES AND ASSOCIATES, INC. (1978). *TSM Planning,* prepared for the North Central Texas Council of Government, Arlington, TX.

WAGNER, F. A., and K. GILBERT (1978). *Transportation System Management, an Assessment of Impacts,* Office of Policy and Program Development. U.S. Department of Transportation, Washington, DC.

EXERCISES

1. A four-lane freeway (two lanes in each direction) connects a city to a large residential suburb. A labor force of 6000 traveling to the city center each day between 7 and 8 A.M. has created congestion problems. Currently, a 5-day workweek is maintained by the employers, but they are willing to adopt the following strategy: (1) 50% of the work force can opt for a 4-day workweek, picking any 4 of the 6 working days except Sundays; (2) 20% of the work force can opt for a 6-day workweek, picking any 6 days of the week; and (3) the balance of 30% maintain the same schedule. Assuming that all workers start work at the same time (8 A.M.), what is the change in traffic flow if the strategy is implemented? (Car occupancy = 1.45.)

2. A company owns a parking facility of 400 spaces for cars and 80 for vans. Its 1000 employees (maintaining a car occupancy of 1.20 and a van occupancy of 3.0) would like to use the parking facility. However, because of a parking shortage, the company pays $10/month for each vehicle parked elsewhere. If a vanpool and carpool program costing $100,000 (which includes the cost of vans and administrative expenses) is implemented (with a van occupancy of 4.0 and a car occupancy of 2.0), will this program work out to be economical? (Assume that the interest rate is 8%.) Is there a time implication in this proposal?

3. A highway is experiencing traffic congestion and the accident rate is also rising. It is planned to introduce staggered work hours in order to spread the peak-hour flow. The existing peak hours are 7 to 9 A.M. with 12,000 vehicles spread evenly over the 2-hour period. The staggered hours proposed and the probable flows are as follows:

   30% of the volume: 6:30–7:00 A.M.
   40% of the volume: 7:00–8:00 A.M.
   15% of the volume: 8:00–8:30 A.M.
   15% of the volume: 8:30–9:30 A.M.

   Calculate the new traffic flows and plot the result.

4. Refer to Exercise 3. If a bus service were introduced to handle 10% of the passenger car occupants (car occupancy 1.2), determine the average headway for the buses during each time period assuming a capacity of 60 passengers/bus.

5. A company wants to determine the economical trip length for a vanpool program, given that the vehicle cost is $10,000 at 10% over 5 years. Insurance is $40/month, fuel is 8 cents/mile, maintenance is 5 cents/mile, working days are 248/year. Determine the relationship between trip cost per person and the economical trip length.

6. An urban arterial has a peak-hour, one-way flow of 1300 veh/hr. The street is currently 56 ft wide and is striped for four lanes. Operating speed is 25 mph and hourly capacity is

about 700 veh/hr. The street is now proposed to be restriped for five lanes with a center reversible lane. The length of the street is 2 miles, car occupancy = 1.2, time value = $5/hr, and speed is 35 mph. If the daily operating cost of the reversible lane is $200/day, what is the saving per year?

7. A planner proposes to improve a 3-mile stretch of an arterial. Currently, the average speed of vehicles is 35 mph, and accidents are 250 per 100 million vehicle-miles. It is estimated that after some basic improvements costing $80,000, the average speed will rise to 45 mph, but so will the accident rate by 20%. If the current flow is 10,000 vehicles, one way, and vehicle occupancy is 1.10, what recommendations would you like to offer? (Time value = $5/hour; accident cost = $20,000 each.)

8. Examine the transportation problems being faced by your university. Which of these are reasonable candidates for applying TSM strategies? Are they related to each other in some way? Draw a diagram showing the connectivity. Write a report on one of these problems.

9. A freeway section with a capacity of 3500 veh/hr experiences two traffic incidents on an average each day during peak hours, creating a bottleneck. Each incident results in a reduction of freeway capacity at the incident site to 2500 veh/hr. The incident occurs randomly when the demand is about 3500 veh/hr. Thirty minutes from the occurrence of the incident the demand is reduced to 1000 veh/hr. Determine (a) the maximum number of vehicles queued and the maximum delay for a vehicle traversing the section of the freeway; (b) the total vehicle-hours of delay resulting from the incidents, per day and per year; (c) whether a sum of $1 million, set apart for improving this section, could be justified, so that these incidents do not occur, assuming that the car occupancy is 1.10 and the value of time is $10/hour.

10. A city served by multilane freeways, which are currently congested during peak hours, has decided either to adopt HOV (high-occupancy-vehicle) lanes or bus lanes, the current modal split for the work trip is 10% bus, 10% carpool, 80% auto, with 1.10 car occupancy. Set up a matrix indicating what advantages and disadvantages each arrangement would have from the point of view of the users and also the operating staff. Discuss your results.

11. The mayor of a large city wants to introduce reserved HOV lanes on all of its congested freeways, with strict enforcement and penalties for misuse. What data would he have to assemble for determining which stretches of freeways would benefit from such action?

12. Select a city planning department or a metropolitan planning organization and investigate whether the ride-sharing and transit use is increasing, decreasing, or has been relatively stable. If there has been a change, investigate the variables that are responsible for this change. How can parking in the central business district affect the change in trip reduction in the future?

13. Compare and contrast TSM with TDM for a problem-ridden corridor in your city. Which types of strategies and actions (institutional/organizational/mandatory/voluntary) have proved successful or unsuccessful?

14. A 15-mile section of an 8-lane freeway in a large metropolitan region is currently operating at capacity with the following characteristics during the 3-hour A.M. and P.M. peak hours: 80% cars, 15% trucks, 5% buses; average car occupancy 1.25; mean free speed 50 mph, and jam density 140 veh/mi. It is proposed to convert one lane of the freeway in each direction for carpool and buses only by encouraging carpooling and transit use. It is estimated that 21% of the drivers will carpool with 2.80 auto occupancy. Assuming Greenshields linear assumption, what travel time savings do you anticipate? (Each truck or bus may be assumed to be equivalent to 4 passenger-car units.)

15. The corporate offices of a company located downtown has a work force of 1000 employees of which 15% use public transport and 5% walk or bicycle to work. The rest of the employees use their own cars and 5% carpool with other employees from the same company. The company has a small parking structure for 350 cars and would like to encourage vanpooling. If a van can accommodate 10 people (including the driver, who rides free) with a one-way average trip length of 8 miles, determine the monthly charge for a passenger assuming the cost of a van is $ 24,000 with a useful life of 7 years, with an interest rate at 8%; operating and maintenance costs are $1.50 per mile. How many vans will be needed for this program?

Chapter 15

Evaluation of Transportation Improvement

1. INTRODUCTION

The term *evaluation* is used in planning and engineering to refer to the merits of alternative proposals. The essence of evaluation is the assessments of the comparative merits of different courses of action. One of these actions may include the "do nothing" alternative (Lichfield et al., 1975).

Engineering and planning alternatives are concerned with problems of choice between mutually exclusive plans or projects (e.g., the examination of alternative investment proposals to alleviate severe traffic congestion in the city center). Evaluation methods may be applied to the problem of choice between sets of independent plans or projects as well as mutually exclusive ones (Mitchell, 1980).

In general, engineering projects in the private sector are built for motives of profit, while in the public sector the motive is, ultimately, the raising of living standards and for social benefit, or for profit, or possibly for political motives. Political motives are not amenable to economic analysis or justification. A value can, however, be put on social benefits and losses (Mitchell, 1980).

Even if a decision maker determines as accurately as possible what consequences will flow from each alternative action, preferences still need to be formulated, particularly when a variety of objectives are involved. In such a case, the crux of the problem is that it is impossible to optimize in all directions at the same time. It is this ubiquitous problem (particularly so in the transportation policy arena), called the *multiattribute problem*, that makes it tough to determine preferences among outcomes (Stokey and Zeckhauser, 1978).

The bottom line seems to be that whereas the basic concepts of evaluation are relatively simple, the actual process of evaluating projects or plans is complex and riddled with controversy. Although engineers and planners try to clarify difficult issues that should be considered by decision makers, it is not uncommon for decision makers to override the results of the analysis. It therefore should be clearly borne in mind that analysis and evaluation are primarily performed by engineers, economists, and planners, whereas the choice of an alternative is done by decision makers (Manheim et al., 1975).

2. FEASIBILITY ISSUES

The solution to a problem should invariably be checked to see whether it is suitable for the situation, acceptable to the decision makers and the public, and one that can be eventually implemented, given the investment. Naturally, at some stage or another, a proposed improvement will be required to satisfy engineering, economic, financial, political, environmental, and social feasibility. For this kind of comprehensive investigation, it is necessary, in complex situations, to obtain the services of economists and sociologists to do justice to the analysis.

Engineers and planners are familiar with the large assortment of analytical tools used in evaluating alternatives. However, in most cases, decision makers demand far more information and justification on the consequences of different alternatives than what is contained in the analysis developed by technicians. Much of this additional information may be qualitative rather than quantitative (Mishan, 1976; Meyer and Miller, 1984).

3. EVALUATION ISSUES

Most evaluation methodologies utilize some form of rating system. Evaluators calculate an index or score, indicating how the welfare of society (or the quality of life) would be affected if a particular alternative were implemented. In effect, they convert all impacts into commensurate units so that they can be added and compared.

Cost–benefit analysis was one of the first evaluation methods used extensively that included a systematic rating procedure and has in the last 25 years been used widely to evaluate all types of public actions. Its rating procedure, although complex, follows the simple arithmetic of placing a monetary value on each impact. The monetary ratings are then aggregated to determine whether benefits exceed costs. The fact that the cost–benefit method has a number of weaknesses led to the development of several other evaluation techniques.

Some evaluation methodologies have established complex procedures for quantifying social welfare ratings. Most of them appear scientific and objective. It does not really matter how sophisticated these methodologies are, because ultimately, the ratings constitute value judgments. Each method has its own weakness. Indeed, each method could possibly lead to a different conclusion (McAllister, 1982).

Several questions naturally stem from this broad description of evaluation methodologies:

- How should evaluation techniques be selected?
- Are economic values sufficient for weighing the costs and benefits of a contemplated public action?

- How are intangible impacts quantified?
- How can equity issues be considered in evaluation?
- How should the concept of time be treated in evaluations?
- How can discount rates be estimated?

Some of these issues are discussed in this chapter.

4. THE EVALUATION PROCESS

The scope of the evaluation process is outlined in Figure 15-1. The evaluation process focuses on the development and selection among several alternatives, and is done in three parts:

1. An evaluation work plan
2. A comprehensive analysis
3. A report containing the results of this analysis

The level of detail and rigor associated with each of these activities will naturally vary from project to project. Figure 15-2 shows a conceptual framework for evaluation of transportation improvements. It identifies the key inputs to the evaluation process, the potential impacts associated with transportation improvements, and the considerations in evaluating the overall merit of improvements. Figure 15-2 also illustrates the interrelationships among the activities constituting the evaluation process, including the iterative nature of evaluation associated with large-scale improvements that often require successive revision and refinement as more critical information is forthcoming about their potential impacts and feasibility.

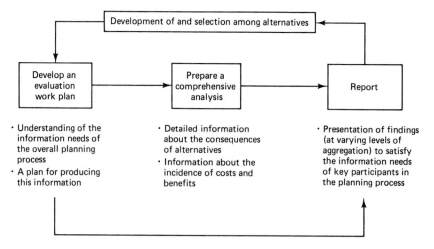

Figure 15-1   Scope of the Evaluation Process (FHWA, 1983).

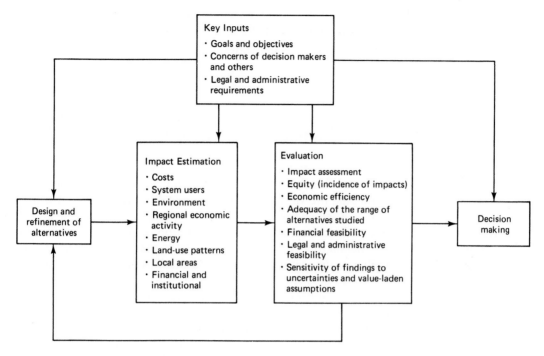

Figure 15-2   Conceptual Framework for Evaluation (FHWA, 1983).

The economic and financial evaluation process of transportation improvements reflected in this section has three major components: (1) the establishment of evaluation criteria; (2) the estimation of costs, impacts, and performance levels; and (3) the overall evaluation of alternative improvements using cost-efficiency and cost-effectiveness techniques as described later in this chapter.

This process is based on the following principles, which underlie good practice in evaluation (Meyer and Miller, 1984):

1. It should be based on a careful examination of the range of decisions to be made and the issues or important considerations in making these decisions.
2. It should guide the generation and refinement of alternatives, as well as the choice of a course of action.
3. Both qualitative as well as quantitative information should be used.
4. Uncertainties or value-laden assumptions should be addressed and explained.
5. All the important consequences of choosing a course of action should be included.
6. Evaluation should relate the consequences of alternatives to goals and objectives established.
7. The impact of each alternative on each interest group should be spelled out clearly.
8. Evaluation should be sensitive to the time frame in which project impacts are likely to occur.

9. The implementation requirements of each alternative should be carefully documented to eliminate the chance of including a "fatal fault" alternative.

10. Decision makers should be provided with information in a readily understandable and useful form.

5. VALUES, GOALS, OBJECTIVES, CRITERIA, AND STANDARDS

The evaluation process involves making a judgment about the worth of the consequences of alternative plans. For example, a large number of metropolitan planning organizations have expressed the view that the planning and development of transportation facilities must be directed toward raising urban standards and enhancing the aggregate quality of life of their communities. Transportation goals have broadened to recognize, for example, the interdependencies of transportation facilities and increased employment opportunities or time savings in travel and an efficient transportation network. The recent literature on urban transportation increasingly demonstrates that transportation improvements constitute a very potent force for shaping the course of regional development.

It may be appropriate, at this stage, to present a brief description of values, goals, objectives, criteria, and standards. A clear statement of their meaning follows. A set of irreducibles forming the basic desires and drives that govern human behavior are called *values*. Because values are commonly shared by groups of people, it is possible to speak of societal or cultural values. Four of the basic values of society are, for example, the desire to survive, the need to belong, the need for order, and the need for security.

A *goal* is an idealized end state; and although they may not be specific enough to be truly attainable, goals provide the directions in which society may like to move toward. An *objective* is a specific statement that is an outgrowth of a goal. Objectives are generally attainable and are stated so that it is possible to measure the extent to which they have been attained.

Criteria result directly from the fact that the levels of attainment of objectives are measurable. In a sense, criteria are the working or operational definitions attached to objectives. They are measures, tests, or indicators of the degree to which objectives are attained. Criteria affect the quantitative characteristics of objectives and add the precision to objectives that differentiate them from goals. One particular type of criterion is a *standard*—a fixed objective [e.g., the lowest (or highest) level of performance acceptable].

The preceding chain is shown in Figure 15-3. A complex structure connects societal values to particular criteria and standards. It may be noted that there may be a high score connected with one criterion as compared to another, in which case trade-offs between criteria may be considered. Of course, decision makers will find it comparatively easy to deal with goals and objectives with a minimum amount of internal conflicts between themselves (Thomas and Schofer, 1970).

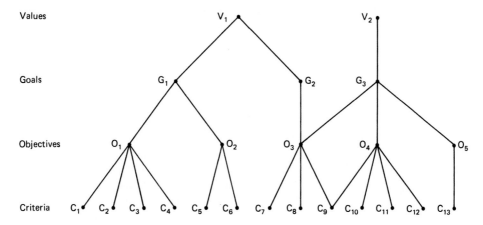

Figure 15-3   Hierarchical Interrelationships among Values, Goals, Objectives, and Criteria.

6. ESTIMATION OF COSTS, IMPACTS, AND PERFORMANCE LEVELS

6.1 Capital, Operating, and Maintenance Costs

Costs play a central role in economic and financial evaluation. When combined with other dollar-valued impacts and compared to other effects, they are helpful in assessing the economic efficiency (i.e., investment worthiness) of the improvement. Detailed knowledge of these costs over the life of the improvement helps to identify and to assign funding sources, thus ensuring that sufficient financial resources are available to implement the improvement.

Existing procedures for estimating costs range from the application of relatively gross average cost measures (e.g., cost per route-mile for two-lane highways) to procedures that use detailed, site-specific information. Achieving a high degree of precision in estimating costs for evaluation of alternatives is important only when similar alternatives are being compared. In such cases, a more detailed cost accounting framework might be required to determine differences between alternatives.

Conversely, when alternatives differ substantially in scale or character, it is more important to establish credible upper and lower limits for the cost of each than to make precise cost estimates based on a single set of detailed assumptions for each alternative (FHWA, 1983).

6.2 User Costs

Transportation system user costs include, for example, changes in travel-time savings, vehicle operating costs, safety, and qualitative changes in the level of service provided by the improvement.

Travel-time savings will accrue from most improvements as congestion is decreased, speed limits increased, or as more convenient routes are provided. Travel-time savings for travelers who take advantage of the improvement and receive the benefit are usually expressed in person-hours of time savings and can be converted to a dollar value using an average value of travel time (i.e., dollars per person-hour).

Vehicle operating cost impacts can accrue if shorter, more direct routes are provided, congestion is decreased, speed limits increased, or if the highway condition improves. Motor vehicle operating costs are usually calculated for the "do-nothing" alternative and each of the improvement alternatives, by estimating the unit cost of the vehicle operation (i.e., cents per vehicle-mile) and the total vehicle miles of travel. The product of these two is the total motor vehicle operating cost.

Safety cost impacts can occur from many types of improvements and can be estimated by applying appropriate "before" and "after" accident rates (e.g., accidents per million vehicle-miles) to the volume of traffic expected to use the proposed improvement. Separate estimates may be developed for fatal, nonfatal injury, and property-damage-only accidents if the potential impact is expected to be large or if the project's safety impact is controversial. Unit costs (e.g., dollars per accident) can then be applied to derive dollar values for accident costs.

Qualitative improvements in the level of service provided can reflect any number of types of effects, including better ride quality from improved pavements, improved transit reliability, or impacts that are difficult to quantify or to assign dollar costs (FHWA, 1983).

6.3 Impacts

A wide range of possible impacts are associated with transportation improvements. These impacts include environmental as well as economic impacts, and are more often described in qualitative or nonmonetary terms rather than in terms of dollars. Such impacts might include air-pollutant emissions from automobiles, noise impacts, water quality impacts, energy consumption, effects on accessibility or land-use patterns, and construction disruption (i.e., for projects that entail substantial amounts of construction activity). To the extent that these impacts represent a significant effect of the improvement, they must be identified and accounted for in the evaluation and factored into the overall assessment (FHWA, 1983).

6.4 Performance Levels

In many cases, formally established criteria for expected economic and financial accomplishments can serve as a valuable evaluation tool. For example, performance levels are often incorporated in the stated goals and objectives for a specific facility to narrow the scope of evaluation and set limits on expected results. Such criteria are also used to monitor implementation, establish budgets, and reallocate financial resources on a periodic basis.

7. EVALUATION OF ALTERNATIVES

The evaluation of alternatives is related to the size, complexity, and number of alternative improvements to be considered. However, answers to the following types of questions, relying on the techniques described in this chapter, should be sought for any evaluation effort:

- Are there other feasible improvements that are not considered and that might compare favorably to those alternatives which were formally evaluated?
- What alternatives were examined and eliminated in the earlier stages of the planning process?

7.1 Impact Assessment

- Have all important differences among alternatives been identified and considered?
- What are the important trade-offs in choosing among alternatives?

7.2 Equity

- What is the distribution of benefits and costs (i.e., among highway users, nearby residents, etc.)?
- Do any groups pay shares of the costs that are disproportionate to the benefits they receive?

7.3 Economic Efficiency

- Does an alternative provide sufficient benefits to justify the required expenditure?
- In comparison with less costly alternatives, does an alternative provide additional benefits to justify the additional funds required?
- In comparison with more costly alternatives, has sufficient weight been given to the benefits that would be forgone in the event that additional expenditure of funds is not justified?
- How do nonmonetary costs and benefits affect conclusions about economic efficiency?

7.4 Financial Feasibility

- Will sufficient funds be available to implement alternatives on schedule? From what sources?
- What is the margin of safety for financial feasibility?
- What adjustments will be necessary if this margin of safety is exceeded?

7.5 Legal and Administrative Feasibility

- Is the implementation of an alternative feasible within existing laws and administrative guidelines? If not, what adjustments might be necessary?
- What approaches will be required in later stages of the planning process? Are problems anticipated in obtaining these approvals? If these problems do materialize, can minor modifications or refinements to alternatives overcome them?

7.6 Sensitivity of Findings to Uncertainties and Value-Laden Assumptions

- Are there uncertainties or value-laden assumptions that might affect the information assembled to help decision makers answer the earlier questions?
- How much would assumptions have to be changed to suggest different answers to these questions?
- What is the likelihood of reasonableness of these changes?

The answer to the many questions listed are subjective in nature; it is quite possible that reasonable individuals might answer them differently even when given the same information.

In a comprehensive evaluation, detailed information about alternatives and their impacts should be summarized in the evaluation report. This report should document the major impacts of the performance of the alternatives analyzed and emphasize the difference in the performance of alternatives with respect to the key impact measures (FHWA, 1983).

8. ECONOMIC AND FINANCIAL CONCEPTS

In determining the economic and financial implications of a transportation improvement, a clear understanding of several basic concepts is essential. These concepts include monetary impacts such as the cost of time, money, equipment, maintenance, and user benefits, as well as nonmonetary costs such as safety, environmental impacts, and effects on the local economy.

Each of these concepts is described briefly in what follows. When applied as part of one of the analysis techniques presented in Section 9, they can assist planners in determining the overall economic efficiency and effectiveness of an improvement (FHWA, 1983).

8.1 Time Value of Money

Benefits and costs will occur at different times during the life of an improvement. However, a cost incurred today cannot be compared directly with the benefits that it will provide in some future year. The funding for an improvement should be considered as an investment that is expected to generate a return. Consequently, the costs and benefits of an improvement must be compared in equivalent terms that account for their timing. These benefits and costs can be either described in equivalent terms by calculating their "present worth" (for the current year or recent year for which data are available) or for an average year over the life of the improvement. Any future cost or benefit expressed in future dollars is properly converted to its corresponding present value by two factors: an inflation rate and a discount rate.

Discount Rate.   The appropriate discount rate is an estimate of the average rate of return that is expected on private investment before taxes and after inflation. This interpretation is based on the concept that the alternative to the capital investment by government is to leave the capital in the private sector.

Inflation Rate.   Inflation can be ignored in performing most aspects of the basic economic evaluation if the evaluation is prepared in present value terms, since inflation has no effect on present values. However, if the costs used are 2 or 3 years out of date, they should be adjusted to a more current year.

The only time when it might be appropriate to project future inflation in evaluating alternative improvements is when there is reasonable evidence that differential future rates of inflation might apply to important components of the evaluation. However, independent of the evaluation of alternative improvements, projections of inflation must be made for a financial feasibility analysis and development of a capital improvement program (FHWA, 1983).

9. ANALYSIS TECHNIQUES

The objective of economic evaluation analysis techniques is to provide sufficient summary information to decision makers and interest groups to do the following:

1. Determine whether the costs of improvements are justified by the anticipated benefits (i.e., whether a proposed improvement is superior to doing nothing).
2. Make comparative overall assessments of different alternative improvements with each other.

In complex situations involving large-scale, costly improvements, it may also be necessary to assess the distribution of benefits and costs among those affected by the improvement, such as user cost, operator cost, and societal cost.

Techniques that support the first preceding purpose come under the category of economic efficiency analysis, or merely *efficiency analysis*, also often called *investment appraisal methods*. These include benefit–cost ratio, present worth, rate of return, equivalent uniform annual value, and other variations of these methods. All involve the translation of impacts (i.e., costs and benefits) into monetary terms.

Techniques that support the second purpose come under the category of cost-effectiveness analysis. These include the use of nonmonetary effectiveness measures either to assess the relative impacts of all alternative improvements in the same terms or to hold constant the requirements that all alternatives must meet. Effectiveness measures are also used in combination with cost values in the form of ratios (FHWA, 1983).

9.1 Economic Evaluation Methods (Efficiency Analysis)

There are four common methods of evaluating transportation projects. These four methods are (1) net present worth (NPW), (2) equivalent uniform annual cost (EUAC), (3) internal rate of return (ROR), and (4) benefit–cost (B/C) analysis. The bottom line in the evaluation of individual projects is: Which project is the most productive? Or which project produces the highest return? Essentially, there are two indices of economic merit that can help answer these questions. The first is the benefit–cost criterion, and the second is the internal rate of return. A brief explanation of each method follows.

By using the concept of equivalence connected with compound interest, the present worth of a single payment F, n years from now, with a discount rate of i is

$$\text{PWSP} = \frac{F}{(1 + i)^n} \tag{1}$$

The present worth of uniform series of equal annual payments is the sum of the present worth of each cost:

$$\text{PWUS} = A\left[\frac{(1 + i)^n - 1}{i(1 + i)^n}\right] \tag{2}$$

Here A is the equivalent uniform annual cost (EUAC). The term in brackets is known as the present worth uniform series factor. Its value has been tabulated for various combinations of i and n. Also,

$$A = \text{EUAC} = \text{PWUS}\left[\frac{i(1 + i)^n}{(1 + i)^n - 1}\right] \tag{3}$$

The term in brackets is called the *capital recovery factor*. For example, if a person borrows $10,000 from a bank and intends to repay the amount in equal yearly installments over an 8-year period at 5% per annum, the equivalent uniform annual payment (EUAC) would amount to

$$\text{EUAC} = 10,000\left[\frac{0.05(1 + 0.05)^8}{(1 + 0.05)^8 - 1}\right]$$

$$= 10,000 \times 0.154722$$

$$= \$1547.22$$

The internal rate of return method has been proposed as an index of the desirability of projects. Naturally, the higher the rate, the better the project. By definition, it is the discount rate at which the net present value of benefits equals the net present value of costs.

Benefit–cost analysis is generally applied to engineering projects to ascertain the extent to which an investment will result in a benefit to society. This analysis can be quite elaborate, and more will be said about this method of analysis in another section. Suffice it to say here that a project that costs less than the benefits derived from the project would be eligible for consideration of being implemented ($B/C \geq 1$).

Example 1

Three alternatives are being considered for improving a street intersection. The annual dollar savings on account of the improvement is shown. Assume that the intersection will last for 25 years and the interest rate is 5%. It is assumed that each of the three improvements is mutually exclusive but provides similar benefits.

Alternative	Total cost	Annual benefits
A	$10,000	$ 800
B	12,000	1,000
C	19,000	1,400

Solution

NPW method: Use the present worth factor for uniform series.

$$\text{NPW(A)} = -10{,}000 + (800 \times 14.094) = \$1275.2$$

$$\text{NPW(B)} = -12{,}000 + (1000 \times 14.094) = \$2094.0$$

$$\text{NPW(C)} = -19{,}000 + (1400 \times 14.094) = \$731.6$$

Therefore, select alternative B with the highest net present worth.
EUAC method: Use the capital recovery factor (crf).

$$\text{EUAC(A)} = -(10{,}000 \times 0.07095) + 800 = 90.50$$

$$\text{EUAC(B)} = -(12{,}000 \times 0.07095) + 1000 = 148.60$$

$$\text{EUAC(C)} = -(19{,}000 \times 0.07095) + 1400 = 51.95$$

Alternative B has the highest EUAC and should be selected.
ROR method: Compare the ROR figure with the "do-nothing" alternative.

$$\text{NPW(A)} = -10{,}000 + (800)(P/A, i, 25 \text{ yr}) = 0$$

$$(P/A, i, 25 \text{ yr}) = 10{,}000/800 = 12.5 = \; > i = 6.25\%$$

$$\text{NPW(B)} = -12{,}000 + (1000)(P/A, i, 25 \text{ yr}) = 0$$

$$(P/A, i, 25 \text{ yr}) = 12{,}000/1000 = 12 = \; > i = 6.7\%$$

$$\text{NPW(C)} = -19{,}000 + (1400)(P/A, i, 25 \text{ yr}) = 0$$

$$(P/A, i, 25 \text{ yr}) = 19{,}000/1400 = 13.57 = \; > i = 5.77\%$$

Alternative B is the least of the three because the rate of return is the highest (and, of course, higher than 5%, so better than the "do-nothing" alternative).

9.2 Cost-Effectiveness Analysis

Cost-effectiveness is a strategy for making decisions rather than establishing a decision rule. This approach provides a general and flexible framework for providing information to aid in the selection of alternative plans. Many of the consequences of proposed transportation plans are difficult to measure if not intangible. Planners resorting to the benefit–cost method place dollar values on benefits and costs, with the motive of being as objective as they can be and in this process oversimplify the complexity of the problem. Cost-effectiveness overcomes some of these snags.

Cost-effectiveness analysis is essentially an information framework. The characteristics of each alternative are separated into two categories: (1) costs and (2) measures of effectiveness. The choice between alternatives is made on the basis of these two classes of information, eliminating the need to reduce the attributes or consequences of different alternatives to a single scalar dimension. Costs are defined in terms of all the resources necessary for the design, construction, operation, and maintenance of the alternative. Costs can be considered in dollars or in other units. Effectiveness is the degree to which an alternative achieves its objective. Decision makers can then make subjective judgments that seem best to them.

Cost-effectiveness analysis arose out of a recognition that it is frequently difficult to transform all major impact measures into monetary terms in a credible manner, and that important evaluation factors can often be stated in more meaningful measures than dollar costs. Basically, the method should be used when preestablished requirements exist regarding the improvement. Usually, these requirements seek to establish the minimum investment (input) required for the maximum performance (output) among several alternative improvements.

One of the common general transportation performance measures that is used in cost-effectiveness analysis is *level of service*. In practice, it is generally desirable to prepare estimates for several cost-effectiveness measures rather than a single measure, because no single criterion satisfactorily summarizes the relative cost-effectiveness of different alternatives. Table 15-1 lists cost-effectiveness indices that might be used for highway and transit evaluation. Many or all of these could be used in a particular cost-effectiveness analysis. Several additional measures could be added, depending on local goals and objectives (Thomas and Schofer, 1970).

Example 2

Citizens (particularly schoolchildren) need a transportation facility to connect their community across a six-lane freeway by means of a footbridge over the freeway or a tunnel under the freeway. The objectives of the project are (1) to provide a safe and efficient system for crossing the freeway, and (2) to maintain and improve (if possible) the visual environment of the neighborhood. The table gives the construction and operating costs of the two alternatives. Annual costs are computed using a 25-year service life and an 8% interest rate. It is anticipated that about 5000 pedestrians will use either of the two facilities per day.

TABLE 15-1   LIST OF TYPICAL COST-EFFECTIVE MEASURES FOR EVALUATION

Highway

Increase in average vehicle speed per dollar of capital investment
Decrease in total vehicle delay time due to congestion per dollar of capital investment
Increase in highway network accessibility to jobs per dollar of capital investment
Decrease in accidents, injuries, and fatalities per dollar of capital investment
Change in air-pollutant emissions per dollar of capital investment
Total capital and operating cost per passenger-mile served

Transit

Increase in the proportion of the population served at a given level of service (in terms of proximity of service and frequency) per dollar of total additional cost
Increase in transit accessibility to jobs, human services, and economic centers per dollar of total additional cost
Increase in ridership per dollar of capital investment
Increase in ridership per dollar of additional operating cost
Total capital and operating cost per transit rider
Total capital and operating cost per seat-mile and per passenger-mile served
Decrease in average transit trip time (including wait time) per dollar of total additional cost

Alternative	Total cost	Annual cost	Annual operating cost	Total annual cost
Footbridge	$100,000	$9368	$100	$9468
Tunnel	$200,000	$18,736	$350	$19,086

The annual difference in cost is $9618. The average cost per pedestrian per trip using the bridge is

$$\frac{9468}{5000 \times 365} = 0.52 \text{ cent}$$

and using the tunnel it is

$$\frac{19,086}{5000 \times 365} = 1.05 \text{ cents}$$

The performance of the two alternatives may be summarized as follows:

1. The tunnel would need long, steep grades and may cause inconvenience to the elderly and handicapped. On the other hand, the bridge will be provided with stairs at either end, and this will eliminate its use by the elderly and handicapped. If ramps are to be provided, long steep grades would have to be provided at three times the cost of the bridge with stairs.
2. The tunnel is safer than the footbridge during the long winter months.
3. From the point of view of aesthetics, the tunnel is preferred to the footbridge.
4. It is possible to predict that the tunnel would be safer than the bridge.

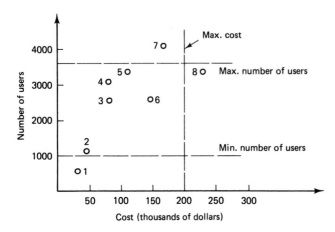

Figure 15-4 Cost-Effectiveness Analysis: Cost versus Users.

Another way to look at cost-effectiveness analysis is to consider it as a way of maximizing the returns (in terms of effectiveness) of expenditures. Consider the following example: A city is considering eight different minibus network configurations, which are mutually exclusive. The only objective is to adopt a configuration that will maximize the number of potential riders in that sector of the city. The city has established the minimum number of potential riders that the alternative will serve and also maximize the amount of money it is willing to spend. Figure 15-4 shows the results of the analysis.

Notice that alternatives 1, 7, and 8 do not qualify under the criteria set. Obviously, by sheer inspection, alternative 5 would be the one that serves the maximum number of users. This problem could get somewhat complicated if an additional criterion were introduced (e.g., number of hours of operation).

9.3 Multicriteria Evaluation Method

The examples that have been given in this chapter so far fall under the general category of single-criterion evaluation methods, because all the benefits and costs have been reduced to monetary terms. The present worth, annual cost, benefit–cost, and rate-of-return methods all fall into this category, because the maximization of net benefits is the single objective the analyst has in mind. However, in the transportation planning process, one usually is dealing with many objectives (or criteria) that reflect the interests of the community. To incorporate several, often conflicting, objectives (or criteria) systematically, planners have developed methods to tackle this situation. One of these methods is generally called the *multicriteria evaluation method*. The steps in applying this method are best illustrated by working through an example.

Example 3

The city of Athens (Greece) is considering the improvement of its transportation system by adopting the following mutually exclusive alternatives: (1) a bus-only system, (2) a light-rail system, (3) a bus + heavy rail system, and (4) a subway system. The city (elected officials and citizens) has recently identified and adopted the following goals and objectives for this improvement project:

1. The alternatives must reduce congestion in the city by capturing the maximum number of commuters in the peak hour.
2. The air-pollution and noise level should be reduced to tolerable acceptance levels through this improvement.
3. The total net revenue to the city should be maximized by adopting one of the alternatives.
4. The rate of deaths and injuries due to accidents should be reduced by at least 50% by adopting one of the alternatives.
5. The alternative chosen should cover the maximum area of the city and should be equitable to all patrons.

The alternatives prepared by consultants to the city are acceptable in principle. However, they all differ in quality, cost, alignment, coverage, and so on. City planners have formulated the following measures of effectiveness for each objective:

Objective 1: percentage reduction of motor vehicles on main arterials and corridors of the city

Objective 2: percentage reduction of pollution

Objective 3: net annual revenue/annual capital cost

Objective 4: probable accident rate/million trips

Objective 5: square miles covered by the proposed network ($\frac{1}{4}$ mile on either side of line).

Objective and MOE	Alternatives			
	Bus	LRS	B + HR	SWS
1. Congestion reduction	20	(27)	25	15
2. Pollution reduction	15	30	22	(35)
3. Revenue	(25)	20	15	10
4. Accidents reduction	15	20	23	(27)
5. Coverage	30	25	(40)	22

Note that each alternative has been assigned a figure representing a particular measure of effectiveness. Also, certain alternatives rank higher than others in a particular measure; for example, the light-rail system reduces congestion the most, by scoring the highest of the row values, 27; and the bus and heavy-rail alternative scores 40 points (the highest) in the row representing "coverage."

If these highest scores are matched to the corresponding relative weighting factor, one can erect another matrix, as shown. Here the highest score for MOE 1, 27, is assigned 20 as the relative weighting factor for the light-rail system; 17 is assigned to the bus + heavy-rail alternative for MOE 5; and so on. By the same token, the other figures are calculated proportionally.

Solution

The city's elected officials (through brainstorming methods and Delphi techniques) have allocated weights on a 10-point scale to each objective as indicated:

Objective	Weight (out of 10)	Relative weighting factor (%)
1	7	20
2	5	14
3	9	26
4	8	23
5	6	17
Total	35	100

This assignment of weights is subjective, but it is one way of assessing the collective opinion of a large group of people who are affected positively (or negatively) by the proposed implementation of the chosen alternative.

Next, planners, engineers, sociologists, economists, and estimators set about collecting data and costs for assigning values to each alternative, as shown in the following matrix. This step is most crucial and involves a lot of work in gathering appropriate data.

Objective and MOE	Alternatives			
	Bus	LRS	B + HR	SWS
1. Congestion reduction	15	(20)	19	11
2. Pollution reduction	6	12	9	(14)
3. Revenue	(26)	21	16	10
4. Accidents reduction	13	17	20	(23)
5. Coverage	13	11	(17)	9
Total	73	81	81	67

It is obvious from the column totals that the light-rail system and the bus + heavy-rail system each score 81 points and can be declared as winners in comparison to the other two alternatives. It is natural that these two alternatives will be subjected to further detailed scrutiny and analysis before a final winner is declared.

If, for example, these two top-ranking alternatives are examined by an investigating team who perform an environmental impact study, it is quite possible that a more calculated decision can be taken. This example illustrates the difficulty in considering situations involving multicriteria application, because the outcome is sensitive to the weights assigned to the criteria.

TABLE 15-2   CAPITAL COST AND EXPECTED BENEFITS

(1) Alternative	(2) Miles	(3) Capital cost	(4) User savings	(5) Savings − cost	(6) Δ (savings) − Δ (cost)
1	5	80	220	140	
2	10	100	300	200	60
3	15	130	340	210	10
4	20	180	370	190	−20
5	25	270	390	120	−70
6	30	380	425	45	−75

9.4 Benefit–Cost Analysis

Public expenditure decisions are generally evaluated using benefit–cost analysis. A simple example of this type of evaluation is demonstrated here. Table 15-2 gives information on the capital cost and expected potential benefits by constructing fom 5 to 30 miles of a light-rail line between downtown and the outskirts of a large city. Column (1) shows the six alternatives, and column (2) indicates the length of each of these six alternatives. The capital cost (in units of millions of dollars per year) is shown in column (3). The present value of user benefits is indicated and this has been derived assuming a money stream for 20 years at 10% per annum in units of millions of dollars. Column (5) is column (4) minus column (3). The last column indicates marginal benefits minus marginal cost. The benefits and costs of the six light-rail systems are plotted in Figure 15-5. The abscissa denotes the length of the light-rail system. The marginal benefits and marginal costs are plotted directly below.

It is obvious, by sheer inspection, that as the light-rail system is extended farther out in the suburbs of the city, the revenues (benefits) from the system become less and less attractive. The question now is: What criteria should be used in evaluating these six alternatives? Because of the nature of the data presented, the economic criterion is used to identify the best course of action.

The figures shown in Table 15-2 are for discrete lengths of the light-rail system. If the cost–benefit relations shown in Figure 15-5 were continuous, the optimum length of light rail would be one where the marginal benefit curve cuts the marginal cost curve. This intersection of curves occurs at 17 miles from the city center. The alternative closest to this intersection is alternative 3 (15 miles).

Applications of Benefit–Cost Analysis.   It is best at this stage to understand the applications of benefit–cost analysis through practical examples. Take the case of a decision maker who is confronted with five projects whose details are provided in Table 15-3.

Case 1.   Assume that the projects are mutually exclusive, but there is no funding constraint. In the absence of a budgetary constraint, it would be clearly advantageous to build project D, which would yield a net benefit of 150.

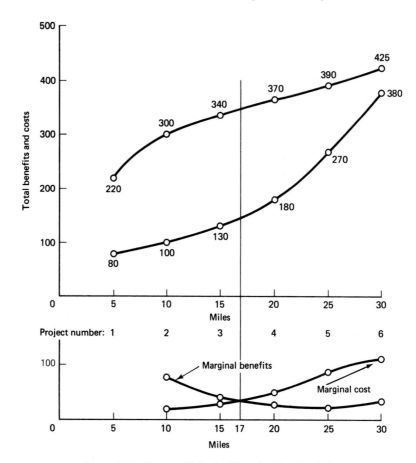

Figure 15-5   Total and Marginal Benefit–Cost Analysis.

Case 2.   Now assume that the projects are no longer mutually exclusive and that a number of projects of the same category (A through E) up to the limit of the budget can be implemented. For example, A through E are alternative forms of bridges. If the budget constraint in this case were 160, then 40 bridges of type A could be built for 160, providing 1280 as benefits and 1120 as net benefits. Similarly, 16 bridges of type B could be built, providing (960 − 160) = 800 as net benefits. Or, four bridges of type C

TABLE 15-3   COST–BENEFIT AND COST OF PROJECTS

Project	Benefits B	Costs C	Ratio B/C	Net benefits, $B - C$	Cumulative initial cost
A	32	4	8	28	4
B	60	10	6	50	14
C	120	40	3	80	54
D	300	150	2	150	204
E	500	450	1.11	50	654

could be undertaken, resulting in 320 as net benefits. Even one bridge of type D could be considered for 150, but this would give only a 150 net benefit. Obviously, the best solution here would be to adopt 40 versions of project A. Ratios are therefore applicable when dealing with a budget constraint.

Case 3.   Assume that the budget is only 10, in which case only projects A and B are eligible. But here again one runs into the problems of indivisibilities, and therefore the only choice is project B.

Case 4.   Assume that a new project C' is added to the list, as follows:

$$C': \quad B = 192, \quad C = 96, \quad B/C = 2, \quad B - C = 96, \quad \text{cumulative cost 150}$$

Also, assume that the budget constraint is 204 and that the projects are not mutually exclusive. Therefore, selecting projects from the top of the list and also taking into account project C', one should undertake $A + B + C + C' = 150$. The question that comes up here is: Which is better—choosing $A + B + C + C'$ or choosing just D, as the choice in case 1, both choices aggregating 150?

Net benefit when $A + B + C + C'$ are considered $= 28 + 50 + 80 + 96 = 254 >$ 150 (if D is considered). This exercise indicates rather clearly how limitations of the budget changes the outcome.

Example 4

An existing highway is 20 miles long connecting two small cities. It is proposed to improve the alignment by constructing a highway 15 miles long costing $500,000 per mile. Maintenance costs are likely to be $10,000 per mile per annum. Land acquisition costs run $75,000 per mile. It is proposed to abandon the old road and sell the land for $10,000 per mile. Money can be borrowed at 8% per annum. It has been estimated that passenger vehicles travel at 35 mph at a cost of 20 cents per mile with a car occupancy of 1.5 persons per car. What should be the traffic demand for this new road to make this project feasible if the cost of the time of the car's occupants is assessed at $10.00 per hour?

Solution

Basic costs:

Construction 15 miles at $500,000 =	$7,500.000
Land acquisition 15 miles at $75,000 =	$ 900,000
	$8,400,000
Deduct sale of land (old alignment)	
20 × 10,000	− 200,000
	$8,200,000

It is assumed that the new road will last indefinitely. Therefore, the investment need not be repaid, only the interest.

Annual cost of initial investment:

$$8\% \text{ of } \$8,200,000 = \$656,000$$

Net annual income: Assume that N vehicles per annum use the road, 5 miles saved at 20 cents per mile.

$$N\left(\frac{5 \times 20}{100}\right) = \$N$$

Time saved per vehicle/trip:

$$5 \text{ miles at 35 mph} = \frac{5}{35} = 0.143 \text{ hour}$$

$$\text{Car occupancy} = 1.5 \text{ persons/car}$$

$$\text{Value of time} = \$10 \text{ per hour}$$

Therefore,

$$\text{Cost saving in time at } \$10/\text{hour} = 0.143 \times 1.5 \times N \times 10 = 2.14N$$

$$\text{Maintenance savings} = 5 \text{ miles} \times \$10,000 = \$50,000$$

$$\text{Total savings} = \$50,000 + \$N + 2.14N = 50,000 + 3.14N$$

Benefit–cost ratio: In order that the project could be justified, the B/C ratio must be equal to or greater than 1. Because the project is assumed to have an infinite life,

$$B/C = \frac{\text{annual benefits}}{\text{annual costs}}$$

or

$$50,000 + 3.14N = 656,000$$

$$N = 192,994 \text{ per annum}$$

$$\text{or about 528 vehicles/day}$$

The project, therefore, would be justified if about 528 vehicles used the new road on a daily basis.

Discussion

The following points should be noted regarding this problem.

1. The social costs of land use have not been considered. For example, the land on either side of the old road would decrease in value, whereas the land adjacent to the new route would increase in value. The net result would raise the value of N.
2. When one works out the B/C ratio, it is generally advisable to use total benefits and total costs. In this problem, B/C was assumed to be 1 and is therefore permissible.

9.5 The Willingness-to-Pay Concept

The willingness-to-pay concept is useful in visualizing the view point of users of a system, such as a bridge or a toll road. In the case of a single-service variable, as shown in Figure 15-6(a) (Manheim et al., 1975), if DD is the demand and the price (p) changes from p_1 to p_2, there will be a corresponding change in volume (v) from v_1 to v_2. The

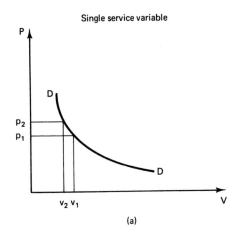

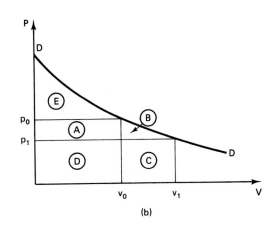

Figure 15-6   Willingness to Pay.

$(v_1 - v_2)$ users who abandoned the system are those who got just enough benefit from the system at price p_1 to use it.

If we consider Figure 15-6(b) using the same argument, reducing the price from p_0 to p_1 increases the number of users from v_0 to v_1. There are three alternative views on the method of measuring the amount of benefit to users.

1. Gross-benefit view

$$\text{Total benefit to users} = Ⓔ + Ⓐ + Ⓓ, \text{ when volume} = v_0$$

$$\text{Total benefit to users} = Ⓔ + Ⓐ + Ⓓ + Ⓑ + Ⓒ, \text{ for } v_1$$

Difference between the two actions $= Ⓑ + Ⓒ$

2. Consumer surplus view: it is the amount of benefit received by users beyond what they actually pay $= Ⓐ + Ⓑ$

3. User-cost view $= Ⓐ - Ⓒ$

APPROXIMATION

$$\text{Gross benefit } B + C = (1/2)(p_0 + p_1)(v_1 - v_0)$$

$$\text{Consumer surplus } A + B = (1/2)(p_0 - p_1)(v_1 + v_0)$$

$$\text{User cost } A - C = v_0 p_0 - v_1 p_1$$

An examination of these three views indicates how different the results can be.

Example 5

A section of a busy road, 1 mile long, has a very high pedestrian accident rate. It is proposed to construct a pedestrian bridge over the road for $18 million and increase the existing speed limit on the road. Details of the existing and proposed conditions are as follows:

Details	Existing	Proposed
Vehicle operating speed	25 mph	35 mph
Peak-hour flow	3000 veh/hr	4000 veh/hr
Length of section	1 mile	1 mile
Accident rate/year	450	25 (estimated)
Cost of driving (veh-mi)	$0.35	$0.20
Cost/accident	3000	3000
Peak hours affected	4 hr	4 hr

Determine the feasibility of this project, if the interest rate is 6% per annum and the design life of the pedestrian bridge is 30 years.

Solution

The direct benefits derived from the construction of the pedestrian bridge can be determined by applying the concepts of consumer's surplus. The following analysis shows the gain in consumer's surplus.

$$\text{Existing price} = \$0.35/\text{veh-mi} \times 1 \text{ mile} = \$0.35/\text{veh}$$

$$\text{Proposed price} = \$0.20/\text{veh-mi} \times 1 \text{ mile} = \$0.20/\text{veh}$$

Annual number of users:

Existing: 3000 veh/hr × 4 hr/day × 365 days = 4.38 million veh/yr

Proposed: 4000 veh/hr × 4 hr/day × 365 days = 5.84 million veh/yr

Consumer's surplus as annual savings

$$= 0.5\,(0.35 - 0.20)\,(4.38 + 5.84) \text{ million} = \$0.7665 \text{ million/yr}$$

Annual savings from accident reduction $= 3000\,(450 - 25)$

$$= \$1.275 \text{ million/yr}$$

Total benefits $= \$2.0415$ million/yr

NPW factor @ 6% for 30 years $= 13.7649$ (see appendix)

$$\text{NPW} = (\$2.0415 \text{ million} \times 13.7649) - 18$$

$$= \$\,10.1041 \text{ million}$$

Because the NPW of this project is greater than zero, the pedestrian project is sound.

10. REPORTING RESULTS

A well-organized summary report can significantly influence the outcome of an economic evaluation effort. In this section, we describe the preparation of a summary report and related materials for use by decision makers.

The objective of good reporting is to satisfy the information needs of decision makers and other people with a wide range of perspectives, technical understanding, and constituents. Properly conceived and executed, written reports and oral presentations should enable decision makers to select, narrow, or reconfigure the alternative improvements, and generally facilitate, rather than complicate, the decision-making process.

A common format for summarizing key evaluation findings has measures or impact categories listed down the left-hand side of a table and alternatives listed across the top, with entries in the table indicating the status of each alternative with respect to each evaluation measure. For those measures that are in common units (e.g., dollars), subtotals can be provided. The normal bias toward quantitative measures can be avoided by including rows summarizing the key difference among alternatives that can only be characterized by a brief phrase. Table 15-4 provides an illustrative example of such a table.

Table 15-5 shows an outline of a typical summary report. The cover, abstract, map, and summary arguments table could be made into a small brochure of about six single-page sheets. A slightly longer version might include one-page summaries of each alternative.

SUMMARY

Evaluation of transportation improvement can have different meanings and concerns for different people, depending on the project under investigation. Only a decade or two ago, for instance, transportation and traffic engineering projects were evaluated taking into account just the economics of benefits and costs to the potential user, purely on an aggregate basis. Today, most evaluation exercises take into consideration the environmental, social, economic, and disaggregate effects of improvements.

There are at least three basic points to be kept in mind in deciding on, and later working with, a particular analysis technique, and these are (1) the evaluator should be fully aware of the goals and objectives of the proposed transportation improvements, and their corresponding measures of effectiveness; (2) the evaluation method adopted must be clear and transparent, even to the layperson; and (3) the evaluation procedure should assist the decision maker to come to a rational determination of what is best for the community or city.

Economic evaluation methods are the most common procedures used in the transportation context. However, their use entails the conversion of measures of effectiveness to dollar units, and this occasionally poses problems. Planners have therefore evolved several other procedures to obviate this problem.

The evaluation of transportation improvements is considered as one of the more difficult tasks in the field of engineering economics. Not only have the principles of

TABLE 15-4 ILLUSTRATIVE SUMMARY FORMAT FOR PRESENTATION OF EVALUATION RESULTS

	Alternative			
	0. "Do nothing"	1. TSM program	2. Arterial improvement	3. Bus improvement program
Mobility	Severe congestion, level of service F by 1986; transit ridership −10%	Reduced congestion, level of service C maintained; transit ridership −10%	Postponed congestion, level of service F by 1988; transit ridership −10%	Reduced congestion, level of service C maintained; transit ridership +80%
Social	Traffic will harm neighborhood	Slightly less traffic disruption	Will make pedestrian movements more difficult	Will strengthen community
Air quality	(Percent reduction in pollutant concentration from 1978 levels) Total daily corridor emissions (1995) — No sites in any alternative are expected to violate 1- or 8-hour national ambient air quality standards			
	CO −60% HC −55% NO$_x$ −40%	CO −65% HC 55% NO$_x$ −40%	CO −65% HC −55% NO$_x$ −40%	CO −70% HC −60% NO$_x$ −40%
Noise	Areas with potential for significant noise impact (80 dB or more)			
	4 areas	4 areas	6 areas	5 areas
Disruption during construction	—	Minor	2 years of modest construction	1 year of modest construction

(All dollar values are present values in thousands of 1978 dollars based on a discount rate of 10%)

Monetary evaluation				
Time	$1550	$850	$1200	$700
Vehicle operating	$400	$250	$350	$200
Accidents	$300	$150	$200	$100
User subtotal	$2250	$1250	$1750	$1000
Transit operating	$650	$750	$750	$500
O&M highways	$800	$900	$850	$550
O&M subtotal	$1450	$1650	$1600	$1050
Capital	$50	$100	$150	$200
Total transportation costs	$3750	$3000	$3500	$2250
Net benefits	—	$750	$250	$1500

(All dollar values are present values in thousands of 1978 dollars based on a discount rate of 10%)

Cost effectiveness				
Capital, O&M costs per daily person miles of travel	$87.7	$101.7	$101.7	$71.4
Capital cost per hour of daily time savings compared to "do nothing"	—	$230	$320	$210
Additional jobs accessible within 30 minutes/$1000				
Via auto	—	200	230	20
Via transit	—	250	90	460
Total both modes	—	450	320	480

657

TABLE 15-5 TYPICAL OUTLINE OF AN EVALUATION SUMMARY REPORT

Cover: Should effectively communicate the following:
  What is being studied
  Location: what part of which area
  That a decision is about to be made
  Who is involved: jurisdiction and/or agencies

Abstract: About two or three pages elaborating on the foregoing items, making clear the choices and the
  major issues

Map: Showing the configuration of the alternatives evaluated

Summary of arguments pro and con for each alternative: A single table highlighting the reasons for
  selection or rejection of each based on the evaluation findings

Purpose of report: Should identify the problem to be addressed by the proposed improvement

The alternatives:
  How they were identified or arrived at
  One-page summary of each, including a more detailed description, benefits, and issues
  More detailed map of each

Evaluation:
  A summary table showing the major differences among options
  More detailed tabular evaluations with text as necessary
  Elaboration on the differences among options
  Explanations for the lack of differences where ones may be expected
  Discussion and clarification of the arguments that may have been advanced by various interests or
    advocates

Appendix:
  Summary of methodology
  Annotated bibliography of available technical references
  Summary explanation of relevant government regulations and programs

---

Source: FHWA, 1983.

capital budgeting need to be understood and applied, but the evaluator needs to have a basic knowledge of microeconomics and equity (Bauer and Gergen, 1968; de Neufville, 1971; Quade, 1975). More recent references on this subject are texts by Pearce (1983) and Schofield (1989). An appendix providing elements of engineering economics may be consulted.

REFERENCES

BAUER, R. A., and K. J. GERGEN (Eds.) (1968). *The Study of Policy Analysis*, W. W. Norton, New York.

DE NEUFVILLE, RICHARD, and J. H. STAFFORD (1971). *Systems Analysis for Engineers and Managers*, McGraw-Hill, New York.

FEDERAL HIGHWAY ADMINISTRATION (FHWA) (1983). *Transportation Planning and Problem Solving for Rural Areas and Small Towns*,

Student Manual, U.S. Department of Transportation, Washington, DC.

LICHFIELD, N., ET AL. (1975). *Evaluation in the Planning Process*, Pergamon Press, Oxford.

MANHEIM, M. L., ET AL. (1975). *Transportation Decision-Making: A Guide to Social and Environmental Consideration*, NCHRP Report 156, Transportation Research Board, National Research Council, Washington, DC.

McALLISTER, D. M. (1982). *Evaluation in Environmental Planning*, The MIT Press, Cambridge, MA.

MEYER, M. D., and E. J. MILLER (1984). *Urban Transportation Planning*, McGraw-Hill, New York.

MISHAN, E. T. (1976). *Cost-Benefit Analysis*, Praeger, New York.

MITCHELL, ROBERT L. (1980). *Engineering Economics*, John Wiley, New York.

QUADE, E. S. (1975). *Analysis for Public Decisions*, Elsevier North-Holland, New York.

SCHOFIELD, J. A. (1987). *Cost-Benefit Analysis in Urban and Regional Planning*, Unwin Hyman, London, UK.

STOKEY, E., and RICHARD ZECKHAUSER (1978). *A Primer for Policy Analysis*, W. W. Norton, New York.

THOMAS, E. N., and J. L. SCHOFER (1970). *Strategies for the Evaluation of Alternative Transportation Plans*, NCHRP Report 96, Highway Research Board, National Research Council, Washington, DC.

EXERCISES

1. A clever community fund-raiser collects funds on the basis that each successive contributor will contribute 1/20 of the total collected to date. The collector makes the first contribution of $100.
   (a) What is expected from the 150th contributor?
   (b) How many contributors does she need to cross the $1 million target?

2. What is the rate of return if $2000 is invested to bring in a net profit of $450 for each of the six succeeding years?

3. A person invests $3000 anticipating a net profit of $200 by the end of the first year, and this profit increasing by 25% for each of the next 6 years. What is the rate of return?

4. A transit system is estimated to cost $2,000,000 with operating and maintenance costs of $28,000 per year after the first year, and 10% incremental increases over the next 12 years. At the end of the sixth year, it is anticipated that new equipment and buses will be needed at a total cost of $3,000,000. Calculate the present worth of costs for this system if the rate of return is $9\frac{1}{2}$%.

5. Estimate the total present sum of money needed to finance and maintain a small transit system, given the following details: initial cost of buses, $2,150,000; annual maintenance cost of buses, $80,000; initial cost of shelters, workshops, etc., $1,500,000; annual maintenance cost of shelters, $40,000; initial cost of equipment, $38,000; annual upkeep of equipment, $13,000; estimated life of buses 7 years; estimated life of shelters 10 years; estimated life of equipment 5 years; interest rate 10%.

6. A small section of a highway can be dealt with in three different ways: (1) it can be left as it is (do nothing), (2) an embankment–bridge combination can be used, or (3) a cutting tunnel combination can be provided. Using whatever method is suitable, determine which alternative has the best financial advantage.

Alternative	Life	Initial cost	Maintenance
1. As is (do nothing)	Perpetual	—	$35,000
2. Embankment–bridge	Perpetual 50 years	$170,000 $200,000	$1000/year $2000/year
3. Cutting tunnel	Perpetual 50 years	$75,000 $300,000	$700/year $2000/year

7. In Example 3, objectives 1 through 5 have certain weights assigned. Suppose that these weights were reassigned as follows:

Objective 1: weight 9 out of 10
Objective 2: weight 8 out of 10
Objective 3: weight 7 out of 10
Objective 4: weight 6 out of 10
Objective 5: weight 5 out of 10

What would be the result of your analysis?

8. A small bus company has worked out the average cost per mile to run buses in the city with the resources it has.

Bus	Average cost per bus/mile
3	$9.00
5	8.40
7	8.00
9	8.00
11	8.70

What is the size of the bus fleet that you would recommend?

9. An old hill road (A) is used by vehicles and the current cost of vehicle and social costs amount to $3.00 million per year. The maintenance cost runs $20,000 per year. Two alternative routes (B and C) are now proposed and their details are given in the table. Discuss which alternative should be adopted, including the one using the old hill road.

Alternative	A	B	C
Interest on construction	—	450	900
Purchase land	—	200	180
Vehicle and social cost	3000	1500	900
Maintenance	20	10	15

10. Different configuration of a highway network for a new town have been proposed. The capital cost and user savings are as shown in the following table. Which configuration would be most economical?

Alternative	Length (mi)	Capital cost (millions of dollars)	User savings (millions of dollars)
A	32	120	130
B	40	121	170
C	45	125	185
D	49	132	195
E	54	140	198
F	59	150	210
G	66	160	212
H	70	172	220

11. A small city has four transportation alternatives for a bus depot based on three criteria: architectural worth, cost, and public opinion. The architectural features have been assessed by a set of 10 architects and urban planners. The cost has been assessed by an independent firm of construction specialists. Public opinion has been obtained through the elected officials of the city council. The results are as follows. The serviceable life of the depot is 40 years.

Alternative	Initial cost (millions of dollars)	Maintenance per year (thousands of dollars)	Architects in favor (out of 10)	Elected officials in favor (out of 25)
A	2.0	9.2	9	18
B	2.8	8.3	7	14
C	2.9	7.8	6	20
D	3.4	6.3	6	17

Provide an analysis for the decision maker to come to a rational decision and discuss your results.

12. A bus company serves passenger transport between two cities having a demand function $p = 1200 - 7q$, where p is the price in dollars/trip and q is the riders/day. Currently, the fare is \$30/trip. The company is planning to buy new buses for a total of \$1 million and reduce the fare to \$25/trip.

(a) What is the current and anticipated daily and annual ridership?

(b) What would be the change in consumer's surplus?

(c) If the design life of the new buses is 7 years and annual interest rate is 8%, what is the feasibility of this investment?

13. The local transportation department has six mutually exclusive proposals for consideration. Their expected life is 30 years at an annual interest rate of 8%. Details follow:

Project	1	2	3	4	5
Cost (millions of \$)	980	2450	6900	750	4750
Annual maintenance cost (millions of \$)	50	70	250	30	180
Annual benefit	150	50	120	95	85

(a) Rank-order the proposals based on increasing first cost, considering maintenance as a cost.

(b) Knowing that the budget of the department is limited, which project would you select?

(c) If the largest net benefit is the criterion, which project would you select?

14. Two highway routes having the following characteristics are being considered for construction:

Routes	A	B
Length (miles)	32	22
Initial cost (millions of $)	15	20
Maintenance cost (millions of $)/yr.	0.16	0.18
Vehicular traffic/day	10.000	15,000
Speed (mph)	40	30
Value of time (dollars/hr)	10	10
Operating cost of veh/mi (cents)	30	40

Compare the two alternative routes and recommend which one should be implemented.

Chapter 16

Transportation Safety

1. INTRODUCTION

One of the unfortunate by-products of modern transportation is the price that society pays for injury and loss of life. Over 60,000 people per year lose their lives in the United States in transportation accidents. About 90% of those fatalities are connected with highways. A rough idea of the distribution of the fatality rate among the modes of transportation in the United States is given in Table 16-1 in terms of 100 million passenger-miles (Morlock, 1978, p. 61).

These figures reveal that air travel is 150 times safer than highway travel. Because highway travel seems to be the most problematic of all the modes, and also because the highway system is the most ubiquitous among transportation modes in the United States, in this chapter, we concentrate heavily on the highway mode.

TABLE 16-1 FATALITY RATE/100 MILLION PASSENGER-MILES IN TRANSPORT IN THE UNITED STATES

Mode	Rate[a]
Autos and taxis	1.90
Motorcycles	17.00
Local transit	0.16
Buses	0.19
Railroads	0.53
Domestic scheduled air carriers	0.13
Water transport	NA

[a] NA: not available.
Source: TAA, 1974, p. 17.

TABLE 16-2   GENERAL STATISTICS ON ACCIDENTS

General statistics	
Deaths	51,500
Disabling injuries	2,000,000
Costs	$34.3 billion

Type of accident	Number of accidents	Number of drivers involved
Fatal	45,400	59,000
Disabling injury	1,400,000	2,400,000
Property damage and nondisabling injury	16,900,000	29,000,000
Total	18,300,000	31,500,000

Source: NSC, 1979.

2. THE HIGHWAY SAFETY PROBLEM

It is not easy to describe the highway safety problem. Engineers like to describe it as a combination of an engineering problem, an education problem, and an enforcement problem. In point of fact, the problem is much more complex, because one could obviously consider it to be a social problem, because of its profound effect on society. It could be considered, at the same time, as an economic problem, and the list could go on. The National Safety Council statistics for accidents in 1978 are shown in Table 16-2.

Regardless of which formula one uses to compute the monetary loss and the physiological and psychological stress due to these accidents, the true cost to society is so great that it defies comprehension. Yet society seems to accept this problem as a part of life.

Statistics can be used to identify areas of potential improvement. A survey of accident experience by accident type provides a preliminary indication of where one should direct resources. For illustrative purpose, 1978 data on fatal accidents nationwide are given in Table 16-3.

In reviewing these statistics, it is evident that there are certain types of accidents that contribute substantially to the highway safety problem. It appears logical to consider the potential effect that may be realized through concerted efforts in these areas.

3. SAFETY RESPONSIBILITY*

There are a large number of transportation-related disciplines that contribute to the safety or lack of safety of a given highway system. In this section, we attempt to enumerate some of these disciplines and the manner in which they influence highway safety.

Persons responsible for the administration of a highway system, whether they be city, county, state, or federal employees, have a very strong influence over the safety of the system. The administration makes decisions relative to the direction of programs,

---

* This section is drawn heavily from U.S. DOT, 1980, pp. 1.1-4 to 1.1-6.

TABLE 16-3 FATAL ACCIDENTS IN THE UNITED STATES

Type of accident		Percent of total fatalities
Pedestrian–vehicle collisions		18
Vehicle–vehicle collisions		44
Head-on	11	
Rear-end	6	
Angle	19	
Sideswipe	8	
Train–vehicle collisions		2
Fixed-object collisions		7
Run-off-the-road; overturn in road (loss of control)		26.8
Bicycle–vehicle collisions		2
Others		0.2

Source: NSC, 1979.

and these programs may be designed to accomplish a wide range of goals and objectives. If a high degree of emphasis is placed on safety within the direction of the programs, certain levels of safety improvement will be achieved.

Perhaps very few people think in terms of the planner as having a direct or substantial influence on the safety of a highway system. On the contrary, safety may begin or end with the decisions made in the planning process. For example, the planner assumes responsibility for street network layout. Planners also assume a role of establishing and controlling land-use policies. Certainly, the distribution and concentration of land use has a great deal to do with the intensity of traffic and consequently, with the safety of a particular street or intersection.

Geometric design is, in a way, the implementation of the planning process. This is the point at which the layouts envisioned by the planner are given dimension. The requirements of design and ultimately of operation must be incorporated at the planning stage. The designer must employ established criteria that are based on operational requirements. The designer must evaluate trade-offs and make decisions as to which alternatives are most practical within the constraints imposed by the administration policy and the basic plan established by the planner. The safety of the facility depends to a great degree on how well the designer utilizes existing technology and how a priority is given to safety criteria.

The construction of the highway is the implementation of planning and design decisions. It is important that construction carry out the intent of the planning and design process as nearly as possible, particularly in areas that relate to operational efficiency and safety.

The construction agency has responsibility for operating traffic concurrently with the reconstruction of a facility. The manner in which detours are designed and operated determines to a great degree the level of safety that will be achieved. Construction agencies consult those responsible for traffic operations and then carry out the directives of this group as closely as possible. Too frequently, detours are thought of in terms of temporary situations. In extensive construction jobs, these should be viewed more as permanent arrangements, as they can last for several months or even several years.

Traffic operations is the implementation of planning and design decisions. It can be viewed as the regulation of the use of a facility. The driver information and traffic control systems must ensure that the safety benefits designed into the facility are achieved. Conversely, traffic control and information systems designed by the operations engineer cannot correct planning and design deficiencies. At best, the effects of a bad condition can only be minimized. To correct these deficiencies requires operations input during the planning and design stages.

Safety is a primary objective in all maintenance. Changing the basic design should not be undertaken without consulting the designer regarding changes, and obtaining necessary concurrence.

The basic objective of enforcement is the safety of the motoring public and pedestrians. Enforcing the regulations in accordance with operational intent of the regulation is most important. Enforcement personnel can add greatly to the safety improvements by identifying trouble spots. Problems in interpreting and negotiating difficult roadway geometry frequently are first detected by enforcement officers. By having a good working relationship between engineering and enforcement, many of these problem areas can be located and corrected before major accidents occur.

With all of these different disciplines involved, highway safety is obviously a cooperative effort. By proper selection of a review team, several different disciplines may be represented, and their input may be most valuable.

4. TYPICAL ACCIDENT CATEGORIES

A "traffic accident" is the commonplace word used to describe a failure in the performance of one or more of the driving components, resulting in death, bodily injuries, and/or property damage. Highway and street accidents can be categorized into at least four categories: (1) multiple vehicle, (2) single vehicle, (3) vehicle–pedestrian, and (4) vehicle–fixed object. The distribution of such accidents varies greatly (Lay, 1986).

In general, high-accident locations in urban areas may be a consequence of high urban density, resulting in high traffic concentrations and congestion. Combinations of factors connected with the driver, the vehicle, and the road (street or highway) are responsible for accidents. Sabey (1980) provides the breakdown shown in Table 16-4.

Accident-based studies involve the development of statistical summaries of the accident data by various characteristics, in order to detect abnormal accident trends. These data are obtained from accident reports. See Figure 16-1 for an accident report form and Table 16-5 for typical accident characteristic categories.

Haddon (1980) has provided a useful framework for analyzing accident situations. His matrix (Table 16-6) indicates countermeasures to cope with dilemma posed by the three components of the highway system.

TABLE 16-4   PRIME CAUSES OF ROAD
ACCIDENTS

Cause	Percent of accidents
Human factors alone	65
Human + road	25
Human + vehicle	5
Road factors alone	2
Vehicle factors alone	2
Human + road + vehicle	1
Total	100

Source: Sabey, 1980.

5. THE HIGHWAY SAFETY IMPROVEMENT PROGRAM*

Highway safety professionals have long recognized the need for an organized approach to the correction of highway safety problems. The recent emphasis on highway safety has led to the availability of funding for the application of procedures to enhance highway safety efforts at the state and local levels. The objectives of these procedures are the efficient use and allocation of available resources and the improvement of techniques for data collection, analysis, and evaluation.

5.1 An Overview

An overview of the Highway Safety Improvement Program (HSIP) is given in Figure 16-2 with respect to the total highway activities of a highway agency. These basic highway activities include the following:

1. Planning and design
2. Construction
3. Safety
4. Operation and maintenance

The safety aspects of highways are handled by maintaining an effective HSIP, consisting of the following three components:

1. Planning
2. Implementation
3. Evaluation

* This discussion is based on U.S. DOT, 1981.

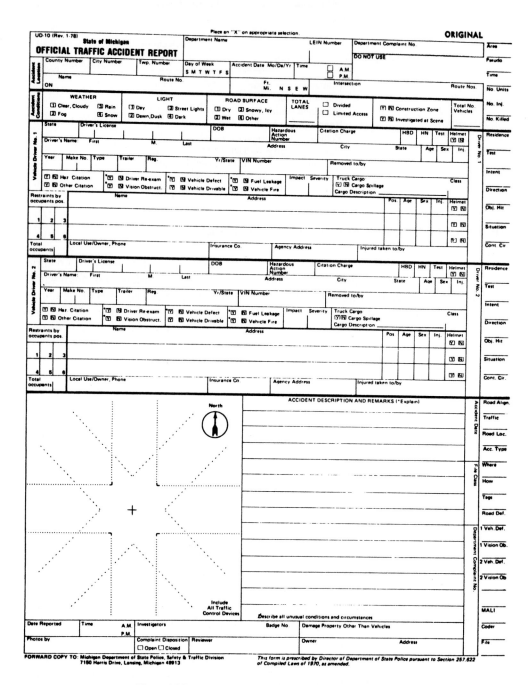

Figure 16-1 Typical Accident Report Form (FHWA, 1981c).

TABLE 16-5   TYPICAL ACCIDENT CHARACTERISTIC CATEGORIES

Description	Categories
Summary by type	1. Left-turn, head-on 2. Right-angle 3. Rear-end 4. Sideswipe 5. Pedestrian-related 6. Run-off-road 7. Fixed object 8. Head-on 9. Parked vehicle 10. Other

Summary by severity	1. Fatal
	2. Personal injury
	  • Incapacitating
	  • Nonincapacitating
	  • Possible injury
	3. Property damage

Summary by contributing circumstances	1. Driving under the influence of alcohol or drugs 2. Reckless or careless driving 3. Ill, fatigued, or inattentive 4. Failure to comply with license restrictions 5. Obscured vision 6. Defective equipment 7. Lost control due to shifting load, wind, or vacuum
Summary by environmental conditions	1. Weather (clear, cloudy, rain, fog, snow) 2. Ambient light (light, dark, dawn, dusk, street lights) 3. Roadway surface (dry, wet, snowy, ice)
Summary by time of day	1. 12:00 midnight–1:00 A.M. 2. 1:00 A.M.–2:00 A.M. 3. 2:00 A.M.–3:00 A.M. 4. 3:00 A.M.–4:00 A.M. ⋮ 24. 11:00 P.M.–12:00 midnight

Source: FHWA, 1981c.

The structure of the HSIP, as shown in Figure 16-3, is established in terms of components, processes, subprocesses, and procedures, and these are defined as follows:

Components: the three general phases of the HSIP: (1) planning, (2) implementation, and (3) evaluation, as shown in the dashed rectangle in Figure 16-3.

Processes: sequential elements within each component. For instance, the four processes within the planning component include (1) collect and maintain data;

TABLE 16-6   HADDON'S COUNTERMEASURE MATRIX

	Before crash	In crash	After crash
Driver	Training, education, behavior (e.g., avoidance of drinking and driving), attitudes, conspicuous clothing (pedestrians, cyclists)	In-vehicle restraints fitted and worn	Emergency medical services
Vehicle	Primary safety (e.g., braking, roadworthiness, visibility), speed, exposure	Secondary safety (e.g., impact protection)	Salvage
Road	Delineation, road geometry, surface condition, visibility	Roadside safety (e.g., no hazardous poles)	Restoration of road and traffic devices
Pedestrian	Education, clothing, behavior, attitude	—	Emergency medical services

Source: Lay, 1986.

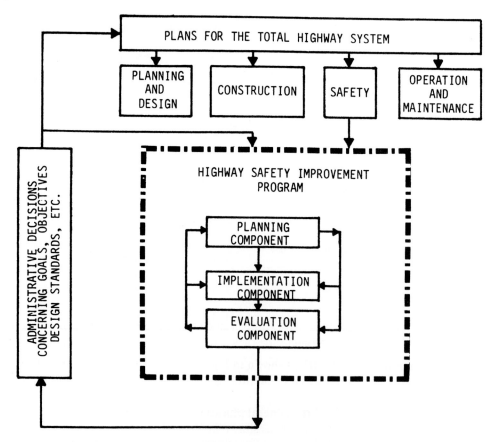

Figure 16-2   Overview of the Highway Safety Improvement Program (FHWA, 1981c).

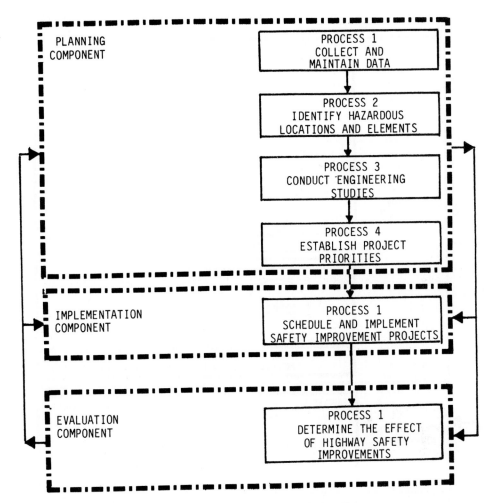

Figure 16-3   Highway Safety Improvement Program at the Process Level (FHWA, 1981c).

(2) identify hazardous locations and elements; (3) conduct engineering studies; and (4) establish project priorities. Process details are shown in Figure 16-3.

Subprocesses: activities that are contained within certain processes. For example, under process 3 of the planning component ("conduct engineering studies"), the three subprocesses are (1) collect and analyze data at hazardous locations; (2) develop candidate countermeasures, and (3) develop projects. Subprocess details are shown in Figure 16-4.

Procedures: possible ways in which each of the processes or subprocesses may be attained. For example, the procedures for identifying hazardous locations and

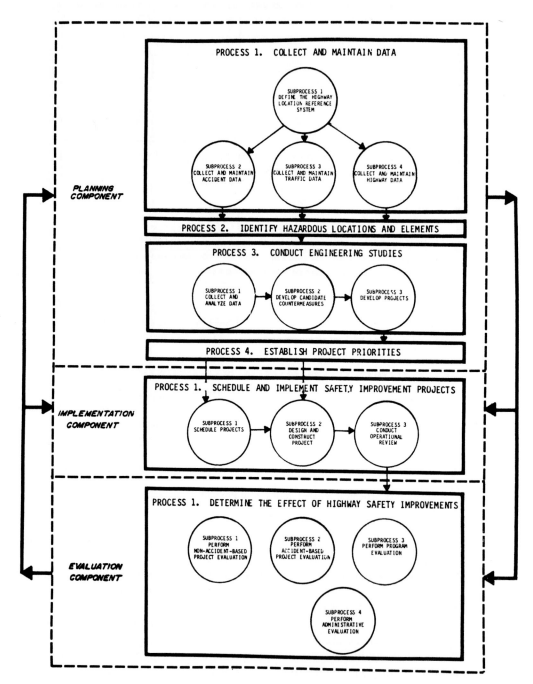

Figure 16-4   Highway Safety Improvement Program at the Subprocess Level (FHWA, 1981c).

elements (process 2) include the frequency method, rate method, rate quality control method, and so on. Details of each process are provided in Figure 16-5.

The planning component consists of the following:

1. A process for collecting and maintaining a record of accident, traffic, and highway data
2. A process for analyzing available data to identify highway locations of features determined to be hazardous on the basis of accident experience or potential
3. A process of conducting engineering studies of hazardous locations and features to develop proposed highway safety improvements
4. A process for establishing priorities for implementing proposed safety improvements

The implementation component entails a process for programming and implementing safety improvements. The evaluation component consists of a process for determining the effect that safety improvements have in reducing the number and severity of accidents and potential accidents. To be included in this process are the following:

1. The cost of, and the safety benefits derived from, the various means and methods used to mitigate or eliminate hazards.
2. A record of accident experience before and after the implementation of a safety improvement.
3. A comparison of accident number, rates, and severity observed after the implementation of a safety improvement with the accident numbers, rates, and severity expected if the improvement had not been made. Note that safety improvements should first be planned, then implemented, and finally, evaluated.

A list of procedures under each process and subprocess is developed based on the following:

1. Widely accepted practices currently in use by various highway agencies
2. Procedures developed and/or used by one or more highway agencies that may offer a useful method under certain conditions
3. New or untested concepts reported in the literature that may offer a worthwhile alternative to existing procedures and deserves further testing for possible future use.

Sixty-four specific procedures are listed in Figure 16-5 as they pertain to the processes and subprocesses of the HSIP. The object in providing this list is to impress on the reader the complexity of the highway safety problem. It is not possible to give details of all 64 procedures, but a few of the more important ones are described in this chapter.
Highway safety improvements may be arranged in the following hierarchy.

PLANNING COMPONENT

<u>Process 1</u> - Collect and Maintain Data

    Subprocess 1 - Define the Highway Location Reference System

        Procedure 1 - Milepost Method
        Procedure 2 - Reference Point Method
        Procedure 3 - Link Node Method
        Procedure 4 - Coordinate Method
        Procedure 5 - LORAN-C Based Method

    Subprocess 2 - Collect and Maintain Accident Data

        Procedure 1 - File of Accident Reports by Location
        Procedure 2 - Spot Maps
        Procedure 3 - Systemwide Computerization of Accident Data

    Subprocess 3 - Collect and Maintain Traffic Data

        Procedure 1 - Routine Manual Collection of Systemwide Traffic
                   Data
        Procedure 2 - Use of Mechanical Volume Counters
        Procedure 3 - Permanent Count Stations
        Procedure 4 - Maintenance of Traffic Data on Maps or in Files
        Procedure 5 - Systemwide Computerization of Traffic Data

    Subprocess 4 - Collect and Maintain Highway Data

        Procedure 1 - Systemwide Manual Collection of Highway Data
        Procedure 2 - Photologging and Videologging
        Procedure 3 - Maintenance of Highway Data on Maps or in Files
        Procedure 4 - Systemwide Computerization of Highway Data

<u>Process 2</u> - Identify Hazardous Locations and Elements

        Procedure 1 - Frequency Method
        Procedure 2 - Accident Rate Method
        Procedure 3 - Frequency Rate Method
        Procedure 4 - Rate Quality Control Method
        Procedure 5 - Accident Severity Method
        Procedure 6 - Hazard Index Method
        Procedure 7 - Hazardous Roadway Features Inventory

<u>Process 3</u> - Conduct Engineering Studies

    Subprocess 1 - Collect and Analyze Data at Identified Hazardous
                Locations

        Procedures 1-5   - Accident Studies
        Procedures 6-14 - Traffic Studies
        Procedures 15-20 - Environmental Studies
        Procedures 21-24 - Special Studies

Figure 16-5   Procedures Used in the Various Processes and Subprocesses (FHWA, 1981c).

Subprocess 2 - Develop Candidate Countermeasure(s)

    Procedure 1 - Accident Pattern Tables
    Procedure 2 - Fault Tree Analysis
    Procedure 3 - Multi-Disciplinary Investigation Team

Subprocess 3 - Develop Projects

    Procedure 1 - Cost-Effectiveness Method
    Procedure 2 - Benefit-to-Cost Ratio Method
    Procedure 3 - Rate-of-Return Method
    Procedure 4 - Time-of-Return Method
    Procedure 5 - Net Benefit Method

<u>Process 4</u> - Establish Project Priorities

    Procedure 1 - Project Development Ranking
    Procedure 2 - Incremental Benefit-to-Cost Ratio
    Procedure 3 - Dynamic Programming
    Procedure 4 - Integer Programming

IMPLEMENTATION COMPONENT

<u>Process 1</u> - Schedule and Implement Safety Improvement Projects

Subprocess 1 - Schedule Projects

    Procedure 1 - Gantt Charts
    Procedure 2 - Program Evaluation and Review Technique (PERT)
    Procedure 3 - Critical Path Method (CPM)
    Procedure 4 - Multiproject Scheduling System

Subprocess 2 - Design and Construct Projects

Subprocess 3 - Conduct Operational Review

EVALUATION COMPONENT

<u>Process 1</u> - Determine the Effect of Highway Safety Improvements

    Subprocess 1 - Perform Accident-Based Project Evaluation
    Subprocess 2 - Perform Non-Accident-Based Project Evaluation
    Subprocess 3 - Perform Program Evaluation
    Subprocess 4 - Perform Administrative Evaluation

Figure 16-5   *(continued)*

1. A *countermeasure* is a specific activity or set of related activities designed to contribute to the solution of an identified safety problem at a single location. Examples of countermeasures are (a) an advance warning sign installation, (b) an impact attenuator installation, or (c) a left-turn prohibition during peak traffic periods at a signalized intersection.

2. A *project* is one or more countermeasures designed to reduce identified safety deficiencies at a highway location. For example, pavement deslicking may be

selected as a single countermeasure to reduce wet-weather accidents at a site, and is called a project. Also, the combination of countermeasures at a site, such as shoulder stabilization, edge lining, and fixed-object removal, is also considered a project.

3. A *program* is a group of projects, countermeasures, and/or activities that are implemented to achieve a common highway safety goal. A program may be applied to numerous locations and may include several types of countermeasures that serve the same purpose.

5.2 Details of HSIP

Now that the framework of HSIP has been described, the broad details of each of the processes, subprocesses, and procedures are examined.

Collect and Maintain Data.   The purpose of this process is to supply the necessary database for the total highway system for use in identifying hazardous locations and elements. The proper identification of such locations is accomplished by use of one of several highway location reference systems. For instance, the milepost method uses a numerical value to represent the distance from a base point to any location on the highway through milepost markers. In recent years, the coordinate method for locating accident sites, using a unique set of plane coordinates, has been utilized. Electronic digitizers are used to produce computerized data files (TRB, 1974).

Accident data are collected, sorted, and processed for use in the identification and subsequent analysis of high-accident locations. The primary source is from police-reported accident reports. Each state has its own system. Reports by location and spot maps are easy to use. Figure 16-1 shows a typical accident report form, and Figure 16-6 is a computerized spot map.

All types of systemwide traffic-related data are routinely collected and maintained for use in the identification of hazardous locations. Because manual data collection on a systemwide basis is very expensive, the use of mechanical volume counters is often preferred. If long-term monitoring of traffic data is desired, one can establish permanent count stations. Continuous traffic volume data from these stations can be used to determine volume fluctuations by time of day, day of week, and so on. Adjustment factors for computing average annual daily traffic (AADT) at other locations, where only short-term volume data are available, can be used. After traffic data are collected, efficient storage and maintenance of such data are most important. Computerized storage and retrieval of large data files are quite common.

The collection and maintenance of highway-related data are accomplished through information on hazardous highway features, such as inadequate vertical and horizontal curves and poor sight distance. Procedures for such data collection are done through manual collection, photo- and videologging, and computerization of highway data. Photologging is a technique of photographing the highway and its environment at equal increments of distance from a moving vehicle, resulting in a visual inventory (Datta and Madsen, 1978).

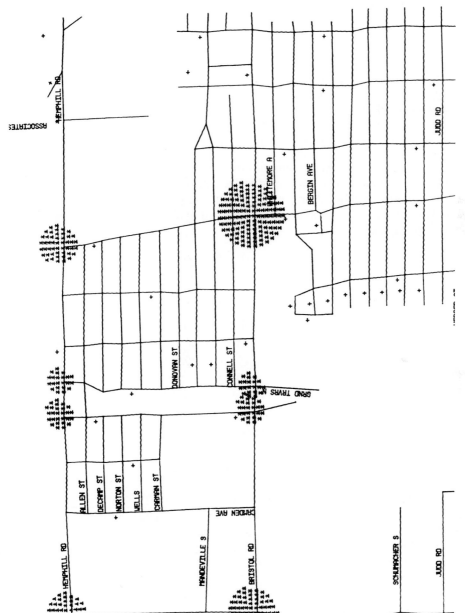

Figure 16-6 Computerized Spot Map (FHWA, 1981c).

Identify Hazardous Locations and Elements. The overall purpose is to identify hazardous spots, sections, and elements based on the accident, traffic, and highway data obtained as described before. Hazardous highway locations need not necessarily be high-accident locations.

There are currently seven procedures in use to accomplish this purpose, ranging in complexity and data usage. Table 16-7 lists the procedures and gives data needs for these seven procedures. Brief descriptions follow.

1. *Frequency method* is used to identify and rank locations on the basis of the number of accidents. A critical value can be established for location selection, such as 10 or more accidents per year (which includes all types of accidents). Highway systems of 2500 miles or less can generally adopt this method. There is no consideration given to exposure (i.e., traffic volumes) or accident severity.

2. *Accident rate method* combines the accident frequency with vehicle exposure (i.e., traffic volume) and is expressed as "accidents per million vehicles for intersections" (and other spot locations) or "accidents per million vehicle-miles of travel" for highway sections. The locations are then ranked in descending order by accident rate. For spot locations:

$$R_{sp} = \frac{A(1,000,000)}{365(TV)} \tag{1}$$

For road sections:

$$R_{se} = \frac{A(1,000,000)}{365(TVL)} \tag{2}$$

where

     R_{sp} = accident rate at a spot (accidents/million vehicles)
     R_{se} = accident rate at section (accidents/million vehicles)
     A = number of accidents for the study period

TABLE 16-7   PROCEDURES TO IDENTIFY HAZARDOUS SPOTS

Data input		Procedure					
	Freq.	Acc. rate	Freq. rate	Rate qual. con.	Acc. sev.	Haz. index	Haz. road. feat.
Accident summaries	×	×	×	×	×	×	
Traffic volume data		×	×	×		×	
Accident severity					×	×	
Average accident experience	×	×	×	×	×	×	
Statistical constants				×			
Other locational data						×	
Roadway features							×

T = period of study

V = AADT during study period; for intersections,

  = sum of entering volumes of all legs

L = length of the section (miles)

Highway systems of 10,000 miles or less could use this method.

3. *Frequency-rate method* is normally applied by first selecting a large sample of high-accident locations based on a "number of accidents" criteria (i.e., frequency method) from which accident rates are computed and the locations are priority-ranked by accident rate. A modified procedure is to plot accident frequency on the horizontal axis and accident rate on the vertical axis. Thus, each accident can be categorized by placing it in a matrix cell. See Figure 16-7 for an example. The upper right-hand corner denotes the most hazardous location.

4. *Rate quality control method* utilizes a statistical test to determine whether the accident rate at a particular location is significantly higher than a predetermined average rate for locations of similar characteristics, based on Poisson's distribution. The critical rate, which is based on the average systemwide accident rate for the highway type, is as follows:

$$R_c = R_a + K \left(\frac{R_a}{M}\right)^{\frac{1}{2}} \tag{3}$$

where

R_c = critical accident rate for a spot (accidents/10^6 veh) or section (accidents/10^6 veh-mi)

R_a = average accident rate for all spots of similar characteristics or on similar road types

M = millions of vehicles passing over a spot or millions of vehicle-miles of travel over a section

K = probability factor determined by the desired level of significance

This K value is determined by the probability that an accident rate is sufficiently large that it cannot be reasonably attributed to random occurrence. Selected values of K are as follows:

P (probability)	0.005	0.0075	0.05	0.075	0.10
K	2.576	1.960	1.645	1.440	1.282

The most commonly used values of K are 2.567 and 1.645 (Zegeer and Dean, 1977).

5. *Accident severity method* is used to identify and priority-rank high-accident locations. Accident severities are classified by the National Safety Council and many states within the following five categories:

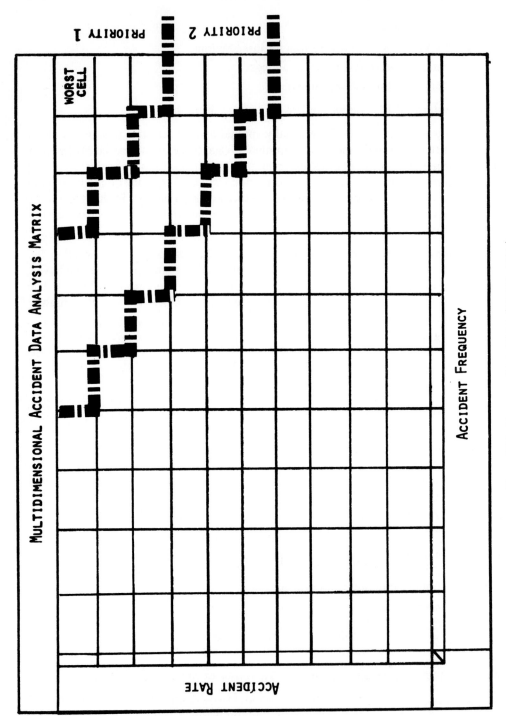

Figure 16-7 Frequency/Rate Matrix (FHWA, 1981c).

680

Fatal accident: one or more deaths (F)

A-type injury: incapacitating accident (A)

B-type injury: noncapacitating accident (B)

C-type injury: probable injury (C)

PDO: property damage only (PDO)

One of many severity methods uses the equivalent property damage only (EPDO) factor, where

$$EPDO = 9.5(F + A) + 3.5(B + C) + PDO \qquad (4)$$

where the letters indicate the numbers of each category. Locations are ranked based on their computed EPDO number (Deacon et al., 1975).

6. *Hazard index method* employs a formula to develop a rating index for each suspect site. Factors used in this method are shown in the example sheet (Figure 16-8). The raw data for each factor are converted to an indicator value through the use of a conversion graph. The indicator value is then multiplied by a weighting factor (based on a survey of professionals in the traffic safety field). The partial hazard indices are totaled to obtain the hazard index for the location (Taylor and Thomson, 1977).

7. *Hazardous roadway features inventory* is based largely on the comparison of existing roadway features with safety and design standards. Examples of such hazardous features are narrow bridges, steep roadside slopes, narrow lanes or shoulders, unprotected bridge overpasses, and so on.

Conduct Engineering Studies.   The purpose of these studies is to collect and analyze data at identified hazardous locations and then to select appropriate safety improvement projects. Subprocesses include the following:

- Analyses of data gathered (described before)
- Developing candidate countermeasures
- Developing projects

Four general categories of studies are needed to analyze locations and these are (1) accident-based studies, (2) traffic-operations-based studies, (3) environmental-based studies, and (4) special studies. A total of 24 studies have been identified within these categories and are shown in Table 16-8. The proper selection and use of these studies provide the information needed to determine the specific safety deficiencies at each location.

Site Number_____ Date_____

Type___Rural Intersection_____

Indicator	Data Value	Indicator Value		Weight		Partial H.I.'s
Number of Accidents	7.67 acc/yr	59	x	0.145	=	8.6
Accident Rate	2.47 acc/MEV	49	x	0.199	=	9.8
Accident Severity	$12,850 dollars	70	x	0.169	=	11.8
Volume/Capacity Ratio	0.17	22	x	0.073	=	1.6
Sight Distance Ratio	>2.0 (wt.avg.)	0	x	0.066	=	0.0
Traffic Conflict	--- conf/hr.)	--	x	0.053	=	---
Erratic Maneuvers	--- e.m./hr.	--	x	0.061	=	---
Driver Expectancy	2.19 (wt.avg.)	37	x	0.132	=	4.9
Info. System Deficiencies	2.79 (wt.avg.)	47	x	0.102	=	4.8
Sums:				0.886 *		41.5

$$H.I. = \frac{\text{Sum of Partial H.I.'s}}{\text{Sum of Applicable Weights}} = \frac{41.5}{0.886} = \underline{47}$$

Relative Strength of Evaluation

Sum of Applicable Weights x 100 =   89 %

Figure 16-8   Hazard Index Method Example (FHWA, 1981c).

Developing specific countermeasures for identified hazardous locations is the next step, based on the known deficiencies of the location. Based on a knowledge of the effectiveness of similar improvements in the past, several candidate improvements can be proposed for the same location. A sample general countermeasure for accident patterns and their probable causes is given in Table 16-9.

To determine the most desirable improvement alternatives at a location, it is necessary to estimate the expected project costs and accident benefits for each alternative. Factors that affect the calculation of accident benefits include accident costs, interest

TABLE 16-8   PROCEDURES AND STUDIES TO ANALYZE AND IDENTIFY HAZARDOUS LOCATIONS

Accident-based procedures (studies)
  Procedure 1: Accident summary by type
  Procedure 2: Accident summary by severity
  Procedure 3: Accident summary by contributing circumstances
  Procedure 4: Accident summary by environmental conditions
  Procedure 5: Accident summary by time of day

Traffic-operations-based procedures (studies)
  Procedure 6: Safety performance studies
  Procedure 7: Volume studies
  Procedure 8: Spot speed studies
  Procedure 9: Delay- and travel-time studies
  Procedure 10: Roadway and intersection capacity studies
  Procedure 11: Traffic conflict studies
  Procedure 12: Gap studies
  Procedure 13: Traffic lane occupancy studies
  Procedure 14: Queue-length studies

Environmental-based procedures (studies)
  Procedure 15: Roadway inventory studies
  Procedure 16: Sight-distance studies
  Procedure 17: Roadway serviceability studies
  Procedure 18: Skid resistance studies
  Procedure 19: Highway lighting studies
  Procedure 20: Weather-related studies

Special procedures (studies)
  Procedure 21: School crossing studies
  Procedure 22: Railroad crossing studies
  Procedure 23: Traffic control device studies
  Procedure 24: Bicycle and pedestrian studies

Source: TRB, 1978; FHWA, 1975; FHWA, 1980.

rates, project service life, accident reduction factors, and traffic growth rates. Sample accident costs are given in Table 16-10.

To accomplish the economic evaluation of improvement alternatives, one can use one or more of the following procedures: cost-effectiveness method, benefit–cost method, rate-of-return method, and the net-benefit method. The first two methods are the ones most used. The cost-effectiveness method is based on the computation of a cost for achieving a given unit of benefit (or effect)—a reduction in accidents. A sample cost-effectiveness analysis worksheet is given in Figure 16-9. The benefit–cost ratio is the monetary accident savings divided by the improvement cost, projects with B/C ratios greater than 1.0 being considered economically sound. A sample benefit–cost analysis worksheet is given in Figure 16-10.

Establish Project Priorities.   The purpose of this task is to establish a priority list of the countermeasures developed in earlier processes. The final outcome should result in the selection of improvements that will culminate in the optimal safety benefits per dollar spent.

TABLE 16-9 GENERAL COUNTERMEASURES FOR ACCIDENT PATTERNS AND THEIR PROBABLE CAUSES

Accident pattern	Probable cause	General countermeasure
Right-angle collisions at unsignalized intersections	Restricted sight distance	Remove sight obstructions Restrict parking near corners Install stop signs (see MUTCD) Install warning signs (see MUTCD) Install/improve street lighting Reduce speed limit on approaches[a] Install signals (see MUTCD) Install yield signs (see MUTCD) Channelize intersection
	Large total intersection volume	Install signals (see MUTCD) Reroute through traffic
	High approach speed	Reduce speed limit on approaches[a] Install rumble strips
Right-angle collisions at signalized intersections	Poor visibility of signals	Install advanced warning devices (see MUTCD) Install 12-in. signal lenses (see MUTCD) Install overhead signals Install visors Install back plates Improve location of signal heads Add additional signal heads Reduce speed limit on approaches[a]
	Inadequate signal timing	Adjust amber phase Provide all-red clearance phases Add multidial controller Install signal actuation Retime signals Provide progression through a set of signalized intersections
Rear-end collisions at unsignalized intersections	Pedestrian crossing	Install/improve signing or marking of pedestrian crosswalks Relocate crosswalk
	Driver not aware of intersections	Install/improve warning signs
	Slippery surface	Overlay pavement Provide adequate drainage Groove pavement Reduce speed limit on approaches[a] Provide "slippery when wet" signs
	Large numbers of turning vehicles	Create left- or right-turn lanes Prohibit turns Increase curb radii
Rear-end collisions at signalized intersections	Poor visibility of signals	Install/improve advance warning devices Install overhead signals Install 12-in. signal lenses (see MUTCD) Install visors Install back plates Relocate signals Add additional signal heads

TABLE 16-9 (*continued*)

Accident pattern	Probable cause	General countermeasure
	Inadequate signal timing	Remove obstacles
		Reduce speed limit on approaches[a]
		Adjust amber phase
		Provide progression through a set of signalized intersections
	Pedestrian crossings	Install/improve signing or marking of pedestrian crosswalks
		Provide pedestrian "walk" phase
	Slippery surface	Overlay pavement
		Provide adequate drainage
		Groove pavement
		Reduce speed limit on approaches[a]
		Provide "slippery when wet" signs
	Unwarranted signals	Remove signals (see MUTCD)
Pedestrian accidents at intersections	Large turning volumes	Create left- or right-turn lanes
		Prohibit turns
		Increase curb radii
	Restricted sight distance	Remove sight obstructions
		Install pedestrian crossings
		Improve/install pedestrian crossing signs
		Reroute pedestrian paths
	Inadequate protection for pedestrians	Add pedestrian refuge islands
	Inadequate signals	Install pedestrian signals (see MUTCD)
	Inadequate signal phasing	Add pedestrian "walk" phase
		Change timing of pedestrian phase
	School crossing area	Use school crossing guards
Pedestrian accidents between intersections	Driver has inadequate warning of frequent midblock crossings	Prohibit parking
		Install warning signs
		Lower speed limit[a]
		Install pedestrian barriers
	Pedestrians walking on roadway	Install sidewalks
	Long distance to nearest crosswalk	Install pedestrian crosswalk
		Install pedestrian actuated signals (see MUTCD)
Pedestrian accidents at driveway crossings	Sidewalk too close to traveled way	Move sidewalk laterally away from highway
Left-turn collisions at intersections	Large volume of left turns	Provide left-turn signal phases
		Prohibit left turns
		Reroute left-turn traffic
		Channelize intersection
		Install stop signs (see MUTCD)
		Create one-way streets
		Provide turning guidelines (if there is a dual left-turn lane)

(*continued*)

TABLE 16-9   *(continued)*

Accident pattern	Probable cause	General countermeasure
	Restricted sight distance	Remove obstacles Install warning signs Reduce speed limit on approaches[a]
Right-turn collisions at intersections	Short turning radii	Increase curb radii
Fixed-object collisions	Objects near traveled way	Remove obstacles near roadway Install barrier curbing Install breakaway feature to light poles, signposts, etc. Protect objects with guardrail
Fixed-object collisions and/or vehicles running off roadway	Slippery pavement	Overlay existing pavement Provide adequate drainage Groove existing pavement Reduce speed limit[a] Provide "slippery when wet" signs
	Roadway design inadequate for traffic conditions	Widen lanes Relocate islands Close curb lane
	Poor delineation	Improve/install pavement markings Install roadside delineators Install advance warning signs (e.g., curves)
Sideswipe collisions between vehicles traveling in opposite directions or head-on collisions	Roadway design inadequate for traffic conditions	Install/improve pavement markings Channelize intersections Create one-way streets Remove constrictions such as parked vehicles Install median divider Widen lanes
Collisions between vehicles traveling in same direction such as sideswipe, turning, or lane changing	Roadway design inadequate for traffic conditions	Widen lanes Channelize intersections Provide turning bays Install advance route or street signs Install/improve pavement lane lines Remove parking Reduce speed limit[a]
Collisions at driveways	Left-turning vehicles	Install median divider Install two-way left-turn lanes
	Improperly located driveway	Regulate minimum spacing of driveways Regulate minimum corner clearance Move driveway to side street Install curbing to define driveway location Consolidate adjacent driveways
	Right-turning vehicles	Provide right-turn lanes Restrict parking near driveways Increase the width of the driveway Widen through lanes Increase curb radii

TABLE 16-9   *(continued)*

Accident pattern	Probable cause	General countermeasure
	Large volume of through traffic	Move driveway to side street Construct a local service road Reroute through traffic
	Large volume of driveway traffic	Signalize driveway Provide acceleration and deceleration lanes Channelize driveway
	Restricted sight distance	Remove sight obstructions Restrict parking near driveway Install/improve street lighting Reduce speed limit[a]
Night accidents	Poor visibility	Install/improve street lighting Install/improve delineation markings Install/improve warning signs
Wet-pavement accidents	Slippery pavement	Overlay with skid resistant surface Provide adequate drainage Groove existing pavement Reduce speed limit[a] Provide "slippery when wet" signs

[a] Spot speed study should be conducted to justify speed limit reduction.

TABLE 16-10   SAMPLE ACCIDENT COSTS

Source	Accident severity	Cost per involvement
NSC (1979)	Fatal	$160,000
	Nonfatal disabling injury	6200
	Property damage (including minor injuries)	870
NHTSA (1975)	Fatality	$287,175
	Critical injury	192,240
	Severe injury—life threatening	89,955
	Severe injury–not life threatening	8085
	Moderate injury	4350
	Minor injury	2190
	Average injury	3185
	Property damage only	520

Source: FHWA, 1981b.

There are at least four methods available: project development ranking, incremental benefit–cost ratio, dynamic programming, and integer programming. A wide range of complexities is involved among the various procedures. Project development ranking is simplest and can be done manually. The other three usually require the use of a computer (Phillips et al., 1976).

Implementation.   The implementation component includes the design, scheduling, and construction of the highway safety improvement selected in the previously

Evaluation No:

Project No:

Date/Evaluator:

1. Initial Implementation Cost, I: $ _____

2. Annual Operating and Maintenance Costs Before Project Implementation: $ _____

3. Annual Operating and Maintenance Costs After Project Implementation: $ _____

4. Net Annual Operating and Maintenance Costs, K (3-2): $ _____

5. Annual Safety Benefits in Number of Accidents Prevented, $\bar{B}$:

Accident Type   Actual   -   Expected   =   Annual Benefit

Total _____ - _____ = _____

6. Service Life, n: _____ yrs.

7. Salvage Value, T: $ _____

8. Interest Rate: _____ % = 0. _____

9. EUAC Calculation:

CR_n^i = _____

SF_n^i = _____

$EUAC = I \, (CR_n^i) + K - T \, (SF_n^i)$

= _____

10. Annual Benefit:

$\bar{B}$ (from 5) = _____

11. $C\text{-}E = EUAC/\bar{B}$ = _____

12. PWOC Calculation:

PW_n^i = _____

SPW_n^i = _____

$PWOC = I + K \, (SPW_n^i) - T \, (PW_n^i)$

13. Annual Benefit

n (from 6) = _____ yrs.

$\bar{B}$ (from 5) = _____ accidents prevented per year

14. $C\text{-}E = PWOC \, (CR_n^i)/\bar{B}$

Figure 16-9 Sample Cost-Effectiveness Analysis Worksheet (FHWA, 1981c).

Evaluation No:

Project No:

Date/Evaluator:

1. Initial Implementation Cost, I: $ _____

2. Annual Operating and Maintenance
   Costs Before Project Implementation: $ _____

3. Annual Operating and Maintenance
   Cost After Project Implementation: $ _____

4. Net Annual Operating and
   Maintenance Costs, K (3-2): $ _____

5. Annual Safety Benefits in Number
   of Accidents Prevented:

   Severity   Actual - Expected = Annual Benefit

   a) Fatal Accidents
      (Fatalities)   _____

   b) Injury Accidents
      (Injuries)   _____

   c) PDO Accidents
      (Involvement)   _____

6. Accident Cost Values (Source _____):

   Severity   Cost

   a) Fatal Accident (Fatality) $ _____

   b) Injury Accident (Injury) $ _____

   c) PDO Accident (Involvement) $ _____

7. Annual Safety Benefits in Dollars Saved, $\bar{B}$:

   5a) x 6a) = _____

   5b) x 6b) = _____

   5c) x 6c) = _____

   Total   = $ _____

8. Services life, n: _____ yrs

9. Salvage Value, T: $ _____

10. Interest Rate, i: _____ % = 0. _____

11. EUAC Calculation:

    CR_n^i = _____

    SP_n^i = _____

    $EUAC = I (CR_n^i) + K - T (SP_n^i)$ = _____

12. EUAB Calculation:

    $EUAB = \bar{B}$

         = _____

13. $B/C = EUAB/EUAC$ = _____

14. PWOC Calculation:

    PW_n^i = _____

    SPW_n^i = _____

    $PWOC = I + K (SPW_n^i) - T (PW_n^i)$ = _____

15. PWOB Calculation:

    $PWOB = \bar{B}(SPW_n^i)$ = _____

16. $B/C = PWOB/PWOC$ = _____

Figure 16-10   Sample Benefit–Cost Analysis Worksheet (FHWA, 1981c).

689

discussed components. Gantt charts, program evaluation and review techniques (PERT), and multiple project scheduling systems (MPSS) provide the necessary tools to control and schedule projects by highway agencies. MPSS includes resource balancing and is a formal method of scheduling and monitoring the status of highway preconstruction and construction activities. One of the primary purposes of MPSS is to achieve optimum utilization of all available financial and personnel resources.

Evaluation.   Evaluation involves obtaining and analyzing quantitative information on the benefits and costs of highway safety improvements implemented. The utilization of estimated benefits and costs reduces the agency's dependence on engineering judgment and helps in selecting future projects with the highest probability of success. This in turn allows for better allocation of scarce safety funds and minimizes expenditures for projects that are marginal or ineffective. For further details, refer to *Highway Safety Evaluation* (FHWA, 1981a).

6. EXAMPLES

Example 1

A state government report declared that in 1985, the motor fuel consumption was 15.75 million gallons, there were 10,342 motor vehicle fatalities, about 1 million motor vehicle-related injuries, 22.825 million motor registrations, and a population of 56,250,000. Assume that fuel consumption = 17 miles per gallon for all vehicles. Calculate the vehicle-miles traveled, and the registration, population, and vehicle-mile death rate.

Solution

$$\text{Vehicle-miles of travel} = 15.75 \times 10^6 \text{ gal} \times 17 \text{ vehicle-miles/gal}$$

$$= 267.75 \times 10^6 \text{ vehicle-miles}$$

$$\begin{array}{c} \text{Registration death rate} \\ \text{(per 10,000 vehicles)} \end{array} = \frac{10{,}342 \times 10^4}{22.825 \times 10^6} = 4.53$$

$$\begin{array}{c} \text{Population death rate} \\ \text{(per 100,000 vehicles)} \end{array} = \frac{10{,}342 \times 10^5}{56.25 \times 10^6} = 18.3$$

$$\text{Death rate/million vehicle-miles} = \frac{10{,}342}{267.75} = 38.63$$

Discussion

Generally, vehicle-miles of travel are difficult to obtain, and so if one knows the total gasoline consumption per year and the average gas mileage for vehicles, one can calculate vehicle-miles of travel. The rates shown before are useful in keeping track of the state's (or region's) overall accident problems.

Example 2

A five-legged intersection has accident-related problems. In 1985, the number of accidents reported was 48 and the average 24-hour volume entering the intersection approaches was 900, 1230, 1560, 1435, and 980 vehicles. Determine the accident rate per million volume.

Solution

Accident rate/million vehicles entering

$$= \frac{48 \times 10^6}{(900 + 1230 + 1560 + 1435 + 980) \times 365} = 21.51$$

Discussion

If the breakdown of the 48 accidents was provided, one could possibly pin down the intersection approach(es) that was contributing to the accidents. A separate accident rate could also be worked out.

Example 3

A 4.5-mile highway section is reported to have had 95 accidents, of which 4 resulted in deaths and 22 in serious injuries. If the ADT on this section is 9500 vehicles, calculate the rate per 10 million vehicle-miles of total, fatal, and serious injuries.

Solution

$$\text{Rate/10} \times 10^6 \text{ vehicle-miles (total)} = \frac{95 \times 10 \times 10^6}{4.5 \times 9500 \times 365} = 60.88$$

$$\text{Rate/10} \times 10^6 \text{ vehicle-miles (fatal)} = \frac{4 \times 10 \times 10^6}{4.5 \times 9500 \times 365} = 2.56$$

$$\text{Rate/10} \times 10^6 \text{ vehicle-miles} = \frac{22 \times 10 \times 10^6}{4.5 \times 9500 \times 365} = 14.10$$

Example 4

Engineers prioritize highway and street improvements based on a systematic and comprehensive study of a large number of candidate locations where accidents are frequent. Hazardous sections of highways and street intersections are generally selected on the basis of numbers of accidents occurring within a stated period (say, one year). However, this method does not allow accidents to be weighted for severity. The severity-weighting process can take several forms. Weighting by the most serious injury to any one person using a severity scale is one way of weighting, where a fatal accident could be weighted 10 and a minor injury could be assigned a weight of 3.

Weighting by accident cost is another method of estimating the cost of accidents at a location. An average cost of $100,000 for a fatality, $5000 for a disabling injury, down to an average of $1000 per property damage could be used. These figures could change from year to year and from city to city.

The number of involvements (motor vehicles, pedestrians, bicycles, etc.) could be used instead of the number of accidents. Rates can be developed for each of these weighting procedures.

Table 16-11 shows a comparison of ranking systems used in practice. Notice that two principal methods of evaluating the risk at locations have been given: number of accidents or involvements and accident or involvement rates. As pointed out before, the rate or risk of accidents is the preferred indicator. This risk or hazard may be expressed as an accident rate.

Example 5

Forecasts of expected reductions of accidents at specific sites can be done from past experience. For example, the numbers of accidents that are expected to be prevented by specific countermeasures can be calculated by using the formula

$$S = \frac{PAT}{365}$$

where

A = number of accidents in T days in the past
P = fractional reduction (% reduction)
S = annual savings in involvements (accidents)

If several improvements $(P_1, P_2, P_3, \ldots, P_n)$ are proposed, the total fractional reduction is

$$P_t = 1 - (1 - P_1)(1 - P_2)(1 - P_3) - (1 - P_n)$$

where P_t is the fractional reduction of the combined improvements (or countermeasures). For example, a hill road may be improved by installing median barrier ($P_1 = 0.24$) and stabilizing shoulder ($P_2 = 0.38$). Here

$$P_1 = 1 - (1 - 0.24)(1 - 0.38) = 0.53$$

If $A = 25$ accidents and $T = 31$ days,

$$S = \frac{0.53 \times 25 \times 31}{365} = 1.125$$

If $V_1 = \text{ADT}_1 = 900$ and $V_2 = \text{ADT}_2 = 1200$, then

$$S = \frac{1.125 \times 1200}{900} = 1.5$$

SUMMARY

Modern transportation technology has helped society to gain mobility and accessibility but at an enormous price. Accidents on highways alone account for an enormous number of lives and property damage. Safety responsibility is shared by a wide range of people: the planner, the engineer, the operations and maintenance personnel, the enforcement officers, and, in a way, the public that uses the facilities.

TABLE 16-11 COMPARISON OF RANKING SYSTEMS

		Section 1	2	3	4	5	Total	Average
				Basic data				
Length (mi)	A	2.4	3.4	2.9	4.9	1.3	14.9	2.98
ADT × 10^{-2}	B	42	37	37	29	27		—
Accidents	C	25	13	9	5	3	55	11
Involvements	D	41	23	17	15	5	101	20.2
Fatal	E	0	0	0	0	1	1	0.2
Injured	F	6	3	3	0	1	13	2.6
Property damage	G	20	9	7	6	2	44	8.8
				Rates				
C/A	H	10.42	3.82	3.10	1.02	2.31	—	3.69
D/A	I	17.08	6.76	5.86	3.06	3.85	—	6.78
$A \times B \times 365 \times 10^{-4}$	J	3.68	4.59	3.92	5.19	1.28	18.66	3.73
Cost[a]	K	50.00	24.0	22.0	6.00	107	209	41.80
R_s[b]	L	7.07	2.61	2.55	1.16	3.12	←Accident rate	
	M	11.15	5.00	4.34	2.90	3.90	←Involvement rate	
				Ranked by:				
Acc. (C)	O	1	2	3	4	5		
Inv. (D)	P	1	2	3	4	5		
Costs (K)	Q	2	3	4	5	1		
Acc./month (H)	R	1	2	3	5	4		
Inv./month (I)	S	1	2	3	5	4		
Acc. rate (L)	T	1	3	4	5	2		
Number rate	U	1	2	0	0	0		

[a] Cost of accidents: $100,000/death, $5000/injury, and $1000/property damage.
[b] $R_s = A \times 10^6/365TVL$; A = accident or involvements.
Source: Baerwald, 1976.

In this chapter, we examined the nature and characteristics of accidents by type, severity, contributing circumstances, environmental conditions, and time of the day. Recognizing the need for an organized approach to solving the highway safety problem, highway engineers have emphasized the need to take up this entire subject in the recently published documents issued by the Federal Highway Administration through the Highway Safety Improvement Program (HSIP). In this chapter, we gave an overview of HSIP and its ramifications. Also, methods of identifying hazards locations and elements of the highway are discussed.

REFERENCES

BAERWALD, JOHN E. (Ed.) (1976). *Transportation and Traffic Engineering Handbook*, Prentice-Hall, Englewood Cliffs, NJ.

DEACON, J. A., ET AL. (1975). *Identification of Hazardous Rural Highway Locations*, Transportation Research Record 543, Transportation Research Board, National Research Council, Washington, DC.

FEDERAL HIGHWAY ADMINISTRATION (FHWA) (1975). *Manual on Identification, Analysis, and Correction of High Accident Location*, U.S. Department of Transportation, Washington, DC.

FEDERAL HIGHWAY ADMINISTRATION (FHWA) (1980). *Safety Design and Operational Practices for Streets and Highways*, Technology Sharing Report 80-228, U.S. Department of Transportation, Washington, DC.

FEDERAL HIGHWAY ADMINISTRATION (FHWA) (1981a). *Highway Safety Evaluation*, Procedural Guide, U.S. Department of Transportation, Washington, DC.

FEDERAL HIGHWAY ADMINISTRATION (FHWA) (1981b). *Highway Safety Engineering Studies*, Procedural Guide, U.S. Department of Transportation, Washington, DC.

FEDERAL HIGHWAY ADMINISTRATION (FHWA) (1981c). *Highway Safety Improvement Program*, FHWA-TS-81-218, U.S. Department of Transportation, Washington, DC.

HADDON, W. (1980). "Advances in the Epidemiology of Injuries as a Basis of Public Policy," *Public Health Reports*, Vol. 95, No. 5, pp. 411–421.

LAY, M. G. (1986). *Handbook of Road Technology*, Vols. 1 and 2, Gordon and Breach, New York.

MORLOCK, E. K. (1978). *Introduction to Transportation Engineering and Planning*, McGraw-Hill, New York.

NATIONAL SAFETY COUNCIL (NSC) (1980). *Accident Facts: 1979*, NSC, Chicago.

PHILLIPS, D. T., ET AL. (1976). *Operations Research: Principles and Practice*, John Wiley, New York.

SABEY, B. (1980). *Road Safety and Value of Money*, TRRL Supplementary Report SR 581, Transportation Research Board, UK.

TAYLOR, J. I., and H. T. THOMSON (1977). *Identification of Hazardous Locations*, Final Report FHWA-77-83, U.S. Department of Transportation, Federal Highway Administration, Washington, DC.

TRANSPORTATION ASSOCIATION OF AMERICA (TAA) (1974). *Transportation Facts and Trends*, TAA, Washington, DC.

TRANSPORTATION RESEARCH BOARD (TRB) (1974). *Highway Location Reference Methods*, NCHRP Synthesis Report 21, TRB, Washington, DC.

TRANSPORTATION RESEARCH BOARD (TRB) (1978). *Priority Programming and Project Selection*, NCHRP Synthesis Report 48, TRB, Washington, DC.

U.S. DEPARTMENT OF TRANSPORTATION (U.S. DOT) (1980). *Safety Design and Operational Practices for Street and Highways,* Technology Sharing Report 80-228, U.S. DOT, Washington, DC.

U.S. DEPARTMENT OF TRANSPORTATION (U.S. DOT) (1981). *Highway Safety Improvement Program*, FHWA-TS-81-21B, U.S. DOT, Washington, DC.

EXERCISES

1. A highway system has six sections of highways that appear to have problems. Their lengths, ADTS, and accidents are given.

	1	2	3	4	5	6
Length (mi)	3.3	2.5	9.2	3.4	5.6	6.8
Average daily traffic $\times 10^{-2}$	45	25	30	35	42	23
Accidents	20	22	24	28	31	8
Involvements	30	32	35	62	35	40
Killed	1	2	2	3	4	0
Injured	9	4	5	8	6	6
Property damage	20	5	8	3	5	8

   (a) Calculate accidents per mile, involvements per vehicle-mile, estimated cost, accident rate, and involvement rate.
   (b) Rank the sections by accidents, involvements, costs (weighted), accidents/mile, involvements/mile, and accident rate.

2. A four-legged intersection reported 23 accidents/year and the 24-hour volumes entering the intersection were 380, 390, 450, and 550 vehicles. Determine the accident rate per million vehicles entering.

3. A large city reported the following facts for one year: motor fuel consumption, 20 million gallons; total population of city, 7.27 million; motor vehicles in city, 12.10 million; percent trucks, 20%; fuel consumption, 15 miles/gal for auto and 12 miles/gal for trucks; fatalities (vehicle-related), 12,314; injuries (vehicle-related), 1.53 million. Calculate the vehicle-miles of travel, registration, population, and vehicle-mile death rate.

4. At a four-lane-by-four-lane signalized intersection, there appear to be a large number of accidents (72 in all) (and near accidents) and their distribution is as follows:

   30% right-angled collisions

   20% rear-end collisions

   22% pedestrian-car accidents (total)

   8% right-twin collisions with pedestrians

   12% left-turn vehicle–vehicle collisions

   9% vehicles running off road in wintertime

   11% sideswipes between vehicles traveling in the opposite direction

(a) What probable causes would you consider?

(b) What general countermeasures would you entertain?

5. A small city with a population of 10,000 people has four intersections having an abnormal number of accidents. What studies would you undertake to investigate the problem?

6. A rural highway is to be improved by means of four separate countermeasures: (1) resurfacing (0.12), (2) widening lanes (0.38), (3) widening roadway (0.08), and (4) installing warning signs (0.52). The current ADT is 1200, but is likely to increase to 1400 after the improvements. The current accidents are 15 fatal and 52 injuries. What is the estimated savings in accidents because of the proposed countermeasure?

Appendix

Elements of Engineering Economics

1. INTRODUCTION

Engineering economics is a branch of economics used by engineers to optimize their designs and construction projects. Project appraisal and cost-benefit analysis require a knowledge of engineering economics. Engineers and planners are concerned with money whether its use or exchange is in the private or public sector. In the language of economics, articles produced, sold, or exchanged are known as goods. At least four sectors of production are necessary to produce a good: labor, land, capital, and enterprise. Capital includes the money, machinery, tools, and materials required to produce a good. The opportunity cost of capital is measured by the interest rate, and the interest on capital is the premium paid or received for the use of money. Here, opportunity cost represents the cost of an opportunity that is foregone because resources are used for a selected alternative and, therefore, cannot be used for other purposes. The interest rate that relates a sum of money at some future date to its equivalent today is called the *discount rate*.

Students needing further information on engineering economics may refer to books listed in the references at the end of this appendix.

2. NOTATION

The following symbols and definitions are used in this appendix.

P = principal, a sum of money invested in the initial year, or a present sum of money

i = interest rate per unit of time expressed as a decimal

n = time, the number of units of time over which interest accumulates

I = simple interest; the total sum of money paid for the use of the money at simple interest

F = compound amount; a sum of money at the end of n units of time at interest i, made up of the principal plus the interest payable

A = uniform series end-of-period payment or receipt that extends for n periods

S = salvage or resale value at the end of n years

3. SIMPLE INTEREST

When one invests money at a simple interest rate of i for a period of n years, the simple interest bears the following relationship:

$$\text{Simple interest } (I) = Pin \tag{1}$$

The sum, I, will be added on to the original sum at the end of the specified period, but will remain constant each period, unless the interest changes.

Example 1

An amount of $2500 is deposited in a bank offering 5% simple interest per annum. What is the interest at the end of the first year and subsequent years?

$$I = (2500)(0.05)(1) = \$125$$

The interest for the second and subsequent years will also be $250. In other words, at the end of the first year, the amount will be $2625; for the second, it will be $2750.

4. COMPOUND INTEREST

However, when interest is paid on the original investment as well on the interest earned, the process is known as *compound interest*. If an initial sum, P, is invested at an interest rate, i, over a period of n years,

$$F = P(1 + i)^n \tag{2}$$

or

$$P = F/(1 + i)^n$$

If interest i is compounded m times per period n,

$$F = P(1 + i/m)^{nm} \tag{3}$$

As m approaches infinity (∞), Eq. 3 can be written as

$$F = Pe^{in} \tag{4}$$

or

$$P = Fe^{-in} \tag{5}$$

Equations 4 and 5 are used when resorting to "continuous compounding," a method frequently used in practice. Factors $(1 + i)^n$ and e^{in} are called compound amount factors, and $(1 + i)^{-n}$ and e^{-in} are called present worth factors for a single payment.

Example 2

What is the amount of $1000 compounded (a) at 6% per annum, (b) at 6% per every quarter, (c) at 6% per annum compounded continuously for 10 years?

Solution

(a) Here, $i = 0.06$, $n = 10$, and $m = 1$. Equation 2 becomes

$$F = 1000\,(1 + 0.06)^{10} = (1000)\,(1.79084) = \$1790.85$$

(b) Here, $i = 0.06$, $n = 10$, and $m = 4$. Equation 3 becomes

$$F = 1000\,[1 + (0.06/4)]^{40} = (1000)\,(1.81402) = \$1814.02$$

(c) Here, $i = 0.06$ and $n = 10$. Equation 4 becomes

$$F = 1000e^{(0.06)(10)} = (1000)\,(1.82212) = \$1822.12$$

Example 3

What is the *effective* interest rate when a sum of money is invested at a *nominal* interest rate of 10% per annum, compounded annually, semiannually, quarterly, monthly, daily, and continuously?

Solution

Assume the sum to be $1 for a period of 1 year. Then the sum at the end of 1 year compounded:

Annually $= 1(1 + 0.1)^{(1)(1)} = 1.10$ Therefore, the interest rate $= 10\%$
Semiannually $= 1(1 + 0.1/2)^{(1)(2)} = 1.1025$ and the interest rate $= 10.25\%$
Quarterly $= 1(1 + 0.1/4)^{(1)(4)} = 1.10381$ and the interest rate $= 10.381\%$
Monthly $= 1(1 + 0.1/12)^{(1)(12)} = 1.10471$ and the interest rate $= 10.471\%$
Daily $= 1(1 + 0.1/365)^{(1)(365)} = 1.10516$ and the interest rate $= 10.516\%$
Continuously $= 1[e^{(0.1)(1)}] = 1.10517$ and the interest rate $= 10.517\%$

Discussion

The comparison illustrates the difference between nominal and effective interest rates. Say, for example, the amount obtained adding interest quarterly is equal to $1.1038. Therefore, $1 + i = 1.10381$, and, therefore, $i = 0.10381$, or 10.381%.

5. UNIFORM SERIES OF PAYMENTS

If instead of a single amount there is a uniform cash flow of costs or revenue at a constant rate, the following uniform series of payment formulas are commonly used in practice.

5.1 Compound Amount Factor (CAF)

The use of the CAF helps to answer the question: What future sum (F) will accumulate assuming a given annual amount of money (A) is invested at an interest i for n years?

$$F = A\left[\frac{(1 + i)^n - 1}{i}\right] = A\left(\frac{x - 1}{i}\right)$$

where $[(x - 1)/i]$ is the uniform series compound amount factor.

5.2 Sinking Fund Factor (SFF)

The SFF indicates how much money (A) should be invested at the end of each year at interest rate i for n years to accumulate a stipulated future sum of money (F). The SFF is the reciprocal of the CAF.

$$F = A\left[\frac{i}{(1 + i)^n - 1}\right] = A\left(\frac{i}{x - 1}\right)$$

where $[i/(x - 1)]$ is the uniform series sinking fund factor.

5.3 Present Worth Factor (PWF)

The PWF tells us what amount P should be invested today at interest i to recover a sum of A at the end of each year for n years.

$$P = A\left[\frac{(1 + i)^n - 1}{(1 + i)^n i}\right] = A\left(\frac{x - 1}{xi}\right)$$

where $[(x - 1)/(xi)]$ is the uniform series present worth factor.

5.4 Capital Recovery Factor (CRF)

The CRF answers the question: If an amount of money (P) is invested today at an interest i, what sum (A) can be secured at the end of each year for n years, such that the initial investment (P) is just depleted?

$$A = P\left[\frac{i(1 + i)^n}{(1 + i)^n - 1}\right] = P\left(\frac{xi}{x - 1}\right)$$

CRF is the reciprocal of PWF. Another way of considering CRF is: If a sum of P is borrowed today at interest rate i, how much A must be paid at the end of each period to retire the loan in n periods?

Example 4

If the interest rate is 5% per annum, what sum would accumulate after 6 years if $1000 were invested at the end of each year for 6 years?

Solution

$$F = A \left[\frac{(1 + i)^n - 1}{i} \right] = 1000 \,(6.80191) = \$6801.91$$

Example 5

A realtor buys a house for \$200,000 and spends \$1000 per year on maintenance for the next 8 years. For how much should she sell the property to make a profit of \$40,000? Assume $i = 12\%$ per annum.

Solution

Future value of \$200,000 in 8 years:

$$F = P(1 + i)^n = 200{,}000 \,(1 + 0.12)^8 = 200{,}000 \,(2.475963) = \$495{,}192.63$$

Future value of annual maintenance:

$$F = A \left[\frac{(1 + i)^n - 1}{i} \right] = 1000 \,(12.299693) = \$12{,}299.69$$

Minimum selling price after 8 years, with a profit of \$40,000

$$= 495{,}192.63 + 12{,}299.69 + 40{,}000$$

$$= \$547{,}492.32$$

Example 6

The city transit system needs to set up a sinking fund for 10 buses, each costing \$100,000, for timely replacements. The life of the buses is 7 years and the interest rate is 6% per annum.

Solution

$$A = F \left[\frac{i}{(1 + i)^n - 1} \right] = (100{,}000)\,(10)\,[0.119135018]$$

$$= \$119{,}135.02 \text{ per year}$$

Example 7

A contractor wants to set up a uniform end-of-period payment to repay a debt of \$1,000,000 in 3 years, making payments every month. Interest rate = 12% per annum. What is the CRF?

Solution

$$i = 0.12/12 = 0.01 \qquad n = 3 \times 12 = 36$$

$$A = P\left[\frac{i(1+i)^n}{(1+i)^n - 1}\right] = 1{,}000{,}000[0.033214309] = \$33{,}214.31$$

6. UNIFORM GRADIENT SERIES

When a uniform series of a number of payments are increasing each year by a similar amount, it is possible to convert them to an equivalent uniform gradient. If the uniform increment at the end of each year is G, then all the compound amounts can be totaled to F.

$$F = \frac{G}{i}\left[\frac{(1+i)^n - 1}{i}\right] - \frac{nG}{i}$$

In order to convert this sum into an equivalent uniform period payment over n periods, it is necessary to substitute the sum for F in the sinking fund factor, giving

$$A = \frac{G}{i} - \frac{nG}{i}\left[\frac{i}{(1+i)^n - 1}\right]$$

Example 8

The maintenance on a transit bus system amounts to $20,000 by the end of the first year, increasing $5000/year for the subsequent 5 years. What is the equivalent uniform series cost each year, with interest at 5% per annum?

Solution

$$A = \frac{G}{i} - \frac{nG}{i}\left[\frac{i}{(1+i)^n\,1}\right] = \frac{5000}{0.05} - \frac{(5)(5000)}{0.05}(0.1809748)$$

$$= 100{,}000 - 90{,}487.4 = \$9512.60$$

Therefore, the uniform series equivalent annual cost

$$= 20{,}000 + 9512.60 = \$29{,}512.60 \text{ for each of the 5 years}$$

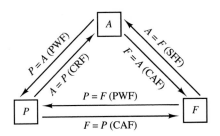

P = F (PWF)

F = P (CAF)

Figure A-1   Relationships among Factors.

7. DISCRETE COMPOUND INTEREST FACTORS

The various factors are summarized in Table A-1 and their relationships in Figure A-1.

8. UNIFORM CONTINUOUS CASH FLOW AND CAPITALIZED COST

The present worth of a uniform continuous cash flow is the amount P, invested at an interest rate i that will produce a cash flow of a per period for n periods. Consider a small time interval t to $t + \delta t$, during which the flow will be $a(\delta t)$, and the present worth of this element is $a(\delta t)(e^{-it})$. The present worth of the entire flow is

$$P = \int ae^{-it}\,dt = -\frac{a}{i}(e^{-it})$$

$$= a\left(\frac{1e^{-it}}{i}\right) = an\left(\frac{1 - e^{-it}}{in}\right) = anf_w$$

TABLE A-1   SUMMARY OF FACTORS

Symbol	Factor	Equation	Symbolic form	Find	Given
A	Single payment: 1. Compound amount 2. Present worth	$F = P[x]$ $P = F/[x]$	$F/P, i, n$ $P/F, i, n$	F P	P F
B	Uniform series: 3. Compound amount 4. Sinking fund 5. Present worth 6. Capital recovery	$F = A[(x - 1)/i]$ $A = F[i/(x - 1)]$ $P = A[(x - 1)/ix]$ $A = P[ix/(x - 1)]$	$F/A, i, n$ $A/F, i, n$ $P/A, i, n$ $A/P, i, n$	F A P A	A F A P
C	Arithmetic gradient: 7. Uniform series equivalent	$A = G\{1/i - (n/i)[i/(x - 1)]\}$	$A/G, i, n$	A	G

Notes: The factor $(1 + i)^n$ is known as the single payment compound amount factor, and is equal to x. In a similar manner, the expression in each equation within square brackets is the factor corresponding to the description, for example, $[ix/(x - 1)] =$ capital recovery factor CRF.

The term $f_w = (1 - e^{-it})/in$ is the present worth factor for uniform flow; it is useful in making comparisons of equipment and services with different service lives, by reducing them to their present values.

Capitalized Cost

Capitalized Cost (CC) is the present value of a series of services or equipment scheduled to be repeated every n periods to *perpetuity*. If P is the amount that must be paid every n periods, where n is the life of the equipment, then the CC is

$$K = P + Pe^{-in} + Pe^{-2in} + \cdots +$$

Summing this geometric series gives

$$K = P\left(\frac{1}{P - e^{-in}}\right) = P\left(\frac{e^{in}}{e^{in} - 1}\right) = P\left(\frac{1}{inf_w}\right)$$

Example 9

The Department of Transportation is faced with a problem. Should it rent trucks for $480 a month each or buy for $33,000 each, assuming the useful life of a truck is 10 years and $i = 14\%$?

Solution

$$\text{Present value of rented truck} = an\left[(1 - e^{-in})/in\right]$$

$$a = 480 \times 12 = \$5760/\text{yr}$$

Therefore,

$$P = (5760)(10)[(1 - e^{-0.14 \times 10})/(0.14)(10)] = (5760)(10)(0.538145) = \$30,997.15 < \$33,000$$

Therefore, it is more advantageous to rent trucks.

Example 10

A comparison between two types of bridges has to be made using the following details.

Type	Steel	Wood
Initial cost	$400,000	$250,000
Paint and maintenance	$400,000/10 yr	$25,000/2 yr
Life	40 yr	20 yr
Interest rate	9%	9%

Solution

Solution	Steel bridge	Wooden bridge
First cost	$K_1 = P[1/(1 - e^{-in})]$ $= 400,000 \, [1/(1 - e^{-0.09 \times 40})]$ $= 400,000 \, (1.028091)$ $= 411,236.51$	$K_1 = 250,000[1/(1 - e^{-0.09 \times 40})]$ $= 250,000 \, (1.19803)$ $= 299,508.41$
Paint and maintenance	$K_2 = 40,000[1/(1 - e^{-0.09 \times 10})]$ $= 67,404.71$	$K_2 = 25,000[1/(1 - e^{-0.09 \times 10})]$ $= 25,000 \, (6.070547)$ $= 151,763.69$
Total	$478,641	$451,272

Based on the calculations shown above the wooden bridge is cheaper.

REFERENCES

De Neufville, R., and J. H. Stafford (1971). *Systems Analysis for Engineers and Managers*, McGraw-Hill, New York.

Jewel, T. K. (1980). *A Systems Approach to Civil Engineering, Planning and Design*, Harper & Row, New York.

Johnson, R. E. (1990). *The Economics of Building*, John Wiley, New York.

Mitchell, B. L. (1980). *Engineering Economics*, John Wiley, Chichester, UK.

Pilcher, R. (1992). *Principles of Construction Management*, 3rd Ed., McGraw-Hill, London.

EXERCISES

1. What is the present worth of a future sum of $3500 in 10 years with interest at 10%?

2. A man deposits $1200, $2000, and $4000 at the end of 1, 2, and 3 years, respectively, at 10% interest per annum. What will be the accumulation at the end of 6 years?

3. A client wants to finance purchase of a house costing $50,000 over a period of 10 years. If the interest is 10% per annum, what will be (a) the annual payment and (b) the monthly payment?

4. A bank offers the following interest rates:
   (a) 6% compounded annually
   (b) 5.9% compounded semiannually
   (c) 5.8% compounded quarterly
   (d) 5.5% compounded monthly
   (e) 5.45% compounded continuously
   Which rate would you select to provide the highest return?

5. I want to double a large sum of inherited money. My bank offers two interest rates: (a) 9% compounded annually and (b) 8.70% compounded continuously. Which one should I select and why?

6. A person buys an automobile, promising to pay $300 per month for 5 years, but after 2 years, when the 24th payment is due, she decides to make a lump sum payment to settle the account. If the interest rate is 10%, what amount will she have to pay?

7. The maintenance for a bus, whose life is 10 years, is $1500 per year starting the fourth year, increasing by $200 for each successive year. What is the present worth of maintenance cost?

Index